Methods in Enzymology

Volume XXII

ENZYME PURIFICATION AND RELATED TECHNIQUES

METHODS IN ENZYMOLOGY

EDITORS-IN-CHIEF

Sidney P. Colowick   Nathan O. Kaplan

Methods in Enzymology

Volume XXII

Enzyme Purification and Related Techniques

EDITED BY

William B. Jakoby

SECTION ON ENZYMES AND CELLULAR BIOCHEMISTRY
NATIONAL INSTITUTE OF ARTHRITIS AND METABOLIC DISEASES
NATIONAL INSTITUTES OF HEALTH
BETHESDA, MARYLAND

1971

ACADEMIC PRESS   New York and London

ACADEMIC PRESS, INC.
111 Fifth Avenue, New York, New York 10003

United Kingdom Edition published by
ACADEMIC PRESS, INC. (LONDON) LTD.
Berkeley Square House, London W1X 6BA

LIBRARY OF CONGRESS CATALOG CARD NUMBER: 54-9110

PRINTED IN THE UNITED STATES OF AMERICA

Table of Contents

Section I. General Methods

Section II. Growth Methods

Section III. Specific Organelles and Extraction Procedures

Section IV. Separation Based on Solubility

Section V. Chromatographic Procedures

Section VI. Separation Based on Specific Affinity

Section VII. Electrophoretic Procedures

Section VIII. Large-Scale Methods

Section IX. Criteria for Homogeneity

Contributors to Volume XXII

Article numbers are shown in parentheses following the names of contributors.
Affiliations listed are current.

CHRISTIAN B. ANFINSEN (31), *Laboratory of Chemical Biology, National Institute of Arthritis and Metabolic Diseases, National Institutes of Health, Bethesda, Maryland*

JAY V. BECK (8), *Department Microbiology, Brigham Young University, Provo, Utah*

GIORGIO BERNARDI (29), *Laboratoire de Génétique Moléculaire, Institut de Biologie Moléculaire de la, Faculté des Sciences, Paris, France*

WILLIAM F. BLATT (6), *Amicon Corporation, Lexington, Massachusetts*

JAMES A. BOYLE (17), *Department of Medicine, University of California, San Diego, La Jolla, California*

ENRICO CABIB (14), *National Institute of Arthritis and Metabolic Diseases, National Institutes of Health, Bethesda, Maryland*

STANLEY E. CHARM (37), *New England Enzyme Center, Tufts Medical School, Boston, Massachusetts*

PAUL W. CHUN (22), *Department of Biochemistry, University of Florida, College of Medicine, Gainesville, Florida*

GEORGE B. CLINE (18), *Department of Biology, University of Alabama at Birmingham, Birmingham, Alabama*

PEDRO CUATRECASAS (31), *Division of Clinical Pharmacology, Departments of Medicine and Pharmacology, The Johns Hopkins University School of Medicine, Baltimore, Maryland*

DAVID R. DAVIES (25), *Laboratory of Molecular Biology, National Institute of Arthritis and Metabolic Diseases, National Institutes of Health, Bethesda, Maryland*

ARNOLD L. DEMAIN (12), *Department of Nutrition and Food Science, Massachusetts Institute of Technology, Cambridge, Massachusetts*

JOHANNES EVERSE (5), *Department of Chemistry, University of California, San Diego, La Jolla, California*

E. H. EYLAR (15), *Merck Institute, Rahway, New Jersey*

MELVIN FRIED (22), *Department of Biochemistry, University of Florida College of Medicine, Gainesville, Florida*

OTHMAR GABRIEL (39, 40), *Department of Biochemistry, Georgetown University Medical School, Washington, D.C.*

JOHN GERMERSHAUSEN (11), *Merck, Sharpe, and Dohme Research Laboratories, Rahway, New Jersey*

ARPI HAGOPIAN (15), *Merck Institute, Rahway, New Jersey*

MINORU HAMADA (30), *Department of Pathological Biochemistry, Atomic Disease Institute, Nagasaki University School of Medicine, Nagasaki-shi, Japan*

S. RALPH HIMMELHOCH (26), *Laboratory of Biochemistry, National Cancer Institute, National Institutes of Health, Bethesda, Maryland*

WILLIAM B. JAKOBY (23), *Section on Enzymes and Cellular Biochemistry, National Institute of Arthritis and Metabolic Diseases, National Institutes of Health, Bethesda, Maryland*

ROBERT J. JOHNSON (1), *Pfizer, Inc., Groton, Connecticut*

H. R. KABACK (13), *Division of Biochemistry, The Roche Institute of Molecular Biology, Nutley, New Jersey*

SEYMOUR KAUFMAN (21), *Laboratory of Neurochemistry, National Institute of Mental Health, National Institutes of Health, Bethesda, Maryland*

MASAHIKO KOIKE (30), *Department of Pathological Biochemistry, Atomic Disease Institute, Nagasaki University School of Medicine, Nagasaki-shi, Japan*

PETER McPHIE (4), *Section on Enzymes and Cellular Biochemistry, National Institute of Arthritis and Metabolic Diseases, National Institutes of Health, Bethesda, Maryland*

CHARLES C. MATTEO (37), *Massachusetts Institute of Technology, Cambridge, Massachusetts*

DAVID E. METZLER (1), *Department of Biochemistry and Biophysics, Iowa State University, Ames, Iowa*

D. JAMES MORRÉ (16), *Department of Botany and Plant Pathology and Department of Biological Sciences, Purdue University, Lafayette, Indiana*

HARVEY S. PENEFSKY (19, 20), *The Public Health Research Institute, of the City of New York, Inc., New York, N.Y.*

D. PERLMAN (10), *School of Pharmacy, University of Wisconsin, Madison, Wisconsin*

E. F. PHARES (36), *Biology Division, Oak Ridge National Laboratory, Oak Ridge, Tennessee*

BURTON M. POGELL (32), *Department of Microbiology, St. Louis University, School of Medicine, St. Louis, Missouri*

J. MICHAEL POSTON (7), *Enzyme Section, Laboratory of Biochemistry, National Heart and Lung Institute, National Institutes of Health, Bethesda, Maryland*

JOHN REILAND (27), *Technical Service Department, Bio-Rad Laboratories, Richmond, California*

R. REPASKE (28), *National Institute of Allergy and Infectious Diseases, National Institutes of Health, Bethesda, Maryland*

CARL RHODES (11), *Department of Biology, The Johns Hopkins University, Baltimore, Maryland*

RICHARD B. RYEL (18), *Department of Biology, University of Alabama at Birmingham, Alabama*

M. G. SARNGADHARAN (32), *Department of Microbiology, St. Louis University, School of Medicine, St. Louis, Missouri*

MORTON K. SCHWARTZ (2), *Memorial Hospital for Cancer and Allied Diseases and Sloan-Kettering Institute for Cancer Research, New York, New York*

J. EDWIN SEEGMILLER (17), *Department of Medicine, University of California, San Diego, California*

D. M. SEGAL (25), *Laboratory of Molecular Biology, National Institute of Arthritis and Metabolic Diseases, National Institutes of Health, Bethesda, Maryland*

LOUIS SHUSTER (34, 35), *Department of Biochemistry and Pharmacology, Tufts University School of Medicine, Boston, Massachusetts*

EARL R. STADTMAN (7), *Enzyme Section, Laboratory of Biochemistry, National Heart and Lung Institute, National Institutes of Health, Bethesda, Maryland*

THRESSA C. STADTMAN (7, 9), *Enzyme Section, Laboratory of Biochemistry, National Heart and Lung Institute, National Institutes of Health, Bethesda, Maryland*

FRANCIS E. STOLZENBACH (5), *Department of Chemistry, University of California, San Diego, La Jolla, California*

S. R. SUSKIND (11), *Department of Biology and The McCollum-Pratt Institute, The Johns Hopkins University, Baltimore, Maryland*

A. L. TAPPEL (3), *Department of Food Science and Technology, University California, Davis, California*

ALEXANDER TZAGOLOFF (19, 20), *The Public Health Research Institute of The City of New York, Inc., New York, N.Y.*

OLOF VESTERBERG (33), *Karolinska Institutet, Department of Bacteriology, Stockholm, Sweden*

C. W. WRIGLEY (38), *CSIRO. Wheat Research Unit, North Ryde, Australia*

MICHAEL ZEPPEZAUER (24), *Department of Chemistry, Royal Agricultural College of Sweden, Uppsala, Sweden*

Preface

The trend in the development of techniques for the purification of enzymes has evolved gradually from a type of trial and error application to that of an art form and, more recently, from art to a form approximating a reasonable scientific basis. While all three approaches remain necessary, it is with marked relief, if not with complete satisfaction, that one can view the contents of this volume and find directions for separation procedures based on such rational parameters as size and shape, charge, and the nature of the ligand. Indeed, advances in the tactics of protein purification have been such that one can reasonably expect that any protein of a given order of stability may be purified to currently acceptable standards of homogeneity. There are a few important ifs and buts in that last statement but there is no question that the opinion represents a higher order of confidence than was appropriate in the past. Recognition of this new plateau in the state of the art has led to the current volume: a group of articles detailing reliable procedures applicable to the purification of enzymes.

Aside from methods related directly to purification, this volume includes a description of peripheral techniques of value in enzyme preparation. For example, much space has been allowed for the methodology of obtaining cells which must be grown in the laboratory, and emphasis has been placed on the isolation of specialized organelles.

Many useful general methods which have been included in previous volumes of this series are listed in the section to which they apply. Such a list is not only particularly useful as a source of direction for isolating specific organelles but also explains why a separate article was not included for the single, most common means of enzyme purification, salting out. The latter was so well recorded in a previous volume that reference to the article should suffice.

xi

There is also a degree of emphasis on the scale of operations. We are at a stage of development at which the enzymologist must both obtain material for enzyme assay from the scrapings of fingernails and prepare a sample sufficient to allow the reaction of substrate with stoichiometric amounts of enzyme. An attempt has been made to cover this range of several orders of magnitude from the clinical biopsy to the pilot plant. The expectation is that the enzymologist will neither rush to build his own pilot plant nor, hopefully, take his own biopsy, but rather that the lessons in methodology will allow a more educated approach to scaling-up preparative procedures.

The reader will note omissions. Thus, methodology for the characterization of the resultant purified enzyme has been limited to a few rapid procedures for the estimation of homogeneity. Similarly, the inclusion of procedures for tissue culture, though highly desirable, was not possible, in part because of the vast number of different techniques that would have to be included. Despite the obvious omissions, and one could go on with such a list, the material covered in this volume will allow the investigator the flexibility to adapt these methods to the varied problems which await.

WILLIAM B. JAKOBY

METHODS IN ENZYMOLOGY

EDITED BY

Sidney P. Colowick and Nathan O. Kaplan

VANDERBILT UNIVERSITY
SCHOOL OF MEDICINE
NASHVILLE, TENNESSEE

DEPARTMENT OF CHEMISTRY
UNIVERSITY OF CALIFORNIA
AT SAN DIEGO
LA JOLLA, CALIFORNIA

METHODS IN ENZYMOLOGY

EDITORS-IN-CHIEF

Sidney P. Colowick   Nathan O. Kaplan

Section I
General Methods

[1] Buffer Preparation[1]

By Robert J. Johnson and David E. Metzler

Buffers of any desired pH can be prepared quickly by mixing together the component stock solutions of a suitable buffer acid and that of its conjugate base in appropriate amounts according to Eq. (1), where the

$$\mathrm{pH} - \mathrm{p}K = \log \frac{[\mathrm{A}]}{[\mathrm{HA}]} = \log \frac{f_d}{1 - f_d} \tag{1}$$

pK represents the *apparent* pK at the ionic strength of the buffer and f_d is the degree of dissociation, i.e., the fraction of total buffer in the dissociated form, A. Certain buffers which have proved to be of use in the spectrophotometric titration of vitamin B_6 derivatives are presented in Table I.[2]

To save time, the chart in Table II can be used. For example, to prepare a buffer having a pH 0.2 unit above the pK of the buffer acid

TABLE I
BUFFER ACIDS SUITABLE FOR USE WITH VITAMIN B_6 COMPOUNDS

Buffer acids	Apparent pK_a	pH Range	Suggested storage form[a]
HCl	—	0–2.5	1 or 2 *N* HCl
Formic acid	3.7	2.7–4.7	1 or 2 *N* Formic acid[b]
Acetic acid	4.6	3.6–5.6	1 or 2 *N* Acetic acid[b]
Cacodylic acid	6.1	5.1–7.1	1 *M* Sodium cacodylate
Dihydrogen phosphate	6.8	5.8–7.8	0.5 *M* KH_2PO_4 and Na_2HPO_4
Triethanolammonium	7.9	6.9–8.9	1 *M* Triethanolamine hydrochloride[c]
Bicarbonate	10.0	9.0–11.0	0.5 *M* $NaHCO_2$
Sodium hydroxide	—	11.5–13	1 or 2 *N* NaOH[d]

[a]HCl, NaOH, formic and acetic acid solutions should be standardized to ±1%.
[b]Stopper tightly; formic and acetic acid solutions lose acid readily and must be titrated quickly during standardization.
[c]Triethanolamine hydrochloride may be prepared by mixing 250 g of triethanolamine in 300 ml of 8 *N* HCl, allowing to crystallize, collecting, and washing with cold methanol or ethanol. Recrystallize by dissolving in hot 0.5 *N* HCl and cooling to room temperature. Wash the crystals with cold methanol, and air dry.
[d]For buffers from 0.02 to 0.2 *M*.

[1]This article is adapted directly from Vol. 18A [77], in which the authors discuss methods for obtaining spectra of vitamin B_6 derivatives.
[2]For other buffers, see Vol. I [16] and N. E. Good, G. D. Winget, W. Winter, T. N. Connolly, S. Izawa, and R. M. M. Singh *Biochemistry* 5, 467 (1966).

TABLE II

THE DEGREE OF DISSOCIATION, f_d, OF A MONOPROTIC ACID VERSUS pH;
$$\Delta = pH - pK = \log_{10} f_d / 1 - f_d$$

Δ	f_d	Δ	f_d
0.00	0.500	0.00	0.500
−0.05	0.471	0.05	0.529
−0.10	0.443	0.10	0.557
−0.15	0.415	0.15	0.585
−0.20	0.387	0.20	0.613
−0.25	0.360	0.25	0.640
−0.30	0.334	0.30	0.666
−0.35	0.309	0.35	0.691
−0.40	0.285	0.40	0.715
−0.45	0.262	0.45	0.738
−0.50	0.240	0.50	0.760
−0.55	0.220	0.55	0.780
−0.60	0.201	0.60	0.799
−0.65	0.183	0.65	0.817
−0.70	0.166	0.70	0.834
−0.75	0.151	0.75	0.849
−0.80	0.137	0.80	0.863
−0.85	0.124	0.85	0.876
−0.90	0.112	0.90	0.888
−0.95	0.101	0.95	0.899
−1.00	0.091	1.00	0.909
−1.05	0.082	1.05	0.918
−1.10	0.074	1.10	0.926
−1.15	0.066	1.15	0.934
−1.20	0.059	1.20	0.941
−1.25	0.053	1.25	0.947
−1.30	0.048	1.30	0.952
−1.35	0.043	1.35	0.957
−1.40	0.038	1.40	0.962
−1.45	0.034	1.45	0.966
−1.50	0.031	1.50	0.969
−1.55	0.027	1.55	0.973
−1.60	0.025	1.60	0.975
−1.70	0.020	1.70	0.980
−1.80	0.016	1.80	0.984
−1.90	0.012	1.90	0.988
−2.00	0.010	2.00	0.990
−2.10	0.008	2.10	0.992
−2.20	0.006	2.20	0.994
−2.30	0.005	2.30	0.995
−2.40	0.004	2.40	0.996
−2.50	0.003	2.50	0.997
−2.70	0.002	2.70	0.998
−3.00	0.001	3.00	0.999

($f_a = 0.61$), one could either mix 39 parts of the acid component of the buffer with 61 parts of the conjugate base or mix 100 parts of the acid with 61 parts of NaOH. If the pH selected is within 0.5 unit of the pK_a, the solution prepared this way should have the desired pH to within 0.1 unit or less (with the exception that at very low pH, the observed pH's of formate buffers will be a little higher than expected, and at high pH the carbonate buffers will yield pH's a little lower than calculated).

[2] Automated Enzyme Analysis—General[1]

By Morton K. Schwartz

During the past decade the use of automated techniques for the performance of enzyme assays has proceeded at a rapid pace. The clinical biochemist has used automation to keep up with the great increase in the number of requested enzyme assays in the hospital laboratory,[2] and the research biochemist has realized that automated enzyme techniques permit him to design his experiment without regard to the number of needed assays.[3,4] Enzyme automation has been described as existing in three stages.[5] In first stage automation, the reaction velocity is continuously recorded, but the reaction mixture must be prepared manually. In second stage automation, both preparation of the reaction mixture and the recording of the reaction velocity are carried out by the machine, but the instrument must be manually calibrated and the results calculated by the operator. Third stage automation, which is still in its development stage, involves all aspects of second stage automation and, in addition, includes automated calculation and presentation of results in a meaningful form with computer-directed feedback devices to control and maintain the operation of the instrument. Thus, third stage automation would fulfill the original definitions of automation.[5]

As enzyme automation has developed, there are three distinct approaches that have been used. These are (1) one- or two-point continuous-flow analysis, (2) constant or variable-time discrete sample

[1] This work was supported in part by Grant CA-08748 from the National Cancer Institute, National Institutes of Health, and Grant T-431J from the American Cancer Society.
[2] D. W. Moss, *Med. Electron. Biol. Eng.* **3**, 327 (1965).
[3] M. K. Schwartz, G. Kessler, and O. Bodansky, *Ann. N. Y. Acad. Sci.* **87**, 616 (1960).
[4] C. Beck and A. L. Tappel, *Anal. Biochem.* **21**, 208 (1967).
[5] M. K. Schwartz and O. Bodansky, *Methods Biochem. Anal.* **11**, 211 (1963).

analysis, and (3) continuous kinetic assays. The purpose of this report is to discuss in general all these techniques and their advantages and their disadvantages, but not to make any attempt to completely survey all the enzyme methods that have been automated or to describe in detail the instruments now available for automated enzyme analysis.

Continuous-Flow Techniques

Single Enzyme Analysis. Most of the automated enzyme assays that have been described are continuous-flow procedures using the Technicon AutoAnalyzer.[6] The AutoAnalyzer is a connected series of modules that allows automated performance of the many steps in a manual enzyme assay. Samples are placed in plastic cups in a rack on a constant-speed turntable equipped with a crook which dips a catheter into the sample and then into wash water at a prechosen rate. The second module is a constant-flow proportioning pump with continually moving steel bars which "milk" sample and reagents by compressing a group of plastic tubes in a forward motion. The tubes include one attached to the dipping catheter on the sampler module and others inserted into reagent bottles. The volume of aspirated liquid is a function of the inner diameter of the individual plastic tubes. A tube with an inner diameter of 0.030 inch will aspirate 0.32 ml/min, and a tube with an inner diameter of 0.081 inch will aspirate 2.00 ml/min. In an enzyme assay plastic tube sizes are chosen so that the final concentration of enzyme, substrate, buffer and necessary cofactors are similar to those in the manual method on which the automated procedure is based.

The samples are diluted if necessary, and the reagents are added in proper sequence by flowing through glass fittings and appropriate lengths of tubing. The reactants are mixed by passage through glass coils with concentric helices that allow the heavier fluids to mix into the lighter fluids as the two reverse top to bottom positions in their passage through the coil. The flowing reactants are interspersed with bubbles of air aspirated into the system through the pump. The air bubbles serve to regulate the flow, maintain the sample as discrete portions during passage through the modules and to act as a "squeegee" to clean the glass–plastic tubing between samples.

The reaction mixture flows into and through a glass coil in a heating bath thermostatically maintained at the temperature desired for the particular enzyme reaction. The time of incubation is controlled by the length of glass coil and the volume of flowing fluid. The standard

[6] M. K. Schwartz, *Automat. Anal. Chem., Technicon Symp. 1967*, Vol. 1, p. 587, Mediad, White Plains, N. Y., 1968.

heating bath coil is 40 feet in length. Since the plastic tubes change in inner diameter with use, the time of incubation may vary from day to day. The precise incubation time is established by aspirating a colored solution such as 4% $K_4Fe_3(CN)_6$ through the sample line and measuring with a stopwatch the time of its transit from the point where the enzyme reaction is started to where it is terminated. If it is necessary to remove protein, the flowing mixture is passed through a dialysis module. The diffusible compounds pass through a semipermeable membrane into a recipient flowing reagent stream, and the remaining solution flows into waste. Dialysis is not complete; usually much less than 50% of the available dialyzable material passes across the membrane. An assumption is made that dialysis of standards and unknowns proceeds at the same rate. Other reagents are added if they are needed, the mixture passes through additional heating modules, if necessary, and then through a constant-flow cuvette in a module chosen to measure the end point of the enzyme reaction. In most continuous flow procedures this is a filter type split-beam colorimeter, an ultraviolet colorimeter, or a fluorometer. The voltage output of the detecting module is applied to operate the pen of a strip chart recorder and the absorbancy or fluorescence of the material in the cuvette registered as a peak. The output may also be fed into a computer. It is possible, by proper choice of AutoAnalyzer modules, to automate any manual enzyme assay. Such techniques have been described for more than 40 enzymes.[7] The versatility of continuous flow technique is emphasized by the independent development of automated ultraviolet, fluorometric, and colorimetric methods for the assay of glutamic-oxaloacetic transaminase activity.[8]

Multiple Enzyme Analysis. Simultaneous analysis of several enzymes in the same sample may also be carried out by continuous flow by constructing plastic–glass tubing manifolds in which the sample is split and entered into several reaction mixtures and then passed through separate incubation and detecting modules. In the SMA 12/60 Auto-Analyzer, suitable lengths of glass-plastic tubing in the flow circuit stagger the arrival to the detecting modules of the several streams and allow the use of one recorder for all the assays.[9] With the SMA 12/60 it is possible to analyze one specimen simultaneously for alkaline phosphatase, glutamic-oxaloacetic transaminase, lactic dehydrogenase, and creatine phosphokinase activity. It is also possible to combine single-

[7] W. E. C. Wacker and T. C. Coombs, *Annu. Rev. Biochem.* **38**, 539 (1969).
[8] M. K. Schwartz, this series, Vol. 17B [261c], p. 866–875.
[9] W. J. Smythe, M. H. Shamos, S. Morgenstern, and L. T. Skeggs, *Automat. Anal. Chem., Technicon Symp. 1967*, Vol. 1, p. 105, Mediad, White Plains, N. Y., 1968.

channel AutoAnalyzer modules into a multiple enzyme system. Levy[10] has devised an AutoAnalyzer manifold (Fig. 1) for the simultaneous assay of lactic dehydrogenase, glutamic-oxaloacetic transaminase, and creatine phosphokinase in the same sample.

Problems in Converting a Manual Enzyme Assay to an Automated System. Although it is possible to automate any manual enzyme assay by Auto-Analyzer techniques, the inherent characteristics of a continuous flow system with a glass–plastic tube manifold may make necessary changes in the incubation time or temperature, the concentration of enzyme or substrate, the method of protein removal, or the reagents or methods used in the estimation of the enzymatically formed products. Some of these problems may be illustrated by the development of a method for the automated determination of L-asparaginase.[11] The assay is based on the incubation of enzyme with L-asparagine buffered at pH 7.8 and the determination of the liberated NH_3.

In the manual L-asparaginase assay, the substrate (10 mM L-asparagine) is dissolved in 0.1 M Tris buffer, pH 8.6, and the enzymatically liberated NH_3 determined in a trichloroacetic acid, protein-free filtrate

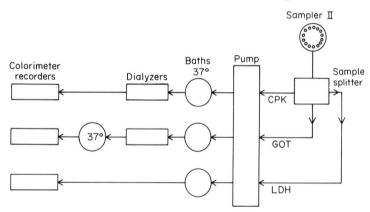

FIG. 1. Composite AutoAnalyzer manifold for the simultaneous analysis of lactic dehydrogenase (LDH), creatine phosphokinase (CPK) and glutamic-oxaloacetic trans-aminase (GOT). (A. Levy, unpublished data, 1969). Lactic dehydrogenase activity is measured by an acid phenylhydrazine method (A. L. Levy, C. Dalmasso, and J. Daly, *Automat. Anal. Chem., Technicon Symp. 1965*, Vol. 1, p. 551. Mediad, White Plains, N. Y., 1966); transaminase by colorimetric coupling of oxaloacetate with 6-benzamide 4-methoxy-*m*-toluidine diazonium chloride [A. L. Babson, P. O. Shapira, P. A. R. Williams, and G. E. Phillips, *Clin. Chim. Acta* 7, 199 (1962)], and CPK by reaction of formed creatine with naphthanol and diacetyl (A. L. Siegel and P. S. Cohen, *Automat. Anal. Chem., Technicon Symp. 1966*, Vol. 1, p. 474. Mediad, White Plains, N. Y., 1967).

[10] A. Levy, unpublished data, 1969.
[11] M. K. Schwartz, E. D. Lash, H. F. Oettgen, and F. A. Tamao, *Cancer* 25, 244 (1970).

by direct Nesslerization, a procedure which does not easily lend itself to automation. The changes in the manual method that were required to automate the assay are listed in Table I. The phenol-hypochlorite reaction was substituted for the Nesslerization procedure in the determination of NH_3. However, Tris buffer interferes with this reaction and prevents color formation. In the automated method, the Tris buffer was replaced with 0.1 M Veronal buffer, pH 8.6, and protein was removed by dialysis rather than by trichloroacetic acid precipitation. Thus, with several changes, the manual method was automated to yield a method with equal precision and accuracy and allowing 60 assays per hour.[11]

Calculation of Enzyme Activity. The calculation of enzyme activity in continuous flow systems is complicated by a number of factors. These include (1) the usual use of a zero time and a single time point for the assay, (2) changes in reaction mixture concentrations as the delivery volumes of the manifold tubes change with use, (3) the effect of partial dialysis on the enzyme activity, (4) variations in incubation time, and (5) the type of standard employed.

The most important consideration in automated enzyme assays by continuous-flow procedures which record activity at a single time, is that the observation be made during the initial zero-order portion of the reaction. The need for this precaution is no greater or less in automated than in manual methods. However, in automated continuous-flow methods, it is possible to build into the procedure the assurance that zero-order kinetics will be obeyed. This is done by adjusting the sensitivity of the method so that full-scale recorder deflection is representative of a substrate change that is within the zero-order portion of the reaction. With this arrangement, all samples with high activity must be diluted and hence are brought into the linear portion of the reaction. L-Asparaginase maintains zero-order kinetics during the hydrolysis of at least the first 25% of the substrate. In the automated method the molarity of L-asparagine in the final reaction mixture is 10 mM or a

TABLE I
DIFFERENCES IN AUTOMATED AND MANUAL
L-ASPARAGINASE METHODS

	Manual	Automated
Buffer	0.1 M Tris, pH 8.6	0.1 M Veronal, pH 8.6
Protein removal	Trichloroacetic acid	Dialysis
Ammonia determination	Nesslerization	Phenol-hypochlorite (Berthelot)

total of 10 μmoles of hydrolyzable ammonia available per milliliter of flow solution. Therefore, liberation of up to 2.5 μmoles ammonia is within the zero-order portion of the reaction. Full-scale deflection of the recorder is obtained with ammonia concentration of 0.75 μmole per milliliter of flow solution, or 7.5% of the total available.

It has been recommended that purified enzymes or commercially available control sera containing high enzyme activity be used as standards in automated enzyme assays. The advantage in the use of such materials is obvious. Since the enzyme standard is treated in the same fashion as the unknown, changes in tube size, incubation time, the effect of dialysis on enzyme activity and other intrinsic automation variables are eliminated. The disadvantages of the use of such materials are the possible lability of the enzyme, the difficulty in establishing the true activity of the "standard" enzyme, and the possibility that the enzyme "standard" is from a different tissue or species than the unknown and acts in a different kinetic fashion. In the L-asparaginase assay, ammonium sulfate solutions have been used to standardize the procedure. This has been necessary because of the unavailability of pure preparations of L-asparaginase and the differences in activity that are observed when purified material is dissolved in water, saline, or protein solutions. For many enzymes it is possible to use commercially available purified enzymes or control sera as standards. In our laboratory we determine glutamic-oxaloacetic transaminase by either an automated fluorometric NAD-NADH technique or by the SMA 12/60 colorimetric procedure. Versatol E, a lyophilized serum with high transaminase activity (General Diagnostics Division, Warner Chilcott), can be used as standard if it is analyzed by a conventional manual method and if appropriate dilutions are used. We have found that dilutions of this material are stable for at least a month when stored in the deep freeze.

The freshly prepared standards were calibrated spectrophotometrically according to the method of Karmen[12] with a Beckman DU Spectrophotometer equipped with thermospacers maintained at 37°. The diluted standards were quite stable during the storage period, and new dilutions of standard prepared from the same lot of lyophilized Versatol E as that used previously checked very well. During a 4-month period, the average deviation of newly diluted standards ranged from 3 to 7%. Values of dilutions of the standard material assayed by the spectrophotometric method[12] were identical with those obtained by the automated fluorometric procedure using standards prepared the previous month (Table II). The data also point out the similarity of values

[12]A. Karmen, J. Clin. Invest. 34, 131 (1955).

TABLE II

Comparison of Transaminase Activity Determined by a
Manual Kinetic Spectrophotometric Method and an
Automated Single Time Point Fluorometric Assay[a]

| Sample | Transaminase activity | | Average deviation (%) |
	Manual	Automated	
1	34	36	2.9
2	65	68	3.0
3	122	117	2.5
4	148	149	0.3
5	201	198	1.0
6	292	266	4.7

[a] Activity is expressed in Karmen Units.

that can be obtained with a single time point assay (automated) and a time-kinetic assay (spectrophotometric).

The use of these enzyme standards in the automated method allowed a direct calculation of results in Karmen units from a standard curve and eliminated the need for measuring incubation time, calculation of factors needed to compensate for changes in inner diameters of plastic tubes and other environmental factors which effect automated continuous flow determinations. However, before enzymes can be used as standards, it is necessary, as was done in this case, to establish that the diluted standards are stable when stored in the deep freeze and then at room temperature during the assay period, and last, that there is a convenient method for ascertaining the precise activity of the standards. It is possible to use molecular extinction coefficients in the calculation of continuous-flow data. However, when this is done it is necessary to measure the exact delivery rates of the manifold tubes and to calculate the precise concentrations of sample in the flowing reaction mixture and the exact time of the enzyme reaction.[13]

An important consideration in the choice of an enzyme for use as a standard is the organ and species from which the enzyme is prepared. It must be determined that the enzyme standard exhibits similar properties and kinetics as the enzyme in the unknown samples. For example, in the acid phosphatase reaction with α-naphthyl phosphate as substrate, it would not be suitable to use as a standard an enzyme prepared from erythrocytes since this enzyme does not exhibit any activity with this substrate.[14] The differences in kinetic properties of

[13] M. K. Schwartz, G. Kessler, and O. Bodansky, *J. Biol. Chem.* **236**, 1207 (1961).
[14] B. Klein and J. Auerbach, *Clin. Chem.* **12**, 289 (1966).

alkaline phosphatase purified from different tissues and different organs has been critically evaluated by Bowers and his associates who point out the difficulties in using purified alkaline phosphatase as an enzyme standard.[15]

Single-Point versus Kinetic Assays. One of the criticisms of continuous-flow enzyme analysis has been that measurement of activity at a single time point yields no information about the constancy of the enzyme reaction during the period of observation. It has been stated, although not established by experimental data, that continuous kinetic recording of enzyme activity has greater accuracy and is a superior technique to single time point assays. The advantages of kinetic assays have been stated to be as follows: (1) blanks are not needed, except for nonenzymatic changes in substrate, (2) short incubation periods allow observation of zero order kinetics, even in the presence of low substrate concentration, (3) short incubation period eliminates the possibility of product inhibition or enzyme denaturation during the reaction, (4) the short incubation period permits more rapid analysis.

As pointed out earlier, continuous flow techniques can be adjusted to assure that the reaction is only run during the zero-order portion of the curve. In addition, it is possible to modify continuous-flow instruments to permit readings at several time points during the reaction. Brown and Ebner[16] built a constant temperature bath to be used on the sample plate of an AutoAnalyzer Sampler II. The modified sampler permitted the immersion of the sample cups in a constant temperature water bath and maintained the samples at a predetermined temperature. In their assay of alkaline phosphatase activity with disodium phenylphosphate as substrate, the enzyme sample and buffered substrate were mixed manually and placed in the sample cups on the turntable. The enzyme reaction mixtures were then sampled at several predetermined times by restarting the sampler and running the samples through several cycles.

The use of kinetic assays for enzymes utilizing ultraviolet change at 340 nm has been widely accepted and semiautomated techniques have been described for many enzyme assays. The most commonly used technique is to carry out the enzyme reaction in a cuvette in a spectrophotometer attached to a Gilford Model 2000 multiple-sample absorbance recording unit. The cuvette temperature is maintained by use of thermospacers. Hess and his associates[17] have utilized such a system for the assay of creatine phosphokinase (CPK) activity. In their system,

[15]G. N. Bowers, Jr., M. L. Kelley, and R. B. McComb, *Clin. Chem.* 13, 595 (1967).
[16]H. H. Brown and M. R. Ebner, *Clin. Chem.* 13, 847 (1967).
[17]J. W. Hess, R. W. MacDonald, G. J. W. Natho and K. J. Murdock, *Clin. Chem.* 13, 994 (1967).

ATP generated during the CPK catalyzed conversion of creatine phosphate to creatine is coupled with glucose and hexokinase to form glucose 6-phosphate. The glucose 6-phosphate is then converted to 6-phosphogluconate in the presence of added NADP and glucose 6-phosphate dehydrogenase. The ultraviolet change during the conversion of NADP to NADPH is continuously recorded. These authors observed that the absorbance change in this reaction followed a sigmoid curve with an initial lag period, a period of linear activity and finally a period of decreased activity due to the occurrence of suboptimal reaction conditions. The time extent of the initial lag period was a variable related to the enzyme concentration. At low activities the lag was as much as 8 minutes and presumably was due to the need to generate enough ATP, and then glucose 6-phosphate, to permit the other portions of the coupled reaction to proceed. It is obvious that single-point assays for such a system would not be acceptable.

Noncontinuous-Flow Techniques

Discrete Sample Systems. Discrete sample analyzers utilize pumps, syringes, moving belts, and other mechanical devices to duplicate exactly the steps in a manual method. Reactions generally are carried to completion and the reactions conducted in plastic or glass vials, cups, or test tubes. Constituents are added as needed and mixed by vibration, shaking, or air jets. If necessary, aliquots are transferred from one tube to another.

Vanderlinde[18] has described these systems as "a mechanized test tube capable of accurately and reproducibly pipetting, incubating and sensing a reaction." The manufacturers of many of these instruments have included enzyme assays in their lists of available tests. Some of these instruments include the Robot Chemist (American Optical Instrument Co.), the Hycel Mark X (Hycel Inc.), the Autochemist (Svenska AB Gasaccumulator [AGA]), Zymat Enzyme Analyzer (Bausch and Lomb Inc.), the Discrete Sample Analyzer–DSA-560 (Beckman Instruments Inc.) and the ACA-Automatic Clinical Analyzer (E. I. du Pont de Nemours and Co.). There is little information in the literature evaluating the enzyme analysis capabilities of these instruments. Proponents of discrete sample analysis claim that the advantages of their approach to automation are that manual procedures can be automated without modification and there is little possibility of interaction between samples. Albert[19] has presented a review of the instruments available for automated analysis.

[18]R. E. Vanderlinde, *Health News*, 44, 4 (1967).
[19]N. L. Alpert, *Clin. Chem.* 15, 1198 (1969).

Automated Techniques for the Study of Enzyme Kinetics

Continuous-flow techniques can be used for the study of enzyme kinetics. With an AutoAnalyzer, changes in reaction parameters such as enzyme or substrate concentration, pH, presence of inhibitors or other factors can be continuously evaluated by equalizing the rates of outward and inward flow from a reagent vessel. The techniques for such studies have been the subject of a recent review.[20] Illingworth and Tipton[21] have described an apparatus for the automated study of enzyme kinetics. This instrument utilizes four power-driven syringes with an integral mixer. The system was described for studies of substrate effects on yeast alcohol dehydrogenase. However, it could be used with any enzyme in which NAD^+ is a cofactor. Enzyme sample containing NAD^+ is pumped at a constant rate from one of the syringes into a mixer, where it is combined with linearly diluted substrate. The enzyme reaction takes place as the solution flows through a temperature-controlled reaction coil and then through a 4-cm flow cell, where the ultraviolet change is recorded in a modified Beckman DU Monochromator and a Holger-Gilford Absorbance Converter. The instrument permits accurate measurement of 0.1 μmole of NADH.

[20]M. K. Schwartz and O. Bodansky, *Methods Biochem. Anal.* **16**, 183 (1968).
[21]J. A. Illingworth and K. F. Tipton, *Biochem. J.* **115**, 511 (1969).

[3] Automated Multiple Enzyme Analysis Systems

By A. L. TAPPEL

Developments pointing toward multiple enzyme analytical systems include many systems for studying enzyme activities and kinetics,[1] and successful applications of many multiple analysis instruments.[2] Automated multiple enzyme analysis systems are applicable to continuous monitoring of column chromatography effluents and effluents from zonal and other centrifugations. Also, these systems are appropriate for the determination of a large number of enzyme activities in individual samples. Several automated multiple enzyme analysis systems have

[1]M. K. Schwartz and O. Bodansky, *Methods Biochem. Anal.* **16**, 183 (1968).
[2]W. J. Smythe, M. H. Shamos, S. Morgenstern, and L. T. Skeggs, *Automat. Anal. Chem.*, *Technicon Symp. 1967*, Vol. 1, p. 105. Mediad, White Plains, N. Y., 1968.

been described.[3-6] The two described here have the capacity to monitor at least 12 hydrolytic enzymes in various applications.[7]

Autoanalysis Systems

Two modes of operation will be described: the slaved sampler mode, and the programmed multichannel valve (PMCV) mode. These two modes differ in the method of delivering the substrates in sequence. The slaved sampler mode was first developed. The development of the PMCV by Eveleigh et al.[8] allowed increased substrate capacities, and this was incorporated into a similar system. AutoAnalyzer (Technicon Corp., Ardsley, New York) components are used except where indicated. The slaved sampler mode is shown in Fig. 1. The master sampler, which

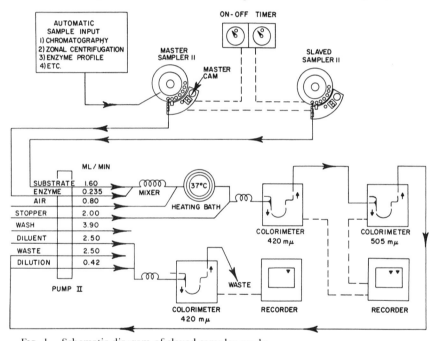

Fig. 1.   Schematic diagram of slaved sampler mode.

[3]H. Pitot and N. Pries, Anal. Biochem. 9, 454 (1964).
[4]H. Pitot, N. Pries, M. Poirier, and A. Cutler, Automat. Anal. Chem., Technicon Symp. 1965, Vol. 1, p. 555. Mediad, White Plains, N. Y., 1966.
[5]C. Beck and A. L. Tappel, Anal. Biochem. 21, 208 (1967).
[6]A. L. Tappel, in "Lysosomes in Biology and Pathology" (J. T. Dingle and H. B. Fell, eds.), Vol. 2, p. 547. North-Holland, Amsterdam, 1969.
[7]D. W. Bradley and A. L. Tappel, Anal. Biochem. 33, 400 (1970).
[8]J. W. Eveleigh, H. J. Adler, and A. S. Reichler, Automat. Anal. Chem., Technicon Symp. 1967,Vol. 1, p. 311. Mediad, White Plains, N. Y., 1968.

is off, thus leaving one channel operationally open. A PMCV distributor block, with a small dead volume to minimize substrate deterioration and mixing, is constructed of plastic and four modified glass connectors (No. K-8, Technicon Corporation). A small refrigerator, which holds the substrates in polyethylene bottles, is adapted to the system by inserting a sixteen-hole plastic port medially in one side.

Substrates and Hydrolytic Enzymes

The substrates and buffers used for assay of hydrolytic enzymes are shown in Table I. Further information regarding these enzymes and their substrates can be found in a recent review.[9] For use in these studies, lysosomes, which contain the major hydrolytic enzymes of animal tissue, were prepared from rat liver. Specific activities of some of these enzymes

TABLE I
SUBSTRATES AND BUFFERS FOR DETERMINATION OF HYDROLYTIC ENZYMES

Enzyme	Substrate	Concentration[a] (mM)	Buffer[b]
α-L-Fucosidase	p-Nitrophenyl-α-L-fucoside	2.50	A, pH 5.2
5′-Phosphodiesterase I (EC 3.1.4.1)	p-Nitrophenyl-5′-phosphothymidine	1.50	A, pH 5.2
5′-Phosphodiesterase IV (EC 3.1.4.1)	Bis-p-nitrophenylphosphate	2.50	A, pH 5.2
α-D-Glucosidase (EC 3.2.1.20)	p-Nitrophenyl-α-D-glucoside	2.00	B, pH 5.0
α-D-Galactosidase (EC 3.2.1.23)	p-Nitrophenyl-α-D-galactoside	2.00	B, pH 4.0
β-D-Galactosidase (EC 3.2.1.23)	p-Nitrophenyl-β-D-galactoside	1.05	B, pH 3.0
β-D-Glucosidase (EC 3.2.1.21)	p-Nitrophenyl-β-D-glucoside	1.00	B, pH 5.0
α-D-Mannosidase (EC 3.2.1.24)	p-Nitrophenyl-α-D-mannoside	6.00	A, pH 5.2
β-D-Glucuronidase (EC 3.2.1.31)	p-Nitrophenyl-β-D-glucuronide	1.90	A, pH 5.0
N-Acetyl-β-D-glucosaminidase (EC 3.2.1.30)	p-Nitrophenyl-N-acetyl-β-D-glucosaminide	2.40	B, pH 4.2 0.1 M NaCl
Arylsulfatase A and B (EC 3.1.6.1)	2-Hydroxy-5-nitrophenyl-sulfate	3.50	A, pH 5.0
Acid phosphatase (EC 3.1.3.2)	p-Nitrophenylphosphate	15.0	A, pH 5.2

[a] Final concentration in reaction mixture.
[b] A = 0.1 M sodium acetate; B = 0.2 M sodium citrate, 0.4 M sodium phosphate.

[9] A. L. Tappel, in "Comprehensive Biochemistry" (M. Florkin and E. H. Stotz, eds.), Vol. 23, p. 77. American Elsevier, New York, 1968.

in liver lysosomes,[6] expressed in millimicromoles of substrate hydrolyzed per minute per milligram of protein, are: acid phosphatase, 580; 5'-phosphodiesterase I, 4; 5'-phosphodiesterase IV, 22; N-acetyl-β-D-glucosaminidase, 860; β-D-galactosidase, 88; and β-D-glucosidase, 6.

Applicability of the Systems

Numerous applications of the slaved sampler mode and the PMCV mode are apparent. One of the more useful applications of autoanalysis is the monitoring of column chromatograms. Continuous monitoring of a chromatogram is performed by periodically sampling an aliquot of the column effluent.[5] Also, enzyme analysis of various tissues and subcellular fractions can be readily performed by an automated multiple enzyme analysis system. Other applications of these two automated modes include the continuous monitoring of zonal centrifugations, and the discontinuous monitoring of density gradient centrifugations, and the determination of selected enzyme kinetic parameters.

In tests of the slaved sampler mode, rat liver lysosomes were analyzed for enzyme activity. The recorder output for 11 hydrolytic enzymes is shown in Fig. 3. The peaks, representing enzyme activity, are well resolved from one another and show definite maxima. Enzyme and substrate blanks are analyzed separately, and the corresponding absorbance values are subtracted from those of the enzymatic reactions. p-Nitrophenol and 2-hydroxynitrocatechol are used for calibration. With the exact reaction time and the protein content of the enzyme sample being known, specific activities can be calculated.

The fundamental knowledge available on these enzymes[9] and the manual methods of their analyses applies directly to these automated methods. Detailed studies of enzyme kinetic parameters, including reaction times, enzyme concentrations, substrate concentrations, pH, and temperature, showed for β-glucuronidase[10,11] and ribonuclease[12] that these automated methods yield valid results. The substrate sampling sequence shown in Fig. 1 was used to analyze various concentrations of the same lysosomal enzyme sample. A plot of enzyme activity vs concentration for enzyme concentrations of 0.04–0.4 mg lysosomal protein per milliliter showed appropriate proportionality. This confirmed the applicability of the system to analysis of enzyme sources with different levels of activity.

[10] A. L. Tappel and C. Beck, *Automat. Anal. Chem., Technicon Symp. 1965*, p. 559. Mediad, White Plains, New York, 1966.
[11] A. L. Tappel and C. J. Dillard, *J. Biol. Chem.* 242, 2463 (1967).
[12] H. Barrera, K. S. Chio, and A. L. Tappel, *Anal. Biochem.* 29, 515 (1969).

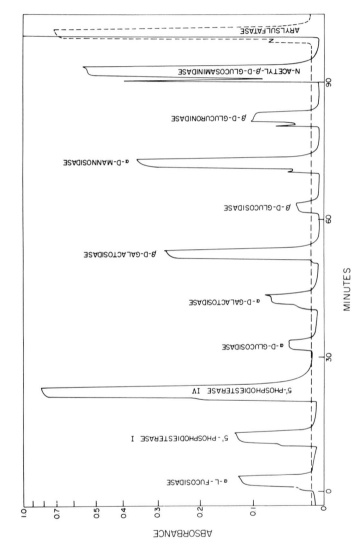

Fig. 3.   Slaved sampler mode: recorder output for eleven hydrolytic enzymes. Absorbance at 420 mμ, ———; absorbance at 505 mμ, ------

In tests of the PMCV modes rat liver lysosomes were analyzed for 8 enzymes and the results are shown in Fig. 4 as a direct recorder readout. In experimental determinations analogous to those shown in Fig. 4, enzymes of lysosomes were analyzed 5 times to test reproducibility. The results in Table II show that reproducibility is good.

In consideration of the relative merits of automated multiple enzyme systems, one must calculate the costs in material and time. With the manual method as a point of reference, three parameters of enzyme analysis are compared. The manual method for a typical glycosidase, N-acetyl-β-D-glucosaminidase, uses approximately 0.03 mg of protein from a lysosome preparation, 1.3 mg of substrate ($6 \times K_m$), and requires approximately 30 minutes of incubation time. The automated method requires 0.20 mg of protein, 10 mg of substrate, and utilizes 5–10 minutes of instrument time. It should be noted that the automated

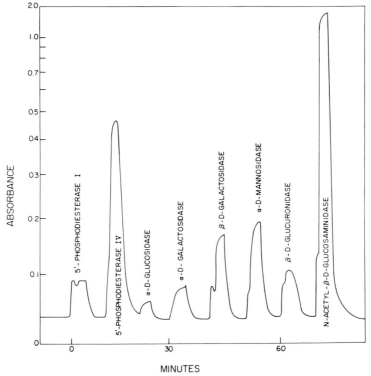

FIG. 4.   PMCV mode: direct recorder readout of eight hydrolytic enzymes. Absorbance is measured at 420 mμ.

TABLE II
REPRODUCIBILITY OF ENZYME ANALYSES

Enzyme and standard	Mean absorbance ± standard deviation
5'-Phosphodiesterase I	0.070 ± 0.007
5'-Phosphodiesterase IV	0.153 ± 0.058
α-D-Glucosidase	0.013 ± 0.005
α-D-Galactosidase	0.052 ± 0.010
β-D-Galactosidase	0.054 ± 0.001
α-D-Mannosidase	0.047 ± 0.008
β-D-Glucuronidase	0.068 ± 0.004
N-Acetyl-β-D-glucosaminidase	0.542 ± 0.043
Acid phosphatase	0.967 ± 0.120
Nitrophenol standard	0.678 ± 0.020

method is performed in a manner analogous to the manual method and that the amounts of enzyme and substrate consumed in the methods are not radically different. By far, the largest investment of operator time in the automated method is in the initial setup of the system. The approximate cost of the equipment arrays shown in Figs. 1 and 2 is $11,000.

Pyridine nucleotide-linked enzyme reactions can be analyzed by measurement of the increase or decrease of absorbance at 340 mμ.[13] Proteases and peptidases also can be determined by automated multiple analysis by measurement of absorbance at 280 mμ, or fluorescence at 350 mμ, or by the Folin-Lowry or ninhydrin analysis of the dialyzate of a reaction mixture.[14] Column chromatography with concurrent on-stream monitoring of the effluent for multiple enzyme activities can be carried out for extended periods of time (>72 hours) with the PMCV mode. Sufficient refrigerated substrate for the analyses can be stored for this period of time.

Acknowledgment

This research was supported by Public Health Service Research Grants AM 05609 and AM 09933 from the National Institute of Arthritis and Metabolic Diseases. The collaboration throughout of Daniel Bradley is gratefully acknowledged.

[13]M. K. Schwartz, G. Kessler, and O. Bodansky, *J. Biol. Chem.* **236**, 1207 (1961).
[14]A. L. Tappel, *Anal. Biochem.* **23**, 466 (1968).

[4] Dialysis

By Peter McPhie

In a previous volume of this series, Craig has discussed the uses of dialysis and ultrafiltration as analytical techniques for the determination and comparison of the diffusional properties of proteins and peptides.[1] His elegant studies have demonstrated complexities in the solution properties of proteins that are not discernible by other techniques. However, by far the most general use of dialysis is in the preparation and purification of macromolecules, by the removal or exchange of low molecular weight contaminants, counterions, and solvent components. Since, over the years, an extensive lore has developed concerning the ways and means of performing dialyses, it may be worthwhile to present the more practicable methods and to assess their relative merits. This chapter will be devoted to preparative methods of dialysis. A discussion of equilibrium dialysis and of electrodialysis is not included.

A number of theoretical analyses of the dialysis process have been attempted,[2] which have predicted the behavior of model systems with varying degrees of success. From a practical point of view, several important factors are self-evident. These include the nature of the dialysis membrane and its selectivity, by which is meant its ability to allow the free flow outward of low molecular weight contaminants, while retaining all the high molecular weight components; the relative surface area of the dialysis membrane to the volume of solution which is being dialyzed; the relative volume of solution inside the dialysis membrane, the retentate, to that of the solvent outside, the diffusate; the concentration of diffusible substances outside the membrane, which will oppose further outward flow. Efforts at improving the dialysis apparatus are usually aimed at optimizing one or more of these parameters.

Dialysis Membranes

Materials suitable for use as dialysis membranes must be in the form of thin films of highly polymerized substances, which, in the presence of solvent, swell to form a molecular sieve, whose pores will allow the

---

[1] L. C. Craig, this series, Vol. X, p. 870.
[2] R. E. Stauffer, in "Physical Methods in Organic Chemistry" (A. Weissberger, ed.), Vol. 3, p. 65. (Wiley Interscience), New York, 1965.

passage of low molecular weight solutes and solvents, while preventing the passage of proteins. The membrane must be chemically inert and without fixed charged groups, which could cause binding of solutes. A number of substances which seem to meet these requirements have been employed, including animal membranes, collodion, colloids deposited in porous pot, polyethylene and Kel·F. At present, cellophane tubing is used almost exclusively for routine work, owing primarily to the fact that tubing of fixed dimensions and of fairly reproducible porosity is commercially available. The tubing is available in six sizes under the trade name "Dialysis Membrane," and its dimensions are given in the table. Other sizes are available by special order.[3]

DIALYSIS MEMBRANE SIZES

Size identity	Visking identity (redundant)	Flat width (inches)	Wall thickness (inches)
8 Dialysis	8/32	0.390	2×10^{-3}
20 Dialysis	20/32	0.984	8×10^{-4}
27 Dialysis	27/32	1.312	10^{-3}
36 Dialysis	36/32	1.734	10^{-3}
1⅞ SS dialysis	1⅞ SS	2.88–3.14	1.6×10^{-3}
3¼ SS dialysis	3¼ SS	4.65–5.10	3.5×10^{-3}

The membranes are not manufactured in a series of graded pore sizes, though experience suggests that the narrower sizes are somewhat more porous than the wider tubing. From the rate of flow of water through the membrane, the manufacturers estimate the average pore radius to be 24 Å. That this is a reasonable estimate is shown by the fact that tubing which will allow the slow passage of ribonuclease A, will not allow the passage of chymotrypsinogen. On the basis of hydrodynamic measurements, these molecules may be regarded as spheres, of radius 19 Å and 24 Å, respectively. Cellulose tubing is available in a much greater variety of widths under trade names other than "Dialysis Membrane." The use of such tubing may be advantageous under certain conditions, especially as it is claimed that this type is of a lower pore radius than Dialysis Membrane. However, the porosity of the cellulose membranes does vary from time to time, since it is manufactured to the requirements of the sausage industry.

Many of the recipes for dialysis that are reported in the literature described the size of tubing employed by the Visking Company's nomen-

[3]The techniques described below were all performed with such tubing, obtained from the Visking Company, a division of Union Carbide Corporation, 6733 West 65th Street, Chicago, Illinois 60638.

clature. A change in ownership of the company was accompanied by a change in nomenclature. Since this may cause some confusion, it seems worthwhile to explain the two systems. The Visking Company measured the size of their tubing by its external diameter in inches when full of water: 8/32, 18/32, and 20/32 inch, etc. Union Carbide Corporation has dropped the fractional denominator of the size identity and refers to these sizes as 8, 18, and 20. When a size is indicated as 36/100, this means that the expanded diameter of the tubing is 36/32 inch and that the tubing is sold in rolls 100 feet in length.

Preparation of Dialysis Tubing

Cellophane tubing is obtained from the manufacturers in rolls 50–1500 feet long, each sealed in a polyethylene bag. The tubing contains glycerin as a plasticizer, traces of sulfurous compounds and heavy metal ions. Some of these contaminants may have rather drastic effects on protein solutions and obviously must be removed before use. The manufacturers recommend that the tubing be soaked in distilled water, in 0.01 N acetic acid, or in a dilute solution of EDTA. However, most workers employ a more vigorous procedure. The method described below has been found successful in a number of laboratories.

Sufficient tubing for the experiments at hand is removed from the roll and allowed to simmer for 1 hour in 2–5 liters of 50% ethanol. The tubing remaining in the roll should be stored in an air-tight container and under refrigeration. During and after preparation, tubing should always be handled with clean surgical gloves. The immersion is repeated sequentially for periods of 1 hour in another equal volume of 50% ethanol, in two changes of 10 mM sodium bicarbonate, 1 mM EDTA solution and in two changes of distilled water. The tubing is stored at 4°, under distilled water. For storage of more than a few days, a preservative, such as sodium azide should be added. Tubing which dries out during storage must be discarded. Before use, the tubing should be thoroughly rinsed with distilled water and then with the solvent which the protein solution is to be dialyzed against. It should be filled with solvent to ensure that no physical damage has occurred during the preparation.

This treatment seems to remove the impurities mentioned above, leaving a dialysis membrane which usually has no adverse effects on protein solutions. However, evidence has been presented that boiled tubing will catalyze the hydrolysis of pyrophosphate to orthophosphate.[4] The effect was ascribed to oxidation of the membrane, pro-

[4]P. R. Watson, J. E. Pittsley, and A. Jeannes, *Anal. Biochem.* 4, 505 (1962).

ducing an increase in the surface charge. A similar phenomenon has been observed on ultrafiltration of oligoribonucleotide solutions through charged synthetic membranes.[5] Consequently, caution must be exercised during attempts to purify or remove such substances by dialysis.

For certain purposes, it may be necessary to change the pore size of the membrane. Mechanical and chemical methods of achieving this have been reviewed by Craig.[1]

Methods of Dialysis

The basic procedure of dialysis, practicable for solution volumes of 1–100 ml, is familiar to all. The solution to be dialyzed is sealed inside a length of tubing, which is then suspended in a much larger volume of solvent. The solvent is agitated with a magnetic stirrer. The procedure is normally carried out in the cold, although, if there is no risk of denaturation or degradation, equilibration will occur more rapidly at room temperature.

The dialysis tubing may be sealed by tieing at least two tight knots directly in the tubing, at each end. Alternatively, the tubing may be closed with unmercerized thread, nylon cord, or dental floss. If the tubing is to be suspended from a support, the upper end may be tied around a rubber stopper. Tappan[6] has described an arrangement which is especially convenient when it is required to withdraw samples from the dialysis bag, during the procedure: the open mouth of the sack is inserted into a rubber sleeve, the sack and sleeve then being closed tightly with a Hoffman style screw clamp. When a large number of dialyses are to be performed, labels may be attached to these clamps, differentiating the samples therein. The clamps are supported by two glass rods which are laid across the top of the dialysis vessel (Fig. 1). On no account should any attempt be made to seal the sack with any form of adhesive.

Whatever the method of sealing, the sacks should include sufficient dead space between the surface of the solution and the seal to allow room for the appreciable flow of solvent into the sack during the course of the experiment. The rate and extent of this inflow depend on several factors, but for an overnight dialysis of a concentrated protein solution, an increase in volume of 50% is not unusual. Since the tubing is fairly resilient, it will expand if such a space is absent, with a concomitant change in the pore size of the membrane. Under extreme conditions, the tubing may burst.

[5]S. R. Jaskunas, C. R. Cantor, and I. Tinoco, *Biochemistry* 7, 3164 (1968).
[6]D. V. Tappan, *Anal. Biochem.* 18, 392 (1967).

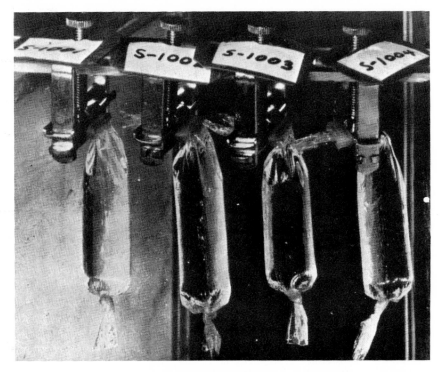

FIG. 1.   The method of suspension of dialysis sacks suggested by D. V. Tappan.[6] Reproduced from *Analytical Biochemistry* [4, 505 (1962)] by courtesy of the author and publishers.

Since dialysis is essentially a diffusion-controlled process, complete removal of dialyzable substances from the sack will be opposed by increasing concentrations of these substances in the exterior. The main effect of stirring is to dissipate solute throughout the diffusate, reducing the effective concentration of solute at the membrane and decreasing the extent of backward flow. A further increase in the rate of dialysis may be achieved by stirring the solution within the sack, thereby preventing the formation of concentration gradients on that side of the membrane.

Perhaps the only systematic investigation into the effect of stirring on dialysis is that of Ogston,[7] who studied the behavior of the indeculator, a device used in his own laboratory. Although the device has met with little popularity, some principles of general relevance were revealed. In the indeculator, the dialysis sack is suspended from a rubber stopper,

[7] A. G. Ogston, *Arch. Biochem. Biophys.* **89**, 181 (1960).

which in turn is mounted on a vertical axle. A system of weights and pulleys serves to rotate this axle in alternate directions, causing stirring of both the inner and outer fluids. As controls, the inner and outer fluids were either stirred separately, by bubbling nitrogen through each, or stirring was omitted. The measured parameter was the flow of sodium chloride from the sack.

Stirring of the diffusate was found to produce some increase in the rate of flow, but not so much as that attained when both diffusate and retentate were stirred. Surprisingly enough, even under optimum conditions, the rate of dialysis was merely double that in the absence of stirring. Without stirring, the viscosity of the solution had little effect on the rate of dialysis. Stirring was found to produce no increase in the rate of dialysis when the viscosity of the retentate was more than twice that of the diffusate, probably as a result of the mechanics of the apparatus.

The Rocking Dialyzer

A popular method of agitating both solvent and solution during dialysis is the rocking dialyzer, an apparatus first described by Kunitz.[8] Here, solvent and sack are contained in a large cylinder, normally of approximately 2 liters capacity. The cylinder is mounted on a horizontal board, which is supported across its middle by an axle. One end of the board is supported on an asymmetric cam which, on rotation causes the board to rock about the axle, so that its plane ranges between 30° above and below the horizontal. The cam is driven by a low power electric motor geared to rotate several times each minute. If the cylinder is not quite filled with solvent, this motion will cause both solvent and sack to flow up and down, thus increasing the speed of the dialysis. A further acceleration results from including two or three glass marbles inside the sack. Agitation must not be too vigorous in order to avoid breaking the membrane. This device seems to increase the rate of dialysis by about a factor of two, when compared with stationary dialysis.[9]

A variation on this theme is the "rock and roll" dialyzer,[10] which is advantageous when a large number of small samples are to be treated. Twelve 150-ml cylinders are clipped about a horizontally mounted drum, of length and diameter 10 inches. The clips are arranged so that the cylinders are at an angle of 10° to the axis of the drum. The rocking motion is produced by mechanical rotation of the drum, at a rate of 28

[8] M. Kunitz and H. S. Simms, *J. Gen. Physiol.* 11, 641 (1928).
[9] M. A. Lauffer, *Science* 95, 363 (1942).
[10] A. M. Stewart, D. J. Perkins, and J. R. Greening, *Anal. Biochem.* 3, 264 (1962).

rpm. Dialysis of 0.1 M sodium chloride required 2–3 hours to reach equilibrium, indicating that the device operates with efficiency similar to that of the rocking dialyzer.

The Rotating Dialyzer

This device had achieved a certain amount of popularity, probably because it is one of the few dialysis systems which can be purchased ready for use.[11] The method was first described by Durrum *et al.*[12] A number of sacks are suspended from a circular rotor, into a large bath of solvent. The method of suspension is to insert the knotted mouths of the sacks through slots in the edge of the rotor. For small-scale dialysis, prepared sacks may be purchased in which the mouth of the tubing is sealed onto a plastic collar while the bottom is sealed by a plastic plug, an arrangement which facilitates loading and withdrawal of samples.

Solvent is agitated by rotating the rotor at 10 rpm, the direction of rotation being reversed several times each minute. The solution in the sack is agitated by including a number of vanes along the bottom of the bath. The height of the rotor is adjusted so that the bottoms of the sacks bump against the vanes.

No evaluation of the performance of this device has been reported, although it seems to be generally satisfactory. The apparatus is available in a range of sizes; a micro model handles 8 sacks, each containing as much as 2.4 ml of sample, while a large capacity device treats 16 samples of 220 ml each. The solvent vessel may be equipped with ports for continuous flow of diffusate.

Circulatory Dialyzers

As stated, the main purpose of stirring during dialysis is to dissipate dialyzable components from the region of the membrane. As dialysis proceeds, i.e., as equilibrium is approached, this becomes less effective and further dialysis can only be achieved by replacing the diffusate with a fresh solution. During a normal dialysis fresh solvent is exchanged two or three times. The logical limit of the procedure is to provide a continuous flow of fresh solvent into the vicinity of the membrane, while solvent containing dialyzable material is removed.

The simplest form of flow device is that described by Hospelhorn,[13] which is shown in Fig. 2. The solution is sealed inside a length of

[11] "The Oxford Multiple Dialyzer" available from Arthur H. Thomas Company, Philadelphia, Pennsylvania.
[12] E. L. Durrum, E. R. B. Smith, and M. R. Jetton, *Science* **120**, 956 (1954).
[13] V. D. Hospelhorn, *Anal. Biochem.* **2**, 180 (1961).

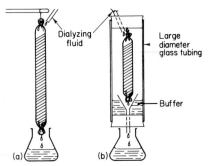

FIG. 2. The circulatory dialyzer of V. D. Hospelhorn. Reproduced from *Analytical Biochemistry* [2, 180 (1961)] by courtesy of the author and publishers. The efficiency of the device and the ease of wrapping the thread around the sack are greatly improved by including a closely fitting glass plug inside the dialysis tubing.

dialysis tubing suspended from a support in the cold room. The diffusate is allowed to drip onto the sack from above at a rate of about 2 ml/min. A small-diameter, untreated cord is wrapped around the tubing in a spiral of low pitch with successive windings about 3 mm apart. The surface tension of the diffusate serves to hold the cord to the tubing. When the complete length of tubing is covered, the flow rate of diffusate is reduced. If the diffusate is water, the experiment may be performed in the open. However, if the diffusate is a buffer, the apparatus must be enclosed so as to prevent changes in buffer composition due to evaporation.

When 50 ml of 0.9 M ammonium sulfate solution were dialyzed in this apparatus through 18/32 tubing and against distilled water, 99% of the salt was removed in 7 hours. The total volume of diffusate was 650 ml.

Lauffer has described a method by which the rocking dialyzer can be converted into a circulatory system.[9] By including two simple check valves in the circuit, the rocking motion is used to drive diffusate through the cylinder, from an external reservoir.

The Dialysis of Large Volumes — Countercurrent Dialysis

Obviously, when the volume of retentate to be treated becomes much larger than 100 ml, the lengths of dialysis tubing and the volumes of diffusate which must be used in any of the above techniques soon become relatively impractical. However, since the treatment of large volumes is a problem often met in the course of large-scale enzyme preparations, a number of investigators have evolved devices aimed at just this capability. These devices all take the form of countercurrent dialysis, in which solvent is allowed to flow along one side of a dialysis membrane while the solution to be treated flows slowly in the opposite direction on the other side. A large number of instruments, of varying degrees of complexity, have been described. Since all require at least some machine shop work, no attempt to describe their construction will be made. These devices are of great use in laboratories where the dialysis of large volumes is a routine procedure, the interested reader is referred to the original articles, where their construction is described in detail.[2,14-16] A form of countercurrent dialysis apparatus is also commercially available.[17]

Dialysis of Small Volumes

The dialysis of large volumes of solution is a problem which frequently occurs in the preparation of biological macromolecules. Once the species under consideration has been obtained in a pure state, it is often desirable to characterize its behavior by various physical chemical techniques. Before these can be performed, one is faced with the opposite problem, that of equilibrating small volumes, i.e., 1 ml or less of protein solution with a solvent of known composition. Several ways of achieving rapid dialysis equilibrium in such systems have been suggested.

The simplest seems to be that of dialysis with the narrowest type of tubing available. However, this is also the most porous tubing (see the table), and its use is precluded when low molecular weight proteins are studied, since appreciable escape through the membrane may occur. Rapid dialysis can only be performed in wider tubing, arranged so as to maintain a reasonable surface area of membrane between the two liquids. A simple arrangement was suggested by Seegers.[18] A glass plug is formed by sealing the ends of a piece of 2-cm diameter glass tubing. The length of the plug is determined by the volume of solution to be dialyzed. For a 1-ml sample, the plug should be 2 cm in length. Before the plug is finally sealed, sufficient mercury is introduced into it so that it will barely float. The plug is slipped into a length of size 27 dialysis tubing, which is sealed just below the plug. The sample is introduced with a Pasteur pipette and the tubing sealed just above the plug. Thereby, the sample is contained in the small volume of the annular space between plug and casing (approximately 0.7 mm wide), but dialysis may occur through the entire surface area of the membrane. Dialysis is allowed to occur in the usual manner.

A second method of dialyzing small volumes, one which seems to be

[14]L. C. Craig and K. Stewart, *Biochemistry* 4, 2712 (1965).

[15]J. G. Davis, *Anal. Biochem.* 15, 180 (1966).

[16]S. Mandeles and E. C. Woods, *Anal. Biochem.* 15, 523 (1966).

[17]"The Webcell Continuous Dialyzer" available from the Scientific Glass Apparatus Company, Inc., Bloomfield, New Jersey.

[18]W. H. Seegers, *J. Lab. Clin. Med.* 28, 897 (1943).

somewhat faster and is also advantageous when a suspension, rather than a solution is involved, is the "flat dialyzer" of Cabib and Algranati.[19] The dialysis membrane is maintained in the shape of a rectangular cell by means of a Plexiglas frame. The frame is made by cutting a square from a 1 mm-thick sheet of Plexiglas. The side of the square must be 2 or 3 mm less than the flat width of the dialysis tubing to be employed. Four triangular pieces are then removed from the square to leave a square frame with diagonal struts. The frame is introduced into a length of dialysis tubing, which is cut to allow about 5 cm of tubing to hang on each side of the frame. One of the ends is wrapped around a narrow plastic rod and tied to the frame with a rubber band. Solution is introduced into the cell, from the other end, by means of a Pasteur pipette. When most of the air bubbles have been squeezed from the cell, the other end is sealed in a like manner. The cell is then mounted as shown in Fig. 3, by being inserted in a rubber band, R. This is tied at one end to a glass rod, G, and, in turn, supported by a clamp. The other end of the rubber band is tied to an untreated cord, S, and suspended by means of the cord from a crank, C. The cord is kept tight by the addition of a small weight, W. The crank is rotated by a motor, at 50 rpm, to produce a wagging motion of the cell.

Using such an apparatus, 96% of the salt was removed from 1 ml of a 2 M ammonium sulfate solution in 30 minutes. In normal suspended dialysis, only 60% of the salt was removed in an equal period of time.

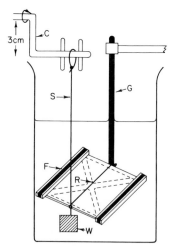

FIG. 3. The rapid dialyzer of E. Cabib and I. D. Algranati. Reproduced from *Nature* (*London*) [**188**, 409 (1960)] by courtesy of the authors and publishers.

[19]E. Cabib and I. D. Algranati, *Nature* (*London*) **188**, 409 (1960).

[5] Lyophilization

By JOHANNES EVERSE and FRANCIS E. STOLZENBACH

Application

Lyophilization, the drying of materials in the frozen state, is a technique that is widely used in the laboratory as well as in industrial processes for the concentration, storage, and distribution of biological materials. While it was hardly more than a laboratory curiosity in 1939, the technique was developed rapidly during World War II in order to preserve and store large quantities of plasma, vaccines, strains of bacteria and viruses. In the years following World War II the technique of lyophilization, or freeze-drying, has found applications in many fields including the bulk-preparation, storage, and distribution of enzymes and the preservation of bacterial cultures.

Many biological materials, which will rapidly deteriorate even in frozen solutions, can be kept in a viable state for many years after lyophilization. Crude extracts, which contain labile active factors, may be freeze-dried to ensure a single homogeneous source of material for future investigations. Usually in these cases the material is distributed over a large number of small containers, and then lyophilized. The containers are sealed after the drying process and stored in the cold. When the need arises for more material, one of the containers is opened. The most useful container for this purpose is the glass ampule. When lyophilization is complete, the vacuum is released by filling the system with an inert gas, and the ampules are sealed at the neck with an oxygen torch. This method is also widely used for the commercial distribution of small amounts of biological materials in a purified form (e.g., bacteria cultures maintained by the ATCC).

Not all biological materials are stable during lyophilization. If lyophilization is being considered as a means of preservation, the method should first be tested with a small amount of the material to ensure that it is not affected by the process.

A more extensive discussion regarding the use of freeze-drying may be found in an old, but still valuable, monograph by Flosdorf.[1] This book also contains considerable information about the lyophilization of medical and bacteriological products.

---

[1] E. W. Flosdorf, "Freeze-drying." Reinhold, New York, 1949.

Techniques and Precautions

Samples to be lyophilized should be aqueous solutions. The presence of organic solvents is highly undesirable, since they will lower the freezing point of water and thereby increase the probability of melting the sample during lyophilization. Melting usually results in extensive foaming and partial denaturation of enzymes. Furthermore, organic solvents with a low boiling point, e.g., ethanol or acetone, possess a sufficiently high vapor pressure at low temperatures to allow most of it to bypass the cold-trap and to condense in the vacuum pump oil. The consequences of such a dilution of the oil may well be disastrous for the vacuum pump involved.

Careful consideration should be given to the lyophilization of buffered solutions. In many cases, the presence of salts in the lyophilized product may not be of any special concern to the investigator, or may be to some advantage in controlling the pH if the sample is to be dissolved again. However, the freezing of buffered solutions can cause considerable shifts in the pH. For example, when a solution of sodium phosphate buffer at pH 7.0 is frozen, the disodium phosphate will crystallize before the monosodium phosphate, and as a result the pH of the solution will be close to 3.5 just before it is completely frozen. In the case of a potassium phosphate buffer the effect will be reversed, since monopotassium phosphate is less soluble than the dipotassium salt, and a pH of 7.5 to 8 may be the final result.[2] The extent to which these shifts occur is related to the rate at which the samples are frozen; the faster the freezing process, the less the shift in pH. Moreover, since slow freezing will result in a partial "freezing out" of the solutes, it becomes important under these circumstances to consider the effects of high salt concentrations on the materials that are to be lyophilized.[3] Many complications may be avoided by a complete desalination of the solutions before freezing.

A variety of containers are suitable for lyophilization. A proper container should withstand the outside pressure during lyophilization under high vacuum, and should be made of a material that allows a reasonable transfer of heat from outside to inside. The most commonly used containers in the laboratory are glass ampules and round-bottom flasks. The size of the container should be such that the solution to be lyophilized occupies not more than 20% of the useful volume.

When very dilute solutions are to be lyophilized, some precautions may have to be taken. The residue from a dilute solution is usually very

[2] O. P. Chilson, L. A. Costello, and N. O. Kaplan, *Biochemistry* 4, 271 (1965).
[3] L. Van der Berg and D. Rose, *Arch. Biochem. Biophys.* 81, 319 (1959); 84, 305 (1959).

light and fluffy, and some of the material may easily be taken along with the flow of water vapor into the cold trap. This may be avoided by placing a small wire screen or glass wool in the neck of the container. The barrier will catch material carried by the water vapor without significantly reducing the flow rate. Another method consists of wrapping the outside of the container in a towel in order to reduce the heat transfer from outside to inside, thereby reducing the flow rate of the water vapor. The latter method is useful also as a precautionary measure, when valuable materials have to be lyophilized.

Samples are to be frozen in a cold bath of acetone and solid CO_2 prior to lyophilization. In order to avoid secondary effects on the biological material it is important to achieve complete freezing of the samples as quickly as possible. Cold baths containing methyl Cellosolve are less effective in this respect than cold baths containing acetone or alcohol, since their increased viscosity does not allow for as rapid a transfer of heat. The container is immersed up to its neck in the cold bath for a few seconds in order to cool the glass and to avoid the formation of undesirable stresses in it. The container is then held at about a 45° angle and slowly rotated to allow the solution to freeze along the container wall in the form of a shell. During this operation the container should be submerged as deeply as possible in the cold bath. After the sample has completely solidified, it is maintained at −70° until it can be connected to the lyophilizing system.

A lyophilizer of sufficient capacity should be selected. If a large number of small samples are to be lyophilized simultaneously, a high flow rate of water vapor will be present in the system, due to the large size of the evaporation surface, i.e., the combined surfaces of all containers. In this case, the capacity of the vacuum pump should be sufficiently high to maintain a proper vacuum, and the cooling capacity of the cold trap should be sufficient to handle the high flow of water vapor. When a few samples of a large volume have to be lyophilized, the volume capacity of the cold trap should be sufficient to contain the total volume of ice, without losing much of its cooling capacity. The flow rate of the water vapor in such cases is generally slow, due to the relatively small evaporation surface.

In preparing the lyophilizer it is important to fill the cold trap with dry ice and solvent, or to start the refrigerating system and have it reach operating temperature before the system is evacuated. Before the vacuum pump is turned on, all connections should be closed to avoid a large flow of air through the cooling system. The lyophilizer is then evacuated and checked for the presence of leaks. When a pressure of 100 μ or less is obtained, the samples may be connected to the system.

In order to safeguard against a possible implosion, it is essential to have the container wrapped in a towel at the moment that the vacuum is applied to it.

During lyophilization the samples are left undisturbed until all ice has evaporated. Removal of the outside ice shell and other movements of the container may result in a partial loss of the sample, due to the sudden increase in flow rate of the water vapor. Thawing of the samples during lyophilization may be indicative of a clogged cold trap, a leak in the vacuum system, a poorly operating vacuum pump, or insufficient capacity of the lyophilizer. Furthermore, the presence of salts or high-boiling organic solvents may allow thawing. In the latter case an ice shell may remain on the outside of the container. Thawed samples must be removed, and the cause of the thawing should be eliminated. The samples need to be frozen again before lyophilization is continued.

When the process is completed, air is slowly admitted into the apparatus before samples are removed from the system. If the lyophilizer is equipped with a diffusion pump, this should be turned off at least 5 minutes before air is admitted to the system. After release of vacuum, samples are taken from the connectors and the vacuum pump is turned off. Finally, the refrigeration unit is turned off or the cold trap is removed. Turning off the vacuum pump before the vacuum has been released can result in a back-up of pump oil into the cold trap, with the possibility of some oil reaching the samples. It is important that cooling be continued until the vacuum pump has been turned off in order to prevent the possibility of water vapor reaching the vacuum pump.

Lyophilizers generally require a minimum of servicing if the instruments are properly cared for after use. The vacuum pump is the heart of the instrument, and should be maintained in good operating condition. Oil has to be changed regularly, at least once a month. The oil should also be changed when there are indications that water, organic solvents, or acids have entered the pump. The cold trap is thawed, emptied, and cleaned after each lyophilization. Allowing water to remain in the system for a prolonged period may result in condensation of water vapor inside the pump, eventually leading to corrosion and malfunction. Any connections between parts should be inspected for proper sealing, and, if necessary, be lubricated with a high vacuum grease.

Equipment

Udder-Type Freeze-Drier. The simplest and most useful freeze-drier is the type which resembles a cow's udder. This may be easily assembled in any laboratory using standard taper glassware, as shown in Fig. 1. The apparatus consists of a large chamber (*a*) and a smaller inner

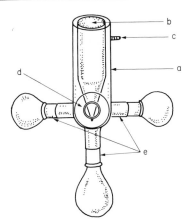

FIG. 1.   Udder-type freeze-drier: (*a*) freezing chamber; (*b*) dry ice trap; (*c*) vacuum outlet; (*d*) manifold; (*e*) standard taper joints.

chamber (*b*). Vacuum is applied at a serrated tubulation (*c*), which is located at the top of the larger outer chamber. At the bottom of the large outer chamber is an attachment (*d*) from which standard taper joints (*e*) are projected. The number may vary, but one must keep in mind the capacity of the freezing chamber as a limit on size and number of ports. A mixture of dry ice and a solvent of choice[4] is then placed in the inner chamber (*b*) to provide subzero temperature for collection of the solvent. Samples are placed on standard taper outlets (*e*) using the proper size standard taper flask as discussed in Techniques and Precautions.

Lyophilization is under way when the vacuum is applied and the unused ports are closed to produce a good working vacuum. A system of this type, as in any system, is only as good as the pump itself. The advantage of a system of this type is the ability to quantitatively collect both the solvent and the freeze-dried sample. This is especially valuable in reclaiming labeled solvents. Another advantage is the ease in handling such a small sample with a minimum of time. It appears that a freeze-drier of this size is a necessary supplement to even the most elaborate type of drier in any laboratory. An additional dry ice trap between the pump and the actual freeze-drier is not always necessary for use, but it is recommended. A more detailed presentation of this apparatus can be found in an article by Campbell and Pressman.[5]

[4]Although ethanol and acetone are good solvents for the purpose, they are volatile and tend to evaporate between uses. Ethelene glycol is preferable because it is not volatile and need not be replenished frequently.
[5]D. H. Campbell and D. Pressman, *Science* 99, 285–286 (1944).

Freeze-Mobile, Small Size. This type of freeze-drier is probably the most widely used because of the relatively low cost and the fact that it is a self-contained and totally portable unit.

The unit consists of a two-tiered lab cart with a pump and a trap on the lower tier. A rubber vacuum tubing then connects the lower trap to a manifold, which is mounted on the upper level. The larger manifold suggested by the manufacturer has 12 ports with a maximum capacity of 100 ml for each port. A limited capacity of only 100 ml is a minor drawback of this manifold. On the side of the units is a McLeod gauge for checking the vacuum in the whole system. The inner trap for the manifold and the lower trap, both have a dry-ice bath capacity of 2.8 liters, which will maintain temperature for up to 12 hours.

In summary, we have found that the convenient features of this unit are its portability, the efficiency of the pump supplied, and the quick-seal valves supplied for the ports of the manifold.

Freeze-Mobile, Large Size with Mechanical Refrigeration. This, in our opinion, is by far the best freeze-drier for versatility and volume. The major drawback is the initial cost, which is slightly over $3000 including the drying chamber and a McLeod gauge. However it is economical to operate because the mechanical refrigeration eliminates the need for constant replenishment of solvents and dry ice.

This particular unit, is completely enclosed in a cabinet with front-opening doors. A condenser, with a capacity of 12 liters of ice and able to reach a temperature of $-65°F$, is contained in the cabinet along with an extremely efficient vacuum pump. A heater-blower combination is also present for defrosting the condenser. The defrosting operation usually requires a maximum of 20 minutes and eliminates the need for dismantling the condensor traps. Controls are mounted on the front of the cabinet along with a McLeod gauge.

The vacuum drum is a manifold with facilities for handling bulk drying. When used as a normal manifold there are a total of 16 ports available, 8 with a ½-inch outside diameter and 8 with a ¾-inch outside diameter. The same chamber may be used for bulk drying by inserting a 3-shelf heat rack which is supplied with the unit. A temperature of 90°F is attainable on each rack, and the racks are adjustable. One of the problems of the other manifolds previously described is overcome in this chamber by using a Lucite top which merely fits on top of the chamber. A thin layer of vacuum grease makes this an efficiently sealed system. Another advantage is that there is no collection of condensate in the actual vacuum chamber. Thus, any volume of material may be used on any port as long as one is within the 12-liter capacity of the condenser.

A unit of this sort allows one to bulk dry or to sample dry in the same versatile chamber. It handles loads of up to 12 liters of condensate and allows freeze drying of 18 samples at one time. The condenser can be totally defrosted and placed back in operation in less than 30 minutes.

[6] Ultrafiltration for Enzyme Concentration

By WILLIAM F. BLATT

Following the pattern of most biochemical procedures, techniques for the enrichment or concentration of enzymes are becoming almost as numerous as the plethora of enzymes themselves. However, the unique properties of specific enzymes are often such that only a given technique out of the multitude available is advisable. For the most part, those techniques to be discussed were originally devised as general preparative methods for protein preparation; while well suited for many enzymes, some possess obvious shortcomings for highly labile species. For purposes of review, and to place ultrafiltration into the general schema of isolation techniques, the following simplified classification has been devised:

1. Precipitative methods
    a. salt and/or solvent concentration
    b. isoelectric precipitation
    c. concentration by ultracentrifugation
2. Solvent removal
    a. flash and vacuum evaporation
    b. lyophilization
    c. freeze-thaw procedures
3. Partition systems
    a. gel exclusion (Sephadex, Biogel, Lyphogel, etc.)
    b. adsorption chromatography
    c. membrane-moderated procedures
        (1) pervaporation
        (2) dialysis (against solutes of high molecular weight)
        (3) electrodialysis
        (4) centrifugally accelerated ultrafiltration
        (5) pressure ultrafiltration

Although suitable for many applications, precipitative and solvent extraction procedures are more apt to be denaturing because of the phase changes common to these techniques. To date, those of the third category, and specifically *membrane-moderated* procedures, provide the most promise.

Membranes capable of selectively admitting or retaining species of given size and shape have been known for some time,[1] and the evolution of ultrafilters with varying degrees of permselectivity have continued to be of interest.[2] In general, most early ultrafiltration membranes were fabricated from collodion (cellulose nitrate) or regenerated cellulose; as such, they suffered from the low water transport common to these classes of materials.

Typifying the earlier approach would be the *pervaporative method* wherein a solution of macrosolute is suspended in Visking tubing held in an air or gas stream. The evaporation of water from the external surface (and the accompanying salt deposition) would lead to further accelerated transport of water from the bulk solution to the outside. Control of the degree of concentration and total recovery of retentate are difficult, and the processing of large volumes is quite unwieldy. But, the procedure does work, and if the adsorptive losses are tolerable, this is certainly one of the least expensive methods available.

Dialysis with cellulosic tubing, utilizing nonpermeating solutes of high molecular weight in the outer dialyzate, can be an effective approach when solution : exchange surfaces are maximized, but basically, it possesses the same disadvantage as exists with all cellophane, i.e., the low water transport characteristic of this material.

Electrodialysis, providing for water removal via salt transport and accompanying endosmotic effects has been used in some flow-through configurations with good results, but concentration of larger volumes requires extensive equipment.

This brings us to the variety of more commonly used ultrafiltrative techniques, those processes dependent on pressure to drive solvent convectively through permselective membranes. Pressure can be provided from exogenous gas sources, by vacuum applied to the underside of the ultrafiltration membrane, or by centrifugal forces in cells of special design. With any of these ultrafiltration methods, water flows through the membrane under the influence of the gradient, carrying with it salts and other low molecular weight, permeable species. Akin to reverse osmosis, pressure ultrafiltration differs in the order of magnitude of the retained solutes and in the lower pressures used in the latter process. In reverse osmosis we are concerned with solutes within

[1] J. D. Ferry, "Ultrafilter Membranes and Ultrafiltration," *Chem. Rev.* **18**, 373 (1935).
[2] W. F. Blatt, B. G. Hudson, S. M. Robinson, and E. M. Zipilivan, *Nature (London)* **216** (5114), 511 (1967).

one order of magnitude of the size of water molecules; in the latter case, retained solutes are of considerably larger size.

Ultrafiltration membranes are generally graded in terms of molecular weight cutoff levels. It is obvious that these are extremely arbitrary designations. Molecular parameters such as size, shape, and charge are more valid criteria to characterize the selectivity of a given membrane. The fact that an ultrafiltration membrane can pass species of a given molecular weight, while being retentive to the solute under study, enables it to serve for more than a simple dewatering step during the overall enzymatic purification procedure. Were only water to be removed during concentration (as is the case with solvent extraction techniques), hyperconcentration of simple and buffer salts occurs. Free passage of these species during membrane ultrafiltration ensures that ionicity and pH are maintained during the procedure, further reducing the possibility of denaturant change. Moreover, if other lower molecular weight contaminating species are removed with the microions, an additional purification step has been afforded during the concentrative process.[3,4]

To improve ultrafiltration, efforts were devoted initially to the selection of membranes that (1) possessed narrow degrees of selectivity ("sharp" vs. "diffuse" cutoff membranes), (2) that would provide rapid ultrafiltration flux, and (3) were minimally absorptive for a broad variety of solutes. These are admirable aims, but membranes clearly meeting all criteria continue to remain elusive. For the most part, some degree of compromise has been necessary. Table I, admittedly incomplete, presents a survey listing of a variety of ultrafiltration membranes, their commercial source, and rated exclusivity limits. The values listed reflect the manufacturers' data both with respect to water flux and to solute retentivity; within the limit of our experience, they appear reasonable. As noted in the table, membranes are available in a broad range of cellulosic and polymeric forms with molecular weight cutoff levels ranging from several hundred through several hundred thousand. Choice of membrane is not always dictated by the solute, in some cases work at extended pH precludes use of cellulosic materials. On the other hand, certain ionic sites in the polymer hydrogels bind strongly coordinated ions, e.g., Diaflo UM membranes are not recommended for use with phosphate buffers above 0.02 M. Autoclavability if desired, as well as solvent compatibilities must be considered in all applications—

[3] W. F. Blatt, M. P. Feinberg, H. B. Hopfenberg, and C. A. Saravis, *Science* 150 (3693), 224 (1967).
[4] A. S. Michaels, "Ultrafiltration," *in* "Advances in Separations and Purifications" (E. S. Perry, ed.). Wiley, New York, 1968.

TABLE I

COMMERCIALLY AVAILABLE ULTRAFILTRATION MEMBRANES

Description	Manufacturer	Type	Water permeability at 100 psi (ml/cm²/min)	Molecular weight cut-off, 80–100% retention
Gel Cellophane	du Pont, Union Carbide	Homogeneous (?) cellulosic	0.004	10,000 (cytochrome c)
PEM Membrane	Gelman	Isotropic cellulosic	0.02	40,000 (ovalbumin)
"P-Membrane"	Schleicher & Schuell	Homogeneous cellulosic	0.08	60,000 (albumin)
CA-Type B	General Atomics	Anisotropic, cellulose acetate	0.007	600 (raffinose)
CA-Type C	General Atomics	Anisotropic, cellulose acetate	0.003	350 (sucrose)
DIAFLO UM-05[a]	Amicon Corp.	Anisotropic, polyelectrolyte complex	0.05	350 (sucrose)
DIAFLO UM-2	Amicon Corp.	Anisotropic, polyelectrolyte complex	0.1	600 (raffinose)
DIAFLO UM-10	Amicon Corp.	Anisotropic, polyelectrolyte complex	0.3	10,000 (Dextran 10)
DIAFLO PM-10	Amicon Corp.	Anisotropic, aromatic polymer	0.5	10,000 (cytochrome c)
DIAFLO PM-30	Amicon Corp.	Anisotropic, aromatic polymer	0.7	30,000 (ovalbumin)
DIAFLO XM-50	Amicon Corp.	Anisotropic, substituted olefin	0.7	50,000 (albumin)
DIAFLO XM-100A	Amicon Corp.	Anisotropic, substituted olefin	0.9	100,000 (7 S globulin)
DIAFLO XM-300	Amicon Corp.	Anisotropic, substituted olefin	1.1	300,000 (apoferritin)
HFA-100	Abcor, Inc.	Anisotropic, cellulosic	0.07	10,000 (Dextran 10)
HFA-200	Abcor, Inc.	Anisotropic, cellulosic	0.40	20,000 (Dextran 20)
HFA-300	Abcor, Inc.	Anisotropic, cellulosic	1.40	70,000 (albumin)
PSAC	Millipore Corp.	Anisotropic, cellulosic	1.2	1,000 (Brom Cresol Green)
PSED	Millipore Corp.	Anisotropic, cellulosic	0.75	25,000 (α-chymotrypsin)
PSDM	Millipore Corp.	Anisotropic, cellulosic	1.0	40,000 (ovalbumin)

[a] Ion exchange membrane.

for the most part, these are available from the manufacturer. Generally, the most efficient and nonplugging ultrafilters are "anisotropic," i.e., composed of a dense, very thin, permselective skin of controlled porosity supported by a spongy understructure. Although water permeability is often used as an expression of the ultrafiltration flux, it enjoys much the same dubious distinction as does molecular weight exclusion in characterizing membrane performance. In the presence of retained solute and, indeed, even in the presence of many salts, considerable decay in flux occurs both initially and as concentration proceeds. Flux reduction by retained solute has been shown to be caused by a phenomenon known as "concentration polarization," i.e., accumulation of macrosolute at the membrane surface which retards solvent flux. Furthermore, if this quasi-gel layer is left undisturbed, it can further consolidate to the extent that the system will cease to flow. Even if ultrafiltrate flow occurs, the passage of permeating solute can be dramatically affected by this "secondary membrane." A more complete discussion of this phenomenon and several approaches to its management are available.[5] To a large extent, control of concentration polarization depends on using the membranes in properly designed equipment to assure good solvent convection in the microscopic vicinity of the membrane. Therefore in applying ultrafiltrative techniques to enzyme concentration and/or partition, the user should not view the process as simply membrane limited, but possibly of greater importance as a function of the equipment used in its application. A greatly abbreviated history and development of equipment devised for membrane-moderated processes, much of it still in current vogue, is schematically shown in Fig. 1. These will be considered in turn.

Dead-End Systems

Where membranes are simply placed on or about supports, and the system pressurized, ultrafiltration will proceed until such time as a consolidated polarization layer of retained solutes reduces the flow to a negligible value. This system has been used with Visking tubing either wrapped about supports with the interior of the chamber exposed to partial vacuum; or placed within a rigid outer matrix and pressurized. In other systems, membranes are placed on sintered supports within vessels which can be pressurized. Despite the limitations imposed by polarization, the systems, though slow, can be useful when dilute

[5] W. F. Blatt, A. Dravid, A. S. Michaels, and L. Nelson, "Solute Polarization and Cake Formation in Membrane Ultrafiltration: Causes, Consequences, and Control Techniques," in Membrane Science and Technology (J. E. Flinn, ed.), Plenum, New York, 1970.

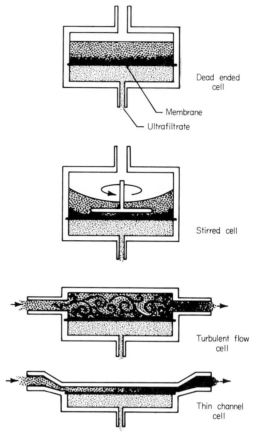

Fig. 1. A comparative view of the evolution of practical ultrafiltration systems.

solutions are employed (generally not to exceed final concentrations of more than 1.0%).

Stirred Systems

Any means of producing agitation at the membrane surface to effect the return of macrosolute to the bulk solution will reduce the polarization layer and help sustain the ultrafiltration flux. For the most part, agitation has been accomplished by motor or magnetically driven stirrers in close proximity to the membrane surface. With these systems, ultrafiltration rates as well as process volume are simply increased by an increase in the membrane area available for exchange. In some cases, mixing (with view toward added shear) has been provided by vibrator-agitated systems—the slight degree of flux improvement is offset to a large extent by the turbulence and frothing when the volume within the cell is reduced to near the level of the stirrer.

Simple stirred pressure systems are by far the most common, and for many applications appear to be satisfactory. Solutes to be concentrated are placed in cells of appropriate size (accessory reservoirs can increase the cell capacity), the system is pressurized, stirring is instituted, and ultrafiltration is allowed to proceed. For most membranes, increasing pressure to above 50 to 60 psi serves no advantage. The general guides for selecting a stirred, pressurized ultrafiltration system are also applicable to all ultrafiltrative processes: selection of a membrane retentive for the species under study; choice of a membrane displaying minimal absorption for the species under study, and stirring or agitator conditions that are minimally denaturing. Cold room operation is certainly recommended for most preparations; as a consequence, a near 50% decrease in flux will be observed with most membranes. Although these systems may not be applicable to all preparations, some studies have indicated that many enzymes may be concentrated successfully using these simple systems, with 80–90% recovery of active material.[6,7]

Turbulent Flow Systems

A more recent development in membrane ultrafiltration introduces the concept of high fluid flow rates in relatively wide channels parallel to the membrane surface, thereby providing turbulence to reduce the polarization boundary layer. Systems incorporating this approach include flat sheet devices, hollow tubes, and even hollow fibers, although the last may properly fit into the final, laminar flow category, discussed below.[8] For the most part, these larger units necessitate high capacity pumps and are more suited to commercial requirements[9] than to the smaller volumes generally considered in laboratory or small pilot operations.

Thin-Channel Laminar Flow Systems

This last approach may, in view of the labile nature of most enzymes, be the most feasible method. High fluid recirculation in thin channels, generally less than 50 mils in height, provides laminar flow at high shear rates and results in sustained ultrafiltrate flux even at relatively high

[6]D. I. C. Wang, T. Sonayama, and R. I. Mateles, *Anal. Biochem.* 26, 277 (1968).
[7]D. I. C. Wang, A. J. Sinskey, and T. A. Butterworth, "Enzyme Processing Using Ultrafiltration Membranes," *in* Membrane Science and Technology (J. E. Flinn, ed.), Plenum, New York, 1970.
[8]E. S. K. Chian and J. T. Selldorf, *Process Biochem.* 4, 47 (1969).
[9]D. I. C. Wang and A. E. Humphrey, *Chem. Eng.* New York, 76, 108 (1969).

solute concentrations. There is evidence to indicate that the shear forces imparted by this technique are more sparing to the enzymes than turbulent flow regimens.[10] Flow over the membrane is generally 10–100 feet/second; the transmembrane pressure for ultrafiltration is provided by external gas. Systems of this design concept vary from laboratory devices of several hundred ml capacity, to units that can process several hundred liters. The operation is basically the same for all units: reservoirs are filled, pressurized, and recirculation through thin channel devices of varying geometry initiated. The geometry of such systems, i.e., channel height, width, length, determines the pressure drop through the unit. Additionally, the viscosity of the solution under study as well as the recirculation rate directly influence the pressure drop. As a consequence, for any particular process, the ultrafiltration flux could be maximized by customizing the equipment. Currently available systems are designed with channel heights of 10–100 mils, for velocities of 10–50 ft/sec (recirculation rates, 100 to 1000 ml/min) which provide pressure drops of the order of 1 atmosphere for most protein solutions. Ultrafiltration fluxes that can be expected are 2- to 3-fold that of stirred cells when dilute solutes are employed, and up to 6-fold when high concentrations are encountered. Efficiency of fractionation, when desired, is considerably increased as well. Some measure of the relative efficiency of each of these processes has been illustrated schematically in Fig. 2. In that the ultrafiltration flow varies considerably with different solutes, no real numbers have been given on the graph and a stylized diagram is employed. In general, with "more polarizing" material (e.g., large macroglobulins) there will be a concurrent shift of all curves to the left; with species "less polarizing" (e.g., albumin), the curves are shifted to the right. As noted, in stirred and laminar flow systems, the ultrafiltrate flux is reduced proportionally to the log concentration of retained solute; the flux may be extrapolated to zero in the region of "gel" concentration of the solute (region of viscosity corresponding to a gel). For dead-ended systems, flow is reduced to a barely perceptible value long before this region is reached.

However, there is evidence to indicate that high rate recirculation is disadvantageous where the lability of the preparation demands the most gentle nondenaturant handling and the shear produced by thin channel recirculation for sustained periods can be denaturing.[11] For this application, a somewhat different approach has been employed,

[10] S. E. Charm and J. Lai, A comparison of ultrafiltration systems for biologicals, Biotechnol. Bioeng., in press.
[11] S. E. Charm and B. L. Wong, Biotechnol. Bioeng., in press.

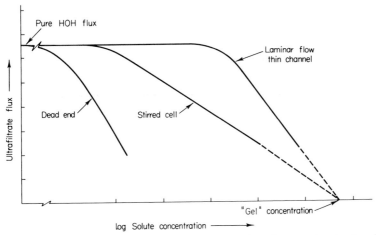

Fig. 2.   Ultrafiltrate flux at varying solute concentrations for nonstirred, stirred, and laminar flow thin channel systems.

wherein the sample is passed only once through the system, and the total concentration is accomplished in this single pass. Low velocities (2–3 feet/minute) and low pressures (<20 psi) are used. A flow-through variable back pressure valve in the egress line of the unit is used to select the desired degree of concentration. Briefly its operation is as follows: Constricting the valve elevates the back pressure to the system which then raises the mean transmembrane pressure. As a consequence, more fluid is expressed through the membrane increasing the selective concentration of the membrane-retained process stream. Concentration ratios to 50-fold can be obtained with single pass operation, using several passes can raise this value considerably. Although the shear forces in this system are minimal, evidently they are of sufficient magnitude to control solute buildup at the membrane interface. Simple gravity flow systems utilizing flow in thin channels, parallel to the membrane surface, are also available; the pressure gradient for ultrafiltration is provided by a reduced pressure in the ultrafiltrate line.

Although, in the former system, there is better control of concentration ratios, either system should be effective in terms of minimum denaturation of the retained macrosolutes.

There are a large number of manufacturers providing UF systems for general application. Table II primarily includes those companies with equipment more suited to laboratory or pilot plant use as categorized in the preceding section.

TABLE II
COMMERCIALLY AVAILABLE LABORATORY ULTRAFILTRATION EQUIPMENT

| Manufacturer | Dead-ended | Stirred[a] | Turbulent flow | Laminar flow | |
				Single pass	Recirculating
Abcor		x	x		
Amicon	x	x		x	x
Biomed				x	
Chemapec		x			
Gelman	x				
Millipore	x	x			
Sartorius	x				
Schleicher & Schuell	x				

[a] Includes vibrator-agitated systems.

Expanded Uses of UF in Enzyme Preparation

Diafiltration and Fractionation. A number of accessory systems are available to complement and extend the uses of ultrafiltration in enzyme preparation. Along with concentration, complete exchange and/or removal of salt may be desirable. Use of a reservoir, containing the dialyzate or exchange solution, with a valve that interconnects the system components with the gas source, enables salt exchange by simple volume replacement. This "diafiltration" procedure is also useful in fractionation schema.

Where contaminating solutes of lower molecular weight are to be removed, choice of a retentive membrane to effect partition using the diafiltration concept can provide an additional degree of fractionation.[12] Higher molecular weight contaminants can be removed by retention in appropriate membrane systems. To be sure, the selectivity of current membranes is not such that it would replace chromatography as a separation tool, but it can be used as an adjunct to this method. Liquid level sensing devices may be integrated with any system to provide automated operation.

Direct Integration of UF with Column Chromatography

Thin channel ultrafiltration allows two laboratory procedures to be merged at a considerable saving of time. In contrast to the usual method of concentrating the column eluate of prepared fractions after collection, it is now possible to directly connect a thin channel (single-pass) unit to a column, affording regulated on-line concentration. With careful attention to the geometry of the system, plug flow can be main-

[12] D. I. C. Wang, A. J. Sinskey, and T. Sonoyama, *Biotechnol. Bioeng.* 11, 987 (1969).

tained, permitting the use of solute-detecting apparatus distal to the concentration device. Such systems do not interfere with further flow into a fraction collector along with concomitant registration of the collected fractions on the recorded pattern.

Comments

A large number of UF systems for direct application to enzyme chemistry have been reviewed. In selecting among these techniques, a number of considerations exist. Choice of membrane should not be entirely dictated by flux, but rather lack of adsorption; systems should be chosen to provide the maximum depolarization, i.e., to achieve maximum flux, commensurate with the need to ensure minimal denaturization. When in doubt for labile proteins, the single-pass system is probably best suited; if the enzyme can withstand the shear, graduation to recirculating systems with the attendant higher flux may be warranted.

[7] An Anaerobic Laboratory

By J. MICHAEL POSTON, THRESSA C. STADTMAN, and EARL R. STADTMAN

The investigation of many aspects of microbiology and biochemistry is hampered by the oxygen lability of the material under study; enzymes, electron carriers, and metabolic intermediates may be rapidly lost in the presence of air. Many bacteria are severely inhibited or may be killed by exposure to air. These properties have made the study of anaerobic mutants particularly difficult since many of the techniques that are well established for use with aerobes cannot be adapted to anaerobic use.

Although any given manipulation can be carried out in conventional anaerobic glove boxes, suitably equipped with remote control devices, experimental operations become extremely difficult, if not impossible, when they involve multistep procedures employing filtration, chromatography, electrophoresis techniques or the use of massive instrumentation such as spectrophotometry, centrifugation, and refrigeration.

To facilitate anaerobic experimentation under conditions that allow great versatility in the utilization of standard laboratory techniques, an anaerobic laboratory chamber was constructed.[1] The chamber is a

[1] The anaerobic facility at the National Institutes of Health, Bethesda, Maryland, was designed by and built under the supervision of the Linde Division of the Union Carbide Corporation.

space of approximately 1400 cubic feet. (40 cubic meters) enclosed by a gas-tight partition. By displacing most of the air with N_2 gas and then removing the remaining oxygen by combining it with hydrogen on a catalyst bed, oxygen tensions of less than 100 ppm can be maintained. The laboratory contains a fume hood, incubators, and a cold room. Most portable laboratory equipment may be used in this chamber. A fairly wide spectrum of experiments have been conducted in its anaerobic environment.

Description

The anaerobic chamber was constructed in a laboratory module and its dimensions were, in part, dependent upon the existing structure. The gas-tight partition is made of $\frac{3}{8}$ inch carbon steel plates supported on a framework of I-beams. All the floor plates are riveted to the concrete floor and all joints and seams are welded. Doors to the gas locks and the emergency doors were fabricated at the site from the same steel as the shell of the chamber. Observation windows are of double-strength safety glass.

The cold room is a commercial, walk-in refrigerator box of sufficient size to accommodate a fraction collector or similar equipment used in cold-room operations. Refrigeration is supplied by liquid nitrogen which sprays as saturated gas at $-195°$ from a perforated pipe (the spray header) mounted at ceiling level (Polar Stream System, Linde Division, Union Carbide Corp.). The cold room is designed to hold any temperature between $-10°$ and $+4°$. Its use is optional. The temperature can be brought from ambient (ca. $25°$) to operating levels in 30–60 minutes.

Nitrogen for the cold room is conveyed directly from a large storage vessel outside the building through insulated copper tubing. The same storage vessel is the source of the nitrogen atmosphere in the chamber itself. It is passed first through a vaporizer assembly that is bathed in water at $21°$ and then through a volume meter before entering the anaerobic chamber.

Operation of the Chamber

Placing the chamber in service once it is stocked with supplies and equipment is a relatively simple matter. Prior to closing the doors, the drain in the waste sink is filled with water. After checking the operation of the emergency system, the chamber is sealed and nitrogen is added through the purge valve. The chamber is vented at the other end of the N_2 circulation loop (Fig. 1). After about an hour, the O_2 level is less than 0.5% oxygen-in-nitrogen (Beckman Oxygen Analyzer,

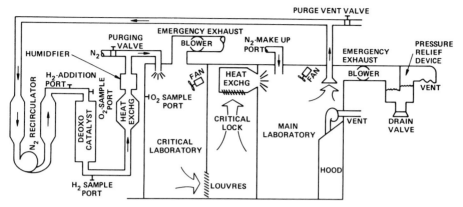

Fig. 1. Schematic illustrating the circulation and purification of gases within the chamber.

Model D2). When this level is reached, the purge is stopped, recirculation is begun, and the addition of hydrogen is started at a rate of about 20 standard cubic feet (scf) per hour. The hydrogen content is monitored with a thermal conductivity analyzer (Model 204A, Analytic Systems Co.) and automatically maintained at 0.5–0.6%. After 4–5 hours of recirculation over a catalyst bed (Deoxo Gas Purifier, Model D-10,000-100, Engelhard Industries), the oxygen level of the hydrogen enriched atmosphere is less than 100 ppm. After 6–7 hours of recirculation, the atmosphere is less than 20 ppm oxygen. The tension continues to drop as recirculation continues and reaches a minimum of 2–3 ppm after about 48 hours.

Oxygen tension is measured electrometrically by an oxygen trace analyzer (ASC Model 306 W, Analytic Systems Co.) in which a continuous stream of the gas to be monitored is passed through a cell containing a lead plate covered by a KOH solution in which a silver screen is suspended. The generation of current is governed by the following reactions:

$$O_2 + 2\ Ag \rightarrow 2\ AgO$$

$$2\ AgO + 2\ H_2O + 4\ e \rightarrow 2\ Ag + 4\ (OH)^-$$

$$2\ Pb + 4\ (OH)^- \rightarrow 2\ PbO + 2\ H_2O + 4\ e$$

A pressure-sensing make-up valve keeps the environment within the chamber at about 1 inch (water column) above the atmospheric pressure. Nitrogen lost through leakage or through operation of the fume hood is replaced by this system. The maximum pressure within the chamber is controlled by the pressure release device (Fig. 1). This is a vented water tank containing a weir so arranged that gas on the internal side will escape under it if the pressure rises above 4 inches of water.

Life Support and Emergency System

The life-support system consists of face masks (two individuals may work within the chamber at once) to which are attached a breathing air line, a vacuum line to remove exhaled air, an oxygen monitor for the breathing air, and an alarm buzzer. Breathing air is supplied to the mask by a compressor housed nearby. If this should fail, a large tank of breathing air (4 hours' supply) is automatically connected to the air line. If this back-up source of air should fail or be exhausted, a pressure activated alarm notifies the person in the chamber to open the valve on the small bottle of breathing air that is carried on the harness of the life-support system. This second back-up source has a minimum of 7 minutes of air and will permit the person to make an orderly exit from the chamber.

During the time when anyone is inside the chamber, a second person is outside observing the operations and well-being of the one inside and monitoring the oxygen level of the breathing air and of the chamber atmosphere. There are observation windows in the wall and each of the doors, as well as strategically placed mirrors so that at no time is the person inside the chamber out of sight or hearing of the monitor.

In the event of a situation in which it is vital to aid the person within the chamber, a button on the control panel activates the emergency system. The emergency doors and the butterfly drain valve on the pressure release device open pneumatically. Large exhaust blowers (Fig. 1) begin to draw fresh air into the chamber through the emergency doors. Within about 7 seconds, the atmosphere inside the chamber will support human life. Even in the event of an electric power failure, the emergency doors can be opened individually by means of the pneumatic system or by manual release of the latches.

Entrance to the Laboratory

Access to the chamber is provided through the personnel lock (Fig. 2) in which the life-support mask is donned. Once the mask is tightly fitted to the face, the worker closes the outside door and the lock is purged. During this time the pressure within the chamber is raised close to its maximum in order to prevent the purge from driving oxygen through the door seals into the anaerobic room. After 15 minutes, the purge is ended and the main laboratory is entered. The inner laboratory may be entered through the critical lock. In practice, it has been found that the inner critical laboratory and the main laboratory are sufficiently

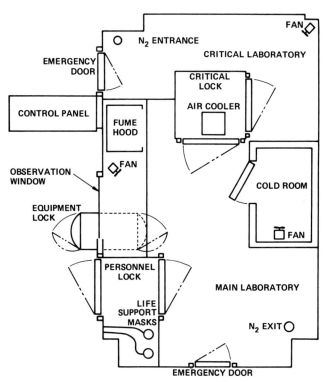

FIG. 2. Floor plan of the anaerobic chamber.

low in oxygen so that this passive-flow lock may be open thereby permitting unhindered passage from one laboratory to the other. The lock is kept in operation during the time when workers are not inside the chamber. There is also a small equipment lock through which items may be sent into or out of the chamber with only a 5 minute purge.

Exit from the chamber is uncomplicated. The doors to the lock between the two laboratories are closed, portable equipment is placed in the personnel lock, the life-support umbilical cord is gathered up, and the door from the main laboratory into the personnel lock is closed. Once the main laboratory is closed off from the personnel lock, the outer lock door may be opened and the mask and its assorted paraphernalia removed.

Consumption of Gases

Typical consumption of nitrogen for bringing the room from 21% to 100 ppm oxygen is about 6000 scf. Recirculation for 24 hours will consume about 4500 scf N_2. An entry purge uses about 1800 scf above the normal recirculation value. Hydrogen consumption is not excessive; a week of operation of the chamber may exhaust one to two cylinders (200–400 scf). If the chamber is not in use for several days, e.g., over a week end or longer, the various gases and the recirculator are turned off. This quiescent stage maintains the chamber at a low level of O_2 since diffusion into the room is slow. This action reduces consumption of N_2 and H_2 and saves time in bringing the room to operational levels when the room is reactivated.

Uses of the Chamber

The anaerobic chamber has been used for purification of oxygen labile enzymes by means of anaerobic chromatographic procedures; preparation and purification of reduced cobalamin derivatives; study of enzymatic reduction and oxidation of various substrates; study of the formation of an enzyme–substrate complex; plating and isolation of anaerobes from enrichment cultures; isolation of mutants of clostridia following mutagenic treatment and replica plating; study of anaerobic mutants of facultative strains; study of the formation of methane by enzyme preparations of methane bacteria.

Section II
Growth Methods

Articles Related to Section II

[8] Enrichment Culture and Isolation Techniques Particularly for Anaerobic Bacteria

By JAY V. BECK

Bacteria having a wide range of metabolic activities are available to the enzymologist largely as a result of successful application during the past eighty years of the enrichment culture or ecological technique first used by Winogradsky[1] but developed and perfected by Beijerinck.[2] It seems most probable that all the interesting and useful bacteria have not been discovered and isolated and that enrichment culture methods will continue to be useful not only for the isolation of new types but also for reisolation of previously described bacteria. The rationale of the method was discussed in an earlier volume of this series by Hayaishi[3] and more recently by Schlegel and Jannasch.[4] A useful outline of some general enrichment procedures was presented by Stanier *et al.*[5] These papers provide adequate sources of information about the general application of enrichment culture methods in isolating aerobic bacteria, but do not fully cover anaerobic procedures. The following section deals with useful methods for the isolation of anaerobic bacteria and presents also procedures for isolation of some interesting aerobic bacteria and blue-green algae.

Anaerobic Bacteria: Enrichment Culture Technique

Anaerobic enrichment culture techniques were extensively used by Beijerinck[2] and other early microbiologists. Many strictly or facultatively anaerobic bacteria were discovered and isolated by application of enrichment methods, and subsequent studies using these bacteria contributed in a major way to understanding many biological phenomena.[6] Successful application of enrichment culture techniques for anaerobic bacteria is dependent upon a somewhat orderly use of a sequence of operations. These include a careful choice of material used for the initial inoculum, a suitable medium prepared and handled so that anaer-

[1]S. Winogradski, "Microbiologie du Sol." Oeuvres Complètes. Masson, Paris, 1949.
[2]M. W. Beijerinck, Verzamelde Geschriften, Vols. 1–6. Nijhoff, The Hague, 1921–1940.
[3]O. Hayaishi, this series, Vol. I [14].
[4]H. G. Schlegel and H. W. Jannasch, *Annu. Rev. Microbiol.* 21, 49 (1967).
[5]R. Y. Stanier, M. Doudoroff, and E. A. Adelberg, "The Microbial World," 3rd ed., p. 87, Prentice-Hall, Edgewood Cliffs, New Jersey, 1970.
[6]H. A. Barker, "Bacterial Fermentations." Wiley, New York, 1956.

obiosis is achieved, favorable incubation conditions, a well-considered transfer schedule, and a final procedure which makes possible the isolation of a pure culture.

Choice of Inoculum. Suitable sources of inocula for the isolation of anaerobic bacteria usually are naturally occurring anaerobic materials. Swamps, black mud, and water containing large amounts of organic substances have been frequently used. However, well aerated surface soil and sediment from normal surface waters contain typical anaerobic bacteria. Rumen contents, animal feces, plant and animal bodies, and other organic material undergoing decomposition have yielded interesting anaerobic bacteria.

Culture Media. Suitable culture media contain the specific substrate of interest, mineral salts and supplements to provide certain growth factors and anaerobiosis. Heavy inocula are usually used in initial enrichment cultures and, since naturally occurring substances usually contain minerals, growth factors and sufficient aerobic bacteria to produce anaerobiosis almost immediately, there is usually little need for first enrichment media to contain anything other than the major carbon or energy source, which is usually supplied at a concentration of 0.5–2.0%. There is no need to sterilize media to which a heavy inoculum of naturally occurring material is to be added.

Media for transfer from the initial enrichment must contain all ingredients essential for microbial growth. Major inorganic substances may be supplied by adding K_2HPO_4, 0.02%; $MgSO_4$, 0.01%; a source of inorganic nitrogen such as $(NH_4)_2SO_4$, 0.05%; and trace elements.[5] A suitable trace element solution contains $CaCl_2$, 0.2%; $MnCl_2 \cdot 4H_2O$, 0.01%; $Na_2MoO_4 \cdot 2H_2O$, 0.01%; $FeSO_4 \cdot 7H_2O$, 0.2%; and $CoCl_2$, 0.01%, which is added to transfer and isolation media at a concentration of 2% (v/v). If a good tap water source is available, the trace element solution may be omitted. An organic source of nitrogen, yeast extract or yeast autolysate, is frequently employed at a concentration of about 0.05% dry weight basis. Excess of organic substance should be carefully avoided to assure that the culture medium encourages only those bacteria that utilize the specific substrate of interest. The first transfer and subsequent enrichment media must be as free of dissolved oxygen as possible. Dissolved oxygen may be removed by vigorously boiling the entire medium and rapidly cooling just prior to inoculation. Media used for transfers are usually sterilized. Sterile eqiupment and aseptic techniques are employed. It may be necessary to add sodium sulfide, sodium hydrosulfite or other substances capable of adjusting the *Eh* of the medium to a suitable level. The hydrogen ion concentration of media and the incubation temperature may be varied to yield specific microbial cultures.

Culture Vessels. The most convenient vessel for anaerobic enrichment culture is a small glass-stoppered bottle of 50–500 ml capacity. This is almost completely filled with the medium; the inoculum is introduced, the bottle is then completely filled, and the stopper is pressed into place. It is advisable to place the bottle in a metal tray or in one half of a petri plate during incubation to prevent contaminating the incubator with fluid that may be expelled from the bottle due to gas formation.

Transfer of Inocula to Subsequent Enrichment Media. Usually two or three transfers are necessary to enable the desired organism to dominate the mixed enrichment culture. The timing of the transfers is important and for some bacteria may be the critical step in the entire isolation procedure. This is because bacterial growth alters the chemical composition of the medium. Obviously, the concentration of the primary substrate is reduced, but more importantly other substances are formed that may be either inhibitory to the anaerobe sought or act as nutrients for other bacteria. Therefore, enrichment cultures should be carefully watched so that inocula for the second and successive transfers may be taken when the desired bacterium is at maximum dominance. Inocula for pure culture isolation are usually taken from the third or fourth enrichment culture.

Pure Culture Isolation. Several general methods are available for isolation of pure culture colonies of anaerobic bacteria.[7] Liquefiable solid media with an additive to poise the *Eh* value and incubation in the absence of atmospheric oxygen achieved by any of several standard procedures may be employed in either the streak or pour plate methods. For some extremely oxygen-sensitive bacteria, plating methods are not successful and other procedures such as the shake-tube method must be used.

In the shake-tube method a small inoculum, 1–2 drops from a suitable enrichment culture, is added to melted, but rapidly cooled agar media in a cotton plugged, soft glass tube about one half full. The inoculated medium is mixed by carefully inverting the tube 2 or 3 times or by rolling the tube between the palms of the hands. A small amount, 1–2 ml, of the medium is then poured into a second tube of the same medium, also about one-half filled. The second tube is mixed, and a small amount is poured into a third tube. This is continued until 8–10 tubes have been inoculated. The tubes are then cooled rapidly in a cold water bath, prepared for removal of oxygen from the tube by using rubber stoppers and a pyrogallol or chromous acid seal, inverted, and incubated at a

[7]L. D. S. Smith and L. V. Holdeman, "The Pathogenic Anaerobic Bacteria," Chapter 2, Thomas, Springfield, Illinois, 1968.

suitable temperature. A few well isolated bacterial colonies usually appear in two or three of the tubes after incubation. The colonies are visually observed for uniformity, and those considered suitable are marked for further study. Cells for microscopic study and for maintenance of pure cultures are obtained by breaking off the bottom of the tube and aseptically transferring the agar plug to a sterile petri plate. The agar plug is cut into thin disks from which material from the colonies can be obtained with an inoculating needle or a Pasteur pipette.

Purine-Fermenting Bacteria (Clostridium acidi-urici and C. cylindrosporum)[8]

Enrichment medium: uric acid, xanthine or guanine (guanine is quite insoluble; 0.2% should be used.), 0.5%; K_2HPO_4, 0.03%; $MgSO_4 \cdot 7H_2O$, 0.01%; and yeast extract, 0.05%. The purine is dissolved in about half the required final volume of hot water, the remaining ingredients are then added, and the pH is adjusted to 7.2 using phenol red indicator.

Transfer medium: enrichment medium with 0.005% $Na_2S_2O_4$
Isolation medium: transfer medium with 1.5% agar added
Inoculum: garden soil or mud from swamp, stream or lake; 0.5–1.0 g in 50- or 100-ml bottle.

Isolation Procedure. Incubate at 30–37° in completely filled, glass-stoppered bottle. Turbidity appears after 24–48 hours and medium becomes more alkaline due to $(NH_4)_2CO_3$ formation. Two or three transfers may be used before inoculating 1 drop of transfer medium culture into agar medium, for serial dilution and isolation by the shake tube method.

Amino Acid–Fermenting Bacteria (Several Clostridia and Peptococcos Glycinophilus)[9,10]

Enrichment medium: specific amino acid 0.5–1.0%, yeast extract, 0.03%; K_2HPO_4, 0.02%; and $MgSO_4 \cdot 7H_2O$, 0.01%; pH 7.0
Transfer medium: enrichment medium with 0.005% $Na_2S_2O_4$ added. Yeast extract content may be increased.
Isolation medium: transfer medium with 1.5% agar added.
Inoculum: soil, mud from swamps or lakes, 0.5–1.0 g for 50–100 ml medium. Many bacteria capable of fermenting amino acids are spore formers. Heating a water suspension of the inoculum at 80° for 10 minutes facilitates isolation of spore-forming bacteria.

[8] H. A. Barker and J. V. Beck, *J. Bacteriol.* 43, 291 (1941).
[9] B. P. Cordon and H. A. Barker, *J. Bacteriol.* 52, 629 (1946).
[10] H. A. Barker, *Enzymologia* 2, 175 (1937).

Isolation procedure: Use inoculum from second or third transfer culture for isolation in a shake tube dilution series.

Methane-Producing Bacteria (Methanobacillus Omelianskii)[11]

Enrichment medium: C_2H_5OH, 1% v/v of 95%; $CaCO_3$, 10% w/v; NH_4Cl, 0.05%; K_2HPO_4, 0.1%; $MgCl_2$, 0.01%; pH 7.0. Tap water.

Transfer medium: Initial enrichment culture medium with added 3% v/v of a 1% $Na_2S \cdot 9H_2O$ and 5% Na_2CO_3 solution. This solution can be sterilized at 120°.

Isolation medium: C_2H_5OH 0.5% v/v of 95%; K_2HPO_4, 0.6%; KH_2PO_4, 0.9%; $(NH_4)_2SO_4$, 0.03%; $M_2SO_4 \cdot 7H_2O$, 0.01%; $FeSO_4 \cdot 7H_2O$, 0.001%; and 1% v/v saturated solution of $CaSO_4 \cdot 7H_2O$; agar, either 0.3% for semisolid or 1.5% for solid medium. Add Na_2S as for the transfer medium.

Inoculum: soil, marine mud, sewer mud, 1–5 g in 50–100 ml bottle.

Isolation Procedure. Incubate at 35–37° in completely filled glass-stoppered bottles for 14 days. After copious gas formation, transfer at least 1 ml of sediment to transfer medium. From the third or fourth transfer culture inoculate a series of tubes using the shake tube method. Development of colonies in the shake tube series is quite slow. Incubation of several weeks may be necessary. Careful selection of isolated colonies is essential, and organism from the selected colony should be immediately recultured and tested for ability to form methane. Descriptions of the bacterial colonies and individual cells have been presented[11].

Aerobic Bacteria

The following procedures, which have been successfully employed in the isolation of aerobic microorganisms, are simply some convenient examples of the many methods that may be found in the literature. Unfortunately, no single source is available that gives good coverage of the many useful techniques that have been employed. The investigator is referred to the general microbiological literature for information on methods used in isolating the many bacteria not listed here.

Pseudomonas Species

Pseudomonads capable of rapidly decomposing any of a large number of simple nitrogenous compounds can be obtained by suitable modification of the following procedure that yields *Pseudomonas fluorescens* when the nitrogen source is aspargine.

[11] H. A. Barker, *Antonie van Leeuwenhoek, J. Microbiol. Serol,* 6, 281 (1940).

Enrichment medium: aspargine, 1%; K_2HPO_4, 0.05%; $MgSO_4 \cdot 7H_2O$, 0.02%; pH 7.0

Transfer medium: enrichment medium with added tracer salt solutions for third or fourth transfer

Isolation medium: transfer medium with added 1.5% agar or nutrient agar.

Inoculum: 0.1 g garden soil.

Isolation Procedure. Inoculate 30 ml of enrichment medium in a 125-ml conical flask (cotton plug). Incubate at 25–30° for 3–5 days. The medium becomes turbid and takes on a typical fluorescent green color. Inoculate transfer medium with 1 drop of active enrichment culture and incubate as above. From second transfer prepare, pour or streak plates using the isolation medium. Colonies tend to spread. Therefore, careful examination of colonies and individual cells is necessary to obtain pure cultures.

Camphor-Decomposing Bacteria (Pseudomonad or Diphtheroid)[12]

The above procedure for general pseudomonads can be modified as follows to yield bacteria capable of utilizing camphor as a carbon and energy source.

Enrichment medium: camphor, 0.1%; NH_4Cl, 0.1%; K_2HPO_4, 0.1%; $MgSO_4 \cdot 7H_2O$, 0.03%

Transfer medium: as above plus trace elements

Isolation medium: transfer medium plus 1.5% agar

Inoculum: garden soil or sewage sludge, 1 g

Isolation Procedure. Add 1.0 g rich soil or 1.0 ml activated sludge to 20 ml enrichment medium in a 500-ml flask. Aerate on shaker at 24–30° for 2–4 days until abundant growth occurs. Then transfer 0.1 ml to a second flask containing 20 ml above medium. Repeat after growth occurs, and finally streak from last flask onto agar with above camphor medium. Colonies should appear in 2–4 days at 25–30°. A second streak plate, inoculated with bacteria from a typical colony on the first streak plate, is recommended.

Iron-Oxidizing Bacteria (Thiobacillus Ferrooxidans)[13]

Enrichment medium: $FeSO_4 \cdot 7H_2O$, 0.5%; $(NH_4)_2SO_4$, 0.05%; KH_2PO_4, 0.02%; $MgSO_4 \cdot 7H_2O$, 0.05%; pH 3.5

[12]W. H. Bradshaw, H. E. Conrad, E. J. Corey, I. C. Gunsalus, and D. Lednicer. *J. Amer. Chem. Soc.* **81**, 5507 (1959), and unpublished observations.

[13]J. V. Beck, *J. Bacteriol.* **79**, 502 (1960).

Transfer medium: Above plus trace elements

Isolation medium: (a) transfer medium plus 1.5% Noble agar or other purified agar

(b) transfer medium plus silica gel prepared according to Kingsbury and Barghoorn[14]

(c) transfer medium (for serial dilution)

Inoculum: these bacteria are limited to areas where iron sulfide minerals are exposed to aqueous environments, i.e., water from exposed coal, copper, lead, iron, or other sulfide mineral deposits.

Isolation Procedure. Add 1.0 ml of inoculum to 50 ml of enrichment medium in a 250-ml conical flask with cotton plug. The CO_2 in the atmosphere is the source of carbon for the bacteria. Growth in the enrichment medium is evidenced by turbidity and precipitation of brownish-red precipitate of ferric iron compounds. Inoculate transfer medium with 1–5 drops of enrichment culture. The first transfer culture can be used as the inoculum for pure culture isolation using either the streak plate method or the serial dilution method into tubes of liquid medium.

Filamentous Blue-Green Algae (Nostocaceae)[15]

Some filamentous members of the blue-green algae (Cyanophyta) are capable not only of light-dependent carbon dioxide fixation, but also of nitrogen (N_2) fixation. This unusual combination of characteristics was recognized by early workers and served as the basis for successful enrichment culture procedures. In spite of the ease with which enrichment cultures may be obtained, pure culture isolation presents a problem because of contaminating bacteria that adhere to the mucoid covering of the filaments of algae. Ultraviolet treatment of blue-green algae enrichment cultures selectively destroys the adhering bacteria, thus making possible the isolation of pure cultures by simple dilution techniques.

Enrichment medium: K_2HPO_4, 0.02%; $MgSO_4 \cdot 7H_2O$, 0.01%; Na_2CO_3, 0.02%; ferric citrate, 0.005%; pH 8.5–9.5

Transfer and isolation media: enrichment medium with trace element solution 1% v/v

Inoculum: soil or surface water.

Isolation Procedure. Inoculate 200 ml of enrichment medium in a liter conical flask with 1 g of soil or 1–5 ml of water. Incubate in light (green-

[14] J. M. Kingsbury and E. S. Barghoorn, *Appl. Microbiol.* 2, 5 (1954).

[15] G. C. Garloff, G. P. Fitzgerald, and F. Skoog, *Amer. J. Bot.* 37, 216 (1950).

house or continuous light) for 2–4 weeks. Algal growth is apparent as a green color in suspension or adhering to the glass surface.

Plating methods are not suitable for pure culture isolation. Instead, serial dilution into sterile medium in small conical flasks is recommended. Cultures from the second or third enrichments are treated in a quartz vessel with 2750A ultraviolet light generated by a mercury vapor lamp. A 20–30-minute treatment is recommended with inocula for preparation of the dilutions being taken at intervals of 5–10 minutes. The higher dilution vessels showing algal growth after 3–6 weeks incubation in light may be considered as possible pure cultures. They should be critically examined for bacterial contamination by the usual plating procedures and for purity of the algae by microscopic observation.

[9] Culture of Anaerobic Bacteria for Biochemical Studies

By Thressa C. Stadtman

For the investigator who wishes to delineate reaction pathways or specific metabolic steps catalyzed by anaerobic bacteria, it is often possible to obtain sufficient cell material by culturing the organism in 10- or 20-liter batches in ordinary Pyrex glass bottles (carboys). Aside from facilities for sterilization of the bottles containing the culture medium to be used, no unusual equipment is needed except for a means of maintaining the cultures, if growth is poor at room temperature, at a temperature above the ambient and a large-capacity centrifuge for collection of the cells.

General Requirements for Laboratory-Scale Culture of Anaerobic Microorganisms

Cultures. Anaerobic bacteria specialized for the fermentation of a particular substrate frequently are isolated from soil or black mud by the enrichment culture technique, especially if strains having the desired properties are not already available in various collections as pure cultures. Alternatively, mutant strains of an available anaerobic organism can be isolated by suitable modification of the standard procedures used in the isolation of mutants of aerobic bacteria. At present, this is not a generally feasible approach, particularly in the case of strictly anaerobic bacteria, because of the difficulty of culturing large numbers of isolates on petri dishes in anaerobic jars. However, in an anaerobic laboratory facility[2] at the National Institutes of Health, it was possible to obtain a number of mutants of *Clostridium sticklandii* by treatment with a chemical mutagen, 1-methyl-3-nitro-1-nitrosoguanidine, and to isolate them on petri dishes exposed to the nitrogen atmosphere of the laboratory in the same fashion that mutants of an aerobic organism are isolated in an ordinary laboratory.[3]

Sterilization of Culture Media. When the microorganism to be cultivated requires a substrate readily fermented by many types of anaerobic bacteria, or complex sources of growth factors, such as yeast extracts, peptones, tryptones, meat extracts, then the medium should be sterilized prior to inoculation with the desired strain. Autoclaves large enough (door opening about 24 by 24 inches) to accommodate standard laboratory-scale Pyrex glass bottles, i.e., 10–20 liters, are most commonly used for this purpose. The heat stable constituents of the medium that can be sterilized together without decomposition are placed in the carboy and autoclaved in about two-thirds of the total volume of water. Enough water to finally fill the culture vessel to the neck is sterilized separately and added after cooling. Constituents of the culture medium which are destroyed by heating at 121° are usually sterilized in separate solutions by filtration through membranes of porosity sufficiently small as to retain bacteria; the sterile filtrate then is added to the sterile culture vessel or to the remaining constituents of the medium that have been heat sterilized and cooled.

For some highly specialized bacteria, i.e., certain species of methane-producing bacteria, sulfate reducing bacteria or *Clostridium kluyveri*, that can be cultured on very simple media under conditions unfavorable to the growth of most types of common contaminants, the medium need not be sterilized. In these cases 50-gallon stainless steel drums have been used as culture vessels. The steel drums are periodically cleaned with high pressure steam or strong detergent solutions.

Removal of Oxygen from Culture Medium. The medium must be freed of dissolved oxygen prior to inoculation in order to culture the strictly anaerobic types of bacteria. Methylene blue, 5–10 drops of a 0.5% aqueous solution per liter of culture medium, (or just enough to give a detectable blue color) added prior to sterilization, serves as a convenient oxidation-reduction indicator. Rapid cooling of the medium following sterilization limits somewhat the reintroduction of oxygen, but, for most species, a relatively nontoxic chemical that reacts rapidly with oxygen

[1] J. V. Beck, this volume [8].
[2] Poston *et al.*, this volume [7].
[3] A. C. Schwartz and T. C. Stadtman, *J. Bacteriol.* 104, 1242 (1970).

must also be added to the medium. Hydrogen sulfide, added as a sterile solution of sodium sulfide, is effective and is tolerated by most anaerobic bacteria. A stock solution of $Na_2S \cdot 9 H_2O$ (50%, w/v, or about 2 M), prepared in advance and stable for several months, is diluted with water and sterilized separately. An amount that will yield a final concentration of 0.02–0.03%, approximately 1 mM, is added to the sterile cooled culture medium just prior to inoculation. This is sufficient to remove most of the dissolved oxygen as judged by the decolorization of the methylene blue, and is within the limits tolerated by anaerobic bacteria isolated from the usual sources. A nonvolatile mercaptan, such as thioglycolic acid ($SHCH_2COOH$), can be added to the culture medium prior to sterilization and is more convenient to use though toxic for some of the highly specialized anaerobes. A final concentration of 0.05–0.075%, calculated as $SHCH_2COONa$, is usually employed. Cysteine at a concentration of about 2 mM (350 mg cysteine·HCl per liter), is sometimes used instead of thioglycolate. Sodium hydrosulfite (dithionite; $Na_2S_2O_4$) which reacts very rapidly with oxygen and almost instantaneously bleaches the methylene blue in the medium is another reducing agent that is sometimes used, particularly for large-scale cultures. The dry compound[4] is added in predetermined weighed amounts, about 30 mg per liter to the final medium just before inoculation. Alternatively, it is introduced, a few grains at a time, until the methylene blue is bleached. It cannot be used with organisms which are sensitive to sulfur dioxide, the oxidation product of dithionite.

Inoculum. The problems of exposure to oxygen and contamination with unwanted organisms during manipulation of unwieldly culture vessels are minimized if relatively large inocula are employed. These are prepared stepwise, often from 10 ml → 100 or 200 ml → 2 liters, in order to inoculate 18 liters of medium in a 20-liter bottle. Use of such large inocula is particularly important for cultivation of markedly oxygen-sensitive anaerobes or in those instances where nonsterile media are employed. An actively growing pure culture of the microorganism (10% by volume) is added to the nonsterile deoxygenated culture medium to ensure rapid development of the desired strain. The most common contaminants encountered in nonsterile media used for cultivation of *C. kluyveri* and certain methane bacteria that grow on simple alcohols or acetate, are sulfate reducers. The latter can be kept to a minimum by limiting the amount of sulfate added to the culture medium, i.e., the use of NH_4Cl rather than $(NH_4)_2SO_4$ as nitrogen source.

---

[4] Exposure of dithionite solutions to air causes rapid oxidation.

By frequent microscopic examination of the cultures, the extent of contamination with undesired species usually can be ascertained.

Anaerobic Seals. Diffusion of oxygen into individual test tube cultures of anaerobic bacteria is prevented by placing at the top a plug of absorbent cotton containing a few drops of a reagent that rapidly reacts with oxygen; immediately thereafter the tube is closed with a tight-fitting rubber stopper.[1] In practice this is accomplished by (1) cutting off the top of the sterile nonabsorbent cotton plug, (2) inserting it further into the head space above the agar or the liquid, (3) packing the space above with a snug-fitting wad of absorbent cotton, (4) introduction of a few drops of the desired reagent, and (5) immediately inserting a moistened rubber stopper. A convenient reagent is alkaline pyrogallol. Separate stock solutions of pyrogallol (50% w/v) and K_2CO_3 (10% w/v), each stable for long periods, are added directly to the absorbent cotton plug. For a gas space of 5–10 ml, the addition of 5–6 drops of the alkaline carbonate solution followed by an equal volume of the pyrogallol solution is sufficient to remove the oxygen present. Use of an excess of potassium carbonate rather than potassium hydroxide to prepare alkaline pyrogallol *in situ* ensures that carbon dioxide, frequently required as a nutrient for anaerobic bacteria, will not be removed from the medium by reacting with the alkali on the plug. Moreover, the amount of carbon dioxide liberated from the plug upon the addition of pyrogallic acid, is often enough to satisfy the CO_2 requirement of a microorganism when the volume of culture medium relative to the head space is small as is the case in test tube cultures.

Another reagent that is sometimes used for this type of anaerobic seal, particularly if carbon dioxide introduction is undesirable, is an aqueous solution of a chromous salt. Solutions of chromous chloride prepared for use as an oxygen adsorbent are commercially available, e.g., from Fischer Scientific Co. The reaction of chromous chloride with oxygen is almost instantaneous and produces a color change from blue to green (chromous to chromic).

Regardless of the reagent used to remove oxygen, the efficacy of the procedure may be readily assessed by inspection when the culture medium contains methylene blue. If the oxidation-reduction indicator remains colorless, the oxygen removal techniques are successful.

Anaerobic bacteria which are only moderately sensitive to oxygen or those which produce considerable amounts of insoluble gaseous products, e.g., H_2 or CH_4, can be protected adequately from oxygen, particularly in culture vessels where the ratio of head space to liquid volume is small, by insertion of a rubber stopper bearing a piece of bent glass tubing, the other end of which is immersed in a tube or flask of water. This type of water trap is routinely used to protect carboy scale cultures from oxygen, and after growth starts the device allows fermentation gases to escape.

Incubation Conditions. Many of the anaerobic bacteria isolated from the soil grow optimally at room temperature or at 30° and may be cultivated successfully without elaborate incubator facilities. Even those with temperature optima in the range of 35° to 40° need not necessarily be cultivated at these higher temperatures. In some instances the rate of multiplication can be decreased by growth at a lower temperature in order to suit the time schedule of the investigator without any observable effect on the ultimate yield of a particular enzyme system.

Visible light may have little effect on many types of nonphotosynthetic anaerobic bacteria but it is generally advisable to protect those types markedly dependent on reactions catalyzed by light-sensitive B_{12} compounds from undue exposure to bright light.

Culture of Certain Specialized Types of Anaerobes. Anaerobic bacteria such as certain physiological types of methane bacteria that require large amounts of carbon dioxide as a cosubstrate for energy yielding reactions, or *Clostridium kluyveri*[5] and *Clostridium* M-E[6] that require carbon dioxide for synthesis of cell constituents, are somewhat more difficult to culture because of technical problems associated with control of pH and carbon dioxide concentration. When sterile solutions are required, the desired amount of sodium or potassium carbonate is sterilized separately in a solution containing a pH indicator such as phenol red. The cooled solution is rapidly neutralized with concentrated HCl which is introduced aseptically prior to its addition to a sterile solution containing the other ingredients of the culture medium. If the pH of the final medium is too high it often can be adjusted by bubbling gaseous carbon dioxide through the solution. To avoid undue loss of carbon dioxide and the concomitant increase in pH, culture media should not be manipulated or stirred more than absolutely necessary after introducing solutions of carbonate. When carbon dioxide-requiring organisms are grown in large-scale fermentors, operations which must be continuously stirred to maintain temperature control,[7] a gas mixture such as one of nitrogen and carbon dioxide is introduced into the head space above the liquid.

For cultivation of species of methane-producing bacteria that are able to grow on carbon dioxide plus gaseous hydrogen, a more elaborate apparatus is required since the electron donor substrate, H_2, is relatively insoluble in the culture medium. Laboratory-scale equipment suitable

[5] E. R. Stadtman and R. M. Burton, this series, Vol. I [84] 1955.
[6] T. C. Stadtman and M. A. Grant, this series, Vol. 17B [168].
[7] E. F. Phares, this volume [36].

for anaerobic growth of microorganisms on this type of gas mixture have been described by Wolfe and his associates.[8]

Harvest and Storage of Cells. A continuous-flow Sharples centrifuge is a convenient device for collection of cells. With anaerobic bacteria it is important to work as rapidly as possible to avoid undue exposure to oxygen. The culture medium should be passed through the centrifuge at the maximum rate compatible with removal of cells. On completion, the cells should be removed immediately from the rotor. The cell paste can be dropped, in small portions, directly into liquid nitrogen or into powdered dry ice. Storage at -10 to $-20°$ preserves the frozen cell paste for many months or at $-80°$ for an even longer period. When needed, the frozen cell material can be thawed in a suitable buffer containing a mercaptan reducing agent and used for cell suspension studies or preparation of extracts.

CELL YIELDS OF SOME ANAEROBIC BACTERIA

Organism	Principal organic constituents of culture medium	Cell yield per liter
Clostridium sticklandii	2–2.25% tryptone + 1% Difco yeast extract + 0.15% H COONa	2–3 g (wet wt)[a] 700 mg (dry wt)[b]
Clostridium M-E	0.5% L-lysine·HCl + 0.5% Difco yeast extract + 0.2% glucose	1 g (wet wt)[a]
Choline and ethanolamine fermenting *Clostridium*	1.25% choline·HCl + 0.6% Difco yeast extract + 0.2% trypticase	1.5 g (wet wt)
	0.88% ethanolamine·HCl + 0.6% Difco yeast extract + 0.2% trypticase	1.5–2.5 g (wet wt)
Clostridium barkeri	0.5% nicotinic acid + 1% Difco yeast extract	2.2 g (wet wt)
	0.5% glucose + 1% Difco yeast extract	3.7 g (wet wt)
Clostridium pasteurianum	2% sucrose	7–8 g (wet wt)
Clostridium kluyveri	2% ethanol + 1% potassium acetate	150–300 mg (dry wt)
Methanosarcina barkeri	Methanol[c]	2–10 g (wet wt)
Methanococcus vannielii	1.5% Sodium formate	0.5 g (wet wt)

[a]Harvested at about two-thirds of maximum growth.
[b]Harvested at beginning of stationary phase.
[c]Cultures started on 1% methanol (v/v) and maintained at about 0.5% level for 5–10 days during which time a total of 3–4% methanol (v/v) is consumed.

[8]M. P. Bryant, B. C. McBride, and R. S. Wolfe, *J. Bacteriol.* **95**, 1118 (1968).

Cell Yields of Some Representative Anaerobes. To give the investigator, unfamiliar with the growth of anaerobic bacteria, some idea of the amount of material obtainable, the average cell yields of several types of anaerobic organisms that have been cultivated routinely in our laboratory, either in 20-liter carboys or in larger cultures, are shown in the table. Cell paste taken directly from the centrifuge rotor, if firmly packed, averages 20–25% solids. Wet weight values in the table are for such cell paste, usually weighed as frozen pellets. In some cases where dry weights are given, cells were resuspended in buffer, washed, and lyophilized.

[10] Molds and Streptomycetes

By D. PERLMAN

Four major considerations should be kept in mind in any program studying the production of specific fungal and streptomycete enzymes: (1) Is the production of the desired enzyme system strain specific as far as the microorganism is concerned? (2) Is the production of the enzyme system related to a particular phase in the growth cycle of the producing organism? (3) Is the production of the enzyme system related to the nutrition of the producing organisms? (4) Is the enzyme level in the cells of the producing organism related to the growth conditions?

Although there have been hundreds of papers and patents published relating to the growth of fungi and of streptomycetes for the preparation of antibiotics, vitamins, and other metabolites,[1] few have been particularly concerned with problems encountered in the preparation of cells of these organisms for specific enzymes. Our attention in this section will focus on the laboratory conditions that influence the growth of molds and of streptomycetes as related to enzyme levels in the cells. Some knowledge of basic microbiological techniques is assumed, and emphasis will be placed on what appears to the author to be technical

[1]General references:

J. R. Norris and D. W. Ribbons, Eds., "Methods in Microbiology." Academic Press, New York, 1969.

W. W. Umbreit, Ed., "Advances in Applied Microbiology," Vol. 1–9. Academic Press, New York, 1959–1967.

W. W. Umbreit and D. Perlman, Eds., "Advances in Applied Microbiology," Vol. 10. Academic Press, New York, 1968.

D. Perlman, Ed., "Advances in Applied Microbiology" Vols. 11, 12. Academic Press, New York, 1969–1970.

points that often determine the success or failure of an experimental program.

Sources and Maintenance of Mold and Streptomycete Cultures

The most important single item in a microbial process is the microorganism whose enzymes are responsible for converting the substrate to a product. It is practically always necessary to use pure cultures. Although all pure cultures were obtained originally—and often can be reisolated—from natural fermentations, soil, diseased tissue, and decomposing material, it is highly desirable to keep pure cultures and maintain them indefinitely. As the applications of microbiology have multiplied, culture collections have increased in number and in size. They are the best sources of the microorganisms one may need, because culture collection organizations are staffed with experts who pay a great deal of attention to purity, identification, and vigor of growth. Such collections are found in many parts of the world; some contain several thousand cultures of diversified types. The names of several of those with large collections of fungal and of streptomycete cultures are listed in Table I.

In requesting cultures from these organizations it is important to specifically mention the strain of the organism. Very frequently two strains of the same species of fungus or streptomycete may have identical morphological characteristics and yet differ widely in enzyme levels. While one strain of *Streptomyces fradiae* will convert Reichstein's compound S to compound F, another of the same species converts this substrate to the isomer, epi-F. One strain of *Fusarium solani* converts progesterone to androstadiene-3,17-dione while 20-β-hydroxyprogesterone is the only metabolite of this substrate by another strain.

TABLE I

SOURCES OF MOLD AND STREPTOMYCETE CULTURES

1. American Type Culture Collection, 12301 Parklawn Drive, Rockville, Maryland 20852
2. ARS Culture Collection, Northern Utilization Research and Development Division (NRRL), Agricultural Research Service, U. S. Department of Agriculture, 1815 North University Street, Peoria, Illinois 61604
3. Centraalbureau voor Schimmelcultures, Baarn, Holland
4. National Collection of Type Cultures, Central Public Health Laboratory, Colindale Avenue, London, N. W. 9, England
5. Institute of Microbiology, Rutgers—The State University, New Brunswick, New Jersey 08903
6. Commonwealth Mycological Institute, Kew, Surrey, England
7. Prairie Regional Laboratory, National Research Council of Canada, Saskatoon, Saskatchewan, Canada
8. Army Research Institute for Environmental Medicine, Natick, Massachusetts 01760

Since most type culture collections have as their primary objective the maintenance of morphological and a few other taxonomic characteristics, it is possible that upon long-term storage some biochemical properties may be lost or at least vary from those present in the original culture. Sometimes it is feasible to cope with such situations by requesting the same strain from another collection where perhaps different methods of culture preservation are used.

There are five methods of preserving fungi and streptomycete cultures which have been found to be useful in many laboratories.

1. *Maintenance on Agar Slants.* In this traditional method the culture is transferred every few weeks on agar media such as those listed in Table II. The spores germinate, form vegetative mycelium, and in time aerial hyphae with spores. The new slants are then stored in a refrigerator (+4°) for several weeks or months, and the cycle is repeated. Storage in a freezer (−20°C) is a useful alternative to the warmer temperature, and the cultures stay viable for longer periods.

The limitation of this method comes from the ability of the culture to mutate. On agar to agar transfer very often the mutants tend to overgrow the rest of the population and thus, after 6 to 10 transfers, the culture (which is usually a mixture of individuals, anyway) contains quite a different population than previously present. This is especially true of cultures which sporulate with difficulty and the asporogenous forms sometimes are "lost" by such agar to agar transfers.

The growth of streptomycetes on the agar media listed in Table I often results in sporulation more rapidly than when the same strains are grown on nutrient agar or some semisolid media. Honey-peptone agar, wort agar, and NRRL sporulation agar are often useful in inducing a fungal culture to sporulate.

2. *Maintenance on Agar under Oil.*[2] This is a variation of method 1 where after the strain has sporulated on the agar slant, sterile mineral oil (autoclaved with 1% water) is layered above the surface of the slant to a depth about 1 cm greater than the top of the agar slant. The covered slants may be stored at refrigerator temperature (+4°) or at room temperature. Since the principle of the method is to prevent the agar slant from drying out, the tip of the slant must not extend above the oil layer; otherwise the slant will lose water, eventually dry out, and the culture will be lost.

The chief disadvantage of this method is the mess that results when it is used; the oil coats everything and is generally difficult to work with.

[2]C. B. Buell and W. H. Weston, *Amer. J. Bot.* 34, 555 (1947).

TABLE II

COMPOSITION OF MEDIA USED FOR GROWTH OF FUNGI AND FOR STREPTOMYCETES

1. Agar media for fungi

Honey-peptone agar: Honey, 60 g; peptone, 10 g; agar, 25 g; water q. s. 1 liter; to pH 6.0 with KOH. Autoclave at 121° for 20 minutes

Saboraud-dextrose agar: neopeptone, 10 g; dextrose, 40 g; agar, 15 g; water q. s. 1 liter. Autoclave at 121° for 20 minutes

Saboraud-maltose agar: neopeptone, 10 g; maltose, 40 g; agar, 15 g; water q. s. 1 liter. Autoclave at 121° for 20 minutes

Wort agar: malt extract, 15 g; peptone, 0.78 g; maltose, 12.75 g; dextrin, 2.75 g; glycerol, 2.35 g; K_2HPO_4, 1 g; NH_4Cl, 1 g; agar, 15 g; water q. s. 1 liter. Autoclave at 121° for 20 minutes

NRRL sporulation agar: glycerol, 7.5 g; cane molasses, 7.5 g; cornsteep liquor, 2.5 g; $MgSO_4 \cdot 7H_2O$, 0.05 g KH_2PO_4, 0.06 g; liquor, 2.5 g; $MgSO_4 \cdot 7H_2O$, 0.05 g; KH_2PO_4, 0.06 g; $CuSO_4 \cdot 5H_2O$, 0.004 g; agar, 25 g; water q. s. 1 liter; pH 5.5–6.0 with KOH. Autoclave at 121° for 30 minutes

2. Liquid media for fungi

Czapek-Dox with glucose: $NaNO_3$, 3 g; KH_2PO_4, 1 g; KCl, 0.5 g; $MgSO_4 \cdot 7H_2O$, 0.5 g; $FeSO_4 \cdot 7H_2O$, 0.01 g; glucose, 40 g; water to 1 liter. Autoclave at 121° for 20 minutes

Raulin-Thom with glucose: glucose, 75 g; tartaric acid, 4 g; diammonium tartrate, 4 g; $(NH_4)_2 HPO_4$, 0.6 g; K_2CO_3, 0.6 g; $MgCO_3$, 0.4 g; $(NH_4)_2SO_4$, 0.25 g; $ZnSO_4 \cdot 7H_2O$, 0.07 g; $FeSO_4 \cdot 7H_2O$, 0.07 g; water q. s. 1 liter. Autoclave at 121° for 20 minutes

Upjohn Medium for Steroid Transformations: Edamin (from Sheffield Farms), 20 g; cornsteep liquor, 3 g; glucose, 50 g; water to 1 liter. Autoclave at 121° for 30 minutes

Cornsteep medium: cornsteep liquor, 40 g; glucose, 20 g; water q. s. 1 liter; pH 6.5 with KOH. Autoclave at 121° for 20 minutes

Calam and Hockenhull's medium: lactose, 30 g; glucose, 10 g; starch, 15 g; acetic acid, 2.5 g; citric acid, 10 g; phenylacetic acid, 0.5 g; $(NH_4)_2SO_4$, 5 g; ethylamine, 3 g; water q. s. 1 liter; pH 6.5 with KOH. Autoclave at 121° for 20 minutes

3. Agar media for streptomycetes

Duggar's agar: asparagine, 0.5 g; glucose, 10 g; beef extract, 2 g; K_2HPO_4, 0.8 g; KH_2PO_4, 0.5 g; agar, 15 g; water q. s. 1 liter; pH 6.5 with KOH. Autoclave at 121° for 20 minutes

Berger's agar: Heinz baby oatmeal, 20 g; Contadina tomato paste, 20 g; agar, 15 g; water q. s. 1 liter. Autoclave at 121° for 20 minutes

Bennett's agar:	yeast extract, 1 g; beef extract, 1 g; NZ amine A (Sheffield Farms), 2 g; glucose, 10 g; agar, 15 g; water q. s. 1 liter; pH 7.3 with KOH. Autoclave at 121° for 20 minutes
Soybean infusion agar:	A 2% soybean meal suspension is boiled for 30 minutes and filtered while hot. The volume is restored and pH adjusted to 7.0 with NaOH. 2 g glucose, 5 g NaCl, and 20 g agar are added. Autoclave at 121° for 20 minutes
Emerson's agar:	glucose, 10 g; yeast extract, 10 g; beef extract, 4 g; peptone, 4 g; NaCl, 2.5 g; water q. s. 1 liter; agar, 15 g. Autoclave at 121° for 20 minutes

4. Liquid media for streptomycetes

Soybean-glucose:	soybean meal, 30 g; glucose, 50 g; $CaCO_3$, 7 g, water q. s. 1 liter. Autoclave at 121° for 30 minutes
Soybean-glycerol:	soybean meal, 30 g; glycerol, 20 g; water q. s. 1 liter. Autoclave at 121° for 30 minutes
Beef extract-peptone:	peptone, 5 g; beef extract, 5 g; glucose, 10 g; NaCl, 5 g; water q. s. 1 liter. Autoclave at 121° for 30 minutes
Perlman's medium:	glucose, 20 g; glycine, 2.6 g; monosodium glutamate, 2.2 g; $K_2HPO_4 \cdot 3H_2O$, 0.5 g; $MgSO_4 \cdot 7H_2O$, 0.5 g; $FeSO_4 \cdot 7H_2O$, 0.025 g; $CuSO_4 \cdot 5H_2O$, 0.012 g; $MnSO_4 \cdot 7H_2O$, 0.016 g; $ZnSO_4 \cdot 7H_2O$, 0.03 g; $CaCl_2 \cdot 2H_2O$, 0.05 g; water q. s. 1 liter. Adjust pH to 7.2 with KOH. Autoclave at 121° for 15 minutes

The advantage lies in the fact that the cultures do not have to be transferred as frequently as with method 1.

3. *Drying of Spores on Soil.*[3] Sterile soil has found wide use for the stock culture maintenance of microorganisms that form spores. In fact, microorganisms that do not form spores also will survive in sterile soil, but they may die out unexpectedly after a period of time. Soil stocks are prepared by mixing enough sand with a rich garden soil to make the soil friable and easy to handle. A small amount of $CaCO_3$ is added, and the mixture is distributed in 16 × 125 mm screw-capped tubes or in tubes plugged with cotton. The tubes are repeatedly autoclaved until random checks of their contents show that sterility has been achieved. A small volume of a thick suspension of spores or of an actively growing nonsporulated culture is then added to the sterile soil, and the moisture is quickly removed by placing the tubes under reduced atmospheric pressure over a drying agent. Soil stocks, thus prepared, are stored at room temperature with the cotton plugs or screw caps protected from dust. All that is necessary

[3] H. C. Greene and E. B. Fred, *Ind. Eng. Chem.* **26**, 1297 (1934).

to start a new culture is to transfer a few grains of soil to the surface of an agar slant and incubate the slant for several days.

The major disadvantage to this method is that many streptomycetes and a few fungi, notably *Ashbya grossypii, Eremothecium ashbya*, and basidiomycetes, do not survive this treatment.

4. *Preservation by Lyophilization*.[4] The most widely used method for culture preservation is lyophilization, also known as freeze-drying. In this procedure, cells in sterile glass ampules are suspended in a carrier or protective agent, such as sterile bovine serum or sterile skim milk, rapidly frozen at low temperature, and dried in a high vacuum. The ampules are then sealed and stored in a refrigerator. If properly prepared and stored, most lyophilized cultures will remain viable for long periods, e.g., 5 to 10 years, without the occurrence of genetic changes. When needed, the cultures are recovered from the ampules by suspending the lyophilized cells in a minimal amount of growth medium and then incubating.

In some laboratories several hundred replicate vials are prepared of a given culture. This stock serves as inoculum for the first vegetative stage of the culture without going through the step of growing the fungus or streptomycete on an agar slant. The major disadvantage to the method is the labor involved and the deleterious effect of the freezing process on the viability of the spores and cells. Usually more than 90% of the cells are killed during freezing and the survivors are those spores and cells that can withstand this physiological shock. Hopefully the survivors contain the enzymes selected for study.

5. *Storage in Liquid Nitrogen*.[5] This relatively new method is based on the long-term viability of spore and cell suspensions in 10% glycerin solution when stored in liquid nitrogen. These suspensions are placed in glass ampules (usually 1 ml) and slowly frozen and then stored in cylinders filled with liquid nitrogen. The process was adapted from that used in the artificial insemination program for cattle. The new vegetative culture can be started by removing a vial or ampule from the liquid nitrogen cylinder and warming it quickly to 37°. The contents are then transferred to agar slants or liquid media, and the inoculated media are incubated as usual. This method can be used for either cells or spores and viabilities of 90% are routinely observed.

Regardless of the method or methods chosen for the preservation of primary-stock cultures, it is of utmost importance that good, descriptive records be kept on these cultures and that the cultures be well labeled.

[4] W. C. Haynes, L. J. Wickerham, and C. W. Hesseltine, *Appl. Microbiol.* 3, 361 (1955).
[5] W. T. Sokolski, E. M. Stapert, and E. B. Ferrer, *Appl. Microbiol.* 12, 327 (1964).

Selection of Media for Growth of Selected Cultures

The choice of a good medium is virtually as important to the success of the experimental program as is the selection of an organism to carry out the fermentation.

Fungi and streptomycetes, in general, grow well in rather simple media. The compositions of several of these are summarized in Table II. Both groups will grow faster in media containing hydrolyzates of proteins, ground seed meals, or the by-product of the cornstarch industry known as "cornsteep liquor." The presence in the media of these nitrogenous materials results in faster cell production, more rapid utilization of carbohydrate, and usually higher levels of enzymes in the cells.

Chemically defined media present distinct advantages for certain types of biochemical studies: The concentration of any one component or of several components can easily be varied to determine its specific effect on cell growth and enzyme levels, and individual components can be deleted or added. These considerations allow the redesigning of a medium to obtain the greatest possible yield of product or enzyme content of the cells. These media may be expensive if unusual ingredients are added. The purity of all ingredients is of utmost importance, e.g., the presence in the water of traces of metals or of descaling amines from the boilers used to produce the steam condensed to form the distilled water, is often deleterious to the growth of the fungus or streptomycete.

The alternative is the nonsynthetic or "crude" medium, which usually allows much higher cell yields and in some instances higher enzyme levels than does a chemically defined medium. An example of a crude medium is one containing soybean meal, cornsteep liquor, glucose, distillers' dried solubles, $CaCO_3$, K_2HPO_4. All fungi and streptomycetes will grow well in this type of medium and the generation time will be much lower than when the same organisms are grown in chemically defined media. On the other hand the levels of desired enzymes in the cells may be considerably lower, especially if the enzymes are involved in the biosynthesis of a nutrient that occurs naturally in these crude materials. An added complication is that many of these natural materials, such as cornsteep liquor, peptones, and seed meals often contain metal ion salts which may markedly affect the enzyme levels in the growing cells. It is possible to reduce the metal ion content of chemically defined media by complexing and extraction with suitable reagents, e.g., the use of dipyridyl to complex iron and extraction the complex with chloroform. Unfortunately, such a procedure cannot be effectively used with media containing crude materials.

The compositions of a few of the types of media used for the growth of fungi are summarized in Table II. The Czapek-Dox and Raulin-Thom media are examples of chemically defined media often used for preparation of cells of fungi. Although the Calam-Hockenhull medium was designed for the production of penicillin by selected strains of *Penicillium chrysogenum*, it will support the growth of many fungi; the phenylacetic acid component can be omitted in these circumstances. Most fungi will grow faster on the Upjohn medium and the cornsteep medium than in the chemically defined medium, and many investigators are satisfied to use these crude media since the amounts of desired enzyme per kilogram of mycelium are significantly more than with the mycelium from the chemically defined medium. The same situation is found with media for the culture of streptomycete.

In pure culture operations it is essential that all media be sterilized before inoculation with the desired microorganism. This can be accomplished by autoclaving, by filtration through bacteriological filters, or by addition of sterilizing chemicals such as β-propionolactone. The most practical method is autoclaving. Among the precautions to be taken when using this method is the avoidance of "overcooking." Some organisms will not grow well in overcooked media, particularly when heat-labile components, such as thiamine and pantothenic acid, or fructose in the presence of phosphate buffer, are required. These heat-labile substances often can be economically sterilized by filtration procedures.

For those who wish to scale-up experiments from small flask-size operations to stirred aerated jars, small fermentors, or other, larger equipment, it is advisable to mention that "overcooking" of media is a common problem as the scale of the operation is increased. Some of the growth-inhibitory effects of "overcooking" can be avoided by sterilizing (autoclaving) the carbohydrate component separately in concentrated solution and adding it to the rest of the medium just before inoculation.

Equipment for Growth of Mold and Streptomycete Cultures

Fungi and streptomycetes are essentially aerobic organisms and for practical purposes require air for optimal cell biosynthesis. They can be grown on the surface of liquid or of semisolid (agar) media, or in aerated flasks or other vessels. Aeration of the latter can be accomplished by either shaking the flask or vessel on a mechanical shaker or introducing sterile, filtered air into the liquid through an aeration device (sparger).

For most small-scale biochemical studies these organisms are grown in shaken flasks. Erlenmeyer flasks, Florence flasks, and Mason jars have all been used. These are usually partially filled with inoculated medium and placed on a shaker. Both reciprocating and rotary shakers have been widely used. The amount of effective aeration on these shakers depends upon the rate of flask displacement, the amount of liquid in the flask, and the viscosity of the liquid. Some data on aeration of aqueous solutions in shaken Erlenmeyer flasks are summarized in Table III. Most cultures of aerobic organisms have shorter generation times at high aeration levels, and in a sense the effective aeration will determine the cell yield. As can be seen from the data in Table III, approximately the same effective aeration can be obtained in a variety of flasks and with shakers operating at different speeds.

Methods Useful in Recovery of Cells as Sources of Enzymes

Since most fungi form filamentous mycelia when grown in the media mentioned in Table II and under aeration levels described in Table III, the recovery of these cells can be efficiently accomplished in the laboratory by vacuum filtration through coarse filter paper. Use of a string-type vacuum filter is sometimes desirable when hundreds of liters are to be treated, and a basket-type centrifuge or a plate and frame press is effective for volumes in the range of 50–150 liters. Fungal mycelium is a perishable product and must be utilized immediately before lysis occurs. If the collected mycelium cannot be treated at once, quick freezing or similar techniques may be used to delay the loss of the enzymes.

Streptomycetes also form filamentous mycelia when grown in aerated culture. However, the strands are usually so fragile and fine that they cannot be collected by simple filtration. Addition of filter aids such as diatomaceous earth (about an equal weight is often used) is sometimes helpful but may lead to complications during recovery of the desired enzymes from the resultant mixture. Collection in a basket-type centrifuge or in a solid bowl high speed centrifuge, e.g., Sharples type, has often been successful for large volumes of cell suspensions. Streptomycete mycelium, like fungal mycelium, is an extremely perishable product and must be utilized immediately or most of the cells will lyse with concomitant decrease in the level of enzyme.

General Precautions

Several general precautions which may be of assistance in preparing fungal and streptomycete cells as enzyme sources: (1) Continuous evaluation of the biochemical characteristics of the producing culture. Since all microbial cultures are mixtures of individuals it is important that the characteristics of the population are kept as constant as possible.

TABLE III

OXYGEN ABSORPTION RATES AS RELATED TO CONTAINER SIZE,
VOLUME OF SOLUTION, AND TYPE OF SHAKING ACTION

Size and type of container	Volume of liquid	Type of shaking action	Oxygen absorption rate (mmoles O_2/l/min)
25 × 150 mm test tube	10	Rotating, 200 rpm[a]	0.45
25 × 150 mm test tube	10	Rotating, 300 rpm	0.70
125-ml Erlenmeyer flask	12.5	Rotating, 200 rpm	0.91
125-ml Erlenmeyer flask	12.5	Rotating, 300 rpm	1.00
125-ml Erlenmeyer flask	25	Rotating, 200 rpm	0.61
125-ml Erlenmeyer flask	25	Rotating, 300 rpm	0.75
125-ml Erlenmeyer flask	12.5	Reciprocating, 100 cpm[b]	0.70
250-ml Erlenmeyer flask	25	Rotating, 200 rpm	0.53
250-ml Erlenmeyer flask	25	Rotating, 300 rpm	0.67
250-ml Erlenmeyer flask	25	Reciprocating, 100 cpm	0.58
250-ml Erlenmeyer flask	50	Rotating, 200 rpm	0.41
250-ml Erlenmeyer flask	50	Rotating, 300 rpm	0.63
250-ml Erlenmeyer flask	100	Rotating, 200 rpm	0.38
250-ml Erlenmeyer flask	100	Rotating, 300 rpm	0.58
500-ml Erlenmeyer flask	50	Rotating, 200 rpm	0.66
500-ml Erlenmeyer flask	50	Rotating, 300 rpm	0.80
500-ml Erlenmeyer flask	50	Reciprocating, 100 cpm	0.68
500-ml Erlenmeyer flask	100	Rotating, 200 rpm	0.55
500-ml Erlenmeyer flask	100	Rotating, 300 rpm	0.69
500-ml Erlenmeyer flask	100	Reciprocating, 100 cpm	0.50
1000-ml Erlenmeyer flask	100	Rotating, 200 rpm	0.68
1000-ml Erlenmeyer flask	100	Rotating, 300 rpm	0.80
1000-ml Erlenmeyer flask	100	Reciprocating, 100 cpm	0.63
1000-ml Erlenmeyer flask	200	Rotating, 200 rpm	0.45
1000-ml Erlenmeyer flask	200	Rotating, 300 rpm	0.59
1000-ml Erlenmeyer flask	200	Reciprocating, 100 cpm	0.36
2000-ml Erlenmeyer flask	200	Rotating, 200 rpm	0.67
2000-ml Erlenmeyer flask	200	Rotating, 300 rpm	0.70
2000-ml Erlenmeyer flask	200	Reciprocating, 100 cpm	0.61
2000-ml Erlenmeyer flask	500	Rotating, 200 rpm	0.47
2000-ml Erlenmeyer flask	500	Rotating, 300 rpm	0.63
4000-ml Erlenmeyer flask	400	Rotating, 200 rpm	0.65
4000-ml Erlenmeyer flask	400	Rotating, 300 rpm	0.70
4000-ml Erlenmeyer flask	400	Reciprocating, 100 cpm	0.55
20-liter fermentor with sparger and baffles	12,000	Two volumes of air/volume of liquid/minute with agitation at 500 rpm	7.2

[a]Flasks on a platform with 1-inch circle.
[b]Flasks on a shaker with a 2-inch stroke.

Although preservation by the methods mentioned above often results in stabilizing the population, none is wholly satisfactory. If the culture has lost its "desirable" characteristics, sometimes a useful strain can be isolated by single spore isolation for the "run-down" population.[6] This is a time-consuming and often frustrating research program. (2) An effort should be made to use ingredients, including the water, of the same quality in preparation of the culture media. If substitutions have to be made, enough of the original ingredient should be retained, or its batch number noted if it is a commercial product, so that comparisons can be made quickly. (3) If the fungal and streptomycete cultures are grown in shaken flasks, the reliability of the shaker action should be checked periodically. Rubber belts stretch, bearings wear, and platform suspensions shift after continued operation of mechanical shakers. All these factors change the effective aeration and indirectly may cause a marked decrease in the growth rate of the organism. (4) Nearly all fungal and streptomycete cultures have rather limited temperature optima for growth and for enzyme synthesis. A shift of 2° may result in as much as a 3- to 10-fold change in enzyme levels. Temperature control can be achieved best by using a water bath shaker. Air incubators are usually controllable to ± 1°, depending upon the temperature control of the room in which they are placed.

[6]F. Reusser, *Advan. Appl. Microbiol.* 5, 189 (1963).

[11] *Neurospora crassa*: Preparative Scale for Biochemical Studies

By CARL RHODES, JOHN GERMERSHAUSEN and S. R. SUSKIND

Strains

A variety of wild-type, auxotrophic, and morphological mutant strains of *Neurospora crassa* have been used for the preparation of enzymes, ribosomes, mitochondria, nuclei, cell walls, and other components of interest to the biochemist. The reader is referred to the individual citations in the text and to the extensive stock list and culture collection maintained by the Fungal Genetics Stock Center, Dartmouth College, Hanover, New Hampshire. The stock lists provide information about the initial isolation and genetic background of the strains, including appropriate references and culture ordering procedures. Additional background information can be obtained from several

[1]Studies on this subject have been supported by grants from the U. S. Public Health Service (GM-16533 and HD-139). Contribution No. 605 from the McCollum-Pratt Institute.

publications dealing with *Neurospora* microbiological and genetic methodology[2,3], a comprehensive bibliography,[4] and a semiannual newsletter, dealing with techniques and research using this microorganism.[5]

A number of wild-type and/or mutant strains have been used for the large-scale preparation of *Neurospora* enzymes, e.g., tryptophan synthetase,[6] tryptophanyl-tRNA synthetase,[7] glutamic dehydrogenase,[8] and nitrate reductase.[9] Conditions which lead to derepression of enzyme synthesis in mycelial cultures have been used to obtain crude extracts with 3–to 10-fold higher specific activities than the repressed level.[6]

Preparations of spores (conidia, ascospores) and mycelia can be assayed for the desired enzyme activity or cell component with significant differences often evident between cultures of a different age as well as between asexual and sexual stages of the life cycle. Methods have been published for determining the cellular localization of enzymes,[10] and for the preparation of mitochondria and mitochondrial components,[11] ribosomes,[12] nuclei and chromatin,[13] and cell walls.[14]

Media

Different media are used for the production of mycelia and conidia and to obtain sexual spores. One medium frequently used for high yields of mycelia or conidia was developed by Vogel and Bonner.[11] It is also used for maintaining stock cultures and for preparing inocula. The medium, made up to 50-fold strength, has the following composition: sodium citrate·$2H_2O$, 131 g; KH_2PO_4, 250 g; NH_4NO_3, 100 g; $MgSO_4$, 4.9 g; $CaCl_2$, 3.8 g; biotin (10 mg/100 ml 50% ethanol, stored in the cold), 2.5 ml; trace element mixture,[15] 5.0 ml; distilled water to

[2]This series Vol. XVIIA [4].

[3]J. R. S. Fincham and P. R. Day, "Fungal Genetics." F. A. Davis, Philadelphia, 1963.

[4]B. J. Bachman and W. N. Strickland, "*Neurospora* Bibliography." Yale Univ. Press, 1965.

[5]B. J. Bachman, ed., *Neurospora* Newsletter. Distributed by Fungal Genetics Stock Center, Humbolt State College, Arcata, California.

[6]R. G. Meyer, J. Germershausen, and S. R. Suskind, Vol. XVIIA [49].

[7]C. Rhodes and S. R. Suskind, (1970).

[8]D. B. Roberts, *J. Bacteriol.* 94, 958 (1967).

[9]R. H. Garrett and A. Nason, *J. Biol. Chem.* 244, 2870 (1969).

[10]J. W. Greenawalt, D. O. Hall, and O. C. Wallis, this series, Vol. X [27].

[11]D. J. L. Luck, this series, Vol. X [53].

[12]F. A. M. Alberghina and S. R. Suskind, *J. Bacteriol.* 94, 630 (1967).

[13]R. S. Dwivedi, S. K. Dutta, and D. P. Bloch, *J. Cell Biol.* 43, 51 (1969).

[14]P. R. Mahadevan and E. L. Tatum, *J. Bacteriol.* 90, 1073 (1965).

[15]Citric acid·$1H_2O$, 5.00 g; Zn SO_4·$7H_2O$, 5.00 g; $Fe(NH_4)_2(SO_4)_2$·$6H_2O$, 1.00 g; $CuSO_4$·$5H_2O$, 0.25 g; $MnSO_4$·$1H_2O$, 0.05 g; H_3BO_3, 0.05 g; Na_2MoO_4·$1H_2O$, 0.05 g. The components are added in order to 95 ml of water at room temperature. Chloroform (1 ml) is added as a preservative, and the solution can be stored at room temperature.

1.0 liter. The components should be added to the water in the order indicated and with continuous stirring. The medium may be stored at room temperature as a 50-fold concentrated stock solution when chloroform (2 ml/liter) is added as a preservative. The medium is diluted to single strength with distilled water before use, and 2% sucrose is added before autoclaving (or is autoclaved and added to the medium separately if large volumes and long periods of autoclaving are necessary). Bacto-Agar (2%, Difco) is added when solid medium is desired. For the obtaining of ascospores, Westergaard and Mitchell's synthetic crossing medium has proved most successful.[16] Other media commonly used for growing vegetative cultures include Ryan's complete medium[17] and Fries minimal medium.[18]

Preparation of Conidia

Slants in screw-cap tubes containing 5 ml of solid medium, prepared as described above, are inoculated with conidia of the appropriate strain and incubated at 30° under fluorescent light illumination for 1 week. About 3 ml of sterile distilled water is added to each slant and a conidial suspension, prepared by vigorously shaking the tube, is used to inoculate a 2.8 liter Fernbach flask containing 500 ml of the medium with 2% agar. Flasks are incubated at 30° with illumination for 7–10 days. The conidia from the agar surface and the walls of each flask are suspended in 250 ml of distilled water by vigorous swirling. The suspension is filtered through four layers of cheesecloth to remove hyphae, and the conidia are concentrated by low-speed centrifugation at room temperature. After several washings with distilled water to remove traces of the medium, the conidia are again centrifuged, the supernatant solution is decanted and the conidial pellet is either used immediately or stored at − 25°. Alternatively, the washed conidia may be lyophilized and stored in a deep freeze. Ten-day Fernbach cultures of wild-type strains yield about 300 mg dry weight of conidia or approximately 10^{10} cells per flask. If the conidia are to be used for extended germination studies, the above procedures should be carried out aseptically.

Preparation of Mycelia

Neurospora may be grown conveniently in liquid culture using 2.5–5-gallon Pyrex or Nalgene carboys. The medium is autoclaved for 40 minutes at 120° and 15 psi. Usually, sucrose is autoclaved separately

---

[16]M. Westergaard and H. K. Mitchell, *Amer. J. Bot.* **34**, 573 (1947).
[17]F. J. Ryan, *Methods Med. Res.* **3**, 51 (1950).
[18]F. J. Ryan, G. W. Beadle, and E. L. Tatum, *Amer. J. Bot.* **30**, 784 (1943).

for 20 minutes to prevent caramelization and is then added to the sterile medium. For inocula we use sterile suspensions of conidia, prepared by adding 3 ml of sterile distilled water to 7-day slant cultures and shaking the slants to suspend the conidia. No effort is made to remove hyphae in these suspensions. Usually one or two slants are used per carboy. The carboy cultures are grown at 30° with vigorous aeration, using compressed air and a sterile glass wool filter trap. When a suspension of wild-type strain, 74-OR23-1A (Fungal Genetics Stock Center No. 987) conidia from a single slant is used to inoculate 20 liters of medium, the culture will be in late log phase at 72 hours; a larger inoculum decreases the growth lag. For tryptophan synthetase preparations, the aerated mycelia are filtered through cheesecloth when they are light pink (60–70 hours). The yield from 10 liters of medium is about 300 g wet weight per carboy, depending on the strain. One should establish optimal growth conditions for each strain and for each purpose.

Larger yields of mycelia are obtained by growth under controlled conditions in a 130-liter fermentation tank (Fermacell FSI-130, New Brunswick Scientific Corporation, New Brunswick, New Jersey). One hundred liters of the medium of Vogel and Bonner without sucrose, is sterilized for 1 hour at 120° and 15 psi, cooled, and maintained at 30° for growth of the culture. A sterile solution of sucrose (2 kg dissolved in 4 liters of distilled water and autoclaved separately) is added through a port, followed by the inoculum, a conidial suspension prepared from 4–6 ten-day-old Fernbach cultures. The culture is stirred at 100 rpm, and aerated with cooled incinerated air at a flow rate of 3 cubic feet per minute. The mycelia are harvested after 20–28 hours. The selection of stirring and sparging rates, the growth temperature, and the age of the culture at harvest will be controlled to a considerable extent by the requirements of the system under study and the properties of the particular strain. For example, the specific activity of tryptophanyl-tRNA synthetase is maximum in mycelia grown for 20–22 hours, and just beginning to turn pink[7]; ribosomes from these young mycelia can be extracted in high yield and are active and stable when assayed for their capacity to support *in vitro* protein synthesis[19]. The specific activity of tryptophan synthetase is highest after 28 hours of growth[6]; cultures grown in the Fermacell for 36 hours yield no appreciable increase in total tryptophan synthetase activity. Crude extracts from older mycelia are less likely to be stable than preparations from younger cultures.

[19] J. Germershausen, H. Rothschild, and S. R. Suskind, (1969).

The culture is harvested from a port by maintaining the fermentor tank under positive pressure. The mycelia are filtered through cheese-cloth, rinsed with tap water, and washed at least twice with distilled water. The wet mycelia are squeezed dry in a stainless steel (19½ × 19¾ × 6⅝ inches) manual press, shredded into small pieces and lyophilized during a period of 2 to 3 days in a large-capacity sublimator (VirTis Co., Gardiner, New York). Alternatively, mycelia are washed with the de-sired buffer, quickly frozen in liquid nitrogen, and stored at − 50°. Immediately after lyophilization, the mycelia are ground to a fine powder in a Waring Blendor at room temperature. Ordinarily, a 100-liter, wild-type culture yields 300–400 g of lyophilized product. Stable, long-term storage of the lyophilized mycelia is best accomplished at − 50°.

Colonial morphological mutants, as well as cultures of spheroplast-like cells grown in the fermentor may be harvested in a Sharples centri-fuge, followed by lyophilization and freezer-storage as described above.[19]

Preparation of Ascospores

When strains of opposite mating type are grown on crossing medium,[16] ascospores are formed in small sacs (asci) within the fruiting bodies (perithecia). Mature spores are ejected from the asci and can be col-lected by scraping the walls of the culture tube.

Preparative quantities of ascospores are considerably more difficult to obtain than either conidia or mycelia. Large numbers of crosses are required to produce a yield of perithecia sufficient for preparative purposes. Washed ascospores have been used as a source of ribosomes[20] and of transfer RNA.[21]

Biochemical Preparations

Crude extracts. Ordinarily all procedures are conducted at 4°. Conidia, ascospores, and quick-frozen mycelia may be ground in a mortar and pestle with 2 volumes of acid-washed sea sand or alumina and 1 volume of buffer. For example, at 1 ml buffer per gram of mycelia, an extract after centrifugation (15,000 g, 20 minutes) contains about 15 mg protein per milliliter. Sonication can be employed instead of grinding and is particularly recommended for conidia, which are relatively difficult to break otherwise.[22] In cases where the preparation is not affected by foaming during extraction, e.g., for transfer RNA prepara-

[20]H. R. Henney and R. Storck, *Science* 142, 1675 (1963).
[21]F. Imamoto, T. Yamane, and N. Sueoka, *Proc. Nat. Acad. Sci. U. S.* 53, 1456 (1965).
[22]S. Linn and J. R. Lehman, *J. Biol. Chem.* 240, 1287 (1965).

tions,[23] a Waring Blendor may be used to disrupt fresh, frozen, or lyophilized mycelia suspended in buffer. This method usually is not recommended for enzyme preparations.

For the large-scale preparation of soluble enzymes, as well as for isolation of cytoplasmic ribosomes,[19] the simplest method involves suspension of lyophilized, powdered mycelia in buffer, followed by manual stirring and centrifugation. In a typical extraction 100 g of lyophilized mycelia are suspended in 1600 ml of 0.1 M potassium phosphate buffer, pH 7.8, and stirred intermittently for 20 minutes. After centrifugation at 15,000 g for 30 minutes, approximately 1400 ml of a supernatant solution containing 15–25 mg of protein per milliliter is obtained. The pellet contains cell wall debris as well as disrupted nuclei and mitochondria. Ribosomes may be prepared from the supernatant fluid by centrifugation at 30,000 g for 30 minutes, followed by ultracentrifugation of the supernatant fraction for 2–4 hours at 105,000 g.[19]

When the preparation of intact mitochondria or nuclei is desired, lyophilized mycelia should not be used. Gentle grinding of fresh mycelia with a mortar and pestle and sea sand or alumina, or the use of a ball or grinding mill avoids disruption of cell organelles.[11] Such methods have also been used for preparation of ribosomes.[12] although larger quantities of ribosomes with higher specific activity can be prepared from lyophilized mycelia.[19] Mitochondrial fractions can be obtained by mycelial disruption with a Gifford-Wood-Eppenbach Micro Mill, followed by sedimentation of nuclei and cell wall debris at 5000 g. Mitochondria are collected by an additional centrifugation at 8000 g, after which the pellet is sedimented through buffered sucrose. If possible, mitochondrial preparations should be monitored by electron microscopy.[24,25]

Removal of Nucleic Acids

Precipitation of nucleic acids or their removal by ion-exchange chromatography is recommended during the initial steps of enzyme purification, especially when heat steps or salt fractionations are anticipated. Precipitation using manganous chloride,[23] streptomycin,[22] and removal of nucleic acids by DEAE cellulose,[26] and phase separation[9] have been reported. Protamine sulfate is quite satisfactory for the removal of high molecular weight nucleic acids from large volumes of crude extract. As described earlier, 100 g of lyophilized mycelia

[23]D. Printz and S. R. Gross, *Genetics* 55, 451 (1967).
[24]D. O. Hall and J. W. Greenawalt, *Biochem. Biophys. Res. Commun.* 17, 565 (1964).
[25]W. E. Barnett and D. H. Brown, *Proc. Nat. Acad. Sci. U. S.* 57, 452 (1967).
[26]W. E. Barnett and J. L. Epler, *Proc. Nat. Acad. Sci. U. S.* 55, 184 (1966).

provides about 1400 ml of crude extract. The pH of this extract is adjusted to neutrality with N acetic acid, and 140 ml of an aqueous 2% solution of protamine sulfate is added. The protamine sulfate solution is freshly prepared at room temperature. The extract is stirred for 20 minutes and centrifuged at 15,000 g for 20 minutes, yielding a white precipitate and about 1500 ml of a clear yellow supernatant solution. Although it does not interfere with subsequent purification steps, the preparation at this stage often contains low molecular weight material, absorbing at 260 mμ, which can be removed by dialysis.

[12] Increasing Enzyme Production by Genetic and Environmental Manipulations[1]

By Arnold L. Demain

The amounts of a particular enzyme produced by a particular microorganism can vary tremendously since enzyme formation is regulated in both positive and negative directions by such control mechanisms as induction, end product repression, and catabolite repression. Catabolic enzymes have been known to vary several thousandfold, and hundredfold changes have been observed in the specific activity of biosynthetic enzymes.[1a] Since each control mechanism is influenced by environmental conditions, traditional factors important in enzyme production are temperature, pH, medium composition, aeration, and the stage of the growth cycle. These conditions are usually determined empirically for each enzyme and each strain and will not be considered further unless they apply to the specific type of control mechanism under discussion. It is thus assumed that these optimal conditions have already been chosen for a particular situation; and I will concentrate on those further manipulations that can be used to modify, to bypass, or to utilize regulatory mechanisms to force "overproduction" of enzymes in the laboratory. It is also assumed that the very important step of screening a wide variety of organisms for the particular enzyme has already been done or that the investigator, for other reasons, has chosen a particular strain as a source of enzyme. Many of the examples of genetic and environmental manipulations quoted below were devised by virtue of detailed genetic knowledge of a particular species, but

[1]Contribution No. 1602 of the Department of Nutrition & Food Science, Massachusetts Institute of Technology, Cambridge, Massachusetts 02139.
[1a]P. H. Clarke and M. D. Lilly, *Symp. Soc. Gen. Microbiol.* **19**, 113 (1969).

there is no reason to expect that they would not work on other micro-organisms. The task of preparing a summary of such manipulations has been facilitated by the excellent survey of Pardee.[2] His contribution forms a basis of the present treatment with additional examples cited to bring the field up to date.

The manipulations to be discussed have been grouped into the following categories:

Environmental:
    1. Addition of inducers
    2. Decreasing concentration of repressors
Genetic:
    1. Mutation to constitutivity
    2. Increasing gene copies

Addition of Inducers

Many enzymes, especially those used by the cell for catabolic pur-poses, are normally repressed and are not produced unless inducers are added to the medium or are formed by metabolism. In many cases, the inducer is the substrate of the enzyme; however, some of the most potent inducers are nonmetabolizable substrate analogs ("gratuitous" inducers), e.g., isopropyl-β-D-thiogalactoside for β-galactosidase, methicillin for penicillin β-lactamase, malonic acid for maleate-fumarate cis-trans-isomerase, and N-acetylacetamide for amidase. The effects of inducers are tremendous in some cases. Certain galactosides increase the specific activity of β-galactosidase in *Escherichia coli* by 1000-fold and result in cells containing several percent of their total protein as this single enzyme.

If an inducer is particularly expensive or not readily available, it may be possible to substitute a compound that can be converted by the organism to the required inducer. For example, in *Pseudomonas fluorescens* and *Neurospora crassa* kynurenine induces kynureninase[3]; the same effect can be obtained by adding tryptophan, a precursor of kynurenine, to the medium.

Decreasing Concentration of Repressors

Many enzymes, especially those with biosynthetic function, are normally formed, but formation can be repressed by the presence or

[2] A. B. Pardee, *in* "Fermentation Advances" (D. Perlman, ed.), p. 3. Academic Press, New York, 1969.
[3] O. Hayaishi and R. Y. Stanier, *J. Bacteriol.* **62**, 691 (1951); W. B. Jakoby and D. M. Bonner, *J. Biol. Chem.* **205**, 699 (1953).

production of end-product repressors or catabolite repressors. Thus, the medium constituents are of utmost importance in production of repressible enzymes. The goal in such cases is a growth medium containing as little of the repressing compound as possible and one that does not lead to extensive internal formation of repressing compounds. Rich, complex media, as well as media containing rapidly utilizable sugars, such as glucose, are to be avoided for maximum enzyme production.

Catabolic enzymes are usually under dual control of induction and repression; repression can be caused by end products or by catabolites of rapidly used carbon sources or by both. Production of urease in *Proteus rettgeri* is induced by urea, repressed by ammonia and also is affected by the catabolite.[4] In *Neurospora*, nitrate reductase is subject to induction by nitrate and repression by ammonia.[5] Aliphatic amidase of *Pseudomonas aeruginosa* is induced by acetamide and catabolically repressed by succinate.[1a] Limitation of repressing compounds has a marked influence on enzyme levels as shown by the following examples. Inorganic phosphate represses phosphatase formation in *E. coli* and nucleases in *Aspergillus quernicus*. By limiting phosphate, the amount of alkaline phosphatase can be increased in *E. coli* from essentially zero up to as high as 5% of the cell protein.[6] Similar treatment of *A. quernicus* results in a 30-fold increase in 5'-nucleotide-forming nuclease, a 50-fold increase in the nuclease producing 3'-nucleotides, and a 40-fold increase in phosphatase.[7] Repression of proteases by ammonia[8] and by amino acid mixtures[9] in bacteria and by sulfate in *Aspergillus niger*[10] can be avoided by limitation of such compounds in growth media. Substitution by pyruvate of glucose as carbon source in *E. coli* leads to a 40-fold increase in phosphoenolpyruvate carboxykinase.[11]

Repression of biosynthetic enzymes by end products can be prevented by manipulations which limit production of these repressing compounds. This can be accomplished in two main ways.

Growth Limitation of Auxotrophic Mutants. Use of growth-limiting amounts of a required growth factor often derepresses enzyme synthesis. For example, histidine limitation of histidine-requiring auxotrophs coordinately derepresses the ten enzymes of the pathway by

[4] I. Magana-Plaza and J. Ruiz-Herrera, *J. Bacteriol.* 93, 1294 (1967).
[5] S. C. Kinsky, *J. Bacteriol.* 82, 898 (1961).
[6] A. Garen and N. Otsuji, *J. Mol. Biol.* 8, 841 (1964).
[7] Y. Ohta and S. Ueda, *Appl. Microbiol.* 16, 1293 (1968).
[8] P. V. Liu and H. C. Hsieh, *J. Bacteriol.* 99, 406 (1969).
[9] J. Chaloupka and P. Kreckova, *Folia Microbiol.* 11, 82, 89 (1966).
[10] G. Tomonaga, *J. Gen. Appl. Microbiol. (Japan)* 12, 267 (1966).
[11] E. Shrago and A. L. Shug, *Arch. Biochem. Biophys.* 130, 393 (1969).

about 25-fold.[12] Leucine limitation in a leucine auxotroph of *Salmonella typhimurium* yields a 40-fold increase in acetohydroxy acid synthetase.[13] Use of glycyl-L-valine, instead of valine, to feed a valine auxotroph of *Hydrogenomonas* growing in a chemostat results in a 5-fold increase in this same enzyme.[14] A thiamine-requiring mutant, grown with limiting thiamine, shows derepressed levels of four enzymes involved in thiamine biosynthesis[15]; the production of one of the enzymes, thiamine phosphate pyrophosphorylase, increases 1500-fold.

If the auxotroph has only a partial requirement, growth in minimal medium will be slow and derepression of the pathway enzymes will result. Thus, growth of a leaky pyrimidine auxotroph in minimal medium leads to a 500-fold increase in aspartate transcarbamylase.[16]

A variation on the above theme is the feeding of a slowly assimilated derivative or precursor of the growth factor to the auxotroph. In *E. coli*, growth of a uracil auxotroph on dihydroorotic acid, instead of on uracil, gives derepressed levels of six enzymes of pyrimidine biosynthesis.[17] Use of djenkolate for a cysteine auxotroph of *S. typhimurium* results in slow growth and high levels of the enzymes involved in sulfate reduction.[18]

Analogs and Inhibitors. In the use of prototrophic microorganisms, production of repressors can be reduced by addition of end-product analogs to the medium. Thus, all ten histidine pathway enzymes are derepressed up to 30-fold by 2-thiazolealanine.[19] Tryptophan synthetase is derepressed by indole 3-propionate,[20] and both inosine monophosphate dehydrogenase and xanthosine monophosphate aminase are derepressed by psicofuranine.[16] In a similar manner, adenine, which inhibits thiamine synthesis, derepresses the enzymes of thiamine biosynthesis.[15]

Mutations to Constitutivity and Hyperproduction

The aim here is to obtain "constitutive" mutants with altered regulatory genes so that the culture no longer requires an inducer or is no

[12] B. N. Ames, R. F. Goldberger, P. E. Hartman, R. G. Martin, and J. R. Roth, *in* "Regulation of Nucleic Acid and Protein Biosynthesis" (V. V. Koningsberger and L. Bosch, eds.), p. 272. Elsevier, Amsterdam, 1967.
[13] M. Freundlich and J. M. Trela, *J. Bacteriol.* 99, 101 (1969).
[14] M. Reh and H. G. Schlegel, *Arch. Mikrobiol.* 67, 110 (1969).
[15] T. Kawasaki, A. Iwashima, and Y. Nose, *J. Biochem. (Japan)* 65, 407 (1969).
[16] H. A. Moyed, *Cold Spring Harbor Symp. Quant. Biol.* 26, 323 (1961).
[17] M. Sheperdson and A. B. Pardee, *J. Biol. Chem.* 235, 3233 (1960).
[18] A. B. Pardee, *J. Biol. Chem.* 241, 5886 (1966).
[19] B. N. Ames and P. E. Hartman, *Cold Spring Harbor Symp. Quant. Biol.* 28, 349 (1963).
[20] D. E. Morse, R. F. Baker, and C. Yanofsky, *Proc. Nat. Acad. Sci. U.S.* 60, 1428 (1968).

longer repressed by end products or by catabolites of a rapidly utilized carbon source. Thus, cultures are obtained that produce "induced levels" of enzyme without inducer or "derepressed levels" in the presence of compounds which normally cause enzyme repression. However, in a few cases, extremely high levels of enzymes are produced, levels much higher than could be obtained by induction or derepression of the parent strain. For example, aeration induces catalase in *Rhodopseudomonas spheroides* so that the enzyme level rises from 0.002% to 1% of the total extractable protein. Selection of mutants resistant to H_2O_2 results in cultures which produce catalase to the extent of 25% of the cell protein.[21] It is possible that hyperproducers result either from mutation at an unlinked regulatory gene, at the operator locus or at the promoter site of an operon. Hyperproducers may also result from increase in gene copies; this will be discussed in a later section.

Removal of Inducer Requirement

Method A. Growth in Chemostat with Limiting Substrate Inducer. When growth of a population of an inducible parent culture is carried out in a chemostat at very low concentrations of a substrate inducer, selection occurs for constitutive mutants that do not require inducer, provided the substrate can be used as carbon source at a concentration that does not induce. For example, growth in a chemostat at a low lactose concentration selects for *E. coli* mutants which form β-galactosidase without inducer and which contain as much as 25% of their protein as β-galactosidase.[22] The technique has also been used to obtain constitutive enzymes of the mandelate pathway in *Pseudomonas putida*.[23]

Method B. Cycling with and without Inducer. Successive growth cycles between a medium not containing inducer and one which does contain inducer, enriches the population in constitutive mutants.[24] For example, the first growth cycle might contain glucose; both the rare constitutive mutants and the predominant β-galactosidase inducible parent cells will grow equally. Transfer of the mixed culture to a lactose medium will favor the constitutive mutant since the inducible parent must take time for induction. Repetition of these cycles of growth will finally allow the constitutive mutant to predominate since each transfer to glucose will return the inducible parent to the uninduced state.

Method C. Poorly Inducing Substrate. When a population is grown in a medium containing as carbon source a compound which is a good

[21]R. K. Clayton and C. Smith, *Biochem. Biophys. Res. Commun.* **3**, 143 (1960).
[22]A. Novick and T. Horiuchi, *Cold Spring Harbor Symp. Quant. Biol.* **26**, 239 (1961).
[23]G. D. Hegeman, *J. Bacteriol.* **91**, 1161 (1966).
[24]G. Cohen-Bazire and M. Jolit, *Ann. Inst. Pasteur* **84**, 937 (1953).

substrate but a poor inducer, constitutive mutants are selected. An example is the use of phenyl β-galactoside to select constitutive β-galactosidase cultures.[25]

Method D. Inhibitors of Induction. Compounds exist that block the induction of certain enzymes. Such compounds are cyanoacetamide in the case of *P. aeruginosa* amidase[1a] and 2-nitrophenyl-β-fucoside in the case of *E. coli* β-galactosidase. Nitrophenylfucoside blocks the induction of a permease for transport of galactosides, e.g., lactose and melibiose, into the cell and has been used to select for constitutive mutants. In this procedure, cells are grown in a mixture of melibiose and nitrophenylfucoside. Since the induction of galactoside permease is inhibited, only mutants that do not require inducer are able to get the galactoside (melibiose) into the cell and to grow.[24] It should be noted that since a separate permease for melibiose exists, this procedure works only in mutants lacking the specific melibiose permease or in wild-type *E. coli* grown at elevated temperature since the specific permease is temperature-sensitive.

Method E. Visible Detection without Enrichment. Methods have been devised to directly visualize a rare constitutive mutant colony among many inducible colonies on agar plates. Enrichment methods as described above are not required in these cases if the mutation frequency is high enough. For example, mutation to β-galactosidase constitutivity can be detected by plating out a mutagenized culture on an agar medium containing glycerol as carbon source. After growth of the colonies, the plate is sprayed with *o*-nitrophenyl-β-D-galactoside. Only constitutive mutants have made β-galactosidase during growth in the absence of inducer. These colonies hydrolyze the colorless reagent releasing the yellow *o*-nitrophenol. The presence of one yellow constitutive mutant colony among hundreds of normal white inducible colonies is easily detected and that clone can be subcultured and maintained.

Resistance to End-Product Repression

Method A. Antimetabolites. The use of toxic antimetabolites to select resistant cultures often yields mutants derepressed in the enzymes of the pathway of the normal metabolite. Trifluoroleucine selects mutants with derepressed levels (as much as 10-fold) of leucine biosynthetic enzymes.[26] Certain canavanine-resistant mutants produce 30 times more of the arginine pathway enzymes than do their sensitive parents.[27] Similarly, resistance to norleucine or 5-methyl tryptophan, or α-

[25] F. Jacob and J. Monod, *J. Mol. Biol.* **3**, 318 (1961).
[26] R. A. Calvo and J. M. Calvo, *Science* **156**, 1107 (1967).
[27] G. A. Jacoby and L. Gorini, *J. Mol. Biol.* **24**, 41 (1967).

aminobutyric acid is accompanied by derepression of enzymes of methionine,[28] tryptophan[29] or valine[30] biosynthesis, respectively. Resistance to valine leads to derepression of three enzymes of isoleucine biosynthesis.[31] Mutants of *Lactobacillus casei* resistant to dichloroamethopterin are derepressed 80-fold in their ability to form thymidylate synthetase.[32] When *Diplococcus pneumoniae* mutants are selected on the basis of resistance to amethopterin, such cultures produce 100 times as much dihydrofolate reductase as the parental culture.[33] In this last case, unlike all others in which the mutation has been mapped, the mutation appears to map in the structural gene for dihydrofolate reductase, rather than in a regulatory gene.

Method B. Visual Detection. Direct detection without enrichment can be used to select mutants not repressible by end products. For example, mutation to phosphatase constitutivity can be detected after growth of colonies in the presence of high phosphate and spraying with *p*-nitrophenyl phosphate.[6] Constitutive mutants are yellow, repressible colonies are white.

Another technique is that of spraying or overlaying colonies with a suspension of bacteria which require a compound for growth, this compound being the end product of the pathway which one desires to derepress. Constitutive mutants often overproduce the end product, excrete it and can cross-feed tester cultures.[34]

Occasionally constitutive mutants can be detected by their colonial appearance. For example, derepressed histidine mutants of *S. typhimurium* have a wrinkled colonial appearance in contrast to the smooth colonies formed by the repressible wild type.[35]

Pigment formation can be used for visually detecting mutants constitutive for enzymes of the purine pathway.[36] The technique uses an auxotroph blocked early in the pathway which forms red colonies due to accumulation of 5-amino-4-imidazole ribonucleotide and its subsequent polymerization into a red pigment. Normally the presence of high levels of adenine in the medium inhibits pigment formation due

---

[28]R. J. Rowbury, *Nature (London)* **206**, 962 (1965).
[29]G. N. Cohen and F. Jacob, *C. R. Acad. Sci.* **248**, 3490 (1959).
[30]M. Kisumi, S. Komatsubara, and I. Chibata, *Amino Acids and Nucleic Acids (Tokyo)* **19**, 1 (1969).
[31]T. Ramakrishnan and E. A. Adelberg, *J. Bacteriol.* **87**, 566 (1964).
[32]T. C. Crusberg and R. L. Kisliuk, *Fed. Proc., Fed. Amer. Soc. Exp. Biol.* **28**, 473 (1969).
[33]F. M. Sirotnak, S. L. Hatchtel, and W. A. Williams, *Genetics* **61**, 313 (1969).
[34]G. A. O'Donovan and J. C. Gerhart, *Bacteriol. Proc.* p. 125 (1968).
[35]J. R. Roth and P. E. Hartman, *Virology* **27**, 297 (1965).
[36]B. Dorfman, *Genetics* **61**, 377 (1969).

to repression of the pathway. After mutation of the auxotroph, the population is plated on adenine-containing agar; derepressed mutant colonies are red.

Method C. Sporulation Mutants. Protease formation by bacilli, as well as sporulation, is inhibited by amino acid mixtures. Isolation of mutants which are able to sporulate in the presence of an inhibitory mixture of amino acids are easily selected by a heating procedure. Such mutants of *Bacillus cereus* are derepressed for protease synthesis producing about ten times as much protease as their parent.[37]

Method D. Reversion of Auxotrophy. Mutation of a culture to auxotrophy and reversion to prototrophy can be used to modify the regulatory properties of the enzyme involved in the nutritional requirement. For example, reversion of mutants lacking homoserine dehydrogenase in one case resulted in a 3-fold increase in enzyme activity compared to the original prototrophic culture.[38]

Resistance to Catabolite Repression

Mutants resistant to catabolite repression are useful when one wants to obtain enzymes from cultures grown in complex media or when non-repressing carbon sources might be too expensive for large-scale fermentations. In addition, some mutants resistant to catabolite repression produce extremely high levels of enzymes. For example, a yeast mutant, in which invertase production is resistant to catabolite repression, produces almost 2% of its cell protein as invertase.[39] Many of the mutants obtained so far are apparently modified in their glucose catabolic pathway since the production of many of their enzymes is no longer repressed by glucose. However, there are some mutants in which resistance specifically applies to catabolite repression of a single enzyme.[40]

Method A. Substrate of Repressed Enzyme as Sole Nitrogen Source. Serial transfer of *Aerobacter aerogenes* in a medium containing glucose plus histidine (but no other nitrogen source) selects mutants resistant to catabolite repression.[41] Normally histidine-degrading enzymes are repressed by glucose, but in this case they are obligatory for growth. The medium thus selects for mutants which produce histidase in the presence of glucose. Such mutants show derepressed levels of histidase,

[37]S. Levisohn and A. I. Aronson, *J. Bacteriol.* **93**, 1023 (1967).
[38]J. C. Patte, G. LeBras, T. Loviny, and G. N. Cohen, *Biochim. Biophys. Acta* **67**, 16 (1963).
[39]J. O. Lampen, N. P. Neumann, S. Gascon, and B. S. Montenecourt, *in* "Organizational Biosynthesis" (H. J. Vogel, J. O. Lampen, and V. Bryson, eds.), p. 363. Academic Press, New York, 1967.
[40]L. A. Chasin and B. Magasanik, *J. Biol. Chem.* **243**, 5165 (1968).
[41]F. C. Neidhardt, *J. Bacteriol.* **80**, 536 (1960).

urocanase, and β-galactosidase when grown with glucose and histidine. A similar technique using a medium containing glucose and N-acetyl lactosamine selects out constitutive mutants of *E. coli*.[42] Such mutants produce high levels of β-galactosidase, β-galactoside permease, tryptophanase and amylomaltase in the presence of glucose.[42,43] Growth of *Pseudomonas aeruginosa* on succinate plus lactamide selects catabolite-repression resistant amidase cultures. The wild-type culture cannot grow since formation of the amidase is repressed by succinate so that nitrogen is not available.[1a]

Method B. Cycling with and without Glucose. E. coli shows a lag in growth when transferred from glucose to a medium containing lactose, maltose, acetate, or succinate.[44] This is evidently due to repression during growth on glucose of enzymes needed for utilization of these other carbon sources. Thus, alternating growth between glucose and succinate, for example, selects mutants resistant to catabolite repression.[44] The selection is also aided by the fact that the generation time in maltose, acetate, or succinate is markedly shorter for the mutant than for the repressible parent.

Method C. Visual Detection. The methods described above for visual detection of other types of constitutive mutants can also be used to detect mutants resistant to catabolite repression. For example, colonies are grown on agar containing glucose and sprayed with O-nitrophenyl-galactoside. If the parent is inducible, an inducer must be included in the agar. If the parent requires no inducer, it may be omitted. Mutants resistant to catabolite repression will be yellow. A similar technique was used recently to isolate mutants of *Aeromonas liquefaciens* resistant to catabolite repression in production of the extracellular enzyme, polygalacturonic acid *trans*-eliminase.[45] The enzyme can be detected after growth of colonies on polygalacturonic acid by flooding of the plate with HCl. Clear zones indicate enzyme production. By replica-plating surviving colonies after nitrosoguanidine mutagenesis on polygalacturonic acid agar plates with and without glucose, repression-resistant mutants were detected by virtue of their ability to produce clear zones on both types of plate.

Method D. Streptomycin Dependency. For some unexplained reason, mutation to streptomycin dependency in *E. coli* decreases the efficiency

[42]W. F. Loomis, Jr., and B. Magasanik, *Biochem. Biophys. Res. Commun.* **20**, 230 (1965).
[43]H. V. Rickenberg, A. W. Hsie, and J. Janacek, *Biochem. Biophys. Res. Commun.* **31**, 603 (1968).
[44]A. W. Hsie and H. V. Rickenberg, *Biochem. Biophys. Res. Commun.* **29**, 303 (1967).
[45]E. J. Hsu and R. H. Vaughn, *J. Bacteriol.* **98**, 172 (1969).

of glucose utilization, resulting in nonspecific resistance to catabolite repression.[46]

Increasing Gene Copies

Occasionally the use of certain of the above techniques to select mutants producing high enzyme levels results in organisms with several copies of the gene dictating production of the enzyme.[22] Since it is generally, but not invariably, true that the greater the number of copies of a gene, the higher the enzyme production, it would be desirable to be able to intentionally introduce extra specific genes into a micro-organism capable of producing a certain enzyme. This already is a reality in *E. coli* as will be described below.

Method A. Episome Transfer. Extrachromosomal DNA segments are present in many enterobacteria. These pieces of DNA, known as episomes,[47] can be introduced into a culture by genetic means. Such a manipulation in *E. coli* with an episome carrying the gene for β-galactosidase has been known to increase enzyme production 3-fold.[25] A method is now available[48] to obtain F'-factors (episomes) containing any part of the *E. coli* chromosome, thus making it possible to specifically increase production of a desired enzyme by episome transfer.

Method B. Phage-Escape Synthesis. When a specialized transducing phage such as λ or $\phi 80$ (but not P1 or P22) carrying structural genes for certain bacterial enzymes is used to lysogenize an *E. coli* strain and the prophage in the lysogenic strain is subsequently induced to replicate, replication of the transducing phage rapidly occurs resulting in an enormous increase in the number of gene copies per cell. Enzymes coded by these genes are synthesized at the same time at rates manyfold higher than the normal rate of synthesis.[49,50] A method is now available for isolation of specialized transducing phages for any *E. coli* gene[51] and this technique should be extremely useful in increasing enzyme production.

Acknowledgment

The author thanks Robert G. Martin for reading the manuscript and for suggesting improvements. The preparation of this article was supported by Public Health Service Research Grant AI-09345 from the National Institute of Allergy and Infectious Diseases.

[46]M. B. Coukell and W. J. Polglase, *Biochem. J.* 111, 279 (1969).
[47]R. P. Novick, *Bacteriol. Rev.* 33, 210 (1969).
[48]B. Low, *Proc. Nat. Acad. Sci. U.S.* 60, 160 (1968).
[49]M. Yarmolinsky, *in* "Viruses, Nucleic Acids, and Cancer," p. 151. Williams and Wilkins, Baltimore, Maryland, 1963.
[50]G. Buttin, *J. Mol. Biol.* 7, 610 (1963).
[51]S. Gottesman and J. R. Beckwith, *J. Mol. Biol.* 44, 117 (1969).

Specific Organelles
and Extraction Procedures

Articles Related to Section III

[13] Bacterial Membranes

By H. R. KABACK

Within the past few years, there has been a burgeoning interest in the structure and function of bacterial membranes. During this time, the membrane has been implicated in transport,[1-10,18] cell division,[11,12,12a] protein synthesis,[13,14] electron transport and oxidative phosphorylation,[15,16] and the binding of nucleotides.[17] The methods used for the preparation of membranes have been quite heterogeneous, and, in many cases, the preparations were not well characterized. Since the techniques used in this laboratory have been extremely useful for both small- and large-scale preparation of bacterial membranes from various gram-positive and gram-negative organisms, they will be described in detail. The membranes prepared by these methods have been extensively characterized with regard to purity, homogeneity, and pertinent physical properties. In addition, they have been shown to be physiologically active with regard to certain integrated membrane functions such as amino acid[1-3,5,8] and sugar[4,6-10,18] transport, phospholipid biosynthesis,[3] and electron transport.[1,18]

In essence, the procedure to be described consists of two steps; (1) the conversion of the organism into an osmotically sensitive form,

[1]H. R. Kaback and E. R. Stadtman, *Proc. Nat. Acad. Sci. U. S.* 55, 920 (1966).
[2]H. R. Kaback and A. B. Kostellow, *J. Biol. Chem.* 243, 1384 (1968).
[3]H. R. Kaback and E. R. Stadtman, *J. Biol. Chem.* 243, 1390 (1968).
[4]H. R. Kaback, *J. Biol. Chem.* 243, 3711 (1968).
[5]H. R. Kaback and T. F. Deuel, *Arch. Biochem. Biophys.* 132, 118 (1969).
[6]H. R. Kaback, *Proc. Nat. Acad. Sci. U. S.* 63, 724 (1969).
[7]H. R. Kaback, *In* "The Molecular Basis of Membrane Function" (D. Tosteson, ed.), p. 421. Prentice-Hall, Englewood Cliffs, New Jersey, 1969.
[8]H. R. Kaback and L. S. Milner, *Proc. Nat. Acad. Sci. U.S.* Kaback and Milner, *PNAS* 65, 1008 (1970).
[9]H. R. Kaback, *In* "Current Topics in Membranes and Transport," (A. Kleinzeller and F. Bronner, eds.), in press. Academic Press, New York, 1970.
[10]H. R. Kaback, *Annu. Rev. Biochem.* 39, 561 (1970).
[11]A. Ryter, Y. Hirota, and F. Jacob, *Cold Spring Harbor Symp. Quant. Biol.* 33, 669 (1968).
[12]B. M. Shapiro, A. G. Siccardi, Y. Hirota and F. Jacob, *J. Mol. Biol.* 52, 75 (1970).
[12a]M. Inouye and A. B. Pardee, *J. Biol. Chem.,* 245, 5813 (1970).
[13]R. W. Hendler, "Protein Synthesis and Membrane Biochemistry." Wiley, New York, 1968.
[14]G. Y. Tremblay, M. J. Daniels and M. Schaechter, *J. Mol. Biol.* 40, 65 (1969).
[15]A. M. Kidwai and C. R. Krishna Murti, *Indian J. Biochem.* 2, 217 (1965).
[16]M. R. J. Salton, "The Bacterial Cell Wall, American Elsevier, New York, 1964.
[17]H. Weissbach, B. Redfield and H. R. Kaback, *Arch. Biochem. Biophys.* 135, 66 (1969).
[18]Barnes, E. M., Jr. and Kaback, H. R., *Proc. Nat. Acad. Sci. U.S.,* 66, 1190 (1970).

and (2) the controlled osmotic lysis of that form in the presence of nucleases and a chelating agent. The structures formed as a result of this procedure consist of intact, "unit-membrane"-bound sacs which are essentially devoid of cytoplasmic constituents.

Regarding nomenclature, the term protoplast is reserved for osmotically sensitive forms derived from gram-positive organisms, and denotes that essentially all the cell wall has been removed. The term is not generally used with reference to osmotically sensitive forms derived from gram-negative organisms, since there is ample evidence that at least part of the lipopolysaccharide component of the cell wall is still present. Instead, the term spheroplast or osmotically sensitive sphere (OSS) is generally employed and will be used here.

Step I. Preparation of Osmotically Sensitive Cells

Various methods used to prepare protoplasts and osmotically sensitive spheres have been reviewed recently by Spizizen.[19] Theoretically, each of these techniques could be used in the preparation of membranes, but, for the purposes of this discussion, only two general methods will be presented. Both methods have been used with excellent results in the preparation of membranes from a variety of bacterial species.

The Penicillin Method

Principle

It has been well established that penicillin specifically inhibits the biosynthesis of the peptidoglycan component of the bacterial cell wall[20] with little or no effect on other metabolic reactions of the cell. Thus, exposure of rapidly growing cells to penicillin results in a form of "unbalanced growth" in which the cell outgrows its peptidoglycan shell. It is the rigid peptidoglycan layer of the cell wall which is responsible for the shape of the bacterium and for its ability to survive in dilute environments,[16] so a cell that has undergone this form of "unbalanced growth" will lyse in a hypotonic environment.

However, in the presence of a suitable osmotic stabilizer, e.g., hypertonic sucrose, viable osmotically sensitive spheres are formed from gram-negative rods.[21] The spheres are capable of induced enzyme

---

[19] J. Spizizen, this series, Vol. 5, p. 122.
[20] J. L. Strominger, K. Izaki, M. Matsuhashi and D. J. Tipper, *Fed. Proc., Fed. Amer. Soc. Exp. Biol.* **26**, 9 (1967).
[21] J. Lederberg, *Proc. Nat. Acad. Sci. U. S.* **42**, 574 (1956).

synthesis and a variety of other biosynthetic reactions and can revert to rods once penicillin is removed.

It should be stressed that in order to obtain optimal effects with penicillin, cells must be growing as actively as possible, i.e., they should be in the early or mid-log phase.

The method to be described is designed to obtain a maximal yield (greater than 99%) of spherical forms as judged by phase contrast microscopy.

Methods

Cell Growth. A starter culture is obtained by inoculation of an enriched medium, e.g. Difco penassay broth or nutrient broth, from a slant and incubation at 37° with shaking in order to provide sufficient aeration. After approximately 10 hours, a small aliquot from this starter culture is used to inoculate a flask of a minimal salts medium, medium A[22] or medium 63,[23] containing 0.5–1.0% glucose, glycerol, succinate, or another carbon source. The culture is grown for 8–10 hours at 30° or 37° under conditions of maximal aeration. The culture is then diluted 1:5 or 1:10 with fresh minimal medium, and incubation is continued until the optical density of the culture just begins to deviate from logarithmic growth.

Spheroplast Formation. At this point, the culture is diluted 1:4 into a medium which has been preequilibrated to 37° and which contains Difco penassay broth, 20% sucrose, 0.2% magnesium sulfate, and 1000 units per milliliter of potassium penicillin G. The preparation is incubated in this medium at 37° for 2–3 hours with intermittent mixing; the suspension is mixed by hand approximately every 30 minutes. The morphological changes that occur in the cells can be followed with a phase contrast microscope and are as follows: *30–60 minutes*—the cells begin to swell at the middle and a small bubble can be observed frequently at the equatorial portion of the rod; *60–90 minutes*—the bubble gradually enlarges and the rounded ends of the original rod deviate toward the midline, giving the appearance of a sphere with two small protuberances at its base; *90–120 minutes*—the sphere gradually enlarges, and the small protuberances begin to become integrated into the much larger spherical mass; *120–150 minutes*—the preparation now consists entirely of spherical forms which have a unique appearance in the phase contrast microscope. When viewed at the proper angle, it can be seen that approximately one-half of each sphere is very dense, and the other half can be seen only by virtue of a very fine limiting membrane. Fur-

[22]B. D. Davis and E. S. Mingioli, *J. Bacteriol.* **60**, 17 (1950).
[23]G. N. Cohen and H. W. Rickenberg, *Ann. Inst. Pasteur* **91**, 693 (1956).

thermore, there is a very sharp line of demarcation between the light and dark halves of the spheres. A schematic representation of this sequence of events is shown in Fig. 1. The progress of the conversion may also be followed by viewing the suspension through a strong light. As the rods become spherical, the strong schlieren effect in the original culture disappears.

The basis for the odd phase contrast appearance of the fully developed spheroplast can be seen in the electron micrograph shown in Fig. 2A. As shown, the ribosomes and "DNA plasm" of the spheroplast are confined to a small portion of the total structure, where they are surrounded by a "unit membrane" structure—the plasma membrane. Enclosing the entire structure is another membranous layer—the grossly enlarged lipopolysaccharide layer of the gram-negative cell wall, the synthesis of which is not inhibited by penicillin. These two membranous structures, i.e., the plasma membrane and the lipopolysaccharide layer of the cell wall, and their relationship are shown in detail in Fig. 2B.

One extremely important point regarding the stability of the spheroplast is that the magnesium concentration used during spheroplast formation and during subsequent manipulations is critical. These structures are labile, even in hypertonic media, if the magnesium concentration is below $10 \, \mathrm{m}M$.

The method can also be used for the preparation of protoplasts from gram-positive organisms such as *Bacillus subtilis* and *Bacillus megaterium*. With these organisms, much lower concentrations of penicillin, i.e., 10–100 units/ml, are effective.

The Lysozyme-EDTA Method

Principle

Lysozyme (muramidase) is an N-acetylhexosamidase, which acts upon the peptidoglycan layer of the cell wall of a variety of organisms.

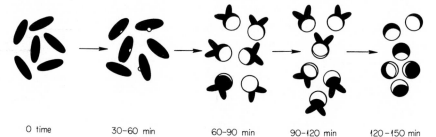

| O time | 30–60 min | 60–90 min | 90–120 min | 120–150 min |

FIG. 1.   Schematic representation of morphological changes occurring during penicillin spheroplast formation of *Escherichia coli*.

Lysozyme treatment, like "unbalanced growth" in the presence of penicillin, exerts its effect on the cell by weakening the rigid peptidoglycan layer of the wall. In this case, however, lysozyme hydrolyzes the peptidoglycan directly. With gram-positive organisms, e.g., *B. subtilis*, *B. megaterium*, or *Micrococcus lysodeikticus*, the peptidoglycan layer is exposed to the external medium, and it is sufficient simply to expose the organism to lysozyme in order to osmotically sensitize the cell. With gram-negative organisms, on the other hand, the peptidoglycan layer of the cell wall is sandwiched between the plasma membrane and the external lipopolysaccharide layer of the cell wall. Thus the peptidoglycan component is not exposed to the external medium, and certain manipu-

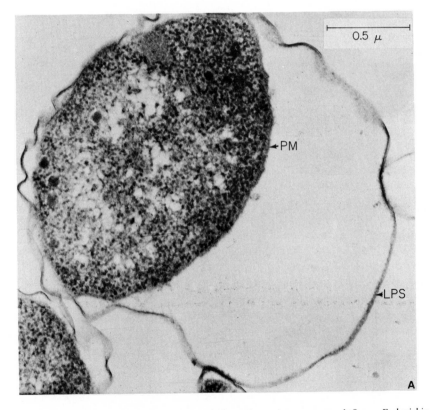

FIG. 2. Electron micrographs of penicillin spheroplasts prepared from *Escherichia coli* W. This micrograph was taken by Dr. Samuel Silverstein of the Rockefeller University. *PM*, plasma membrane; *LPS*, lipopolysaccharide. B approximately × 125,000.

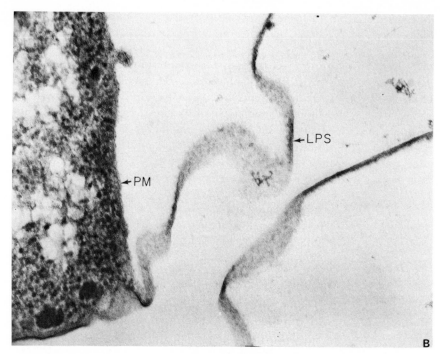

FIGURE 2B

lations (i.e., EDTA,[24] alkaline pH,[25,26] or a combination thereof[27]) are necessary in order to expose the peptidoglycan layer to lysozyme. It is also noteworthy that with many gram-negative organisms or under certain growth conditions, the cell does not become spherical on treatment with lysozyme. However, this observation apparently is not related to osmotic fragility. In the author's hands, many lysozyme-treated gram-negative organisms which do not convert to spheres lyse effectively on osmotic shock yielding spherical membrane vesicles. It should be obvious from the above discussion that the ability and efficiency of the lysozyme-EDTA method with any particular set of conditions is dependent on the strain of organism and its physiological state. The method to be described has been used to prepare membranes from a number of gram-negative (*Escherichia coli* ML, K12, W, and B,

[24]M. C. Karunairatnam, J. Spizizen and H. Gest, *Biochim. Biophys. Acta* 29, 649 (1958).
[25]N. D. Zinder and W. F. Arndt, *Proc. Nat. Acad. Sci. U. S.* 42, 586 (1956).
[26]J. Spizizen, *Proc. Nat. Acad. Sci. U. S.* 43, 694 (1957).
[27]H. R. Mahler and D. Fraser, *Biochim. Biophys. Acta* 22, 197 (1956).

Salmonella typhimurium LT, *Pseudomonas aerogenosa, Micrococcus denitrifi-cans,* and *Azotobacter vinlandii*) and gram-positive (*B. subtilis, B. mega-terium,* and *Clostridium thermoaceticum*) organisms grown under a variety of conditions with excellent results.

Method

Cell Growth. Cultures, 8- to 9-hour, of the appropriate organism, grown under the conditions indicated above for the penicillin method, are diluted 1:8 to 1:10 with fresh medium (equilibrated to 37°), and the incubation is continued until the cells have reached approximately mid- or late-log phase (140–150 Klett units with a No. 60 filter or approximately 0.5 mg/ml, dry weight).

Cell Harvest. The culture is centrifuged at approximately 16,000 *g* until the supernatant fluid is clear, and the pellet is washed twice with 10 m*M* Tris·HCl, pH 8.0, at 0°. The cells are resuspended (1 g, wet weight, per 80 ml) at room temperature in 30 m*M* Tris·HCl, pH 8.0, containing 20% sucrose, and the suspension is swirled by means of a magnetic stirrer.

Spheroplast Formation. Potassium EDTA, pH 7.0, and lysozyme (Worthington, crystalline) are added to final concentrations of 10 m*M* and 0.5 mg/ml, respectively, and the suspensions are incubated for 30 minutes at room temperature.

With organisms which are very sensitive to lysozyme (*E. coli* W and K12), the addition of high concentrations of the enzyme results in gross clumping and sedimentation of the bacteria. This phenomenon, rather than being a matter for concern, is an excellent criteria of the sensitivity of an organism to lysozyme.

The lysozyme procedure has certain advantages over the penicillin method. It is much more convenient for large-scale preparation of membranes (from 20 to 40 liters of cells), since the method does not require the dilutions inherent in the penicillin technique. Perhaps more importantly, the lysozyme method obviates the necessity of exposing the cells to an undefined growth medium. On the other hand, spheroplasts prepared by the penicillin method are much more homogeneous morphologically and are probably more sensitive to osmotic shock. It is also noteworthy that penicillin spheroplasts retain so-called "pericytoplasmic enzymes"[28] which are released during treatment with lysozyme-EDTA.[29]

[28]L. A. Heppel, *Science* 156, 1 (1967).
[29]L. A. Heppel and H. R. Kaback, unpublished information (1967).

Step II. Preparation of Membranes

Principle

Protoplasts or spheroplasts formed by treatment with penicillin or lysozyme-EDTA release their intracellular contents when placed in hypotonic media. Furthermore, the membrane reanneals by an unknown mechanism, yielding closed, empty membrane vesicles which can be easily sedimented. Assuming that the membrane becomes transiently discontinuous during lysis and is thereby able to release the intracellular constituents, the intramembranal milieu would be expected to equilibrate with the external medium during this period. Thus, the greater the lysis ratio, i.e., the volume of protoplasts or spheroplasts to the volume of the lysis solution, the more dilute the intramembranal contents should be. Since much of the intracellular DNA which is released on lysis adheres to the membranes and makes the preparations difficult to handle, DNase is added to the lysates. RNase, which is also usually added, can be omitted.

It has been found that EDTA, in addition to causing the release of lipopolysaccharide,[30] facilitates the release of RNA from the membranes. Membranes prepared without EDTA contain approximately 25% of the RNA of the whole cells whereas membranes prepared in the presence of EDTA contain less than 2% of the RNA of the whole cells. Since DNase activity requires the presence of magnesium, the addition of EDTA inhibits DNase at the same time that it facilitates the further release of nucleic acid. Therefore, it is necessary to add magnesium back to the lysates after they have been exposed to EDTA in order to reactivate DNase. The remainder of the procedure to be described consists essentially of differential and density centrifugation coupled with extensive homogenization of the membrane vesicles. These final manipulations are designed to fractionate the few residual whole cells and/or partially lysed elements from the preparation.

Method

Centrifugation. The protoplast or spheroplast suspensions prepared as described above are centrifuged at approximately 16,000 g until the supernatant fluid is clear.

Homogenization. The protoplast or spheroplast pellet is resuspended in the smallest possible volume of 0.1 M potassium phosphate buffer, pH 6.6, containing 20% sucrose and 20 mM magnesium sulfate using a Teflon and glass homogenizer. Homogenization is facilitated by the

---

[30]L. Leive, *Biochem. Biophys. Res. Commun.* **21**, 290 (1965).

addition of DNase (Worthington, crystallized once) and RNase (Worthington, crystallized three times) to final concentration of 3–5 mg/ml (such that the final concentrations of DNase and RNase in the total lysate, see below, will be approximately 10 μg/ml). Homogenization is carried out using a motordriven plunger until the suspension is uniformly dispersed.

Lysis. The homogenized, concentrated protoplast or spheroplast suspension is poured directly into 300–500 volumes of 50 mM potassium phosphate buffer, pH 6.6, which has been equilibrated to 37°. The lysate is incubated for 15 minutes at 37° with vigorous swirling. Potassium EDTA, pH 7.0, is then added to 10 mM final concentration, and the incubation is continued for 15 minutes. Shortly after the addition of EDTA, the turbidity of the suspension decreases and the viscosity increases. Finally, magnesium sulfate is added to a final concentration of 15 mM and the incubation is continued for another 15 minutes at 37°; during this period the viscosity decreases.

Isolation of Membranes. The lysates are centrifuged at a minimum of 16,000 g for 30 minutes or until the supernatant is clear. The pellet is resuspended by vigorous homogenization in a solution of 0.1 M potassium phosphate buffer, pH 6.6, containing 10 mM EDTA at 0°. Occasionally, the membrane pellet is viscous. In this case, magnesium sulfate, DNase, and RNase are added to give 20 mM, 100 μg/ml, and 100 μg/ml final concentrations, respectively; and the suspensions are incubated at 37° for 30 minutes with shaking. The preparation is then centrifuged at approximately 45,000 g until the supernatant is clear, and the pellet is resuspended by homogenization in 0.1 M potassium phosphate–10 mM EDTA as described above.

Removal of Cell Debris. The sample is centrifuged at approximately 800 g for 30 minutes and the yellowish, milky, supernatant fluid is carefully decanted and centrifuged at 45,000 g until it is clear. The low speed centrifugation removes essentially all the whole cells and partially lysed cells, and is repeated until the supernatant contains less than one rod per oil immersion field in the phase contrast microscope (1000× magnification). Low speed centrifugation can be repeated on the sediment after extensive homogenization in order to obtain maximal yield of membranes.

Washing of Isolated Membranes. The high speed pellet which now is almost free of whole cells is washed 4–6 times by resuspension and vigorous homogenization in 0.1 M potassium phosphate–10 mM EDTA. This step is followed by centrifugation at 45,000 g until the supernatant fluid is clear (usually approximately 30 minutes).

Quantitative Removal of Whole Cells (Optional). In order to quantatively remove whole cells and partially lysed forms from the membrane preparation, the high speed pellet obtained as described above can be resuspended in a solution of 0.1 M potassium phosphate, pH 6.6, containing 10 mM magnesium sulfate and 20% sucrose. It is important in this step to homogenize the preparation extensively so that there is a minimum of aggregation. The suspension is carefully layered on top of 60% sucrose (60 g in a total of 100 ml) containing 0.1 M potassium phosphate, pH 6.6, and 10 mM magnesium sulfate and is centrifuged at 64,000 g for a minimum of 90 minutes in a Spinco Model L centrifuge using a SW 25.1 swinging-bucket rotor. Centrifugation can be carried out overnight, since this is a density equilibrium method. The thick layer of membrane remaining at the interface is carefully aspirated, diluted with 0.1 M potassium phosphate-10 mM EDTA, centrifuged at 45,000 g until clear, and the pellet is washed 3 to 4 times by homogenization and centrifugation in 0.1 M potassium phosphate-10 mM EDTA.

Storage. After the last wash, the membranes are resuspended by homogenization in 0.1 M potassium phosphate, pH 6.6, at a concentration of 5–10 mg/ml, dry weight (corresponding to 4–7 mg of protein per milliliter), and frozen in small aliquots of 2–3 ml in liquid N_2. The frozen preparations may be stored in liquid N_2 for an indefinite period with little or no loss of activity with regard to sugar or amino acid transport and phospholipid biosynthesis.

Step III. Characterization of the Membrane Preparations

A. Purity.

The purity of membrane preparations can be established by both direct and indirect methods as follows:

Direct Methods

Electron Microscopy. Figures 3 and 4 are electron micrographs of membranes prepared from *E. coli* ML 308-225 and *E. coli* W, respectively. As shown in Fig. 3, membranes prepared from *E. coli* ML 308-225 consist predominantly of intact "unit membrane"-bound sacs varying from 0.5 to 1.5 μ in diameter. The great majority of these sacs are surrounded by a single trilaminar membrane layer which is 65–70 Å thick. The sacs appear to be empty and without internal structure. It is also noteworthy that membranes prepared using either the lysozyme-EDTA or penicillin methods appear to be morphologically identical (compare Fig. 3A and B). Membranes prepared from *E. coli* W treated with either lysozyme-

EDTA or penicillin are shown in Fig. 4A and B, respectively. In this case, the structures seen also consist of intact "unit membrane"-bound sacs but the membranes are much more heterogeneous. The diameters of these sacs vary from 0.1 to 1.5 μ; the sacs are surrounded by one to five or six trilaminar membrane layers. As with *E. coli* ML 308-225, the

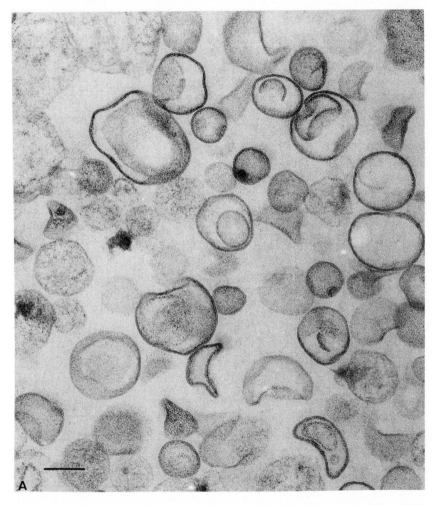

FIG. 3. Electron micrographs of membrane preparations from *Escherichia coli* ML 308-225. (A) Membranes prepared from lysozyme-EDTA-induced spheroplasts. Micrograph taken by Dr. V. Marchesi of the National Institute of Arthritis and Metabolic Disease. (B) Membranes prepared from penicillin-induced spheroplasts. Micrograph taken by Dr. Samuel Silverstein of the Rockefeller University. The markers in the lower left-hand corners of these micrographs are 0.4 μ.

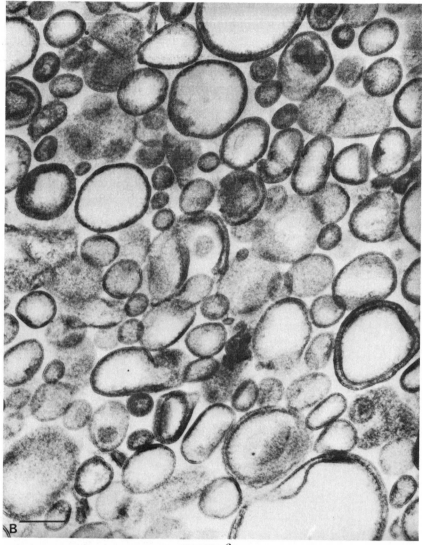

FIGURE 3B

membrane vesicles are also empty and devoid of internal structure, and each trilaminar membrane layer is also 65–70 Å thick. Although there appear to be more multilayered forms with the lysozyme-EDTA preparation as compared to the penicillin preparation, this cannot be stated with certainty. Membranes prepared from other strains of *E. coli* (i.e., K12, W2244, and K₂It) and from *S. typhimurium* are essentially indistinguishable from *E. coli* W. The reason for this striking morphological

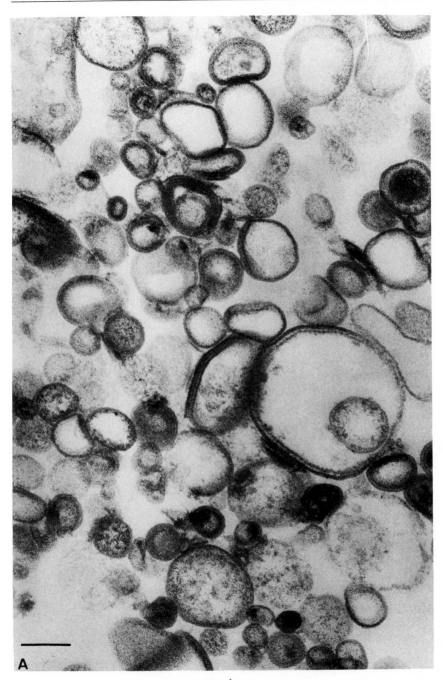

FIGURE 4A

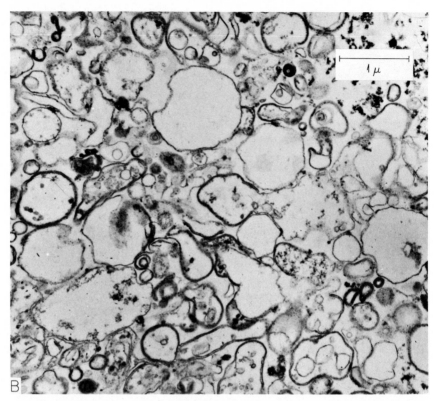

FIG. 4. Electron micrographs of membrane preparations from *Escherichia coli* W. (A) Membranes prepared from lysozyme-EDTA-induced spheroplasts. Micrograph taken by Dr. V. Marchesi of the National Institute of Arthritis and Metabolic Disease. (B) Membranes prepared from penicillin-induced spheroplasts. Micrograph taken by Dr. Samuel Silverstein of the Rockefeller University. The marker in the lower left-hand corner of A is 0.4 μ.

difference between ML membranes and membranes prepared from other strains of *E. coli* and *S. typhimurium* is unknown at present; however, it may be related to the observation that ML 308-225 membranes contain negligible quantities of lipopolysaccharide whereas the other gram-negative membrane preparations mentioned above contain significant quantities of this cell wall component (see below). Miura and Mizushima[31] have presented a method by which the plasma membrane and lipopoly-saccharide fractions of *E. coli* K12 lysates can be resolved. Futhermore, Okuda and Weinbaum[32] have recently obtained evidence for an enzyme

[31] T. Miura and S. Mizushima, *Biochim. Biophys. Acta* **150**, 159 (1969).
[32] S. Okuda and G. Weinbaum, in preparation (1970).

that degrades lipopolysaccharide. Either or both of these techniques may prove useful for the preparation of more highly purified plasma membrane fractions from gram-negative organisms other than *E. coli* ML.

Membranes prepared from lysozyme-EDTA-treated *B. subtilis*, a gram-positive organism, appear to be very similar morphologically to the gram-negative membrane preparations discussed previously.

On scanning many such fields as those shown in Figs. 3 and 4, less than one ribosome-containing structure is seen for every 200–300 large, empty membrane vesicles.

Phase Contrast Microscopy. Direct observation of the membrane preparations by this technique in a bacterial counting chamber shows less than one bacterium per 500 membranes and essentially no unlysed spheroplasts or protoplasts.

Density Centrifugation. When membranes, whole cells (or spheroplasts), or a mixture of membranes and whole cells are layered on top of buffered 60% sucrose and centrifuged as described previously, the membranes do not penetrate the sucrose barrier, whereas the whole cells sediment to the bottom. As a measure of cytoplasmic constituents, DNA and RNA can be measured in fractions collected from these tubes (shown schematically at the bottom of each bar in Fig. 5). Although a small pellet is recovered by centrifugation of the membranes (Fig. 5, top bar graph), this is not significantly enriched with DNA or RNA, indicating the absence of whole cells or unlysed spheroplasts. Furthermore, on microscopic examination this pellet is found to contain adherent clumps of membranes and almost no whole cells. With whole cells (Fig. 5, middle bar graph) or whole cells mixed with membranes (Fig. 5, bottom bar graph), essentially all the DNA and RNA sediments with the pellet despite a thick layer of membranes remaining at the interface.

Viability Assays. The membrane preparations contain less than 0.05% of the viable cells contained in the culture from which they were obtained.

Indirect Methods

Measurement of Cellular Constituents. When various intracellular constituents are used as a measure of cytoplasmic contamination, the membrane preparations are found to contain less than 3% of the DNA and RNA, approximately 15% of the protein, and at least 70% of the lipid of the whole cells from which they were derived. Less than 1% of the activities of glutamine synthetase, fatty acid synthetase, leucine-activating enzyme, or β-galactosidase, and 2% or less of each of the so-called "pericytoplasmic enzymes" (alkaline phosphatase, 5'-nucleotidase, ribosomal ribonuclease, etc.)[28] are found in the preparations.[29]

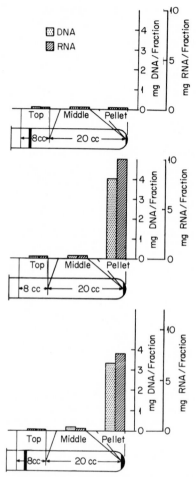

Fɪɢ. 5.   DNA and RNA content of fractions from density barrier centrifugation of membranes, whole cells, or membranes mixed with whole cells. Top bar graph, membranes; middle bar graph, whole cells; bottom bar graph, membranes plus whole cells. The tubes from which these fractions were obtained are represented schematically below each bar graph.

Disc Gel Electrophoresis. Although approximately 15% of the total protein remains in the membranes, the great majority of the cytoplasmic proteins are lost. When sonicated whole cells, sonicated membranes, or a 100,000 *g* sediment from sonicated membranes are subjected to disc gel electrophoresis, numerous intensely staining bands are obtained with the sonicated whole cell preparation, all but two of which are missing from the sonicated membranes. Since these two bands do not decrease in intensity when the 100,000 g sediment from sonicated membranes is subjected to the procedure, even these two proteins are probably not cytoplasmic contaminants. Thus, the quantity of cytoplasmic protein contaminating the membranes which will migrate on disc gel electrophoresis is vanishingly small.

Cell Wall Constituents. Analyses for contaminating cell wall constituents reveals that membranes prepared from penicillin-induced spheroplasts contain only 10% or less of the diaminopimelic acid of the spheroplasts. Regarding the lipopolysaccharide content of membranes prepared from gram-negative organisms, membranes prepared from *E. coli* GN-2, W2244, K$_2$lt, and from *S. typhimurium* SB 102 contain approximately 7-17% lipopolysaccharide (by dry weight), whereas membranes prepared from *E. coli* ML 308-225 contain less than 3%.[33]

ORD and CD Studies. Optical rotatory dispersion (ORD) and circular dichroism (CD) studies[34] of membranes prepared from *E. coli* ML 308-225 and K12 are essentially identical to those found with *B. subtilis* membranes and human red blood cell membranes[35] and Ehrlich's ascites cell membranes.[36] The magnitude and position of the ORD and CD spectral bands are typical of "membrane" protein.

Composition

The membranes are approximately 60-70% protein, 30-40% phospholipid, and approximately 1% carbohydrate. The predominant phospholipid is phosphatidylethanolamine (65-70%), but there are also significant quantities of diphosphatidylglycerol (approximately 15%), phosphatidylglycerol (10-15%), phosphatidic acid (5-10%), and phosphatidylserine.

The membrane proteins can be characterized by disc gel electrophoresis after solubilization with sodium dodecyl sulfate or phenol-acetic acid according to the methods of Maizel[37] or Rottem and Razin,[38] respectively.

Physical Properties

Although the sectioned material presented in Figs. 3 and 4 gives the impression that the vesicles are enclosed by a continuous membrane,

[33]These determinations were carried out independently by Dr. M. J. Osborne of the University of Connecticut, Storrs, Connecticut, and Dr. George Weinbaum of the Albert Einstein Medical Center, Philadelphia, Pennsylvania.
[34]A. Gordon, D. F. H. Wallach and H. R. Kaback, unpublished information (1969).
[35]J. Lenard and S. J. Singer, *Proc. Nat. Acad. Sci. U. S.* 56, 1828 (1966).
[36]D. F. H. Wallach and P. H. Zahler, *Proc. Nat. Acad. Sci. U.S.* 56, 1552 (1966).
[37]J. V. Maizel, *Biochem. Biophys. Res. Commun.* 28, 815 (1967).
[38]S. Rottem and S. Razin, *J. Bacteriol.* 94, 359 (1967).

it is only from techniques other than those in which there is a good possibility for sampling errors due to the thinness of the sections used that such conclusions may be drawn. The electron micrographs presented in Fig. 6A and B were obtained using negative staining so that the surface of the vesicles could be observed. Figure 6A shows a typical *E. coli* ML 308-225 membrane, and Fig. 6B shows membranes prepared from *E. coli* W. In both cases, there are no gross defects in the surface of the vesicles. It is also significant that the stain used does not penetrate the interior of the vesicles.[39]

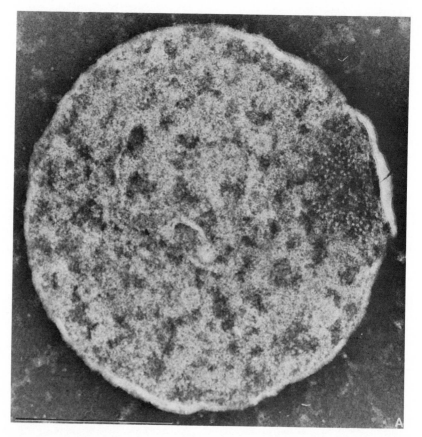

FIG. 6. Electron microscopy of negatively stained membrane preparations. (A) *Escherichia coli* ML 308-225; (B) *E. coli* W. This study was carried out by Dr. V. Marchesi of the National Institute of Arthritis and Metabolic Disease. The markers in the lower left-hand corner of A and in the lower right-hand corner of B are 0.4 μ.

[39] L. Rothfield and R. W. Horne, *J. Bacteriol.* **93**, 1705 (1967).

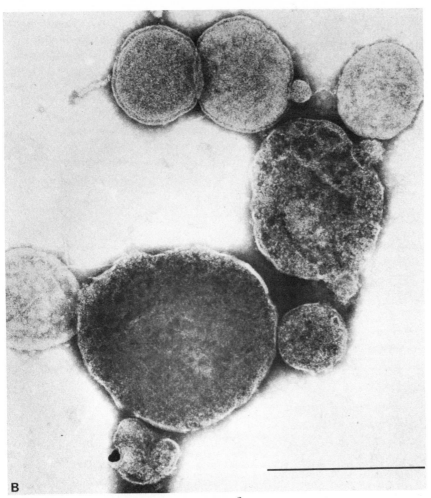

B

FIGURE 6B

Further evidence for continuity is provided by electron microscopic observation of membranes prepared by freeze etching.[40] As shown in Fig. 7A, *E. coli* ML 308-225 membranes, when examined by this technique, have a continuous, convex outer surface with a fine granular surface exposed within the cleavage plane. Moreover, as shown in Fig. 7B, when the concave, or inner, surface of the vesicle is visualized, the membrane is also continuous, and the granularity is much coarser. All the membranes seen in preparations such as those shown have this

[40]D. Branton, *Annu. Rev. Plant Physiol.* **20**, 209 (1969).

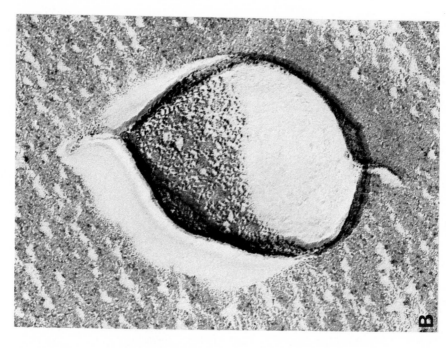

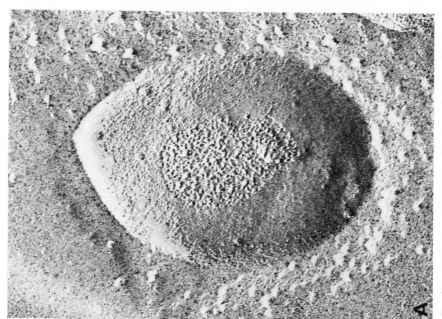

Fig. 7A. Electron microscopy of freeze etched membrane preparations from *Escherichia coli* ML 308–225. The micrographs were taken by Drs. T. Tillack and V. Marchesi of the National Institute of Arthritis and Metabolic Disease. Approximately × 140,000.

appearance (i.e., a smooth outer convex surface with fine granularity within the cleavage plane; and a coarsely granular concave surface). Thus, it is unlikely that any of the vesicles turn inside out during lysis, since, in such membranes, the convex surface within the cleavage plane should appear coarsely granular and the concave surface should appear finely granular.

Finally, evidence for continuity is provided by experiments which demonstrate that the vesicles behave as osmometers (i.e., the vesicle membrane is semipermeable). When the external osmolarity is increased, water leaves the vesicles, causing them to shrink. This shrinkage can be measured directly as a decrease in the intramembranal volume[9] or indirectly as a decrease in light scattering.[5]

Sugar and Amino Acid Transport by Isolated Bacterial Membrane Preparations

P-Enolpyruvate-P-transferase-Mediated Sugar Transport[4,6-10]

Aliquots (50 μl) of the membranes prepared as described above are diluted to a final volume of 100 μl containing, in final concentrations, 50 mM potassium phosphate buffer, pH 6.6, 10 mM magnesium sulfate, 0.3 M lithium chloride, and 0.1 M P-enolpyruvate (Boehringer-Mannheim, trisodium salt, adjusted to pH 6-7 with sodium carbonate). The samples are incubated at a given temperature for 15 minutes. After this period, radioactively labeled sugar (i.e., glucose, methyl-α-D-glucopyranoside, fructose, or mannose) is added at a final concentration of 3.64×10^{-5} M, and the incubation is continued for an appropriate period of time. To terminate the reaction, each sample is rapidly diluted 25- to 50-fold with 0.5 M lithium chloride at room temperature, immediately filtered through a Millipore HA filter (Millipore Filter Corporation, Bedford, Massachusetts), and washed once with an equal volume of 0.5 M lithium chloride. The dilution, filtration, and washing procedures are conducted in less than 30 seconds; the filters are immediately removed from the suction apparatus, mounted on planchets with tight-fitting stainless steel rings that cover the periphery of the filters, dried, and counted in a gas flow counter at 10-20% efficiency. Each set of samples is corrected for a control obtained by diluting the samples before adding the radioactive substrate and omitting P-enolpyruvate. With the use of the assay conditions described above and methyl-α-D-glucopyranoside-^{14}C at a specific activity of 73.4 mCi/mmole, control values are 400 cpm or less.

The temperature at which this assay is carried out is crucial.[4,9] Although the initial rate of vectorial phosphorylation[9,10] is maximal at

46°, the membranes are "leaky" at this temperature and will lose methyl-α-D-glucopyranoside-^{14}C-P if the incubation is carried out for a prolonged period of time (i.e., usually over 10 minutes).[4,9] The temperature at which a particular membrane preparation begins to "leak" can be determined as described in references 7 and 9.

Dehydrogenase-Coupled Amino Acid and Sugar Transport[1-3,8,18]

Aliquots (50 μl) of membranes prepared as described above are diluted to a final volume of 100 μl containing 50 mM potassium phosphate buffer, pH 6.6, and 10 mM magnesium sulfate. The samples are incubated at 30° for 15 minutes. At this time, sodium D-(−)-lactate and, immediately thereafter, a radioactive amino acid or β-galactoside (e.g., proline-^{14}C or lactose-^{14}C) are added to yield final concentrations of 20 mM and approximately 10 μM or 0.2 mM, respectively, and the incubation is continued for an appropriate period of time. To terminate the reaction, each sample is rapidly diluted and washed as described above for sugar transport. In this case, however, 0.1 M lithium chloride (rather than 0.5 M lithium chloride) is used to terminate the reactions and wash the samples. The remainder of the procedure is carried out as described above for sugar transport.

[14] Yeast Spheroplasts

By ENRICO CABIB

Principle

Yeast cells are treated with a mercaptan compound and EDTA, whereby the cell wall is made more sensitive to the attack of lytic enzymes.[1] The cell wall is then digested with snail gut juice.[2] An osmotic stabilizer must be present during incubation to prevent bursting of the spheroplasts.

Reagents

Yeast cells, preferably from the logarithmic phase of growth
EDTA, 0.1 M
2-Mercaptoethanol
Snail intestinal juice, either obtained as outlined in Vol. 1 [20], or

[1] E. A. Duell, S. Inoué, and M. F. Utter, *J. Bacteriol.* **88**, 1762 (1961).
[2] A. A. Eddy and D. H. Williamson, *Nature (London)* **179**, 1252 (1957).

purchased commercially (two widely used preparations are "glusu-lase," from Endo Laboratories, Garden City, N. Y. and "Suc d'Helix pomatia" from Industrie Biologique Francaise, Gennevilliers, France). To the crude juice an equal volume of 0.2 M KCl, containing 2 mM EDTA is added, and the solution is filtered through a Sephadex G-25 column, previously equilibrated with 0.1 M KCl, containing 1 mM EDTA. The filtered extract is stable for many months if kept at $-10°$.

Citrate-phosphate buffer, pH 6.1 (see Vol. I [16])

Mannitol, 10% w/v

Mannitol, 20% w/v

Mannitol, 10% w/v containing one-twentieth its volume of citrate-phosphate buffer, pH 6.1

Aureomycin

Procedure.

For each gram of yeast (wet weight) 1.4 ml of 0.1 M EDTA and 0.024 ml of 2-mercaptoethanol are added, and the volume is brought to 3.5 ml with distilled water. The mixture is incubated for 30 minutes at 30° with shaking. The suspension is centrifuged in the cold for 10 minutes at 20,000 g, and the pellet is washed once with 5 ml of 10% mannitol. To the final pellet, the following additions are made: 0.57 ml of citrate-phosphate buffer, pH 6.1; 0.067 ml of 0.1 M EDTA; snail gut juice (see Comments below); a volume of 20% mannitol equal to the sum of those of buffer, EDTA and snail enzyme; 10% mannitol to complete the volume to 6.7 ml. The suspension is incubated at 30° in an orbital shaker-incubator, at the minimal speed sufficient to prevent the cells from sedimenting.

The progress of spheroplast formation can be followed under the phase contrast microscope, in two ways. If the yeast cells of the strain utilized are elongated, the conversion from an oval to a spherical shape is observed. If the yeast cells are round, a small portion is diluted in 10–20 volumes of distilled water: any spheroplast which had formed will lyse, while the cells with undigested cell walls will remain intact. A small number of spheroplasts usually lyse despite the presence of osmotic stabilizer, giving rise to "ghosts." Intact spheroplasts can be easily distinguished from ghosts, since the latter have lost the luminous halo which is clearly visible under phase contrast.

Once the conversion is practially complete, the spheroplast suspension is centrifuged for 10 minutes at 1000 g. The pellet is resuspended as gently as possible in 15 ml of 10% mannitol, containing citrate-phosphate, and is recentrifuged in the same way. After another washing with 7.5 ml of the mannitol-buffer mixture, the spheroplast pellet is resuspended in enough of the same mixture to obtain the desired dilution.

The spheroplasts may be stored for several days at 4° without appreciable lysis. In this case, Aureomycin is added to a final concentration of 50 μg/ml, to prevent bacterial contamination.

Comments.

The pretreatment with SH compounds and EDTA[1] greatly facilitates the subsequent attack by the snail enzymes.[3] Thus, at least with some strains, such as *Saccharomyces carlsbergensis* strain 74, National Collection of Yeast Cultures, England, or *Saccharomyces cerevisiae*, strain A364A,[4] it is no longer necessary to use cells from the logarithmic phase of growth. Cells from the stationary phase will also be converted completely to spheroplasts, although it is advisable in this case to increase the amount of snail enzyme by about one-third. This is by no means general, and difficulties have been experienced in our laboratory with other strains, such as *S. cerevisiae* αS288C.[5]

In general, the appropriate amount of enzyme must be ascertained for each strain of yeast. For logarithmic phase cells of *S. carlsbergensis*, strain 74, or *S. cerevisiae*, A364A, a volume of enzyme corresponding to about 0.3 ml of the original "glusulase" was usually sufficient to convert 1 g of yeast (wet weight) into spheroplasts in 45 minutes.

Mannitol can be substituted with other substances, such as sorbitol, rhamnose or 0.6 M KCl.[6] In our experience, however, the use of KCl with *S. carlsbergensis* gave rise to a variable amount of lysis, and the spheroplasts were not well preserved after storage in the refrigerator.

The spheroplast preparation may be contaminated with numerous bud scars, at least if "glusulase" is used as source of lytic enzymes. The bud scars consist mainly of chitin[7] and are resistant to the attack of "glusulase," which has only weak chitinase activity.

---

[3]However, H. T. Hutchison and L. H. Hartwell [*J. Bacteriol.* 94, 1697 (1967)] reported that spheroplasts from yeast, pretreated with SH compounds, showed marked impairment in RNA synthesis. Thus, for some applications it may be preferable to omit the pretreatment.
[4]Kindly furnished by Dr. L. H. Hartwell.
[5]Kindly provided by Dr. G. Fink.
[6]H. Holter and P. Ottolenghi, *C. R. Trav. Lab. Carlsberg* 31, 409 (1960).
[7]J. S. D. Bacon, E. D. Davidson, D. Jones, and I. Taylor, *Biochem. J.* 101, 36C (1966).

[15] Isolation of Plasma Membranes from Mammalian Cells

By E. H. Eylar and Arpi Hagopian

Much of our present knowledge of membrane biochemistry has been derived from microscopic, chemical, and enzymatic studies of membranes from convenient source materials, such as mitochondria,[1] cell stroma,[2-4] bacterial membranes,[5-8] and myelin.[9] In recent years methods for the preparation of plasma membranes have been described for liver cells,[10,11] bladder epithelial cells,[12] intestinal brush border cells,[13,14] cultured fibroblasts,[15] and Ehrlich ascites carcinoma cells.[16]

Our main objective in the preparation of plasma membranes was to preserve as much as possible the membrane integrity and biological properties. In order to avoid scrambling and disruption of the plasma membranes, minimum trauma during cell rupture was desired. For many cell types, the Dounce homogenizer is appropriate for this purpose. However, in the case of cells more resistant to breakage, such as the Ehrlich ascites tumor cell, less gentle techniques must be employed even under hypotonic conditions; use of the French pressure cell in this case produces small fragments of the plasma membrane.[17,18] For HeLa cells and mouse myeloma S194 cells, use of the Dounce homogen-

[1]D. E. Green, *Isr. J. Med. Sci.* 1, 1187 (1965).
[2]S. Bakerman and G. Wasemiller, *Biochemistry* 6, 1100 (1967).
[3]T. A. J. Pranherd, "The Red Cell", Blackwell, Oxford, 1961.
[4]J. T. Dodge, C. Mitchell, and D. J. Hanahan, *Arch. Biochem. Biophys.* 100, 119 (1963).
[5]M. R. J. Salton and J. H. Freer, *Biochim. Biophys. Acta* 107, 531 (1965).
[6]C. Weibull and L. Bergstrom, *Biochim. Biophys. Acta* 30, 340 (1958).
[7]G. D. Shockman, J. J. Kolb, B. Bakay, M. J. Conover, and G. Toinnier, *J. Bacteriol.* 85, 168 (1963).
[8]S. Razin, J. Morowitz, and T. T. Terry, *Proc. Nat. Acad. Sci. U. S.* 54, 219 (1965).
[9]J. S. O'Brien, *J. Theor. Biol.* 15, 307 (1967).
[10]D. M. Neville, Jr., *J. Biophys. Biochem. Cytol.* 8, 412 (1960).
[11]P. Emmelot and C. J. Bos, *Biochim. Biophys. Acta* 58, 374 (1962).
[12]R. M. Hays and P. Barland, *J. Cell Biol.* 31, 209 (1966).
[13]J. Overton, A. Eichholz, and R. K. Krane, *J. Cell Biol.* 26, 693 (1965).
[14]G. G. Forstrer, S. M. Sabesin, and K. J. Isselbacher, *Biochem. J.* 106, 381 (1968).
[15]L. Warren, M. C. Glick, and M. K. Mass, *J. Cell. Physiol.* 68, 269 (1967).
[16]V. B. Kamat and D. F. H. Wallach, *Science* 148, 1343 (1965).
[17]G. M. W. Cook, M. T. Laico, and E. H. Eylar, *Proc. Nat. Acad. Sci. U. S.* 54, 247 (1965).
[18]J. F. Caccam and E. H. Eylar, *Arch. Biochem. Biophys.* 137, 315 (1970).

izer releases the plasma membranes as ghosts or large fragments clearly discernible by phase contrast microscopy.

Alternatives to this procedure include the methods of Warren et al.,[15] which, while maintaining the structure of the plasma membrane by fixation, inactivate biological activities. In the decompression technique of Kamat and Wallach,[19] enzyme activities are maintained but severe fragmentation into vesicles occurs.

As models for membrane preparations, HeLa cells and the S194 myeloma cells were chosen—the former as an example of a cell which does not secrete macromolecules except possibly collagen, and the latter as an example of a cell secreting a high percentage of its total protein as immunoglobulin IgA.

After cell rupture, the nuclei and unbroken cells are removed by light centrifugation; the plasma membranes are subsequently derived from the supernatant fluid. The plasma membranes from liver[10,20] or kidney and intestine[21] are obtained from the nuclear pellet. This result suggests that the intercellular interaction of cells in tissues promotes aggregation of plasma membranes during fractionation and hinders subsequent isolation. By contrast, membranes are obtained from HeLa cells and cells in suspension in high yield and purity as indicated by microscopic, enzymatic, and chemical characterization.[22]

Negative factors that modify the yield and quality of plasma membranes include trauma during cell breakage, leading to dissociation, and aggregation with other membranes and cellular components. These factors are probably amplified by damage to nuclei, which release DNA, and rupture of lysozomes, which release enzymes that degrade proteins or lipids. In order to maximize the yield of membranes, therefore, attention should be paid to use of dilute solutions, isotonicity, concentration of metal ions such as Ca^{2+} or Mg^{2+}, and potential use of EDTA. It is quite possible that no scheme generally applicable for preparation of membranes can be devised, but that for each cell a different set of parameters must be sought.

Preparation of Plasma Membranes

Reagents

Calcium acetate, 1 mM in buffered saline, pH 7.1; buffered saline: 0.14 M NaCl, pH adjusted with Tris buffer

[19]D. F. H. Wallach and V. B. Kamat, this series, Vol. 8, p. 164.
[20]P. Emmelot, C. J. Bos, E. L. Benidetti, and P. Rumke, Biochim. Biophys. Acta 90, 126 (1964).
[21]R. Coleman and J. B. Finean, Protoplasma 63, 172 (1967).
[22]H. B. Bosmann, A. Hagopian, and E. H. Eylar, Arch. Biochem. Biophys. 128, 51 (1968).

Tris·HCl buffer, 50 mM, pH 7.1.

TKM buffer: Tris-HCl buffer, 50 mM, pH 7.1; MgCl$_2$, 5 mM; KCl, 25 mM

Sucrose solutions in Tris buffer, 50 mM, pH 7.1, at 0.80, 0.98, 1.175, 1.37, 1.48, 1.59, and 1.60 M.

Cell Sources. HeLa cells are grown in spinner culture at 37° in Eagle's spinner medium supplemented with 10% calf serum.[23] Myeloma cells[24] are grown in Eagle's medium (modified by Dulbecco) supplemented with 20% horse serum. The cells are harvested in the logarithmic phase of growth and counted. The volume of packed cells is determined by centrifugation for 20 minutes at 2000 rpm.

Plasma Membranes from HeLa Cells

Cell Rupture. All operations are carried out at 2°-4°. The packed HeLa cells (2 to 6 × 10^9 cells) are washed 4 times with 1 mM calcium acetate-buffered saline pH 7.1. The washed cells are resuspended in 5-10 volumes of 50 mM Tris buffer, pH 7.1, and are broken in a tight-fitting Dounce homogenizer (pestle B). The number of homogenizer strokes is important; if too many or too few, the membranes do not sediment reproducibly in the sucrose gradient. The number of strokes necessary for rupture varies from one cell population to another, therefore, it is important to monitor constantly with the phase microscope: 20-30 strokes generally proved satisfactory. The optimal degree of rupture occurs when only 5-10% of the cells are visible under the phase microscope. Free nuclei and plasma membrane "ghosts" are easily discernible in the homogenate along with cytoplasmic material.

Removal of Nuclei and Whole Cells. The broken cell homogenate is centrifuged for 10 minutes at 4000 g to remove whole cells, nuclei, and other contaminants, such as mitochondria. Some plasma and smooth membranes are lost at this stage, but much of the membranous material can be recovered by washing the 4000 g pellet.

Sucrose Gradient. The 4000 g supernatant fluid (4 S) is adjusted to 1.59 M sucrose; 20 ml is placed at the bottom of a cellulose nitrate centrifuge tube used with the Beckman SW 25.2 rotor. Thirteen milliliters of 1.175 M sucrose is then carefully layered onto the 20 ml of the sample containing 1.59 M sucrose, followed by 13 ml of 0.98 M sucrose over which is layered 10 ml of 0.80 sucrose. Finally, 2 ml of 50 mM Tris buffer, pH 7.1, is added. The discontinuous gradient is

[23]H. Eagle, *J. Cell. Med.* **102**, 592 (1955).

[24]K. Horibata and A. W. Harris, *Exp. Cell Res.* **60**, 61 (1970). We are indebted to Dr. M. Cohn, The Salk Institute, for the myeloma 194 cell line.

then centrifuged at 70,000 g (23,500 rpm) for 16 hours at 3°; five membranous bands and a pellet are observed with the unaided eye, as shown schematically in Fig. 1. The plasma membrane is found as a large translucent pellet at the bottom of the tube (Fraction S7). The fractions are then diluted 4:1 with 50 mM Tris buffer, pH 7.1, and centrifuged at 70,000 g for 1 hour at 3°C to remove the sucrose and adsorbed proteins. The pellets are made up to a suitable volume with 50 mM Tris at pH 7.1, and analyzed.

Final Purification of HeLa Plasma Membranes. In order to achieve further purification of the plasma membrane fraction (fraction S7), this material is resuspended in 1.59 M sucrose, placed on a gradient prepared in the same way as the initial gradient, and centrifuged as described above.

Preparation of Myeloma Plasma Membranes

Cell Rupture and Removal of Nuclei. The S194 myeloma cells are washed twice in 0.14 M NaCl and broken in 0.25 M sucrose in TKM buffer. For this purpose, the ratio of solution to cell volume is 10:1 or 20:1. The cells were broken with 40–60 strokes of the Dounce homogenizer. The degree of cell rupture was carefully monitored with the phase contrast microscope. Nuclei and unbroken cells were removed by centrifugation at 1000 rpm for 3 minutes in a clinical centrifuge. The pellet was washed once. The two supernatant fluids were combined, then centrifuged at 5000 g for 10 minutes. Most of the plasma membranes remain in the supernatant fluid.

Sucrose Gradient. The supernatant fluid is applied to the top of the discontinuous sucrose gradient shown in Fig. 1. The gradient is centrifuged at 25,000 rpm in the Beckman SW 27 rotor for 4 hours. The plasma membranes are found at the 0.25–1.37 M sucrose interface.

Myeloma	HeLa
0.25	0.80
1.37	0.98
1.48	1.175
1.60	1.59

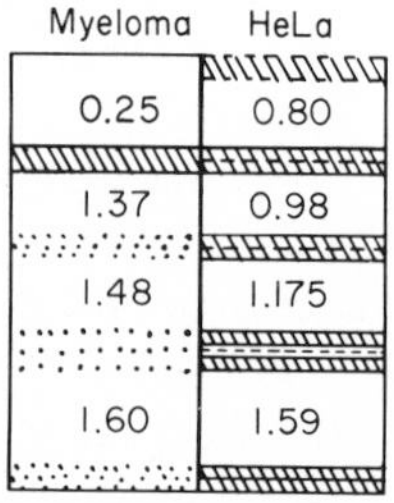

Fig. 1. Schematic representation of the membranous fractions derived from the HeLa and myeloma S194 cells after centrifugation in a discontinuous sucrose gradient at 70,000 g. The dashed lines show the initial sucrose interfaces, crosshatched and dotted areas show the membranous material, and the figures refer to the sucrose molarity. The plasma membrane from the myeloma cells is found at the 0.25–1.37 M sucrose interface; the plasma membrane from the HeLa cells is found in the pellet (fraction S7).

The membranes are carefully removed, diluted 1:4 with the TKM buffer, and centrifuged at 70,000 g for 90 minutes. The pellet is resuspended in TKM buffer.

Properties of the Purified Plasma Membranes

Yield. The yield of plasma membranes was 40–50% based on the recovery of appropriate enzymatic activities. Washing of the nuclear pellet partially released bound plasma membranes so that the yield could be increased to 60–80%.

Microscopic Observation. The gradient and regradient preparations of HeLa cell plasma membranes are morphologically homogeneous, consisting of large membranous structures and aggregates of the same material devoid of subcellular contamination. Under the phase microscope, these preparations resemble ghosts; when viewed with the electron microscope, no gross fragmentation or cytoplasmic elements are seen. The plasma membranes from the myeloma cells appear as broken sacules or large membranous fragments in the phase microscope.

Enzymatic Characterization. One of the questions presented in membrane fractionation is whether specific enzyme markers exist that are characteristic of particular membranes. Since many enzymes appear to be only loosely bound, and may be released from membranes, at least two criteria should be met before an enzyme may be considered a specific membrane marker; (a) the specific activity be greatly increased in the purified membrane fraction, and (b) the distribution of the enzyme be predominantly in a specific membrane fraction. It is evident that an enzyme must be strongly bound for it to be used as a membrane marker. Loosely bound enzymes may be released during fractionation procedures, and become bound to other membranes. In the case of HeLa cells, the specific activity of 5′-nucleotidase (EC 3.1.3.5) shows an increase of 120-fold (Table I) and is obtained in approximately 40–60% yield.[22] Since this enzyme is not present in significant quantities in the other membrane fractions, the results clearly demonstrate that it is predominantly found in the plasma membrane and can be considered to be a plasma membrane marker. In the myeloma plasma membranes, the 5′-nucleotidase is purified 6-fold and is found in 50% yield. Thus, it appears that the 5′-nucleotidase is perhaps an exclusive membrane marker of the plasma membranes of HeLa and myeloma cells, a property it shares with the collagen:glucosyl and collagen:galactosyl transferases[25,26,27] in the case of HeLa cells (Table I).

[25] A. Hagopian, H. B. Bosmann, and E. H. Eylar, *Arch. Biochem. Biophys.* **128**, 387 (1968).
[26] H. Bosmann and E. H. Eylar, *Biochem. Biophys. Res. Commun.* **30**, 89 (1968).
[27] H. B. Bosmann and E. H. Eylar, *Biochem. Biophys. Res. Commun.* **33**, 340 (1968).

TABLE I
ENZYME ACTIVITIES ASSOCIATED WITH SPECIFIC CELLULAR MEMBRANES
OF HeLa CELLS

| Enzyme | Increase[a] in specific activity of enzymes in | |
	Plasma membranes	Smooth internal membranes (Golgi)
1. 5'-Nucleotidase	120	—
2. Collagen:glucosyl transferase	145	—
3. Collagen:galactosyl transferase	160	—
4. Alkaline phosphatase	20	—
5. ATPase (Na⁺, K⁺, Mg²⁺ stimulated)	30	—
6. Phosphodiesterase	6	—
7. Glycoprotein:galactosyl transferase	—	28
8. Glycoprotein (PSM):fucosyl transferase	—	15
9. Glycoprotein (fetuin): fucosyl transferase	—	14
10. Polypeptide:N-acetyl-galactosaminyl transferase	—	47
11. Glycoprotein:sialyl transferase	—	15

[a]The increase refers to the ratio of specific activities in the purified fraction to that in the cellular homogenate.

The collagen:glycosyl transferases are not found in the smooth membranes, but appear to be associated only with the plasma membranes.[25] The gradient centrifugation procedure for preparation of the plasma membranes results in purification of these two enzymes which is the same order of magnitude as achieved with the 5'-nucleotidase. These enzymes are strongly bound to the plasma membrane; none of the other membrane fractions, including the smooth membranes, contained collagen:glycosyl transferase activities. The collagen:glucosyl transferase, an enzyme of high specificity, attaches glucose from UDP-glucose onto galactose residues in collagen, completing formation of an α-glucosyl-(1-2)-galactose disaccharide attached to the hydroxylysine moiety.[26,27]

Three other enzymes, alkaline phosphatase, ATPase, and phosphodiesterase, are also found in HeLa cell plasma membranes and were enriched by 20-, 30-, and 6-fold, respectively, during the course of purification (Table I). It has been reported that these three enzymes, as well as the 5'-nucleotidase, are enriched in the plasma membranes of liver cells,[19] intestinal mucosa, and kidney.[21]

The absence of particular enzymes may also be used in some cases as a criterion for purity. Thus, esterase activity, a microsomal enzyme,[28] succinic dehydrogenase, a mitochondrial enzyme,[29] and the UDPase which appears to be concentrated in the Golgi membranes,[30] were not detected in the plasma membrane fraction. Likewise absent from the plasma membranes were the multienzyme group of glycosyl transferases, which have been found in HeLa cells (Table I). They appear predominantly, if not exclusively, in the smooth internal membranes (Golgi).[25] One of these enzymes, the polypeptide: N-acetylgalactosaminyl transferase, an enzyme that forms the protein-carbohydrate linkage in membrane glycoprotein, is widely distributed in many cell types and may be considered a marker for the smooth internal membranes. Galactosyl transferase, another member of the multienzyme group of glycosyl transferases involved in the assembly of the carbohydrate units of glycoproteins, can be considered a marker for the smooth membranes (Golgi). This enzyme has recently been found in liver in the Golgi membrane.[31]

Chemical Composition. Because the distribution of enzymatic activities may vary with cell type,[32] Coleman and Finean[21,32,33] have suggested the use of the molar ratio of cholesterol/phospholipid as a characteristic plasma membrane marker; a high ratio may be a general property common to many plasma membranes (Table II). It is seen from Table II that this ratio is approximately 20-fold higher in the plasma membranes as compared to the smooth internal membranes. The value for the HeLa cell plasma membranes (0.9 to 1.05) agrees well with values reported[33] for the cholesterol/phospholipid ratio in the plasma membrane fractions of erythrocyte (1.17) and intestinal epithelium (1.1). The question arises, however, whether this criterion may be used as an indicator for degree of purity of a particular plasma membrane preparation since it is known that cholesterol may be lost from the plasma membranes prior to or during fractionation. The smooth internal membranes, by contrast, are characterized by lipids having relatively high proportions of phospholipid.

[28] A. I. Lansing, M. L. Belkhode, W. E. Lynch, and I. Lieberman, *J. Biol. Chem.* 242, 1772 (1967).
[29] E. C. Slater and W. D. Bonner, *Biochem. J.* 52, 185 (1952).
[30] H. B. Novikoff and S. Goldfischer, *Proc. Nat. Acad. Sci. U.S.* 47, 802 (1961).
[31] B. Fleischer, *Fed. Proc., Fed. Amer. Soc. Exp. Biol.* 28, 404 (1969).
[32] R. Coleman and J. B. Finean, *Biochim. Biophys. Acta* 125, 197 (1966).
[33] J. B. Finean, R. Coleman, and W. A. Green, *Ann. N.Y. Acad. Sci.* 137, 416 (1966).

TABLE II

CHEMICAL COMPOSITION OF PLASMA MEMBRANES FROM HeLa CELLS

Component	Plasma membranes	Smooth internal membranes
Protein	55–60[a]	32
Total lipid	40	67
Phospholipid	17.4	45.5
Cholesterol	9.5	1.4
Sialic acid	0.3	0.1
Neutral lipid	13.1	20.4
Cholesterol Phospholipid (molar ratio)	0.9–1.05	0.06

[a]Values are expressed as percent dry weight, except for phospholipid.

The amino acid composition of the protein found in the smooth internal membrane and the plasma membrane fractions from HeLa cells[22] show a high degree of similarity; they also resemble structural protein prepared from many membrane sources.[34] Whether this result is simply fortuitous or a significant indication of a major protein component in all membranes remains a question.

[34]D. O. Woodward and K. D. Munkres, in "Organizational Biosynthesis" (H. J. Vogel, J. O. Lampen, and V. Bryson, eds.). Academic Press, New York, 1967.

[16] Isolation of Golgi Apparatus

By D. JAMES MORRÉ

The isolation of Golgi apparatus allows a direct approach to the study of the physical, chemical, and enzymatic properties of this cell component provided that the preparations are free of other cell fractions and representative of the state of the Golgi apparatus *in situ*. In its most familiar form, the Golgi apparatus (singular or plural) is that part of the cell's cytoplasmic membrane system consisting of regions of stacked cisternae (dictyosomes) which lack ribosomes.[1,2] Dictyosomes, in turn, are composed of heterogeneous elements, i.e., platelike regions

[1]D. J. Morré, H. H. Mollenhauer, and C. E. Bracker, *in* "Results and Problems in Cell Differentiation" (T. Reinert and H. Ursprung, eds.), Vol. 2, pp. 82. Springer, Berlin, 1970.

[2]D. J. Morré, T. W. Keenan, and H. H. Mollenhauer, *Proc. 1st Int. Symp. Cell. Biol. Cytopharmacol., Venice, Italy, 1969*, in press.

(the saccules) of various types, several kinds of tubules, intercisternal regions, coated vesicles, a variety of secretory vesicle types and the contents of cisternae.[1,2] The procedures described fractionate decisively, yet are sufficiently mild to prevent gross modification or loss of individual Golgi apparatus components. The electron microscope has been, and will continue to be, indispensable in the qualitative assay of isolated Golgi apparatus fractions. Quantitative estimates of yield and purity are provided by assays for enzyme activities concentrated in Golgi apparatus and other cell fractions.

Centrifugal forces are calculated for the middle of the tube. All solutions are prepared in distilled water. Specific activities of enzymes are given as micromoles of substrate transformed per minute per milligram of protein.

Rat Liver: Purification Procedure

Reagents

a. Buffer: Mix 50 ml of a 0.2 M solution of Tris acid maleate (24.2 g of Tris + 23.2 g of maleic acid or 19.6 g of maleic anhydride in 100 ml) and 37 ml of 0.2 N NaOH, dilute to a total of 200 ml, and adjust to pH 6.4 with NaOH.

b. Sucrose, 2 M, prepared in buffer (a)

c. Dextran,[3] 10%

d. Magnesium chloride, 10 mM

e. Mercaptoethanol, 0.1 M

f. Sucrose, 1.25 M

Procedure. Male rats, 200–250 g, 50 days old of the Holtzman strain[4] provided with standard diet and drinking water ad libitum[5] are killed by decapitation and drained of blood. Alternatively, animals are anesthetized by intraperitoneal injection of 0.5–1 ml of pentobarbital (Nembutal) solution (20 mg/ml), and the livers are drained of blood by first clamping the portal vein and hepatic artery and then severing the inferior vena cava just below the diaphragm at the junction of the hepatic veins. The livers are removed, weighed, and minced rapidly at room temperature with scalpels or single-edged razor blades. All other operations are at 0–4°. The minced tissue, in lots of approximately 10 g each, is mixed with 20 ml of chilled homogenization medium. The homogenization medium is prepared by combining 50 ml

[3]Average molecular weight 225,000, Sigma Chemical Company, St. Louis, Missouri.

[4]The Holtzman Company, Madison, Wisconsin.

[5]H. H. Mollenhauer, D. J. Morré, and C. Kogut, *Exp. Mol. Pathol.* 11, 113 (1969).

of reagent a, 25 ml of reagent b, 10 ml each of reagents c and d, and 5 ml of reagent e. Fixatives are not used. Homogenization is for 40–80 seconds at 5000–10,000 rpm using a Polytron 20 ST homogenizer.[5,6] The homogenate is squeezed through a single layer of Miracloth[7] to remove unbroken cells and connective tissue. The homogenate is centrifuged for 15–30 minutes at 2000–5000 g using a rotor of the swinging-bucket type to concentrate the Golgi apparatus. The lipid on top of the tube and the supernatant fluid are removed by suction. The yellow-brown phase of the pellet (usually upper one-half to one-third) which lies above the red to pink and dark brown layers containing whole cells, nuclei, and fragments of plasma membrane is resuspended in a portion of the lipid-free supernatant (final volume of about 6 ml for each liver homogenized), layered on 1.5–2 volumes of 1.25 M sucrose (reagent f), and centrifuged for 30 minutes at 90,000–150,000 g using a rotor of the swinging-bucket type. The Golgi apparatus are collected at the 1.25 M sucrose-homogenate interface while mitochondria and endoplasmic reticulum enter the sucrose layer. The band containing the Golgi apparatus is removed from the top using a Pasteur pipette fitted with a rubber aspirator or a syringe fitted with a curved 16-gauge stainless steel cannula. This fraction is resuspended in the clear red supernatant fluid (for optimal preservation of morphology), in the homogenization medium or in distilled water (for highest fraction purity), and the Golgi apparatus are collected by centrifugation at 2000 g for 30 minutes. The supernatant fluid is removed, and the surface of the pellet is carefully rinsed with distilled water. The pellets are finally resuspended in distilled water or an appropriate buffer solution at a concentration of 5–10 mg of Golgi apparatus protein per milliliter. The yield is 5–10 mg Golgi apparatus protein per 10 g of liver with a fraction purity of at least 70–80%.[2,8–13] Further removal of contaminating cell components and a fraction purity of 90% or more is achieved by repeated resuspension in distilled water or isolation medium and centrifugation at 2000 g for 30 minutes.[11] Although dictyosomes remain intact, washing removes some of the

[6]Kinematica, Lucerne, Switzerland.

[7]A porous, cellulosic fabric, Chicopee Mills, New York.

[8]D. J. Morré, R. Cheetham, and W. Yunghans. *J. Cell Biol.* 39, 96a (1968).

[9]D. J. Morré, L. M. Merlin, and T. W. Keenan, *Biochem. Biophys. Res. Commun.* 37, 813 (1969).

[10]D. J. Morré, R. L. Hamilton, H. H. Mollenhauer, R. W. Mahley, W. P. Cunningham, R. D. Cheetham, and V. S. LeQuire, *J. Cell Biol.*, 44, 484 (1970).

[11]R. D. Cheetham, D. J. Morré, and W. N. Yunghans, *J. Cell Biol.*, 44, 492 (1970).

[12]W. N. Yunghans, T. W. Keenan, and D. J. Morré, *Exp. Mol. Pathol.* 12, 36 (1970).

[13]T. W. Keenan and D. J. Morré, *Biochemistry* 9, 19 (1970).

peripheral tubules of the cisternae through vesiculation and reduces total yield of Golgi apparatus protein. Alternatively, purification of Golgi apparatus membranes is accomplished by centrifugation in either a continuous or discontinuous 0.9 to 1.25 M sucrose gradient[14] with attendant morphological changes similar to those accompanying repeated resuspension-centrifugation.

Comments. When large-volume swinging-bucket rotors are employed for the centrifugation steps (Servall HB-4 at 2000 g; Spinco SW 27 at 90,000 g; or rotors of comparable capacity and design), it is possible to process livers from 10–12 rats in a single run of less than 3 hours. With batch isolations, the Miracloth filtration step becomes time consuming and is eliminated.[15,16] The procedure is easily adapted for use with zonal rotors[17,18] but, with conventional methods of loading, the Golgi apparatus are held so long in the total homogenate before sedimentation that unstacking occurs along with loss of peripheral tubules and changes in cisternal morphology. In addition, isolations can be carried out more efficiently in less time with large-volume, swinging-bucket rotors.

Golgi apparatus isolations are influenced by diet of the animal[5] and the physiological state of the tissue. With the Holtzman strain, animals are fed normal diets of Purina Laboratory Chow and are not fasted prior to sacrifice.

Yield and purity of the isolated fractions are critically dependent upon the procedures of homogenization. When properly carried out, intact portions of the Golgi apparatus (dictyosomes) which sediment at low centrifugal force are obtained in high yield.[2,5,10,15] Extensive homogenization yields smaller membrane pieces which sediment with the microsome fraction. Even with gentle homogenization, it is necessary to establish the appropriate centrifugal forces and time periods necessary to sediment the Golgi apparatus quantitatively for each combination of tissue and conditions of homogenate preparation. The Polytron homogenizer appears to be superior to other types of homogenizers commonly used for cell fractionation. However, preparations of useful quantity and purity have been obtained using loose-fitting, all-glass homogenizers[2] or glass-Teflon homogenizers,[19] but

[14]Unpublished procedure of L. Ovtracht and D. J. Morré, (1969).
[15]R. W. Mahley, R. L. Hamilton, and V. S. LeQuire, *J. Lipid Res.* **10**, 433 (1969).
[16]S. N. Nyquist, R. Barr, and D. J. Morré, *Biochim. Biophys. Acta* **208**, 532 (1970).
[17]D. J. Morré, H. H. Mollenhauer, R. L. Hamilton, R. W. Mahley, and W. P. Cunningham, *J. Cell Biol.* **37**, 157A (1968).
[18]B. Fleischer, S. Fleischer, and K. Ozawa, *J. Cell Biol.* **43**, 59 (1969).
[19]H. Schachter, I. Jabbal, R. L. Hudgin, L. Pinteric, E. J. McGuire, and S. Roseman. *J. Biol. Chem.* **245**, 1090 (1970).

these devices tend to yield preparations in which the Golgi apparatus are unstacked and fragmented as well as contaminated with plasma membrane fragments.[2] Other details of the procedure appear less critical. The dextran retards unstacking of dictyosomes.[14] Ions and sulfhydryl compounds are included for preservation of enzymatic activities.[9,10]

A procedure developed by Schachter et al.[19] involves use of a glass homogenizer of the Potter-Elvehjem type (4 gentle strokes of a slowly rotating, motor-driven Teflon pestle) and centrifugation of the total homogenates. Rat liver is homogenized in 4 volumes of 0.5 M sucrose-Tris buffer-magnesium-dextran medium and diluted to 16 volumes per gram of liver. The homogenate is filtered through 4 layers of gauze and layered onto a buffered sucrose gradient consisting of 1.7 M sucrose, 1.3 M sucrose, and 0.7 M sucrose in a ratio to sample of 2.5:2:2 followed by centrifugation at 25,000 rpm (Spinco SW 25.1) for 45 minutes. The Golgi apparatus collect at the 0.7/1.3 M sucrose interface. Golgi apparatus prepared in this manner tend to be unstacked, and the preparations consist largely of single cisternae and cisternal fragments. However, for many types of analyses, such as isolation of highly purified glycosyl transferases, this type of preparation[19] as well as those from procedures employing zonal rotors,[18] are adequate if sufficiently characterized.

Rat Liver: Assay Procedures

Morphology

The identification of isolated Golgi apparatus is based on their morphology which is so characteristic that it serves as a reliable marker.[1,2,5,10,15,20] After positive staining of thin sections, fields of stacked cisternae are observed (Fig. 1A); each stack consisting of 3–4 cisternae surrounded by numerous small vesicular profiles and secretory vesicles. Appropriate planes of section show continuity between the parallel cisternae and the peripheral tubules which usually appear in section as small vesicles (Fig. 1A). Preparatory to fixation for electron microscopy, the bottom portions of lusterloid centrifuge tubes containing the pellets are removed using a sharp blade, sliced into wedge-shaped sectors and dropped into fixative (for details of fixation, see Fig. 1A). The plastic from the centrifuge tube does not interfere with fixation and is dissolved during the dehydration step before embedding. The pellet sectors retain the shape of the centrifuge tube which facilitates their orientation prior to sectioning.

[20]R. L. Hamilton, D. J. Morré, R. Mahley, and V. S. LeQuire, *J. Cell Biol.* **35**, 53A (1967).

Golgi apparatus fractions are assayed most rapidly and accurately by the technique of negative staining.[21,22] A 1% aqueous solution of phosphotungstic acid (PTA) is neutralized to pH 6.5–7 with KOH. A drop of the resuspended cell fraction is placed on a carbon-stabilized, collodion-coated grid and mixed with a drop of the PTA solution. The excess solution is removed slowly with absorbent paper and, after drying, the preparations are ready for examination in the electron microscope. Alternatively, and particularly with preparations containing sucrose, a droplet of the PTA solution is placed on a small piece of parafilm and a small amount of the Golgi apparatus suspension (the quantity adhering to a needle) is spread on the surface of the PTA. The carbon-coated grid surface is then touched to the surface of the droplet. Excess solution is removed as before.

The tubular nature of the Golgi apparatus cisternae is most evident in negatively stained preparations (Figs. 1B and 1C). Negative staining also emphasizes the 300–1000 Å electron translucent lipoprotein particles of the secretory vesicles and of the peripheral tubules that are characteristic of the hepatocyte Golgi apparatus.[2,10,15,23] The extensive system of peripheral tubules is continuous with the platelike central regions of the cisternae; the latter corresponding to the parallel cisternae seen in thin section. Often, the tubules are highly anastomosed and closely spaced near the periphery of the plate. In some cisternae, the platelike regions predominate; others consist mainly of tubular elements. Both types of cisternae exist within a single stack or dictyosome. Tubules are a consistent feature of all cisternae in isolated Golgi apparatus preparations from a wide range of animal and plant sources and many of the small Golgi apparatus vesicles seen in thin sections correspond to cross sections through these tubules.[1,2,10,22,24,25]

Enzymatic Activities

Levels of contaminating cell components are estimated from assays of 5′-nucleotidase (plasma membrane), glucose-6-phosphatase (endoplasmic reticulum), and succinic-INT-reductase (mitochondria). Glycosyl transferases and thiamine pyrophosphatase are used to monitor the fate of Golgi apparatus and Golgi apparatus fragments. Assays are

[21]R. W. Horne, *in* "Techniques for Electron Microscopy" (D. H. Kay, ed.), p. 341. Blackwell, Oxford, 1965.

[22]W. P. Cunningham, D. J. Morré, and H. H. Mollenhauer, *J. Cell Biol.* **28**, 169 (1966).

[23]R. W. Mahley, R. L. Hamilton, D. J. Morré, and V. S. LeQuire, *Fed. Proc., Fed. Amer. Soc. Exp. Biol.* **27**, 666 (1968).

[24]H. H. Mollenhauer and D. J. Morré, *Annu. Rev. Plant Physiol.* **17**, 27 (1966).

[25]H. H. Mollenhauer, D. J. Morré, and L. Bergman, *Anat. Rec.* **158**, 313 (1967).

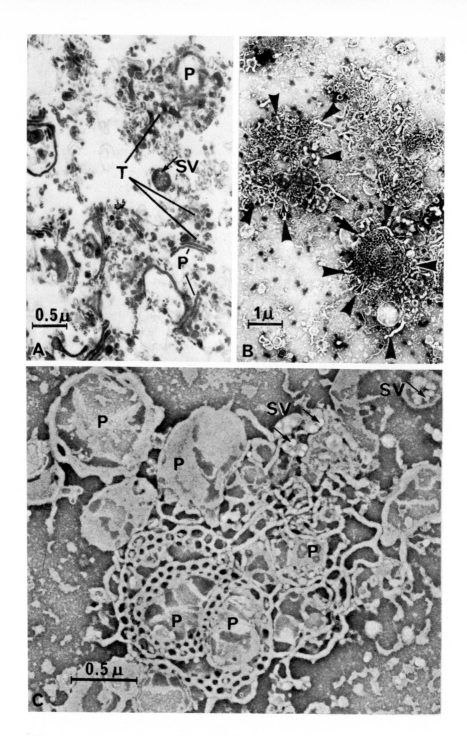

carried out as soon as possible after preparation of the fractions. The assay medium is preincubated for 5 minutes at 37° for temperature equilibration. Assays are initiated by adding Golgi apparatus fractions diluted immediately before assay, and are at 37° with gentle shaking. Protein is determined by the Lowry procedure[26] with bovine serum albumin as the standard.

5'-NUCLEOTIDASE

Reagents

Substrate: 5.5 mM magnesium chloride, 55 mM Tris, 11 mM AMP, final pH adjusted to 8.5 with HCl
Stopping reagent: 10% w/v trichloroacetic acid
Molybdate·H_2SO_4: 2.5% w/v ammonium molybdate in 5 N H_2SO_4
Reducing agent: Aminonaptholsulfonic acid reagent according to Fiske and SubbaRow[27]

Procedure.[28] To 0.9 ml of substrate is added 0.1 ml of Golgi apparatus suspension (0.2–0.5 mg protein). The reaction is allowed to proceed for about 15 minutes so that an amount of inorganic phosphate is formed within the range of 0.1–1 μmoles, and is then stopped by the addition of 1 ml of 10% trichloroacetic acid. Protein is removed by centrifugation and 1 ml of the cleared supernatant is used for phos-

[26]O. H. Lowry, N. J. Rosebrough, A. L. Farr, and P. J. Randall, *J. Biol. Chem.* **193**, 265 (1951); see also this series, Vol. 3 [73].
[27]C. H. Fiske and Y. SubbaRow, *J. Biol. Chem.* **66**, 375 (1925); see also this series, Vol. 3 [115].
[28]See also P. Emmelot and C. J. Bos, *Biochim. Biophys. Acta* **120**, 369 (1966).

FIG. 1. Electron micrographs showing the appearance of Golgi apparatus isolated from rat liver. (A) Positively stained [E. S. Reynolds, *J. Cell Biol.* **17**, 208 (1963)] (lead citrate) thin section of a fraction fixed in glutaraldehyde (2.5% in 0.1 M sodium phosphate buffer, pH 7.2) for 1.5–2 hours at 4° followed by osmium tetroxide (1% in 0.1 M sodium phosphate, pH 7.2) at 4° for 16 hours; dehydrated through an acetone series and embedded in Vestopal [A. Ryter and E. Kellenberger, *J. Ultrastruct. Res.* **2**, 200 (1958)]. Dictyosome at the upper right was sectioned tangentially and shows the platelike portion of a cisterna (*P*) and attached tubules (*T*) in face veiw. Other dictyosomes were sectioned transversely and show these same cisternal features in vertical section. SV, secretory vesicles. (B) Negative contrast (PTA) showing single, relatively intact dictyosomes consisting of stacked cisternae with tubular peripheries. The large, electron transparent tubules at the dictyosome periphery (arrows) often are distended due to the presence of lipoprotein particles. (C) Negative contrast (PTA) of a dictyosome partially unstacked to show both the system of peripheral tubules and the central platelike portions (*P*) of cisternae. The form of the cisternae varies from typically platelike with a few peripheral tubules to almost entirely tubular. Secretory vesicles (*SV*) contain the very low density lipoprotein particles (arrow) of the secretory product.

phate determination by adding 0.5 ml of the molybdate·H_2SO_4 solution and 0.2 ml of the reducing agent for color development. Distilled water is added to a final volume of 5 ml. Absorbance is read after 20 minutes at 660 mμ. Sodium-potassium tartrate (10 mM) may be included in the assay medium to inhibit acid phosphatases.[29]

GLUCOSE-6-PHOSPHATASE

Reagents
    Substrate: 55 mM Tris, 11 mM glucose 6-phosphate, 11 mM mercaptoethanol, final pH 6.6
Other reagents as for 5'-nucleotidase

Procedure.[30] Golgi apparatus and reference fractions are prepared in the presence of mercaptoethanol. The incubation and assay procedures are the same as for 5'-nucleotidase except for the substrate.

SUCCINIC-INT-REDUCTASE

Reagents
    Substrate: 55 mM potassium phosphate, 0.11% (w/v) 2-(p-indophenyl)-3-(p-nitrophenyl)-5-phenyl tetrazolium (INT); 55 mM sodium succinate, 25 mM sucrose
    Stopping reagent: 10% w/v trichloroacetic acid

Procedure.[31] To 0.9 ml of substrate is added 0.1 ml of Golgi apparatus suspension containing 0.2–0.5 mg protein. The reaction is allowed to proceed for 15 minutes. One milliliter of the trichloroacetic acid solution is added and the reduced dye is extracted with 4 ml of ethyl acetate in a stoppered glass tube. The extinction coefficient is 20.1 $\times$ 10³ liter·mole^{-1}·cm^{-1} at 490 mμ.

THIAMINE PYROPHOSPHATASE

Reagents
    Substrate: 33 mM sodium barbital, 15 mM calcium chloride, 3.3 mM thiamine pyrophosphate, final pH 8.0
    Stopping reagent: 10% w/v trichloroacetic acid

Procedure.[32] The assay is initiated by adding 0.1 ml of Golgi apparatus suspension (0.2–0.8 mg protein) to 2.9 ml of substrate. After 20 minutes,

[29]R. H. Mitchell and J. N. Hawthorne, *Biochem. Biophys. Res. Commun.* **21**, 333 (1965).
[30]See also this series, Vol. 2 [83].
[31]R. J. Pennington, *Biochem. J.* **80**, 649 (1961).
[32]J. M. Allen and J. J. Slater, *J. Histochem. Cytochem.* **9**, 418 (1961).

or after 0.3–3 μmoles inorganic phosphate are released, the reaction is terminated by addition of 1 ml of the trichloroacetic acid solution. Precipitated protein is removed by centrifugation, and 2 ml of the cleared supernatant fluid is used for determination of inorganic phosphate according to the method of Berenblum and Chain as modified by Martin and Doty.[33] Formation of precipitates prevents use of the Fiske and SubbaRow method for determination of inorganic phosphorous.

UDP-Galactose: N-Acetylglucosamine Galactosyltransferase

Reagents

a. Tris·HCl, 80 mM, pH 7.5, containing 0.2 M mercaptoethanol

b. Magnesium chloride, 40 mM, containing 20 mM manganous chloride

c. N-Acetylglucosamine, 18 mM

d. UDP-D-galactose, 1.2 mM, to which UDP-D-galactose-^{14}C has been added to contain approximately 10^6 dpm/ml

e. Triton X-100,[34] 1% (v/v)

f. EDTA, 0.25 M

g. AG_1-X_2 anion exchange resin (Cl⁻form)[35] prepared in distilled water

Procedure. The activity is estimated from the increase in the rate of UDP-galactose hydrolysis which occurs in the presence of a suitable acceptor by an adaptation of the method of Palmiter.[9,36] Golgi apparatus and reference fractions are prepared in the presence of mercaptoethanol. Two series of measurements are required for each assay. The conditions are as follows. Series A. Equal volumes of each of solutions a through d are mixed. Series B. Identical to series A except in that distilled water is substituted for solution c. Solution d should be freshly prepared to ensure low levels of free galactose. Golgi apparatus and other cell fractions are dispersed in 1% Triton X-100 (e). For each pair of analyses, 0.1 ml of enzyme solution (0.1–0.2 mg Golgi apparatus protein) is added to each of two shell vials containing 0.1 ml of series A or B, respectively, with a final volume of 0.2 ml. Incubations are for 10 minutes. The reaction is stopped by addition of 0.1 ml of 0.25 M EDTA (f), and the mixture is placed on an anion exchange column con-

---

[33]See O. Lindberg and L. Ernster, *in* "Methods of Biochemical Analysis" (D. Glick, ed.), Vol. 3, p. 1. Wiley (Interscience), New York, 1956.

[34]Octylphenoxypolyethoxyethanol, Rohm and Haas Company.

[35]200–400 mesh, Bio-Rad Laboratories, Richmond, California.

[36]R. D. Palmiter, *Biochim. Biophys. Acta* **178**, 35 (1969).

sisting of about 1 ml of packed resin (g) contained in a Pasteur pipette plugged with glass wool. A second Pasteur pipette is used to effect the transfer. The products of the reaction, N-acetylaminolactose and free galactose, are eluted from the columns with three washes of 0.4 ml each of distilled water, at the same time rinsing the reaction vial and transfer pipette. The washes containing the radioactive products are collected in scintillator vials, and radioactivity is determined by liquid scintillation counting. To correct for nonspecific hydrolysis of UDP-galactose, the value for series B is subtracted from the value for series A.

Comments. All enzyme assays should be conducted under optimal conditions where product formation or disappearance of substrate is proportional to time of incubation and to the quantity of protein present. Normally, at least two protein concentrations are used for each assay.

With Golgi apparatus fractions taken directly from the sucrose gradients (Table I), the combined levels of 5'-nucleotidase, glucose-6-phosphatase, and succinic-INT-reductase show the fractions to be at least 80% Golgi apparatus-derived material.[2,8,9,11] The Golgi apparatus itself appears to contain significant levels of the first two of these enzymes.[14] Thus, the actual purity of the fractions may be somewhat higher than 80%.

Using the galactosyl transferase (Table I), it is possible to estimate the yield of Golgi apparatus as between 40 and 50% (as much as 70%) since 40–50% of this activity is recovered in the Golgi apparatus fraction.[9] Similar results are obtained with an N-acetylglucosaminyl-glycoprotein transferase and its endogenous acceptor.[9,37] That the unrecovered activity contained in the microsomal fraction represents Golgi apparatus fragmented during homogenization remains to be established.

Several other sugar transferase activities are localized in Golgi apparatus fractions from rat liver. In studies by Schachter *et al.*,[19,38–40] transferases of rat liver Golgi apparatus were assayed in the presence of suitable mucopolysaccharide acceptors. For each transferase, the Golgi apparatus fractions exhibited the highest specific activities when compared to other cell fractions, and, in each instance, the activity of the Golgi apparatus fraction accounted for approximately 40% of the total activity of the whole homogenate (Table II). Although representing marker enzymes for the Golgi apparatus of rat liver, the use of the

[37]R. R. Wagner and M. A. Cynkin, *Biochem. Biophys. Res. Commun.* 35, 139 (1969).
[38]H. Schachter, E. J. McGuire, and S. Roseman, *Proc. Can. Fed. Biol. Soc.* 10, 78 (1967).
[39]H. Schachter, I. Jabbal, and S. Roseman, *Proc. Can. Fed. Biol. Soc.* 12, 77 (1969).
[40]H. Schachter, personal communication, (1969).

TABLE I
ENZYMATIC ACTIVITIES OF GOLGI APPARATUS FRACTIONS ISOLATED FROM RAT LIVER[a–c]

	Specific activity	Relative specific activity[d]
5'-Nucleotidase		
Golgi apparatus	0.053	1.9
Plasma membrane	0.900	32.0
Homogenate	0.028	—
Glucose-6-phosphatase		
Golgi apparatus	0.028	0.7
Endoplasmic reticulum	0.200	5.0
Homogenate	0.040	—
Succinate-INT-reductase		
Golgi apparatus	0.003	0.1
Mitochondria	0.385	12.8
Homogenate	0.030	—
Thiamine pyrophosphatase		
Golgi apparatus	0.053	10.6
Endoplasmic reticulum	0.010	2.0
Plasma membrane	0.005	1.0
Homogenate	0.005	—
UDP-galactose:N-acetylglucos-amine-galactosyl transferase		
Golgi apparatus	3.81×10^{-3}	95.0
Endoplasmic reticulum	0.06×10^{-3}	1.5
Plasma membrane	Not detected	—
Homogenate	0.04×10^{-3}	—

[a] D. J. Morré, L. M. Merlin, and T. W. Keenan, *Biochem. Biophys. Res. Commun.* 37, 813 (1969).
[b] R. D. Cheetham, D. J. Morré, and W. N. Yunghans, *J. Cell Biol.* 44, 492 (1970).
[c] R. D. Cheetham and D. J. Morré, unpublished results (1969).
[d] Relative specific activity = $\dfrac{\text{units of enzyme activity of fraction}}{\text{units of enzyme activity of homogenate}}$

sialyl-, galactosyl-, and N-acetylglucosaminyl-transferases for this purpose is presently limited by the need for preparation of suitable high molecular weight acceptors. In contrast, the UDP-galactose: galactosyl-N-acetylglucosamine transferase is relatively uncomplicated in its estimation,[9,18] as is the sialyl transferase which catalyzes the incorporation of sialic acid from CMP-N-acetylneuraminic acid to lactose or N-acetyl lactosamine with the formation of the corresponding trisaccharide.[41] It is not yet possible to consider any of the above glycosyl transferases as universal marker enzymes applicable to Golgi apparatus of all cell types.

[41] See this series, Vol. 7 [62].

TABLE II

GLYCOSYL TRANSFERASE REACTIONS OF A GOLGI APPARATUS-RICH FRACTION ISOLATED FROM RAT LIVER[a]

Enzyme	Sugar nucleotide	Glycosyl transferase reaction		Specific activity ($\times 10^3$)		% Total activity in Golgi apparatus
		Acceptor	Product	Homogenate	Golgi apparatus	
Sialyl transferase	CMP-NANA	AGP⁻(SA)	AGP	0.83	7.04	44
Galactosyl glycoprotein transferase	UDP-Gal	AGP⁻(Gal,SA)	AGP⁻(SA)	0.18	2.25	42
N-Acetylglucosaminyl transferase	UDP-GlcNAc	AGP⁻(GlcNac,Gal,SA)	AGP⁻(Gal,SA)	0.40	3.65	43
Galactosyl-N-acetylglucosamine transferase	UDP-Gal	GlcNAc	Gal-GlcNAc	0.10	1.06	40

[a] Exogenous glycoprotein acceptors were prepared by stepwise pretreatment of orosomucoid (α_1-acid glycoprotein-AGP), a glycoprotein containing an oligosaccharide side chain terminating in a trisaccharide unit: sialic acid (SA) $\rightarrow$ galactose (Gal) $\rightarrow$ N-acetylglucosamine (GlcNAc) $\rightarrow$ glycoprotein, with purified sialidase [AGP⁻(SA)], β-galactosidase [AGP⁻(Gal,SA)], and β-N-acetylglucosaminidase [AGP⁻(GlcNAc, Gal, SA)]. Livers of male Wistar rats were carefully homogenized in 4 volumes of 0.5 M sucrose in Tris–Mg²⁺–dextran and Golgi apparatus were recovered from the 0.7/1.3 M sucrose interface of a discontinuous sucrose gradient (SW 25.1, 25,000 rpm, 45 minutes). These data were provided through the courtesy of H. Schachter, Department of Biochemistry, University of Toronto, Canada.

Thiamine pyrophosphatase activity has been shown by cytochemical procedures to be concentrated in the Golgi apparatus of rat liver.[42,43] However, the levels of activity which occur in endoplasmic reticulum, plasma membrane (Table I), and other cell fractions limits this enzyme as a specific *in vitro* marker for rat liver.[44]

Other Animal Tissues: Purification Procedures

The procedures described for isolation of Golgi apparatus from rat liver have been successfully modified for use with rat mammary gland[45] and for mucopolysaccharide-secreting glands of a snail (*Helix pomatia*).[46] However, no general scheme of fractionation can be given since sedimentation behavior of Golgi apparatus varies from one source to another. For example, the large Golgi apparatus of the snail sediment quantitatively at 730 *g* for 6 minutes.[46] Nevertheless, the homogenization medium, the method of tissue homogenization, and the sucrose gradient procedure seem to be applicable to most tissues.

In exploring the use of this method of isolation with tissues other than liver, it is important to guard against postmortem changes. Golgi apparatus of cells such as adrenal cortex seem particularly susceptible, and failures of the isolation procedure with this tissue were traced to loss of recognizable Golgi apparatus while the tissue was being collected prior to homogenization. For organs containing large amounts of connective tissue, more vigorous homogenization is required, but at the expense of the integrity of cytoplasmic components.

An interesting modification has been used with rat testes.[47] Tissue is homogenized in a medium as described for rat liver except with a final sucrose concentration of 1 *M*. The Golgi apparatus are obtained by flotation centrifugation for 30 minutes at 100,000 *g*, thereby eliminating one centrifugation step. Testes are unique in that endoplasmic reticulum is sparse and that sufficient mitochondria are attached to the plasma membrane to cause it to sediment in 1 *M* sucrose. The same procedure when applied to rat liver resulted in fractions heavily contaminated with plasma membrane.

The first successful isolation of Golgi apparatus was from epididymus

[42]A. B. Novikoff, E. Essner, S. Goldfischer and M. Heus, *in* "Interpretation of Ultrastructure" (R. J. C. Harris, ed.), p. 149. Academic Press, New York, 1962.

[43]E. Essner and A. B. Novikoff, *J. Cell Biol.* 15, 289 (1962).

[44]R. Cheetham, D. J. Morré, C. Pannek, and D. S. Friend, unpublished results (1969).

[45]T. W. Keenan, D. J. Morré, and R. Cheetham, Nature 228, 1105 (1970).

[46]L. Ovtracht, D. J. Morré, and L. M. Merlin, *J. Microsc. (Paris)* 8, 989 (1969).

[47]W. P. Cunningham and H. H. Mollenhauer, unpublished procedure, (1969).

using a medium containing 0.34 M NaCl.[48,49] Addition of NaCl has not proved beneficial for isolation of Golgi apparatus from rat liver, nor has the NaCl procedure yielded consistent results with epididymus or other plant or animal tissues in our laboratory.

Other Animal Tissues: Analytical Methods

UDP-galactose; N-acetylglucosamine galactosyl transferase is concentrated in all animal Golgi apparatus fractions thus far examined.[2,9,18,19,40,45,46,50] Thiamine pyrophosphatate has been localized within the Golgi apparatus for a large number of cell types by ultrastructural cytochemistry,[42,43,51,52] and is found in preparations of isolated Golgi apparatus.[11,44,46] Additionally, extensive cytochemical studies by Novikoff and co-workers[42,51,52] show certain nucleoside diphosphatase activities to be specifically localized in Golgi apparatus membranes. Although concentrated in the Golgi apparatus, the presence of these enzymes in other cell fractions should be checked for each tissue type before these enzymes are used as specific *in vitro* markers for Golgi apparatus.

Onion Stem: Purification Procedure

Reagents
Isolation medium[53] 0.5 M sucrose, containing 10 mM sodium phosphate, pH 6.8, and 1% (w/v) dextran[3]

Plant Material. Green onions (*Allium cepa*) purchased locally are stored at 4°. Stem explants are harvested by cutting roots and lignified stem regions from the onion base. A cone of tissue, 0.5–1 cm diameter at the base and 0.5–1 cm high, is removed from the central portion of the onion bulb using a scalpel fitted with a narrow blade (Bard Parker No. 11). Included in the explant are stem, meristematic region, and leaf bases. Scale leaves and the green portions of the onion are discarded.

---

[48] E. L. Kuff and A. J. Dalton, *in* "Subcellular Particles" (T. Hayashi, ed.), p. 114. Academic Press, New York, 1959.
[49] W. C. Schneider and E. L. Kuff, *Amer. J. Anat.* **94**, 209 (1954).
[50] B. Fleischer, *Fed. Proc., Fed. Amer. Soc. Exp. Biol.* **28**, 404 (1969).
[51] A. B. Novikoff and S. Goldfischer, *Proc. Nat. Acad. Sci. U. S.* **47**, 802 (1961).
[52] S. Goldfischer, E. Essner, and A. B. Novikoff, *J. Histochem. Cytochem.* **12**, 72 (1964).
[53] D. J. Morré, and H. H. Mollenhauer, *J. Cell Biol.* **23**, 295 (1964). Homogenization in coconut milk (liquid endosperm from coconuts purchased locally and cleared by centrifugation for 1–3 hours at 90,000 g)[54] gives better morphological preservation of onion dictyosomes than does the synthetic homogenization medium. However, the possibility of introducing extraneous constituents into the preparation from the coconut milk arises.

Procedure. Approximately 5 g of stem explants from 30–50 onions are washed and blotted. Homogenates prepared by using the Polytron are satisfactory, but best results are obtained by chopping the tissue with razor blades. The motor-driven, cam-operated chopping device diagrammed in Fig. 2 is fitted with ten double-edged razor blades spaced 5 mm apart. The chopping surface is a Plexiglas trough. The stem explants are combined with 5 ml of homogenization medium on the chopping surface. The tissue is then chopped for about 5 minutes at a rate of 30 chops per second using the motor-driven device. The homogenate is filtered through a single layer of Miracloth, squeezing by hand, in order to remove cell walls and debris.

Homogenates are centrifuged at 10,000 g for 30 minutes to remove nuclei, plastids, and mitochondria. The entire 10,000 g supernatant is layered onto a sucrose cushion composed of a bottom layer of 0.25 ml of 1.8 M sucrose and a top layer of 0.5 ml 1.6 M sucrose contained in a 5.4 ml lusterloid centrifuge tube. The Golgi apparatus are centrifuged onto the sucrose cushion by centrifugation for 30 minutes at 35,000 g. After centrifugation, supernatant liquid is removed by suction and replaced by additional sucrose layers as follows: 1.0 ml 1.5 M; 1.0 ml 1.25 M, and 1.0 ml 0.5 M sucrose. Purification is achieved by centrifugation in the sucrose gradient for 3 hours at 100,000 g. Individual dictyosomes band at the 0.5 M/1.25 M sucrose interface and are removed using a Pasteur pipette or a syringe fitted with a curved 16-gauge stainless steel cannula. Dictyosomes are resuspended in distilled water and concentrated by a final centrifugation at 35,000 g for 30 minutes.

For optimal preservation of morphology, fixative solution[55] is added to the homogenization medium, immediately before homogenization, at a final concentration of 50 mM.[22,56] The procedure is the same as without fixatives except that centrifuge speeds are adjusted for a more rapid sedimentation rate and protective gloves should be worn. The initial Golgi apparatus fraction is collected between 4000 and 12,000 g.

Comments. Plant Golgi apparatus isolated in the absence of coconut milk or glutaraldehyde exhibit a marked tendency to unstack. Purification beyond the first differential centrifugation step in the absence of coconut milk or glutaraldehyde results in preparations that consist largely of single, separated cisternae or aggregations of dictyosomes. Dictyosomes isolated in the presence of glutaraldehyde, and to a lesser

[54]D. J. Morré, J. Horst, S. Nyquist, and W. Yunghans, *Proc. Indiana Acad. Sci. 1967* 77, 154 (1968).
[55]Glutaraldehyde: A 50% stock solution (Fischer Scientific, Biological Grade) to which 0.16 mg/ml activated coconut charcoal is added to remove impurities.[56]
[56]D. J. Morré, H. H. Mollenhauer and J. E. Chambers, *Exptl. Cell Res.* 38, 672 (1965).

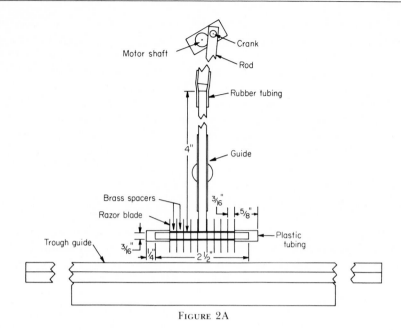

FIGURE 2A

FIGURE 2B

degree in the presence of coconut milk, have the same appearance as those of whole tissue fixed in glutaraldehyde.[22,54,57] Onion dictyosomes collect at a density of about 1.125 in linear sucrose gradients (Fig. 3) and account for about 2% of the total particulate nitrogen recovered from the homogenates. Preparations are obtained in quantity and are useful for morphological studies as well as certain types of chemical analyses and *in vivo* labeling studies with radioactive metabolites.[58] Enzyme activities are generally altered (inhibited or activated) by glutaraldehyde.[57] However, partial glutaraldehyde stabilization has been used to localize a phosphorylcholine-cytidyl transferase in dictyosomes of onion stem.[59]

Other Plant Tissues: Purification Procedures

The procedures described here for onion stem have been used with varying degrees of success for etiolated stems of mung beans (*Phaseolus aureus*), snap beans (*Phaseolus vulgaris*)[60] and garden peas (*Pisum sativum*); storage roots of radish (*Rhaphanus sativus*)[24]; inflorescence of cauliflower (*Brassica oleracea* var. *botrytis*)[22] and for an alga (*Acetabularia*).[61] A procedure described by Ray *et al.*[62] for etiolated stems of peas involves razor blade chopping of 2–4 g of 8-mm long segments cut from the third internode of etiolated pea seedlings. Homogenates are prepared at ice bath temperature in 4 ml of a medium containing 40 mM Tris, pH 8.0; 0.4 M sucrose; 10 mM KCl; 1 mM EDTA; 0.1 mM MgCl$_2$; 1 mM dithiothreitol; and 0.1% bovine serum albumin. The homogenate

---

[57]D. J. Morré and H. H. Mollenhauer, *Proc. Inidana Acad. Sci. 1968* **78**, 167 (1969).

[58]D. J. Morré, *Plant Physiol.* **45**, 791 (1970).

[59]D. J. Morré, S. Nyquist, and E. Rivera, *Plant Physiol.,* **45**, 800 (1970).

[60]H. H. Mollenhauer, unpublished results, (1969).

[61]G. Werz, unpublished results, (1969).

[62]P. M. Ray, T. L. Shininger, and M. M. Ray, *Proc. Nat. Acad. Sci. U. S.,* **64**, 605 (1969).

FIG. 2.   Diagram for construction of razor blade chopping device. The device consists of a horizontal steel bar holding ten single-edge razor blades separated by brass spacers. The blades and spacers are retained by a short length of plastic tubing at each end of the bar. The bar is soldered at the center to a length of 1/4 inch o.d. brass tubing which slides into a 5/16 inch o.d. bronze tube used as a guide. The 1/4-inch tubing is connected to a 3/8 × 1/8 inch copper rod which, in turn, is attached to a 5/16-inch offset crank pin. The motor, equipped with a 1:6 gear reducer (Palo Laboratory Supplies, Inc. New York, N. Y.) is wired in series with a potentiometer for further speed reduction. The motor and guide are supported by 1/2-inch brass tubing attached to a support rod on a 7 × 10 inch ring stand base. Closed at one end, the 2 3/8 × 14 inch trough is constructed from 1/4-inch Plexiglas and slides on a 19-inch Plexiglas guide bolted to the ring stand base. Dimensions are not critical except for the trough width and crank offset. This unpublished design is reproduced here by the courtesy of Ronald Gamble, Purdue University.

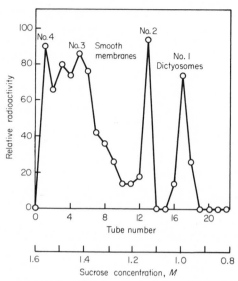

FIG. 3.   Linear sucrose gradient fractionation of the 4000–12,000 g pellet from 5 g onion stem labeled prior to homogenization for 4 hours with choline-1,2-[14]C. Glutaraldehyde[55,56] (50 mM) was added to the homogenate to stabilize the dictyosomes.

is squeezed through cheese cloth and cleared of debris by centrifugation for 2 minutes at 1000 g. Purification is achieved by rate zonal centrifugation followed by isopycnic sucrose gradient centrifugation. Fractions are first centrifuged in a 15–35% linear sucrose gradient for 25 minutes at 10,000 g. Material that has traveled about half as far as the mitochondrial band is then collected and layered on a 25–45% linear sucrose gradient followed by centrifugation for 120 minutes at 40,000 g. Dictyosome membranes band midway between the center of the gradient and the gradient-sample interface. The method yields preparations consisting largely of single cisternae plus aggregates of two to several cisternae.

Plant Tissues: Assay Procedures

Positive identification of isolated plant dictyosomes is presently restricted to examination of the preparations with the electron microscope either with positively stained thin sections of fixed and embedded pellets or by negative staining.[24] The gross morphology of plant dictyosomes is similar to that of animal dictyosomes already described.[25] Cytochemical studies suggest IDPase to be concentrated in plant Golgi apparatus.[63] Both IDPase and a glycosyl transferase are concentrated in the fractions described by Ray et al.[62]

[63]M. Douwalder, W. G. Whaley, and J. Kephart, *J. Cell Sci.* 4, 455 (1969).

[17] Preparation and Processing of Small Samples of Human Material

By James A. Boyle and J. Edwin Seegmiller

Introduction and Scope

The ever increasing number of human hereditary diseases now being precisely characterized in terms of a deficiency of a specific gene product provide the basic tools for greatly expanding our understanding of the biochemistry and genetics of diploid cells. Similar well characterized genetic and biochemical markers have been used in expanding our knowledge of bacterial biochemical genetics in the past. Search of the phylogenetic scale for a comparable group of well defined mutations for extending such studies to diploid organisms leads to the human species as the one in which the largest number of these have been characterized.

The detailed study of the "experiments of nature" presented in the human inborn errors of metabolism has already made substantial contributions to our understanding of a variety of biological processes through the perterbations created by a deficiency in the product of a single gene. The detailed characterization of the precise defect in gene product therefore provides the touchstone not only for making precise medical diagnoses but also for converting many puzzling hereditary diseases of obscure clinical interest into subjects of value for major scientific investigation.

The study of human enzyme defects is often beset at the very outset by the major problem of obtaining sufficient material either to assay the enzyme under study or to purify it. Recent years have witnessed increasingly good use being made of the little material that is available by the adoption of one or more of three approaches: (1) by the development of assays sensitive enough to measure enzymes in very small samples of human tissues obtained either by biopsy or by other means; (2) by the purification of the enzyme-containing tissue from other tissues thus concentrating the sample; (3) by the technique of tissue culture whereby the original sample may be grown *in vitro*, thus increasing the magnitude of available material manyfold.

We herein survey some aspects of the applications of these three approaches. It is clearly impossible to cover all areas of this field as it relates to the study of human enzymes. We have therefore concentrated on a description of methods with which we ourselves are familiar and which have been used or are in use in our own laboratory.

Small Samples of Human Tissue

The methods for obtaining small samples of human tissue are too numerous to discuss in detail here. Suffice it to say that samples weighing from 1 to 100 mg can be obtained from many body organs by biopsy. These organs include the liver, kidney, large and small bowel, stomach, brain, conjunctiva, lacrimal, and salivary glands, the mucous membrane of the lip, the thyroid gland, the lung, peripheral nerve, bone, skeletal muscle, and skin. Moreover, many enzymes may be conveniently studied in blood which can be obtained in relatively large amounts from one individual over a period of time. We shall discuss the preparation of samples of tissue obtained by intestinal biopsy, hepatic biopsy, or from clippings of the stratum corneum around the fingers; the technique of assay of stable erythrocyte enzymes using a filter paper collection technique will also be described.[1]

Intestinal Biopsy

The activity of a number of enzymes in human intestinal mucosa has been studied increasingly in recent years,[2] and the ease and safety with which such tissue can be obtained by peroral biopsy indicates that it will be used more extensively for biochemical studies in the future. Jejunal mucosa may be sampled with a Rubin or a Crosby[3] intestinal biopsy capsule. Samples of tissue weighing from 20 to 100 mg may be so obtained. The sample is removed from the capsule and placed on moistened filter paper in a petri dish on ice; excess fluid is blotted away, and the sample is weighed. If the enzyme in question is known to be stable when stored frozen, the sample may then be wrapped in Parafilm and stored at $-20°$ in a small plastic or glass tube with a snap on or screw on lid. The Parafilm ensures against weight loss from the stored sample probably by decreasing the air space around the biopsy to a minimum, thereby preventing sublimation of tissue water into the relatively large and dehydrated atmosphere of the container. If this simple precaution is taken, tissue weight can be as accurate a basis of measurement as protein content,[4] although a protein determination is usually performed as well. The treatment of the sample thereafter depends on

---

[1] W. Y. Fujimoto, M. L. Greene, and J. E. Seegmiller, *J. Pediat.* 73, 920 (1968).
[2] A. Dahlqvist, *Anal. Biochem.* 7, 18 (1964).
[3] R. B. W. Smith, H. Sprinz, W. H. Crosby, and B. H. Sullivan, Jr., *Amer. J. Med.* 25, 391 (1958).
[4] W. M. Walter, Jr., and G. M. Gray, *Gastroenterology* 54, 56 (1968).

the nature of the enzyme. We shall outline as an example the preparation of tissue for the assay of xanthine oxidase.[5]

The specimen of jejunal tissue is placed in cold 0.154 M potassium chloride and homogenized in a TenBroeck glass tissue grinder (2 ml capacity) in an ice bath, the final dilution (w/v) being in the range 1:25 to 1:40. The homogenates are centrifuged for 60 minutes at 105,000 g at 4° and the supernatant fluid is dialyzed 1 hour against 500 volumes of 0.154 M KCl at 4°.

The xanthine oxidase activity of the homogenate may be determined with hypoxanthine-8-^{14}C as the substrate: incubate portions of the homogenate (0.1–0.2 ml) in stoppered glass tubes with 0.07 ml of 0.07 M Tris buffer pH 8.1 containing 0.34 μCi (0.45 μmole) of hypoxanthine-8-^{14}C for 1 hour at 37°. Stop the reaction by immersing the reaction tubes in boiling water for 2 minutes. Remove the denatured protein by centrifugation and separate the reaction products (xanthine and uric acid) by high voltage electrophoresis (4000 V for 2–3 hours at 25°; after applying 50 μl with 3 μg of carrier purines on Whatman 3 MM filter paper in 0.05 M borate buffer, pH 9.0, with 1 mM EDTA). Localize the purines on the paper under UV light. Cut them out and count in 17 ml of toluene phosphore containing 2,5-diphenyloxazole (0.45% w/v) and 1,4-bis-2(5-phenyloxazolyl)benzene (0.01% w/v). This radiochemical method allows the measurements of very low activity of xanthine oxidase activity in small amounts of crude tissue homogenates.[5]

Hepatic Biopsy

The advent of the thin-beveled Menghini needle,[6] has reduced the intrahepatic phase of liver biopsy to 1 second, thus minimizing the possibility of complications, yet allowing adequate tissue for some enzymatic studies. The technique can be used safely with pediatric patients as well as with adults[7]; 10–15 mg of hepatic tissue usually can be obtained.

The sample should be immediately separated from blood and visible connective tissue, blotted dry and weighed on a torsion balance. Many enzymes in the liver are stable for long periods if the sample is covered with parafilm and kept frozen at −20° in a vessel with the least possible air space.

As with jejunal tissue, the further handling of the biopsy material depends on the enzyme activity that is to be assayed. Since the amount

[5]K. Engelman, R. W. E. Watts, J. R. Klinenberg, A. Sjoerdsma, and J. E. Seegmiller, *Amer. J. Med.* 37, 839 (1964).
[6]G. Menghini, *Gastroenterology* 35, 190 (1958).
[7]W. A. Walker, W. Krivit, and H. L. Sharp, *Pediatrics* 40, 946 (1967).

of hepatic tissue obtained may be somewhat less than the amount of jejunal tissue, it is homogenized in a TenBroeck homogenizer of 1–2 ml capacity. For example, 2 ml of cold 0.154 M KCl is used per gram of liver. Spin the homogenate at 12,000 g for 20 minutes at 4° and dialyze the supernatant liquid with mechanical agitation for 2 hours against the same solution in the cold. Xanthine oxidase activity may be determined by assaying small aliquots of the liver homogenate in the presence of xanthine-6-^{14}C and excess uricase in pyrophosphate buffer (pH 8.3, 0.1 M) and measuring the amount of $^{14}CO_2$ produced.[5]

Stratum Corneum

Anyone who has used a pair of nail clippers is familiar with the observation that small pieces of keratinized skin (stratum corneum) may be removed painlessly from around the ends of the fingernails (Fig. 1). Approximately 25 mg of clippings can be removed in this way from an adult hand and this can serve as a source of tissue enzymes whose activity may not be demonstrable in blood. For example, stratum corneum of normal skin is a rich source of histidine α-deaminase.

Procedure. The assay of histidine α-deaminase is based on the accumulation of urocanic acid with an absorption maximum at 277 mμ at pH 9.2.[8] However, because of the turbidity of homogenates of human stratum corneum, direct spectrophotometry cannot be used.

Homogenates are prepared by grinding 20 mg of the tissue with 1.0 ml of 10 mM sodium pyrophosphate buffer pH 9.2 in a glass-walled TenBroeck-type tissue grinder (2-ml capacity).[10] One milliliter portions of the homogenate are dialyzed at 4° against 1 liter of 5 mM sodium pyrophosphate buffer pH 9.2 containing 0.1 ml of 0.1M glutathione (305 mg dissolved in 10 ml of 0.1M sodium pyrophosphate buffer) for 1 hour. Under these conditions over 90% of the urocanic acid is removed.

Incubate both control and experimental tubes with 0.50 ml of 0.1 M sodium pyrophosphate buffer, pH 9.2, and 0.05 ml of 0.1 M glutathione (prepared as described above). The experimental tubes receive 0.05 ml of 0.1 M L-histidine (neutralized with 1 N NaOH) and the control tubes 0.05 ml 0.1 M sodium pyrophosphate buffer, pH 9.2. Add 0.4 ml of the homogenate of the stratum corneum to both the control and experimental tubes, and adjust the total volume of each tube to 1.2 ml with

[8]B. N. LaDu, R. R. Howell, G. A. Jacoby, J. E. Seegmiller, and V. G. Zannoni, *Biochem. Biophys. Res. Commun.* 7, 398 (1962).
[9]H. Tabor and A. H. Mehler, in this series, Vol. II, p. 228.
[10]V. G. Zannoni and B. N. LaDu, *Biochem. J.* **88**, 160 (1963).

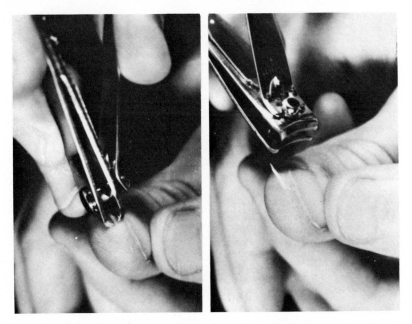

FIG. 1.   Removal of strateum corneum from the finger tips with a pair of nail clippers. The operation is painless. Approximately 25 mg of tissue can be obtained from each hand in this manner.

water. Remove 0.2-ml samples at zero time and after incubation at 37° for about 2 hours, the time depending on the amount of enzyme activity present.

Deproteinize the samples immediately by the addition of 0.4 ml of 7% (w/v) perchloric acid and maintain in an ice bath for 10 minutes. Centrifuge at 3000 g for 10 minutes and measure the supernatant fluid at 277 mμ. One microgram of urocanic acid per milliliter in a 10-mm light path gives a reading of 0.260. Under the conditions of the assay it is possible to detect accurately the formation of less than 0.25 μg of urocanic acid per hour.

Filter Paper Technique for Collection of Blood Samples

Some of the erythrocyte enzymes that are relatively stable may be measured in samples of blood collected directly onto filter paper. This technique is particularly useful for preliminary screening tests to be performed on infants or in screening tests of inmates of mental institutions since the procedure does not require a venipuncture. Blood is obtained from a single finger stick (or heel stick in infants), spotted directly onto a Whatman No. 3 filter paper and allowed to soak completely through the paper to obtain a spot approximately 1.5 cm in diameter. Multiple separate spots may thus be obtained, but each spot should be the result of only a single application of the finger to the paper. The spots are allowed to dry and may then be carried or mailed without refrigeration to the laboratory.[1] If the assay is to be delayed, the sample may be stored in a plastic bag at $-5°$ to retard deterioration of the enzyme activity.

Replicate, uniform-sized disks are cut out from the center of each blood spot using a 3-mm skin biopsy punch, a paper punch, or a cork borer. These may then be stored until the assay is to be performed. This method has been applied, for example, to the measurement of hypoxanthine phosphoribosyltransferase (inosinic acid pyrophosphorylase) in individuals suspected of having diminished activity of this enzyme. Each small disk is placed in a disposable test tube (12×75 mm Falcon Plastics Corporation, Los Angeles, California) and soaked in 50 μl of 0.01 M Tris buffer, pH 7.4, for 30 minutes. Add 50 μl of a previously prepared substrate mixture containing 10 μl 0.5 M Tris buffer, pH 7.4, 5 μl of 0.1 M $MgCl_2$, 10 μl of 9 mM 5-phosphoribosyl 1-pyrophosphate solution, 15 μl of 4.56 mM hypoxanthine-8-^{14}C solution (3.8 mCi/mmole) and 10 μl of water. The mixture is incubated in a shaking water bath at $37°$ for 30 minutes and the reaction is terminated by the addition of 20 μl 0.1 M EDTA, pH 7.4, to each tube. Gas-tight syringes, originally designed for gas chromatography, and repeating dispensers (Hamilton Company, Whittier, California) are used to make all additions to the assay tubes. The product, inosinic acid (IMP), is separated from hypoxanthine either by applying 20 μl over a 3 μg IMP marker on Whatman 3 MM paper using electrophoresis at 4,000 volts in 50 mM borate buffer, pH 9.0 containing 1 mM EDTA for 30 minutes, or by thin-layer chromatography, applying 3 μl over a 3 μg IMP marker on cellulose (Eastman Chromagram Sheet 6065, Eastman Kodak, Rochester, New York) using 1.6 M LiCl for development. The spots corresponding to the IMP markers are located under UV light, cut out, and counted in a liquid scintillation counter.

Concentration of the Enzyme-Containing Tissue

The human tissue to which this approach is applied most often is blood. Serum or plasma contain the enzymes to be assayed for the diagnosis of a number of human genetic defects, and no detailed description will be given of these procedures. Many enzymes are present in the lymphocyte or granulocyte that are not present in the red cell. Although a 20-ml sample of blood may contain sufficient white cells with enough total enzyme for measurement by the assay available, the activity may be so diluted by the large volume of plasma and erythrocytes that these must be discarded. In many cases, it is desirable to have only a particular fraction of blood for study.

Fractionation of Whole Blood

The recent interest in the chemistry of the human lymphocyte, granulocyte and platelet has been facilitated by the development of simple procedures for the isolation of these elements from blood.

Isolation of Leukocytes from Whole Blood — Survey of Methods and Principle

The leukocyte population of heparinized blood consists of a mixture of granulocytes, lymphocytes, and monocytes. A variety of techniques have been used to separate the white cells from whole blood. The most useful procedures utilize differences in the sedimentation behavior between leukocytes, platelets, and erythrocytes. The sedimentation of the red cells is accelerated by the use of agents which agglutinate them. Dextran in a high molecular weight form has been widely used for this purpose,[11] but other agents, such as phytohemagglutinin,[12] fibrinogen,[11] and polyvinylpyrrolidone,[13] have also been employed. More recently Ficoll[14], hydroxymethylcellulose or methylcellulose[15] have also been used to aggregate red cells. Most of the techniques for the isolation of white cells from whole blood using gravitational techniques with or without procedures causing selective erythrocyte destruction, have been reviewed by Maupin,[16] by Walford,[17] and more recently by Böyum[18]. We shall describe here the dextran mixing method of Skoog and Beck[11] and the technique of Böyum[18] which is based on the agglutination of erythrocytes by dextran and their differential sedimentation through a layer of Hypaque.

DEXTRAN MIXING METHOD[11]

Materials

Dextran, molecular weight 275,000 (Mann Research)

NaCl, 0.9%

Heparinized fresh human blood collected in siliconized glass or in plastic syringes. All glassware should be siliconized.

[11] W. A. Skoog and W. S. Beck, *Blood* 11, 436 (1956).
[12] J. G. Li and E. E. Osgood, *Blood* 4, 670 (1949).
[13] K. Ulrich and G. E. Moore, *Acta Haematol.* 35, 338 (1966).
[14] A. W. Richter, *Acta Physiol. Scand.*, 59, Suppl. 213, 130 (1963).
[15] A. Böyum, *Nature (London)* 204, 793 (1964).
[16] B. Maupin, *Rev. Hematol.* 14, 250 (1959).
[17] R. L. Walford, "Leucocyte Antigens and Antibodies," Chapter I, Grune and Stratton, New York, 1960.
[18] A. Böyum, *Scand. J. Clin. Lab. Invest.*, Suppl. 97, 31 (1968).

A 3% dextran solution is made with 0.9% saline. Solutions should be prepared fresh because dextran, although stable, is an excellent medium for mold growth. Alternatively, solutions may be made up using sterile technique and stored at 4°C prior to use.

A 2:1 dextran solution:blood mixture is made in a plastic or siliconized glass syringe. The final volume may be as little as 5 ml. Mix thoroughly by inverting repeatedly. Avoid foaming which causes hemolysis of the red cells and traps leukocytes. If foaming does occur, expel the froth through the needle. Mixing and all other procedures may be carried out either at 0° or at 25° because dextran-induced agglutination of the red cells is not temperature dependent within this temperature range. Stand the syringe vertically with needle upright for 20 minutes. The course of red cell sedimentation may be followed if the syringe is placed in a rack with back lighting. As soon as the red cells have sedimented, expel the overlying plasma-dextran suspension of leukocytes through a needle bent to an angle of 120°, taking care not to disturb the sedimented cells. The leukocytes may now be centrifuged at 250 g for 5 minutes. The clumps of cells in the pellet may then be broken with a fine pipette using a small volume of appropriate physiological salt solution. A gentle stream of fluid is directed at the cells, which may be aspirated gently back and forth into the pipette.

The dextran method which yields platelets as well as leukocytes affords about a 70% yield of leukocytes and removes approximately 99% of the erythrocytes. The mean cell sedimentation time is of the order of 11 minutes when the final volume of the red cell–dextran mixture is 5 ml. This figure increases to 18 minutes if the final volume is 90 ml. If a better yield of leukocytes is desired, virtually 100% recovery can be obtained by using a preparation of bovine fibrinogen containing 40–50% citrate by weight (Sigma Chemical Company) as the red cell agglutinating agent. A 6% solution of fibrinogen in 0.9% saline is used, the ratio of fibrinogen solution:blood being 1:1 or 2:1. Erythrocyte sedimentation rates are faster than with dextran, but the degree of contamination of the pellet with red cells is about the same; the disadvantage of the fibrinogen method is that it cannot be used at 0°. Because of the increase in the viscosity of fibrinogen which occurs at the lower temperature, sedimentation of the cells does not take place.

Hypaque-Dextran Two-Phase Method[15]

One advantage of this method is that it is suitable for smaller volumes of whole blood of the order of 2 ml. Moreover, since the dextran is not mixed with blood its concentration is about 1/15 to 1/20 that obtained with the mixing method. Another advantage to the Hypaque-dextran two-phase system is the more efficient removal of the red cells. The principle of both methods is essentially the same although, in the two-phase system, erythrocytes are not mixed with the agglutinating agent but are layered on top of it and are aggregated by Hypaque-dextran at an interphase.

Reagents

    Hypaque (sodium *N*-methyl-3,5-diacetamido-2,4,6-triiodobenzoate) (Winthrop Laboratories, New York)

    Dextran (Mann Research)

A 33% solution of Hypaque is made with distilled water. This may be stored at 4° as a stock solution. Ten parts of this are added to a 6% dextran solution containing 0.9% NaCl. The volume of the Hypaque-dextran mixture should be at least 75% of the volume of blood. Venous blood anticoagulated with EDTA (1 part of 10% EDTA, pH 7.4, to 10 parts of blood) is layered gently onto the Hypaque-dextran mixture in tubes with an internal diameter of 5–10 mm. The mixture has a density of about 1.070 g/ml. It thereby prevents sinking of whole blood, whose density is approximately 1.055 g/ml, but does permit the erythrocytes with their greater density to sink. Red cells are clumped by dextran at the interface and fall to the bottom of the tube. With time, the erythrocytes sediment as one large clump while the white cells remain in the overlying plasma layer. When the red cell mass has just passed through the interface between plasma and Hypaque-dextran, the leukocyte-rich layer of plasma may be removed with a pipette. The yield of white cells is about 40%.

Separation of Lymphocytes from Granulocytes

Many methods of separating lymphocytes from granulocytes have been described. Most of the techniques are somewhat complicated. Some approaches have been based on the difference in behavior between lymphocytes and granulocytes in an albumin density gradient[19] or have utilized a gelatin sedimentation technique.[20] Other workers have explored the ability of granulocytes to ingest iron particles in order to remove these cells from lymphocytes by reason of their acquired increase in density[21] or by reason of their behavior in a magnetic field.[22,23]

[19]A. I. Spriggs and R. F. Alexander, *Nature (London)* **188**, 863 (1960).

[20]A. S. Coulson and D. G. Chalmers, *Lancet* **1**, 468 (1964).

[21]B. Cassen, J. Hitt, and E. F. Hays, *J. Lab. Clin. Med.* **52**, 778 (1958).

[22]J. Hastings, S. Freedman, O. Rendon, H. L. Cooper, and K. Hirschhorn, *Nature (London)* **192**, 1214 (1961).

[23]K. Carstairs, *Lancet* **1**, 829 (1962).

The demonstration that granulocytes can be retained on columns of siliconized glass beads[24,25] has led to the use of a variety of column techniques utilizing glass wool,[26] nylon,[27] polystyrene,[28] or cotton[29] to retain granulocytes while allowing the passage of lymphocytes. We shall discuss here the use of a packed nylon wool column. The method is a slight modification of that proposed by Brittinger and co-workers.[30]

Reagents

Dextran, 6% in 0.9% saline

Eagle's Minimal Essential Medium (Grand Island Biological Company)

Nylon wool (Leucopak, Fenwall Laboratories)

Procedure. Twenty-five milliliters of blood are mixed with an equal volume of 6% dextran in a 0.9% sodium chloride solution and allowed to settle at 37° for approximately 45 minutes in a siliconized glass centrifuge tube or using a syringe as described above under Dextran Method. The supernatant liquid is withdrawn and is mixed with an equal volume of Eagle's Minimal Essential Medium. The mixture is allowed to pass slowly over a 10 × 1 cm column of tightly packed nylon wool. To minimize the number of lymphocytes rendered nonviable by this procedure, the nylon wool should be washed repeatedly in distilled water for a week prior to use and rinsed with isotonic saline. In addition the packed column should be preincubated at 37° to ensure that it is at the same temperature as the leukocyte-rich plasma. The column is washed with a half to one bed volume of medium for 1 hour at 37° after the application of the plasma. The total eluate is combined and centrifuged at 200 *g* in 50-ml plastic centrifuge tubes (Falcon Plastic Corporation, Los Angeles) at 4°. The cell pellet is washed four times with cold 0.9% sodium chloride. Trypan blue staining may be used as one index of cell viability following this procedure. Three drops of cell suspension and one drop of 1% trypan blue (Grand Island Biological Company) are incubated at 37° for 30 minutes. Nonviable cells will

[24]J. E. Garvin, *J. Exp. Med.* 114, 51 (1961).

[25]Y. Rabinowitz, *Blood* 23, 811 (1964).

[26]L. Brandt, J. Börjeson, A. Nordén, and I. Olsson, *Acta Med. Scand.* 172, 459 (1962).

[27]K. Hirschhorn and C. S. Ripps, in "Isoantigens and Cell Interactions" (J. Palm, ed.), p. 57. Wistar Institute Press, Philadelphia, 1965.

[28]A. E. R. Thomson, J. M. Bull, and M. A. Robinson, *Brit. J. Haematol.* 12, 433 (1966).

[29]J. O. Lamvik, *Acta Haematol.* 35, 294 (1966).

[30]G. Brittinger, R. Hirschhorn, S. D. Douglas, and G. Weissmann, *J. Cell Biol.* 37, 394 (1968).

stain with the dye. In our hands[31] this technique gives a population of lymphocytes with a granulocyte contamination of not more than 2%. Over 99% of the cells fail to stain with trypan blue, suggesting that the bulk of the cell population is viable. The eluent contains more red cells than white cells.

Removal of Contaminating Erythrocytes

An erythrocyte lysing procedure has been described which, it is claimed, has no apparent effect on white cells.[32] Quickly add 6 ml of cold distilled water to a 2-ml suspension of cells in 0.9% sodium chloride. Stir vigorously for exactly 30 seconds and add 2 ml of 3.5% aqueous NaCl solution for each original 2 ml of cell aliquot to restore isotonicity. The cells may be centrifuged at 380 g for 10 minutes; the supernatant liquid, containing hemoglobin and debris from the hemolysate, is discarded. A variation of this lytic procedure applied directly to whole blood achieves a primary separation of leukocytes from erythrocytes.[33]

Isolation of Granulocytes

Granulocytes may conveniently be isolated by the procedure described by Böyum.[34]

Reagents

Hypaque (Winthrop Laboratories); a stock solution of 33% is used
Ficoll (Pharmacia Fine Chemicals Inc., New Jersey); a sucrose polymer is used as a 9% aqueous solution
Dextran (Mann Research) used as a 4.5% solution in 0.9% NaCl

Procedure. Venous blood is collected in heparin to give a final heparin concentration of 10 USP units per milliliter of blood. Eight milliliters of a mixture of anticoagulated blood and 0.9% NaCl (1 part blood to 3 parts saline) are layered carefully onto 3 ml of a mixture of 33% Hypaque and 9% Ficoll (10:24) in a siliconized or heparinized round-bottomed glass tube with an internal diameter of at least 13 mm. The layering may conveniently be accomplished with a syringe to which is attached thin plastic tubing. The orifice of the tubing should remain against the side of the tube, approximately 10 mm above the meniscus,

---

[31] J. D. Schulman, V. G. Wong, T. Kuwabara, K. H. Bradley, and J. E. Seegmiller, *Arch. Int. Med.* 125, 660 (1970).
[32] H. J. Fallon, E. Frei, J. D. Davidson, J. S. Trier, and D. Burk, *J. Lab. Clin. Med.* 59, 779 (1962).
[33] J. A. Schneider, K. Bradley, and J. E. Seegmiller, *Science* 157, 1321 (1967).
[34] A. Böyum, *Scand. J. Lab. Clin. Invest.*, Suppl. 97, 77 (1968).

during addition of the Hypaque-Ficoll solution. The tube is then centrifuged at 20° for 40 minutes at 400 g. After centrifugation, lymphocytes and monocytes (mononuclear layer) will be seen as a white ring at the interface between the Hypaque-Ficoll mixture and the plasma; the bottom fraction contains red cells and granulocytes. The mononuclear layer may be removed with the plasma and about half of the Hypaque-Ficoll mixture using a Pasteur pipette. To remove these cells entirely, it is necessary to move the pipette over the whole cross area of the tube because many of the mononuclear cells are located on the wall of the tube. To obtain the granulocytes, remove the remainder of the Hypaque-Ficoll down to 1–2 mm above the erythrocyte-granulocyte pellet. One milliliter of cell-free plasma and 0.4 ml of 4.5% dextran in 0.9% NaCl solution are added to the pellet; the contents of the tube are mixed thoroughly but gently by inversion, taking care to avoid foaming. The mixture is transferred to another tube, and the erythrocytes are allowed to settle at 4°. The plasma supernatant liquid, which now contains the granulocytes, is removed with a Pasteur pipette or using a syringe for sedimentation as described above under Dextran Method. There is about one red cell per white cell in this preparation. If contamination with red cells is considered undesirable these can be removed by hypotonic lysis as described above.

The yield of granulocytes obtained by this procedure is approximately 50% using heparin as the anticoagulant and may be increased to 60% if EDTA is substituted for heparin in a ratio of 1 ml 10% EDTA, pH 7.4, to 50 ml blood. Contamination with mononuclear cells is of the order of 1%. The yield of mononuclear cells in the first step is approximately 98%. There is less than 0.1% contamination with granulocytes and less than 10% with red cells using EDTA. If heparin is used, this value is increased to 15%. If centrifugation is carried out at 4°, the purity of both fractions is decreased.

Tissue Culture Techniques

Tissue culture undoubtedly constitutes one of the most powerful techniques for the most efficient utilization of small samples of human tissue. General methods for the colonial growth of isolated mammalian cells have previously been discussed in this series.[35] In this section we confine ourselves to a description of methods for setting up a primary culture of human fibroblasts, the technique of human amniotic cell culture, a description of a method of cloning single fibroblasts and outline a procedure for the establishment of a long-term suspension culture

[35] R. G. Ham and T. T. Puck, this series, Vol. 5, p. 90.

of a strain of euploid human lymphoid cells. In the discussion that follows we assume familiarity with general tissue culture and aseptic techniques and the need to check cultures at intervals for contamination with mycoplasma. These methods allow in theory, and indeed in practice, the capability of obtaining relatively enormous quantities of material from a very small piece of human tissue weighing in many instances less than 0.1 g despite the limited life span of 20–60 generations for euploid cultured human fibroblasts.

The Establishment of Fibroblast Cultures from a Skin Biopsy

After cleansing of the skin with Phisohex, followed by alcohol, and infiltration with a local anesthetic, a small disk of skin is incised with a sterile 3-mm biopsy punch. Alternatively, a tent of anesthetized skin may be raised with toothed forceps and snipped off with sterile scissors or scalpel. The sample is immediately placed in a tube containing 10 ml of medium 199 with 20% nondialyzed fetal calf serum (Grand Island Biological Company). This may be kept at room temperature until it reaches the laboratory. In many cases it is quite feasible to send the sample via airmail. Successful cell explants have been obtained after as long as 6 days in transit, from donors who have been dead as long as 24–48 hours, or by using human serum from AB positive blood type individuals in lieu of the culture medium described above.

On arrival in the laboratory, the sample is placed in a sterile watch glass or petri dish and covered with 1–2 ml of medium 199. It is minced with sterile scissors and forceps into a number of very small pieces (10–20). These are transferred with a pipette to a screw-top plastic flask of 30-ml capacity (Falcon Plastics No. 3024). The flask is incubated at 37° and is initially tipped on its side to allow the medium to drain freely from the fragments of minced tissue. This procedure favors fixing of the tissue to the flask. After a period of 90 minutes on its side, 20 ml of medium 199 with 20% fetal calf serum and 50 μg of streptomycin and 50 units penicillin per m is introduced gently down the side of the flask, care being taken to avoid dislodging the newly affixed tissues.

After 3–4 days, microscopic examination will show an outgrowth of epithelial cells from several of the fragments that have become fixed. These are round, oval, or irregularly shaped cells. Some 3 days later, characteristically spindle-shaped fibroblasts appear (Fig. 2). The medium should be changed gently at this time and twice weekly thereafter. After 3–4 weeks from the time of establishment of the primary explant, the culture may be passed into another identical flask. The trypsinization of the cells at this stage should be extremely gentle, as

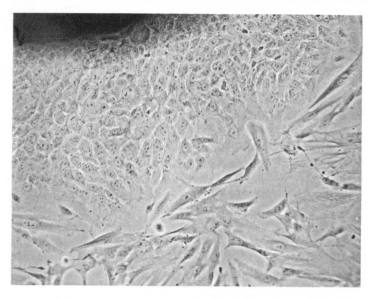

FIG. 2.   Outgrowth of fibroblasts and epithelial cells from a skin biopsy. The main mass of the biopsy is at the top left-hand corner of the picture. The cells nearer the mass are epithelial cells. The spindle-shaped elongated cells farther away are human fibroblasts. Phase contrast; × 100.

the aim is to dislodge the fibroblasts that have grown out while leaving the epithelial cells (which tend to die in this medium in any case) and the main mass of fixed tissue behind. Once this has been done, more medium is added to the original flask to allow further outgrowth of fibroblasts from the primary explants. If the cells are growing properly with a generation time of 24 hours and have been harvested correctly they may be expected to reach confluency within 3 days of the first passage. At this time they should be passed into 2 flasks of 250-ml volume. These in turn may each be transferred into 2 flasks within 3–4 days. Once the culture is firmly established, the strain may be transferred to 32-oz glass prescription bottles by adding the contents of 2 flasks into one bottle. The average time from the "taking" of the skin biopsy to the establishment of the strain in glass bottles is 6–7 weeks.

Cloning Diploid Mammalian Cells

In some instances in order to gain a better understanding of the functional, genetic or morphological differences between individual cells it becomes necessary to develop strains grown from a single parental cell, i.e., to clone a single cell. Cloning techniques also offer the capability of developing strains derived from cells which, in conventional mass cultures, may be overgrown by more hardy or more rapidly growing cells.

A number of methods for single cell cloning have been described, notably by isolation of single cells in a capillary tube,[36] by the implantation of single cells in drops of medium under mineral oil,[37] by isolation of a single cell or clone within a glass or plastic well that separates it from the rest of the culture and permits its removal by trypsinization,[38] by isolation of single cells on glass squares,[39] or by selection of cell clones from sloppy agar.[40] These methods have been successful in cloning aneuploid cell lines with a high rate of efficiency (approximately 100% in most cases); the efficiency in cloning euploid strains has been somewhat lower. Here, we describe the technique of Freeman et al.[41] as practised in our laboratory with human diploid fibroblast strains.

Reagents

Cloning medium: 1 part medium 199; 1 part NCTC 109 with 30% nondialyzed fetal calf serum (all from Grand Island Biological Company); 100 μg of neomycin per milliliter; 10 units of mycostatin per milliliter

Dulbecco's buffered phosphate saline, Grand Island Biological Company (GIBCO)

Trypsin, 0.25%, in GIBCO solution A

Trypan blue, 0.4% in Hanks' basic salt solution (Grand Island Biological Company)

Glass chips prepared from Corning No. 1 coverslips

Plastic petri dishes (60 × 15 mm) (Falcon Plastics)

Procedure. Fibroblasts in 250-ml Falcon flasks are considered ready for cloning if cells have been undergoing division for the previous 2–3 days as judged by microscopy. The growth medium is removed on the day before cloning and is replaced with 20 ml of cloning medium. To clone, the medium is decanted and the cells are washed twice with 2 ml of Dulbecco's phosphate-buffered saline. One to 2 ml of 0.25% trypsin is introduced into the flask and swirled gently. The excess is poured off and the flask is incubated for 30–60 seconds at 37°. Examine the cells

[36]K. K. Sanford, A. B. Covalesky, L. T. Dupree, and W. R. Earle, *Exp. Cell Res.* 23, 361 (1961).
[37]A. Lwoff, R. Dulbecco, M. Vogt, and M. Lwoff, *Virology* 1, 128 (1955).
[38]T. T. Puck, S. J. Cieciura, and H. W. Fisher, *J. Exp. Med.* 106, 145 (1957).
[39]D. M. Schenck and M. Moskowitz, *Proc. Soc. Exp. Biol. Med.* 99, 30 (1958).
[40]I. MacPherson and L. Montagnier, *Virology* 23, 291 (1964).
[41]A. E. Freeman, T. G. Ward, and R. G. Wolford, *Proc. Soc. Exp. Biol. Med.* 116, 339 (1964).

under the microscope for signs of detachment (rounding up) and then hasten the process by striking the edge of the flask vigorously a few times against a hard edge. It is important not to overtrypsinize the cells. Five milliliters of cloning medium is then added to stop the action of the trypsin, and the cells are vigorously dispersed by sucking the medium back and forth into a pipette. We then put the cell suspension into a glass centrifuge tube. From this moment on, speed is essential because 50% of the cells will attach to the wall of the tube within 5 minutes. Since there is usually a sufficient number of cells, it is not necessary to rinse the flask. The cells are again vigorously dispersed in the centrifuge tube and allowed to stand for 1 minute to allow any clumps that may have escaped dispersion to settle. At this point we make a quick count of the viable cells in the following way: 0.2 ml of the cell suspension is added to 1.8 ml of saline. Add 1 ml of trypan blue. After mixing, place a drop under both ends of a hemacytometer and count the number of viable cells (those which are not stained) in the 4 outside large squares and in the middle square at both ends of the counting chambers (total of 10 squares counted). The number of cells per milliliter of undiluted medium is given by $C \times 15 \times 1000$ where C is the total cell count in the 10 squares. The counts at each end of the hemacytometer should be within 20% of each other, and less than 10% of the cells should be dead. After counting, dilute the cells in cloning medium to yield a concentration of 5×10^2 per milliliter. About 60–70 cells per 10 squares, a total of 10^6 cells per milliliter, requires three serial dilutions, 1:10, 1:10, and 1:20 to obtain the desired concentration. It is important to have this approximate cell density: too few cells reduces the chances of cloning considerably and too many reduces the chance of finding a glass chip with only one cell on it. One milliliter of cell suspension is then placed in each of 5 separate Falcon plastic petri dishes (60 × 15 mm) containing a single layer of glass chips and 4 ml of cloning medium are added to each dish. The chips are prepared by grinding No. 1 cover glasses (Corning) in a mortar and pestle to a size which is sufficient to pass through a size 10, but not a size 20, mesh screen. The chips are placed in a small glass specimen jar and washed 3 times with distilled water, 3 times with 70% ethanol and once with ether. After ether has evaporated, the chips are thoroughly dried in an oven and sprinkled into a series of petri dishes so as to provide a single layer. The plates are autoclaved.

Petri dishes containing the cells are incubated in 5% CO_2 and air in a humidified incubator, the chips are checked in 24–48 hours by microscopy. Chips with only one cell attached are selected for cloning. With care and practice it is possible to obtain 20 or so satisfactory chips from each dish. The chip is placed in a fresh plate to which medium has previously been added so that the chip does not stick to the forceps. Twenty to 30 plates are seeded with single glass chips in this manner. Because the cells are transparent and thus difficult to see against the glass, the location of each cell on the chip is marked on a drawing. This aids in the rapid scrutiny of the chip on successive days but is not an absolute safeguard since the cell may migrate widely. Cells should show some evidence of division one day after plating. Human fibroblasts have a generation time of approximately 24 hours: if 4 or more cells can be seen at this early stage the culture should be discarded since this is evidence that the products are not derived from a single cell.

Amniotic Fluid Cell Culture

The increasing use of the technique of amniocentesis affords the opportunity of obtaining material which may be used for studies of enzyme development in man. This knowledge has been applied to the prenatal detection of an increasing number of serious hereditary diseases that have a basis in biochemical aberrations.[42-44] Moreover, a knowledge of fetal sex, which is of importance in sex-linked human disorders, may be rapidly obtained from the examination of amniotic cells for heterochromatin bodies but should be confirmed by karyotype; karyotyping for detection of Down's syndrome and other chromosomal abnormalities is also feasible.[45,46] Some of these approaches require large quantities of cells necessitating the culture of human fetal cells derived from amniotic fluid.[47] Rather than deal with all the variations in techniques that have been reported for culturing cells, we shall describe the approach in use in our laboratory.

The amniotic fluid is maintained at room temperature while it is being transferred to the laboratory; if chilled, cells tend to grow less well. When the volume of the sample is approximately 10 ml, transfer it into a 250-ml Falcon flask with an equal volume of a mixture of 3 parts medium 199, 3 parts NCTC-135, and 4 parts fetal calf serum with 100 μg of neomycin per milliliter. If the volume of the fluid is more than 10 ml, it is centrifuged at 20° for 30 minutes at 250 g and the cell pellet is resuspended in 10 ml of the supernatant liquid. Smaller samples,

[42]H. L. Nadler, *Biochem. Genet.* 2, 119 (1968).
[43]H. L. Nadler and A. B. Gerbie, *N. Engl. J. Med.* 282, 596 (1970).
[44]J. A. Boyle, K. O. Raivio, K. H. Astrin, M. Graf, J. E. Seegmiller, and C. B. Jacobsen, *Science* 169, 688 (1970).
[45]A. P. Amarose, A. J. Wallingford, and E. J. Plotz, *N. Engl. J. Med.* 275, 715 (1966).
[46]C. Valenti, E. J. Schutta, and T. Kehaty, *J. Amer. Med. Ass.* 207, 1513 (1969).
[47]C. B. Jacobson and R. H. Barter, *Amer. J. Obstet. Gynecol.* 99, 796 (1967).

such as 1–2 ml, are placed in a 30-ml capacity Falcon flask (No. 3012) and receive an equal volume of medium. Incubation is at 37°, the medium being changed after 5 days. Some of the amniotic cells float free in the growth medium so that the fluid to be replaced must be decanted into a sterile round-bottomed plastic tube (Falcon No. 2070) and centrifuged at 250 g for 10 minutes. The pellet of cells is added back to the flask with fresh medium or used for heterochromatin determination. The medium is changed every third day. Cells are usually ready for transfer toward the end of the second week. More vigorous mechanical agitation of the flasks is necessary during trypsinization of amniotic cells since these strains appear to have an inordinate tendency to stick fast to this plastic surface. Once the strain is established, it is possible to grow cells for many purposes including enzyme assays that may require a relatively large amount of material.

Establishment of Long-Term Lymphocyte Lines from Small Blood Samples

The observation that human euploid fibroblast cultures die after 20–60 cell generations, together with the fact that these cells grow only in monolayer culture, imposes, to some extent, restrictions on their use. The use of mutant human cells in studies of somatic cell genetics is greatly facilitated by the possession of cells with a virtually limitless life span. Such cells would also be useful for the isolation of relatively large amounts of a variety of subcellular components. Aneuploid cell lines without this limitation can be produced by viral transduction of human fibroblasts. Alternatively human lymphocytes may be grown in suspension culture on a long-term basis and the strains, once established, appear to be permanent if not immortal. These cells may thus be useful for the study of human genetic variation. The persistence of host enzymes in "long-term strains" has been demonstrated using genetically polymorphic enzyme markers.[48] The purification and comparative study of enzymes in long-term cultures of cells from normal individuals and from those with metabolic defects therefore becomes possible. Until quite recently such cultures could be established only from individuals with a variety of benign or malignant lymphoproliferative disorders or from normal subjects and only by the use of very large amounts of blood.[49–51]. The use of small quantities of a specially purified phytohemagglutinin (PHA), a phytomitogen from the red kidney bean *Phaseolus vulgaris* which stimulates normal lymphocytes to transform into blast-like cells

---

[48]J. H. Conover, P. Hathaway, P. R. Glade, and K. Hirschhorn, in press.
[49]G. E. Moore, E. Ito, and K. Ulrich, *Cancer* 19, 713 (1966).
[50]P. R. Glade, I. M. Paltrowitz, and K. Hirschhorn, *Bull. N. Y. Acad. Sci.* 45, 647 (1969).
[51]G. E. Moore, R. E. Gerner, and H. A. Franklin, *J. Amer. Med. Ass.* 199, 519 (1967).

and then to undergo mitosis and cell division, has allowed the establishment of long-term suspension cultures of lymphoid cells from 20-ml samples of human blood.

The method that we use is essentially that of Glade and his colleagues.[52,53] Twenty to 25 ml of blood (from an adult) or 10–15 ml (from a child) are taken into a heparinized plastic syringe (100 units per milliliter of whole blood). The needle is replaced with a clean one, which is then capped. The syringe is placed upright, and the blood is allowed to sediment under gravity at 37° for 1–2 hours or until just before the appearance of a definite buffy coat. Using the sterile needle cap, the needle is bent to an angle of 120° from the vertical to permit decanting of the plasma into a heparinized glass centrifuge tube from the upright syringe. After centrifuging at 250 g for 30 minutes at room temperature, plasma is poured off gently and the cell pellet is suspended in 20 ml of Roswell Memorial Institute 1640 medium (Grand Island Biological Company) with 20% fetal calf serum, 100 μg of neomycin per milliliter, and highly purified PHA (No. E-118 Burroughs Wellcome) 0.025 μg/ml. The lymphocyte concentration should be close to 1 to 2×10^6 cells per milliliter. Incubate at 37° in a 250-ml Falcon flask, disturbing the cells as little as possible. At 1 week the medium is changed by standing the flask on end, allowing the cells to sediment by gravity, and decanting the top one-third of the medium. Replenish this with fresh medium and repeat the process once weekly to three times weekly as needed thereafter. If desired the culture can be mailed to another laboratory at the end of the first week. The tube should be filled with culture medium to prevent excessive shaking and foaming of the contents in transit. At about the third to fourth week, fibroblasts and epithelial cells will be seen growing out but these will die by the next week. At about the sixth week, sometimes a little later, the lymphocyte culture should be becoming established if the attempt is going to be successful. Signs of such success include the finding of lymphoblasts clinging to the vessel wall, the formation of large cell aggregates and a distinct tendency for the medium to become acidic. The generation time of these cells in established culture is about 30 hours.

The success rate of this technique in our hands is currently 20%. Hopefully, future modifications of the method may improve this value and thus afford a simple and reliable means of growing long-term suspension cultures of "normal" human cell strains. However, the observation that the success in obtaining a viable cell line is enhanced if the

[52] P. R. Glade, J. A. Kasel, H. I. Moses, J. Whang-Peng, P. F. Hoffman, J. K. Kammermeyer, and L. N. Chessin, *Nature (London)* 217, 564 (1968).
[53] S. W. Broder, P. R. Glade and K. Hirschhorn, *Blood* 35, 539 (1970).

donor is experiencing a viral infection,[54] suggests the possibility that even these "normal cells" may be harboring a virus that contributes to their longevity in culture. This concept receives further indirect support from the marked increase in success in establishing primary cultures by the addition to the initial growth media of 0.5 ml of a cell lysate from an established lymphocyte line.[55] These considerations suggest that appropriate safety precautions should be taken at least until the benign nature of such a presumed virus is established. These precautions could include the use of a biohazard laminar flow hood and appropriate waste disposal system as well as avoiding growth of lymphocyte strains derived from personnel currently working in the laboratory.

[54]P. R. Glade, Y. Hirshaut, S. D. Douglas, and K. Hirschhorn, *Lancet* 2, 1273 (1968).
[55]K. W. Choi and A. D. Bloom, *Science* 170, 89 (1970).

[18] Zonal Centrifugation

By George B. Cline and Richard B. Ryel

Zonal centrifuge rotors are coming into general use as highly efficient tools for the isolation and purification of a variety of particles and macromolecules.[1,2] Although the terms "zonal centrifuge" and "zonal ultracentrifuge" are commonly used when referring to a specific zonal rotor in a preparative-type centrifuge, the terms are perhaps misnomers except in the cases where special centrifuges have been built to spin specific rotors (i.e., Model K[3] and RK systems). However, since an entire system is required for any zonal separation, we will use the systems approach and call any centrifuge a zonal centrifuge whenever any zonal rotor is being used in it. All zonal centrifuge rotors use a liquid density gradient for the stablization of sedimenting zones of particles or molecules.

A zonal centrifuge combines most of the functions of the analytical centrifuge with the functions of the preparative centrifuge by the collection of analytical data while obtaining preparative quantities of products. Thus it is apparent that different approaches can be made to zonal separations. First, separations can be made for analytical data only.

[1]N. G. Anderson, *J. Phys. Chem.* 66, 1984 (1962).
[2]N. G. Anderson and G. B. Cline, *in* "Methods in Virology" (H. Koprowski and K. Maramorosch, eds.) 137. Academic Press, New York, 1967.
[3]N. G. Anderson, D. A. Waters, C. E. Nunley, R. F. Gibson, R. M. Schilling, E. C. Denny, G. B. Cline, E. F. Babelay, and T. E. Perardi, *Anal. Biochem.* 32, 460 (1969).

One example is that which involves the separation of a particle-bound enzyme or the free enzyme itself on a rate or isopycnic basis with subsequent calculations of sedimentation coefficient, banding density, concentration or specific activity of the material in the zone. Second, separations can be made for the purpose of collecting fractionated products which are used as starting materials for additional physical, chemical, or ultrastructural studies. While it may be somewhat artificial to separate these approaches, it is important to consider both since the design of protocols can be tailored for different rotors and for the type of results desired.

Zonal centrifuges can be classified as either high speed or low speed depending on the type of zonal rotor being used. Low speed systems are limited to 6000 rpm while high speed systems range up to 60,000 rpm. The rotors can be further divided into two categories: (1) batch-type rotors where a fixed volume (batch) of starting sample is fractionated and (2) the continuous-sample-flow-with-banding type (flo-band) in which fresh sample is continuously pumped over the centripetal surface of the density gradient. Nomenclature can be confusing because flo-band rotors can be used for batch separations also, but the batch-type rotors cannot normally be used for continuous sample flow.

Zonal rotors provide versatility in selecting approaches to fractionation problems. Figure 1 shows three approaches to separations in batch-type zonal rotors when beginning with a mixture of particles. Since the middle approach is the most useful for beginning studies when a variety of cell components is to be isolated, step by step instructions are included as a protocol for preparing a starting sample, assembling the rotor and its associated components, preparing and building a discontinuous density sucrose gradient in the B-XIV zonal rotor and suggested ways to handle collected fractions. Although the B-XIV is used as the reference rotor, the protocol can be easily varied for application to any zonal rotor. A partial vocabulary of terms to describe zonal separations is presented as an appendix since zonal systems involve equipment and operations that are not necessarily common to other better known separations procedures. However, before presenting the protocol, it is important to briefly consider the theoretical aspects of zone sedimentation through density gradients as well as some of the more important factors that affect the separation and operation of a zonal centrifuge system.

Practical Considerations

Factors Affecting Particle Separations in Liquid Gradients

Subcellular particles, viruses, and molecules differ in size, density,

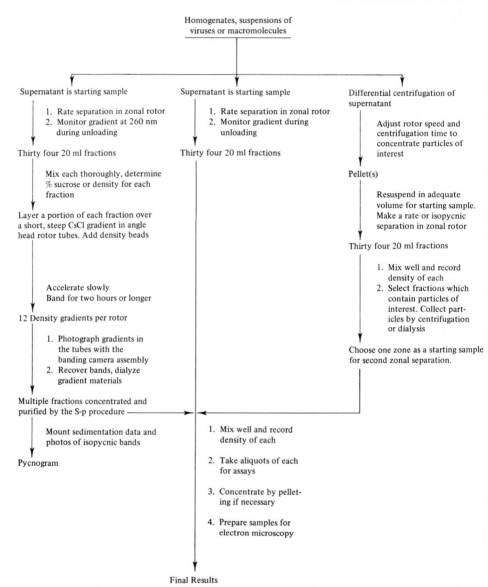

Fig. 1.   Three approaches to separation of subcellular particulates, viruses, or macromolecules by combining the processes of rate separation, isopycnic banding, differential centrifugation, and various sample analysis procedures.

permeability to the gradient solute, rate of permeability to gradient solute, sensitivity to pH ions, organic solvents, heavy metals, enzymes, temperature, lack of water (dehydration), and perhaps hydrostatic pressure. Thus liquid gradients present to sedimenting particles a constantly changing environment of density, viscosity, dehydration, and hydrostatic pressure.

Factors such as density and viscosity of the gradient and dehydration of a particle slow the rate of sedimentation while increased radius means increased gravitational force which tends to move the particle faster. When gradient shape and viscosity factors just offset acceleration due to increased gravitational force, the particle moves with constant velocity (isokinetically).

The gravity-time factor is important whether or not sedimentation rates are to be computed. Electronic integrators, which sum the gravity-time factor, are available from each zonal rotor supplier and are required for exact reproducibility of any two separations (All other conditions remain unchanged.) The device converts the rotational speed to ω^2 and sums the revolutions constantly. Since gravity $\times$ time equals some number, the $\omega^2 t$ value can be reached by low rotor speed for a longer time or high rotor speed for a shorter time. Diffusion of gradient and zones may affect resolution in separations of long duration. The $\omega^2 t$ value is required for the computer program. It can be derived from Table I for approximate values or can be determined accurately by an electronic integrator.

TABLE I

(Part a)

$\omega^2 t$ VALUES FOR ROTOR SPEED AND TIME[a]

RPM	$\omega^2 t$ 10 min, 600 sec.	$\omega^2 t$ 20 min, 1200 sec.	$\omega^2 t$ 30 min, 1800 sec.	$\omega^2 t$ 1 Hour, 3600 sec.	$\omega^2 t$ 2 Hours, 7200 sec.
500	1.6446×10^6	3.2883×10^6	4.9329×10^6	9.8658×10^6	1.9732×10^7
1,000	6.5785×10^6	1.3153×10^7	1.9732×10^7	3.9463×10^7	7.8926×10^7
2,000	2.6314×10^7	5.2611×10^7	7.8925×10^7	1.5785×10^8	3.157×10^8
3,000	5.9207×10^7	1.1838×10^8	1.7758×10^8	3.5517×10^8	7.1034×10^8
5,000	1.6446×10^8	3.2883×10^8	4.932×10^8	9.8658×10^8	1.9732×10^9
8,000	4.2102×10^8	8.4178×10^8	1.2628×10^9	2.5256×10^9	5.0512×10^9
10,000	6.5785×10^8	1.315×10^9	1.9732×10^9	3.9463×10^9	7.8926×10^9
15,000	1.4802×10^9	2.5954×10^9	4.4396×10^9	8.8792×10^9	1.7758×10^{10}
20,000	2.6314×10^9	5.2611×10^9	7.8925×10^9	1.5785×10^{10}	3.157×10^{10}
30,000	5.9207×10^9	1.1838×10^{10}	1.7758×10^{10}	3.5517×10^{10}	7.1034×10^{10}
40,000	1.0526×10^{10}	2.1045×10^{10}	3.15705×10^{10}	6.3141×10^{10}	1.2628×10^{11}
50,000	1.6446×10^{10}	3.2883×10^{10}	4.9329×10^{10}	9.8658×10^{10}	1.9732×10^{11}
55,000	1.9901×10^{10}	3.9789×10^{10}	5.969×10^{10}	1.1938×10^{11}	2.3876×10^{11}
60,000	2.3683×10^{10}	4.7352×10^{10}	7.1035×10^{10}	1.4207×10^{11}	2.8414×10^{11}

$$\omega^2 t = \left(\frac{2\pi \text{ rpm}}{60}\right)^2 t(\text{sec})$$

TABLE I
(Part b)

RPM	$\omega^2 t$ 4 Hours 14,400 sec	$\omega^2 t$ 6 Hours	$\omega^2 t$ 8 Hours	$\omega^2 t$ 12 Hours	$\omega^2 t$ 18 Hours	$\omega^2 t$ 24 Hours
500	3.9463×10^7	5.9195×10^7	7.8926×10^7	1.1839×10^8	1.7758×10^8	2.3678×10^8
1,000	1.5785×10^8	2.3678×10^8	3.157×10^8	4.7356×10^8	7.1033×10^8	9.4711×10^8
2,000	6.314×10^8	9.471×10^8	1.2628×10^9	1.8942×10^9	2.8413×10^9	3.7884×10^9
3,000	1.4207×10^9	2.1310×10^9	2.8414×10^9	4.2620×10^9	6.3931×10^9	8.5241×10^9
5,000	3.9463×10^9	5.9195×10^9	7.8926×10^9	1.1839×10^{10}	1.7758×10^{10}	2.3678×10^{10}
8,000	1.010×10^{10}	1.5154×10^{10}	2.0205×10^{10}	3.0307×10^{10}	4.5461×10^{10}	6.0614×10^{10}
10,000	1.5785×10^{10}	2.3678×10^{10}	3.1570×10^{10}	4.7356×10^{10}	7.1033×10^{10}	9.4711×10^{10}
15,000	3.5517×10^{10}	5.3275×10^{10}	7.1034×10^{10}	1.0655×10^{11}	1.5983×10^{11}	2.1310×10^{11}
20,000	6.314×10^{10}	9.471×10^{10}	1.2628×10^{11}	1.8942×10^{11}	2.8413×10^{11}	3.7884×10^{11}
30,000	1.4207×10^{11}	2.131×10^{11}	2.8414×10^{11}	4.262×10^{11}	6.3931×10^{11}	8.5241×10^{11}
40,000	2.5256×10^{11}	3.7885×10^{11}	5.0513×10^{11}	7.5769×10^{11}	1.1365×10^{12}	1.5154×10^{12}
50,000	3.9463×10^{11}	5.9195×10^{11}	7.8926×10^{11}	1.1839×10^{12}	1.7758×10^{12}	2.3678×10^{12}
55,000	4.7752×10^{11}	7.1628×10^{11}	9.5504×10^{11}	1.4326×10^{12}	2.1488×10^{12}	2.8651×10^{12}
60,000	5.6828×10^{11}	8.5242×10^{11}	1.1366×10^{12}	1.7048×10^{12}	2.5573×10^{12}	3.4097×10^{12}

[a]Calculated by R. E. Canning and W. Rasmussen, Molecular Anatomy (MAN) Program, ORNL.

Gradient Materials

A gradient material should meet several requirements. Sucrose is used for most rate separations and cesium chloride is used for many isopycnic separations. Thus it is apparent that there is no ideal all-purpose gradient material. The one basic requirement is that the gradient permit the desired type of separation. The additional factors can be itemized in the form of questions:

1. Is the density sufficient to band particles isopycnically?
2. Will it affect biological activity?
3. Is it hyper- or hypoosmotic?
4. Will it interfere with the assay technique?
5. Can it be removed from the purified product?
6. Is it opaque in the ultraviolet range?
7. Is it cheap and readily available?
8. Can it be recovered for reuse?
9. Can it be sterilized?
10. Is it corrosive to the rotor?

Table II contains the most commonly used gradient materials, some of their solvents, and densities at 20°C. Density will sometimes decrease at lower temperatures by a decrease in solubility. Characteristics of many of these materials are discussed elsewhere.[4,5]

Gradient Shape and Zone Capacity

The mass of sample which can be fractionated into zones in a density gradient depends on: (a) the concentration of the starting sample, (b) the shape of the density gradient curve, (c) the mass of material in each zone, and (d) the resolution required. All these factors are related and must be considered whenever maximal amounts of sample are to be fractionated at one time. Zone capacity studies have been reported[6-8] and present details that cannot be covered here.

The function of a density gradient in zonal separations is to provide a stabilizing medium to keep particles in zones. When the first particles sediment out of a starting zone, they encounter a higher gravitational force but also meet a higher density and viscosity environment. The particles in the trailing edge of the zone can be sedimented at the same rate as particles at the leading edge because they are in a lower density

[4] J. B. Ifft, D. H. Voet, and J. Vinograd, *J. Phys. Chem.* **65**, 1138 (1961).
[5] A. S. Hu, R. M. Bock, and H. O. Halvorson, *Anal. Biochem.* **4**, 489 (1962).
[6] S. P. Spragg and C. T. Rankin, Jr., *Biochim. Biophys. Acta.* **141**, 164 (1967).
[7] V. N. Schumaker, *Advan. Biol. Med. Phys.* **11**, 245 (1967).
[8] A. S. Berman, *Nat. Cancer Inst. Monogr.* **21**, 41 (1966).

TABLE II
TYPES OF GRADIENT MATERIALS[a]

Materials	Solvent	Maximum density
Sucrose (66%, 5°C)	H_2O	1.33
Sucrose-D_2O (65%)	D_2O	1.37
Silica sols	H_2O	1.30
Diodon	H_2O	1.37
Glycerol	H_2O	1.26
CsCl	H_2O	1.91
CsCl-D_2O	D_2O	1.98
Cs formate	H_2O	2.1
Cs acetate	H_2O	2.0
RbCl	H_2O	1.49
Rb formate	H_2O	1.85
RbBr	H_2O	1.63
KAc	H_2O	1.411
K formate	H_2O	1.57
K formate-D_2O	D_2O	1.63
Na formate	H_2O	1.32
Na formate-D_2O	D_2O	1.40
LiBr	H_2O	1.83
LiCl-D_2O	D_2O	1.33
Polyvinylpyrrolidone	H_2O	—
Albumin	H_2O	—
Sorbitol	H_2O	—
Ficol	H_2O	1.173

[a] Compiled from various sources.

and viscosity environment though one with less gravitational force. Zones of particles of similar size thus tend to stay together; but whether they stay together, are concentrated or diluted, depends on additional factors.

Zonal centrifuge rotors are divided into sector-shaped compartments. A zone of particles sedimenting from a smaller radius to a larger radius is diluted even when a zone width remains the same and diffusion is minimal. Thus zonal rotors generally do not provide for a concentration of particles but only for a separation and purification (and dilution). By choosing the proper gradient shape, radial dilution problems can be minimized. In fact, if only a single component is to be isolated, a gradient shape can be designed for zone concentration.[9]

Three basic density gradient curves or shapes are: (a) linear with rotor radius, (b) convex with radius, and (c) concave with radius. The shape of

[9] S. P. Spragg, R. S. Monod, and C. T. Rankin, Jr., *Separation Sci.*, 4, 467 (1969).

the gradient in the rotor is the important factor, and when designing gradients, the conversion from shape as provided by the gradient pump to shape in the rotor is sometimes critical. Convex gradients offer maximal capacity down the sedimentation path. The steepest part of the curve is under the starting sample where it is needed most and less steep toward the periphery of the rotor where radial dilution has decreased the concentration within a zone. Linear gradients offer the same capacity throughout. Concave shaped gradients offer increasing capacity with increasing radius and can be used to reduce zone width to keep concentration constant. Maximal steepness offers maximal capacity and reduced resolution while shallow gradients offer reduced capacity but higher resolution.

The gradient curve may be continuous or discontinuous. The continuous gradients are built with a gradient engine[10] from two solutions of different density while the discontinuous gradients are built from several solutions of different density. Only a simple pump or air pressure is required to move the solutions. The discontinuous gradients are the easiest to design and build and offer almost limitless variations in shape and range. These gradients offer additional advantages of high resolution and moderate capacity. The interfaces between adjacent solutions provide a short steep gradient to narrow zone width while the shallow areas between interfaces permit zones to sediment rapidly and be separated easier from each other.

Figure 2 shows the shape of the gradient before being built into the B-XIV zonal rotor as well as the shape once it is in the rotor. The effect of sedimentation on zone width is also shown to indicate the changing shape which a zone has when passing through a shallow area and an interface (Fig. 3). The interfaces act to reshape and sharpen zones into Gaussian shaped peaks. Resolution of separation appears better with discontinuous gradients than with continuous gradients when working with the whole cell homogenates.

Resolution

It is often easier to isolate a single component from a mixture than to separate or resolve several different types of particles from a mixture in one step. For a single component, a measure of the resolution attained during the sedimentation can be calculated from measured or computed sedimentation coefficients of particles across the zone.[11] This gradient resolution (GR) is defined by:

---

[10]Available from each of the zonal centrifuge manufacturers.
[11]N. G. Anderson, *Quart. Rev. Biophys.* 1, 217 (1968).

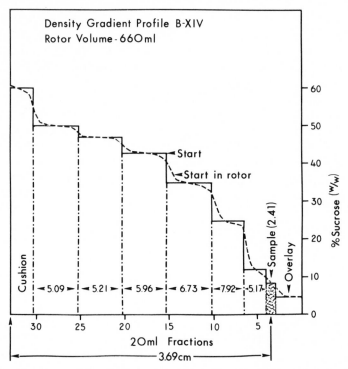

FIG. 2. The shape of the discontinuous sucrose density gradient before loading into the B-XIV zonal rotor (solid lines) and after loading into the rotor with the sample (dashed line). The 20-ml sample occupies a 2.41 mm zone width when 60 ml of overlay is used. The radius occupied by each sucrose solution in the rotor is shown in millimeters.

$$GR = \frac{s_{20,w}}{Z_{s_{20,w}}} \qquad (1)$$

where $s_{20,w}$ is the sedimentation coefficient of the center of the zone in Svedberg units under standard conditions and $Z_{s_{20,w}}$ is the width of the zone at half height at standard conditions.

Ifft[4] and co-workers have approached zone resolution from the parameters of radius and standard deviations of concentration of particles in two adjacent zones such that,

$$A = \frac{r}{\sigma_1 - \sigma_2} \qquad (2)$$

where r is the radial distance between the centers of the zones and σ_1 and σ_2 are standard deviations of concentrations in each zone. Price[12] has substituted gradient volume for radial distance in Eq. (2).

[12]C. A. Price, in "Manometric Techniques" (W. W. Umbreit, R. H. Burris, and J. F. Stauffer, eds.), 5th ed. in press.

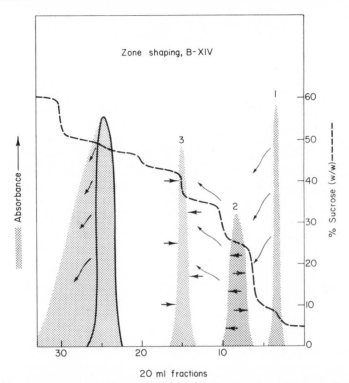

20 ml fractions

Fig. 3. Various shapes which a zone of particles may have during sedimentation through a discontinuous gradient. *1* represents the starting zone, *2* the broadening of the zone in a "shallow" area of the gradient, and *3* the narrowing of a zone when it meets an interface. The zone at the left is overloaded, the dark line denoting the shape a maximally loaded zone would have.

Computations

Calculation of sedimentation rates of some subcellular particles through liquid gradients can be made difficult because of the particle's reaction to changes in osmotic pressure. However, for calculation of sedimentation rates in density gradients, the osmotic factors are generally ignored in favor of simplicity of computation. The same factors described for sedimentation of particles in a homogeneous medium[13] are used to calculate particle sedimentation in gradients. The rate of sedimentation is dependent upon the particle size, shape, and density as well as the viscosity and density of the medium and the amount of the applied gravitational force. For a nongradient situation where the suspending medium *m* is homogeneous, the sedimentation coefficient *s*, at temperature *T* is expressed by:

[13]T. Svedberg and K. O. Pederson, "The Ultracentrifuge" Oxford Univ. Press (Clarendon) London and New york.

$$s_{T,m} = \frac{(dR/dt)}{\omega^2 R} \tag{3}$$

where R is the distance of the particle from the axis of rotation; ω is the angular velocity in radians per second and dR/dt is particle velocity. For comparison purposes, sedimentation coefficients are normalized to the velocity in water at 20° so that Eq. (3) is arranged to correct for any solution:

$$s_{20,w} = \frac{(dR/dt)}{\omega^2 R} \frac{\eta_{T,m} (\rho_p - \rho_{20,w})}{\eta_{20,w} (\rho_p - \rho_{T,m})} \tag{4}$$

where $\eta_{T,m}$ is the viscosity in poise of the medium at temperature T, ρ_p is the density of the particle, $\rho_{20,w}$ is the density of water at 20°, $\eta_{20,w}$ is the viscosity of water at 20° and $\rho_{T,m}$ is the density of the medium at the temperature used.

Equation (4) can be used to calculate sedimentation rates of particles through gradients by dividing the gradient into a number of incremental fractions, determining the viscosity and density of the gradient solution in each fraction as well as the position of that fraction in the rotor. Computations are made from one fraction to the next adjacent fraction and result in a series of laborious steps. Numerous methods have been described for evaluating rates of particles in sucrose gradients.[7, 14–17] Such repetitive calculations are most easily done by computer, and the program of Bishop[18] has been widely used. This program divides the gradient into a number of aliquots and uses an adaptation of the trapezoidal integration technique described by Martin and Ames.[16]

If Eq. (4) is rewritten as

$$s_{20,w} \, \omega^2 dt = \frac{(\rho_p - \rho_{20,w})}{(\rho_p - \rho_{T,m})} \frac{\eta_{T,m}}{\eta_{20,w}} \frac{dR}{R} \tag{5}$$

then the left side of the equation can be integrated to give

$$\int_0^t s_{20,w} \, \omega^2 dt = s_{20,w} \, \omega^2 t \tag{6}$$

and the right side of Eq. (5) may be approximated by the expression

[14] C. de Duve, J. Berthet, and H. Beaufay, *Progr. Biophys.* **9**, 325 (1959).
[15] R. J. Britten and R. B. Roberts, *Science* **131**, 32 (1960).
[16] R. G. Martin and B. N. Ames, *J. Biol. Chem.* **236**, 1372 (1961).
[17] J. F. Thomson and E. T. Mikuta, *Arch. Biochem.* **51**, 487 (1954).
[18] B. S. Bishop, *Nat. Cancer Inst. Monogr.* **21**, 175 (1966).

$$\frac{(\rho_p - \rho_{20,w})}{\eta_{20,w}} \int_{Ri}^{Ri} F(R)\,dR = \frac{(\rho_p - \rho_{20,w})}{\eta_{20,w}} \sum_{Rs}^{Rs} F(R_i)\,(R_i - R_{i-1}) \quad (7)$$

where i represents a given aliquot of the gradient, R_i the maximal radius of that given aliquot, $(R_i - R_{i-1})$ is the width of the aliquot in the rotor, and R_s is the radius of the mass center of the starting sample zone in the rotor. $F(R)$ is the sum expression of values relating to viscosity and density to aliquot number. See Barber[19] for details. The values of $F(R)$ are computed from equations relating the density and viscosity of sucrose as a function of concentration and temperature. See Bishop[18] for details on computer programming for zonal centrifuge rotors.

Tables III and IV give the volume-radius relationships for rotors B-XIV and B-XV. Values are used in computing sedimentation coefficients.

Types of Zonal Rotors

Since most commercially available zonal rotors have the same dimensions as the original rotors produced by the Molecular Anatomy Program of the Oak Ridge National Laboratory, they carry the same identifying

TABLE III

OAK RIDGE—TYPE B-XIV ZONAL ROTOR:
VOLUME VS RADIUS[a]

Volume (ml)	Radius (cm)	Volume (ml)	Radius (cm)
20	0.	340	4.988
40	0.		5.110
60	2.665		5.228
80	2.906	400	5.343
100	3.126		5.456
	3.328		5.566
	3.517		5.674
	3.695		5.775
	3.863	500	5.878
200	4.024		5.979
	4.177		6.080
	4.324		6.181
	4.466		6.282
	4.603	600	6.385
300	4.735		6.490
	4.863		6.601

[a]Calculated by E. L. Rutenberg, Molecular Anatomy (MAN) Program, ORNL.

[19] E. J. Barber, *Nat. Cancer Inst. Monogr.* 21, 219 (1966).

TABLE IV
Oak Ridge—Type B-XV Zonal Rotor:
Volume vs Radius[a]

Volume (ml)	Radius (cm)	Volume (ml)	Radius (cm)
40	0.	880	6.594
80	2.667		6.724
120	3.023		6.851
160	3.333	1000	6.976
200	3.610		7.099
	3.863		7.219
	4.098		7.336
	4.318		7.452
	4.525	1200	7.566
400	4.722		7.685
	4.909		7.794
	5.089		7.902
	5.262		8.010
	5.428	1400	8.117
600	5.589		8.223
	5.745		8.330
	5.896		8.438
	6.043		8.547
	6.186	1600	8.658
800	6.325		8.774
	6.461	1665	8.890

[a]Calculated by E. L. Rutenberg, Molecular Anatomy (MAN) Program, ORNL.

numbers. This is important for comparison of data from one laboratory to another. However, the numbering system does not suggest what the complete function of the rotor is. The prefix "A" means that the rotor is one operating at low speed (up to 6000 rpm) while the prefix "B" means that the rotor is capable of higher speed (up to 60,000 rpm). Various manufacturers have additional prefixes or suffixes which indicate design modifications. The original identifying numbers were given to new rotors and to modifications of rotors in the order in which they were conceived and built[11]. Table V is a list of the most widely used rotors. Rotors which are not widely available commercially are not included, and perhaps some rotors which are now becoming available are unintentionally omitted.

Gradients in early zonal rotors were recovered centrally by displacement with heavy piston solution (core unloading). Gradients in more recent rotors such as the B-XXIX (and B-XIV and B-XV with special cores) can be recovered from the wall by displacement with water (edge unloading).

TABLE V
ZONAL ROTORS AVAILABLE FOR PREPARATIVE CENTRIFUGES[a]

Rotor	(rpm) Top Speed	Type of centrifuge	Gradient path length (cm)	Approx. rotor volume (ml)	$g \times 10^3$ max
A-XII	5,000	IEC[b] PR-6	12	1300	7
	6,000	MSE[c] Mistral 6-L	12	1300	10
Z-15[e]	8,000	IEC B-20	8	700	
B-IV	40,000	Spinco[d] L-4	4	1725	90.9
		MSE Superspeed 50	4	1660	91
B-IX[f]	40,000	Spinco L-4	1	750	90.9
B-XVI[f]	40,000	Spinco L-4	1	750	
B-XIV al[g]	32,000	Spinco Model L	5.3	650	91.3
		IEC B-60	5.3	660	91.3
	30,000	MSE Superspeed 50	5.3	650	67.1
B-XIV ti[h]	48,000	Spinco L-3-50, L2-65-B	5.3	660	171.8
		MSE Superspeed	5.3	660	171.8
	50,000	IEC B-60, B-75	5.3	660	171.8
	60,000	IEC B-60, B-75	5.3	650	247
B-XIVa-ti		Same as B-XIV but with edge unloading	—	—	—
B-XV al	22,000	Spinco L series	7.5	1665	48.1
B-XV ti	35,000	Spinco L series	7.5	1665	122
		IEC B-60	7.5	1670	122
		MSE Superspeed 50	7.5	1670	122
B-XXIX ti		Same as for B-XV ti	7.5	1430	122

[a]Rotors made by one manufacturer should be operated only in designated centrifuges.
[b]International Equipment Company, Needham Heights, Massachusetts.
[c]Measuring and Scientific Equipment LTD., London, England.
[d]Spinco Division of Beckman Instruments, Palo Alto, California.
[e]Can be used for continuous sample flow with a 110 ml gradient volume.
[f]Continuous sample flow.
[g]Aluminum.
[h]Titanium.

Rate and Isopycnic Separations

   Samples fractionated in zonal centrifuges range from serum proteins[20] to whole cells.[21] Since particles sediment on the basis of size (shape) and density, separations can be designed to exploit either one or both of these parameters for either analytical data or preparative amounts of products. The following list covers the common procedures of where the sample is put with respect to the gradient and the type of separation to be performed in the B-XIV rotor.

[20]W. D. Fisher and R. E. Canning, *Nat. Cancer Inst. Monogr.* 21, 403 (1966).
[21]C. W. Boone, G. S. Harell, and H. E. Bond, *J. Cell. Biol.* 36, 369 (1968).

Sample position relative to gradient position	Sample volume to gradient volume	Separation type	Typical products of rationale of separations
On top	20/640	Rate	Serum proteins, viruses, cells and components
		Isopycnic	Mitochondria, lysosomes, nuclei, membranes, viruses, whole cells
On top	500/160	Isopycnic	Cell membranes, viruses
On bottom	20/640	Rate or isopycnic	Any particle with density less than gradient, e.g., membranes, serum, lipoproteins
In middle	20/640	Rate or isopycnic	Lighter material goes up, heavier goes down, sample can remain in position
Throughout	660/660	Isopycnic	Proteins, cells, membranes, viruses

Since size is the most obvious and variable physical parameter among subcellular components, it is this dimension that can be exploited most easily for separations purposes. Figure 4 shows a display of most of the major cell components within the coordinates of size and density (in CsCl). The size ordinate is on a log base since most components differ by orders of magnitude in sedimentation coefficient (size). Most beginning zonal methods involve rate separations. Although separations can be made in density gradients in any angle head or swinging-bucket rotor, the large sector-shaped compartments of zonal rotors make them most ideal for analytical and preparative studies.

Figure 5 shows a series of diagrams representing the sequential steps in loading a B-XIV or B-XV type of zonal rotor as compared with separations in a swinging bucket tube. Careful analysis of these diagrams will aid a prospective operator in executing the protocol outlined below (see *Zonal System Assembly and Operation*).

It is apparent from Fig. 4 that while many particles may have the same size, they generally do not share a common density. For highest resolution of components, a two-dimensional sedimentation approach is warranted. This is the S-ρ procedure which is presented below as a method of processing zonal fractions.

Zonal Fractionation Procedure (Protocol)

Gradient Preparation

Sucrose is inexpensive, readily available, and meets many of the criteria for a gradient material established above. Use the highest grade of

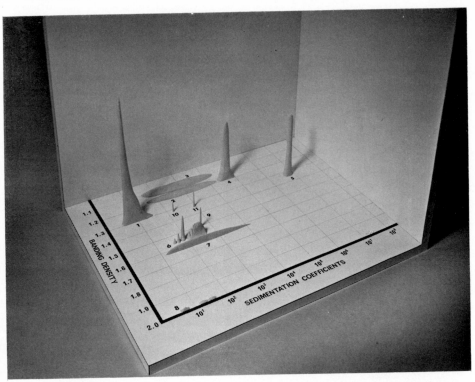

Fig. 4. Diagrammatic presentation of distribution of subcellular components as a function of sedimentation rate and banding density. Components are: *1*, soluble proteins; *2*, membranes with attached particles; *3*, smooth membranes; *4*, mitochondria; *5*, nuclei; *6*, ribosomal subunits, ribosomes, and polysomes; *7*, glycogen; *8*, RNA; *9*, poliovirus; *10*, T3 bacteriophage; *11*, adenovirus, type 2. Banding densities for most particles are those for cesium chloride. Note that viruses fall in a clear space—the "virus window." Reprinted from *Science*, 154, 103 (1966).

sucrose available. However, even reagent grade sucrose often contains ion and enzyme contaminants, so assays should be made on each lot of sucrose to determine whether it is usable. Sucrose can be purified by passing it through heated activated charcoal columns. Purified sucrose solutions are available commercially.[22,23] Do not use ordinary table sugar except as piston solution. Its use as a gradient material yields results completely different from results with reagent grade sucrose.

Sucrose discontinuous gradient solutions should be made just prior to fractionation and cooled to 0–5°. Do not store dilute sucrose solutions

[22]Elanco, Eli Lilly and Co., Indianapolis, Indiana.
[23]Harshaw Chemical Co., Cleveland, Ohio.

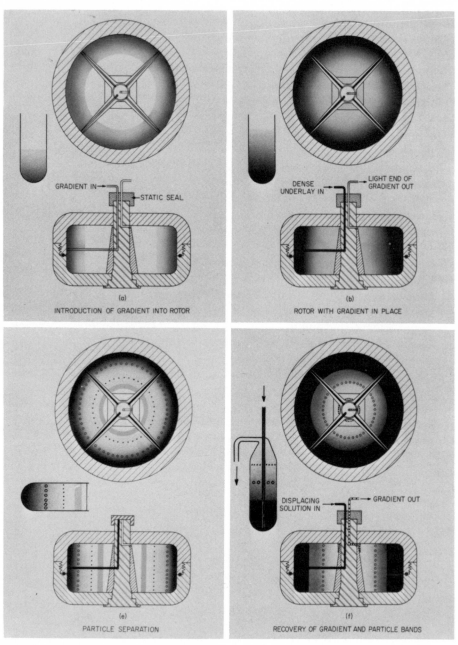

FIG. 5. Schematic diagrams of operation that apply to both B-XIV and B-XV zonal centrifuge rotors. Rotor shown at various stages of loading and unloading in top and side view. (a) Start of gradient introduction into rotor spinning at low speed, (b) completion of loading of gradient into rotor, (c) movement of sample layer into the rotor through the

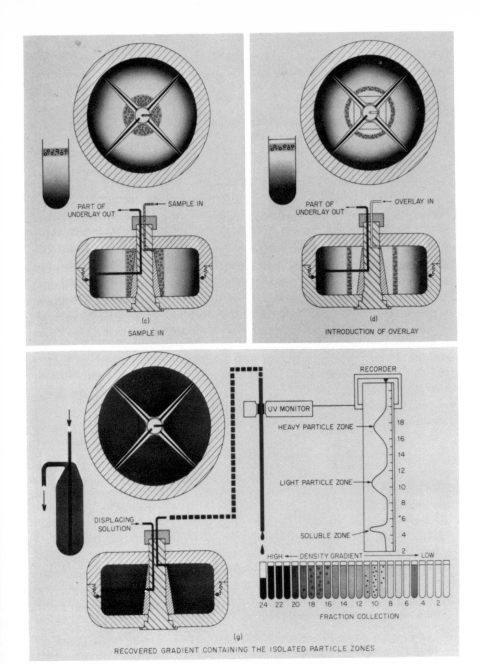

center line, (d) introduction of overlay into rotor to move the starting zone away from the core faces, (e) separation of particles at high speed, (f) displacement of separated zones out of rotor at low speed, and (g) completion of unloading and collection of sample tubes. Note that the swinging-bucket equivalent of each step is also included. Reprinted from *Anal. Biochem.* **21**, 235 (1967).

for long periods even in the refrigerator since bacteria and mold growth may develop. A stock solution of 66% sucrose (w/w) can be made by dissolving 5 pounds of high grade sucrose in 1200 ml of distilled water. A homogenizer is required.[24] Such a high concentration in the stock solution is a natural preservative, and batches can be maintained at 0–5° for long periods without the development of biological contaminants. Solutions for the discontinuous sucrose gradient outlined in the *Operation* section are made from the stock solution according to dilution factors from Table VI. Dilutions should be made with appropriate buffers and ions to adjust the working solutions to the desired concentrations and pH. Agents such as ethylenediaminetetraacetate (EDTA, a chelating agent) or dioctyl sodium sulfosuccinate (DOSSS, surface-active agent) may be added to the gradient solutions at this time. One milliliter of a 0.0075% solution of DOSSS per 100 ml of gradient solution is effective in preventing the aggregation of cell organelles and membranes.

Make the following solutions and store them at ice cold temperatures in graduate cylinders or beakers:

100 ml	0.066 M phosphate buffer (or equivalent) for overlay
100 ml	9% sucrose
75 ml	25%
75 ml	35%
100 ml	43%
100 ml	47%
100 ml	50%
300 ml	55%
800 ml	56% (for piston)

Record all data on a zonal fractionation protocol sheet such as that shown in Table VII.

Sample Preparation

Reproducible preparation of the sample is critical for reproducible results. The method and degree of homogenization vary with the type of tissue used. For rat liver preparations, sacrifice the animal quickly by cervical dislocation or decapitation. The use of barbiturates or ether is not advised if the ultrastructure of cell components is important. Excise the liver quickly, wash it in ice cold 0.25 M sucrose (8.5% w/w) or buffer which has been adjusted to the proper osmolarity. Using forceps, blot the liver on filter paper and excise from 1 to 1.5 g into a tared beaker containing cold sucrose or buffer. Mince the liver with

[24]Large size Waring Blendor, Lourdes, etc.

TABLE VI

CORRELATION BETWEEN STOCK 66 WEIGHT PERCENT SUCROSE AND FINAL CONCENTRATION

Dilute the following ml of 66% sucrose to 1 liter		Final conc. (%)	Dilute the following ml of 66% sucrose to 1 liter		Final conc. (%)
4°C	25°C		4°C	25°C	
15.15	14.25	1.00	545.46	516.06	36.00
30.30	14.90	2.00	560.61	531.21	37.00
45.46	16.06	3.00	575.76	546.36	38.00
60.61	31.21	4.00	590.91	561.51	39.00
75.76	46.36	5.00	606.06	576.66	40.00
90.91	61.51	6.00	621.21	591.81	41.00
106.06	76.66	7.00	636.36	606.96	42.00
121.21	91.81	8.00	651.62	622.12	43.00
136.36	106.96	9.00	666.67	637.27	44.00
151.52	122.12	10.00	681.82	652.42	45.00
166.67	137.27	11.00	696.97	667.57	46.00
181.82	152.42	12.00	712.12	682.72	47.00
196.97	167.57	13.00	727.27	697.87	48.00
212.12	182.72	14.00	742.43	713.03	49.00
227.27	197.87	15.00	757.58	728.18	50.00
242.43	213.03	16.00	772.73	743.33	51.00
257.58	228.18	17.00	787.88	758.48	52.00
272.73	243.33	18.00	803.03	773.63	53.00
287.88	258.48	19.00	818.18	788.78	54.00
303.03	273.63	20.00	833.33	803.93	55.00
318.18	288.78	21.00	848.49	819.09	56.00
333.33	303.93	22.00	863.64	834.24	57.00
348.49	319.09	23.00	878.79	849.39	58.00
363.64	334.24	24.00	893.94	864.54	59.00
378.79	349.39	25.00	909.09	879.69	60.00
393.94	364.54	26.00	924.24	894.84	61.00
409.09	379.69	27.00	939.39	910.00	62.00
424.24	394.84	28.00	954.55	925.15	63.00
439.39	409.99	29.00	969.70	940.30	64.00
454.55	425.15	30.00	984.85	955.45	65.00
469.70	440.30	31.00	1000.00	970.60	66.00
484.85	455.45	32.00	—	—	—
500.00	470.60	33.00	—	—	—
515.15	485.75	34.00	—	—	—
530.30	500.90	35.00	—	—	—

scissors before homogenizing in an ice cold Potter-Elvehjem homogenizer. Use 15 double strokes and count as a stroke the one in which the tip of the pestle is first worked to the bottom of the chamber. "Over" homogenization will break many of the large subcellular components while "under" homogenizing will give a low yield.

TABLE VII
Zonal Fractionation Protocol Sheet[a]

Rotor _____   Sample ID _____   Z-UAB Run # _____

Sample weight _____   Date _____

     volume _____   Operator _____

     buffer _____   ROTOR SPEED _____

density (%w/w) _____   TIME AT SPEED _____

    preparation _____   Start decel. @

_____ $\omega^2 t \times 10^6$ _____

_____   Start unload @

_____ $\omega^2 t \times 10^6$ _____

_____ $\omega^2 t \times 10^6$ during

Gradient solution _____ ml _____ % or ρ banding _____

		% SUCROSE
_____	_____	
_____	_____	
_____	_____	
_____	_____	1 _____
_____	_____	2 _____
_____	_____	3 _____
_____	_____	4 _____

Gradient loading rate _____   5 _____

Sample loading rate _____   6 _____

Overlay loading rate _____   7 _____

     volume _____   8 _____

   composition _____   9 _____

Gradient unloading rate _____   10 _____

Spectrophotometer wave length _____   11 _____

Recorder chart speed _____   12 _____

No. fractions collected _____   13 _____

     volume _____   14 _____

Assays or tests:	fractions	starting samples		
	_____	_____	15	_____
			16	_____
density	_____	_____	17	_____
total protein	_____	_____	18	_____
RNA	_____	_____	19	_____
DNA	_____	_____	20	_____
Cytochrome			21	_____
oxidase	_____	_____	22	_____
Acid phosphatase	_____	_____	23	_____
Catalase	_____	_____	24	_____
_____	_____	_____	25	_____
_____	_____	_____	26	_____
_____	_____	_____	27	_____
_____	_____	_____	28	_____

Zonal system problems _____   29 _____

_____   30 _____

_____   31 _____

_____   32 _____

Suggestions for change _____   33 _____

		34 _____
Pertinent reference _____		35 _____
		36 _____
		37 _____
		38 _____
		39 _____
		40 _____

[a] Used at the University of Alabama at Birmingham.

Transfer the homogenate to a cold centrifuge tube for clarification at 500 rpm for 8 minutes in the No. 256 head in an International PR-2 or PR-6 refrigerated centrifuge. Whole cells, nuclei, and aggregates will be in the pellet. Follow by light microscopy. The supernatant fluid is used as the starting sample.

Zonal System Assembly and Operation

1. Determine that all passages in the vanes of the rotor septa are free from materials that might stop or impede the flow of the gradient being introduced into the rotor. Do this by running a stream of water from a squirt bottle through the vane passages. All O-rings should be lubricated to assure proper sealing of rotor parts. Assemble the rotor and cool it to refrigerator temperature.

2. Connect the gradient pump to the rotor edge line of the seal assembly. Use small-diameter tubing (1/8 inch i.d.). Put a bubble trap between them to catch bubbles before they get into the rotor. Keep the tubing as short as practical. Fill the tubing with some of the 9% sucrose solution to displace all air. The tubing leading upward from the bubble trap (escape line) and downward to the seal assembly are clamped off when they are full (Rochester Pean 6¼-inch hemostats are suggested for this purpose).

3. Mount the cooled rotor on the centrifuge drive and install the Plexiglas safety plate. Turn on the refrigeration. Turn on coolant water to seal assembly. Turn the $\omega^2 t$ integrator to standby.

4. The seal assembly can be attached either before or after accelerating the rotor to 2000 rpm (after is preferred). Unclamp the line leading downward from the bubble trap to the seal assembly. A Plexiglas refrigeration plate, if available, should cover the chamber opening to prevent condensation forming on the chamber cold wall. The configuration of the components is shown in Fig. 6.

5. Pump 50 ml of the 9% sucrose solution to the wall of the spinning rotor by way of the edge line as the first step of the discontinuous gradient. Stop the pump; put the pump input line into the next beaker containing 25% sucrose. Turn on pump and slowly load in 75 ml. Stop

B-XIV, B-XV Rotor assembly

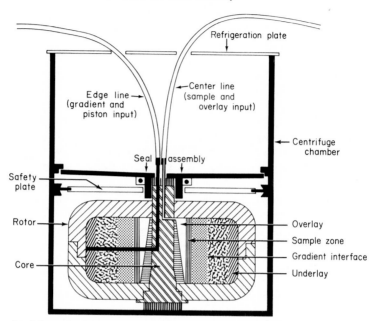

FIG. 6.   Schematic diagram of the B-XIV or B-XV rotor assembly showing the relationship between the rotor, seal assembly, safety plate, and refrigeration plate during loading and unloading.

pump and repeat for remaining solutions in the order of increasing density. A loading rate of from 30 to 50 ml per minute is recommended. Do not introduce bubbles into the rotor at any time.

6. After the gradient solutions are all loaded, pump in the cushion solution. The rotor is filled when the 9% sucrose emerges from the center line. The gradient is now ready for sample loading. If the rotor is kept refrigerated, the discontinuous gradient can be held at 2000 rpm for about 30 minutes while the sample is being prepared. Continuous gradients may be held much longer, especially if they are relatively shallow.

It is best to prepare the sample while the gradient is being loaded into the rotor. This can best be done by two people or by one experienced operator. If one person must do it, load the gradient first and prepare the sample thereafter.

7. Carefully draw off the supernatant fluid with a 20-ml syringe and a 16-gauge hypodermic needle. The supernatant fluid represents the starting sample for the zonal separation and is kept ice cold until loaded into the rotor.

8. Connect the syringe with the sample to the rotor by way of the center line to the seal assembly. A three-way Luer-Lok valve is useful, but there are other ways equally convenient.

9. To load the sample into the rotor, first release the clamp on the bubble trap escape line and switch the 3-way valve the right way and slowly (5–10 ml/min) push the sample into the rotor. Cushion solution will be displaced from the edge line and lost through the bubble trap. Unless fluid is being put into the rotor, at least one line to the rotor should always be clamped off. The seal acts as its own pump and will move fluid and air into or out of the rotor on its own. Switch integrator to "Count."

10. When all sample is out of the syringe, switch the valve to draw in 5 ml of overlay solution. Push this into the rotor. This solution "washes" the syringe and keeps most of the sample in a reasonably narrow starting zone. Small bubbles can be tolerated in the sample.

11. Follow the 5 ml "wash" solution with at least 55 ml more of overlay solution. A minimum of 60 ml should be used to displace the starting fraction away from the flat surface of the rotor core into a higher gravitational field.

12. Clamp off edge and center lines to the seal assembly and remove it from the spinning rotor.

13. Accelerate the rotor to 4000 rpm and securely attach the vacuum cap to the rotor. This increase in speed causes the rotor to expand slightly and take in a small amount of air. The trapped air at the center of the rotor will aid in cap removal at 2000 rpm at the end of centrifugation. Vented caps can be put on at 2000 rpm.

14. After the cap is on, close the centrifuge lid, wait a minute or two for the vacuum to increase in the chamber and then accelerate the rotor to 30,000 rpm for 35 minutes. Make sure that both the vacuum and refrigeration are functioning properly.

15. Decelerate the rotor with the brake to 5000 rpm, then turn the brake off and let the rotor coast to 2000 rpm. Be sure that drive is turned on. Bleed the vacuum from the centrifuge.

16. Open the lid of the centrifuge, remove the vacuum cap from the rotor and replace the seal assembly on the spinning rotor. Make sure the seal coolant is on. Do NOT let the rotor stop. Attach the center line from the seal assembly to a flow cell in a spectrophotometer or any other appropriate absorbance or light scatter detector. Use a chart recorder for a continuous record of absorbance change. A bubble trap may be used in the center line between the rotor and the flow cell to trap small bubbles coming out with the gradient. Bubbles going through the flow cell can interfere with detection. Baseline the spectrophotometer with water in the cell.

17. Clamp the exit line from the bubble trap. Unclamp the line from the pump to the rotor wall and begin pumping ice cold underlay or piston solution to the rotor wall. Monitor the gradient at 260 nm or 280 nm and collect 20-ml fractions. Switch integrator to "standby."

18. Mix each fraction thoroughly and determine the weight percent of sucrose refractometrically (Bausch and Lomb Abbe 3-L) or gravimetrically. Record the data for each fraction. Keep all fractions ice cold.

Gradient Fractionation and Zone Analysis

Generally thirty-four 20-ml fractions are collected from the B-XIV sized rotors and forty-two 40-ml fractions are collected from the B-XV sized rotors. Larger zonal rotors such as the K-X yield seventy samples of 100 ml each.

The simplest and most foolproof method is to collect the fractions by hand. For routine separations, it is sometimes easier to use a fraction collector and collect on a time basis.

The large number of fractions and the volume of each fraction can present problems for enzymatic assay. After collection, each gradient fraction should first be thoroughly mixed and then assessed for density by refractive index measurements. Fractions must be kept cold and analyzed rapidly to suppress or reduce the effects of microbial growth and to preserve enzymatic activity. The originator of the zonal centrifuge, Dr. Norman G. Anderson, has also provided a rapid kinetic analyzer system which solves the problem of multiple assays on large numbers of fractions within a very short time. The details of the analyzer are presented elsewhere,[25,26] but a brief description is given here because of its vital role in making the zonal rotors more useful.

The GeMSAEC analyzer (named for the National Institutes of General Medical Sciences and the Atomic Energy Commission) is a centrifugal spectrophotometer in which special rotors (which contain either 15, 16, 30, or 45 cuvettes, depending on the commercial supplier)[27-29] spin past a fixed light source and a photodetector. The output of the photodetector as well as the spinning of the rotor are interfaced with a small digital computer. The maximum reaction volume is 500 μl, and in many reactions only 10-μl samples are required. A sample of each gradient fraction and reagent is pipetted either manually or auto-

[25]N. G. Anderson, *Anal. Biochem.* **23**, 207 (1968).
[26]N. G. Anderson, *Science* **166**, 317 (1969).
[27]Electro-Nucleonics, Inc., Fairfield, New Jersey.
[28]American Instrument Co., Inc., Silver Spring, Maryland.
[29]Union Carbide Corp., New York, New York.

matically into special wells aligned radially in a central distributor disk. The disk fits into the center of the flat rotor which contains the cuvettes in the periphery. When the rotor is accelerated, the sample and the reagent are centrifuged from their respective wells and are "dumped" into their respective cuvettes at about the same time. The cuvettes are siphons, so that air can be automatically drawn up through the mixture immediately after dumping. The air cycle triggers the computer to start collecting data on each successive cuvette for any given number of readings. Since the rotor speed is about 800 rpm, absorbance measurements collected from all cuvettes in one revolution occur almost simultaneously. Collected data are rapidly reduced and processed by computer programs and either displayed in standard ordinates on a cathode ray tube (CRT) and/or are printed on paper by a teletypewriter. The dumping of the sample and reagent as well as the progress of each reaction can be followed visually on an oscilloscope. While the reactions from the first set of samples are being collected and reduced by computer, a second and third distributor disk are readied with the remaining zonal fractions. After all the data is collected from one set of samples, the cuvettes are emptied by air pressure, washed several times, dried with air, and the rotor stopped. The second distributor disk is inserted and another set of data is collected. This analyzer is solving one of the major problems of separations in zonal rotors—the time involved in doing multiple assays on all the collected fractions.

The S-ρ Procedure

It is apparent from Fig. 4 that there are several components from rat liver cells which have the same sedimentation rate in sucrose but different densities. To exploit differences in banding density, the sucrose fractions can be centrifuged in a second gradient material to yield a 2-dimensional gradient separation. Resolution of separation with this S-ρ technique is quite high if the particulates can withstand the second gradient material. A significant concentration is achieved with this procedure.

1. Line up 12 polycarbonate tubes for the No. 30 Spinco rotor in a rack. Label each with the corresponding gradient fraction number.

2. Put into each of the above tubes 2 each of a series of plastic beads[30] which have various discrete densities. For isolation of glycogen the density must exceed 1.63, so add beads ranging from 1.1 to 1.65.

3. There is a choice here. Either add dry CsCl or a solution of CsCl. If dry material is used, add it to each tube and pour in enough sucrose

[30]Spinco Division, Beckman Instruments, Palo Alto, California.

gradient fraction to fill to the neck of the tube. If dry CsCl is used, do step 7 next.

4. If saturated CsCl (buffered with phosphate, pH 7.0) is used, first transfer 20 ml of each gradient fraction to the No. 30 tubes. Put a short (5 inch) Pasteur pipette into each tube, small end down.

5. Fill a syringe with the saturated CsCl solution. Use a 20-gauge needle.

6. Carefully run 7 ml of the CsCl from the syringe down the pipette to the bottom of the tube. The CsCl will layer nicely under the sucrose fraction. Pull the pipettes out.

7. Screw on the tube caps and carefully put each tube into the No. 30 rotor. Remember, top rotor speed must be reduced because of the high density solution.

8. Very slowly accelerate the rotor on the following schedule. Verify top rotor speed for your system: 5 minutes at 2000 rpm; 5 minutes at 5000 rpm; 10 minutes at 10,000 rpm; 2 hours at 25,000 rpm. Longer periods may be required for small particles or to get dry CsCl into solution.

The programmed speed increase permits the larger particles to orient themselves in the gradient before hitting the outer tube wall. For very large components the acceleration program should be lengthened to keep the particles from hitting the tube wall.

9. Brake the rotor to 2000 rpm, turn off the brake, and let the rotor decelerate to rest. Bleed the vacuum and carefully remove the rotor. Do NOT shake, spin, or wobble the rotor. Remove the tubes gently and put them in a rack. Bands should be visible to the unaided eye. Too rapid deceleration, especially from about 3 rpm to rest, will create large numbers of very thin bands. The same "rapid" deceleration will also resolve two bands which had been very close together.

10. Get a permanent record of the position of the bands with respect to the position of the reference beads. You can either draw what you see or get a photographic record of light scattered from the bands. The photographic record is best obtained using a banding camera.[31] This device will give both polaroid pictures and 4×5 negatives. Time exposures will often show zones that are not readily apparent in normal room light.

11. Recover the bands by using either a Band Recovery Apparatus[32] (preferable) or by hand with syringes or pipettes.

12. Recovered bands are dialyzed free of CsCl and sucrose against buffer by using small diameter tubing (½ inch diameter). Suspend

[31]Electro-Nucleonics, Inc., Fairfield, New Jersey.
[32]Electro-Nucleonics, Inc., Fairfield, New Jersey.

the sacs just under the surface of the buffer in a 250-ml graduated cylinder. The CsCl will stream to the bottom of the cylinder and stay there. The buffer will not normally need to be changed if only a few samples are being dialyzed. The CsCl in the bottom stays reasonably concentrated and facilitates recovery and reuse.

13. Photographs from above can be cut into strips and localized beneath the zonal fractions for the production of the pycnogram.[33]

Preparing Zonal Fractions for Electron Microscopy

Sedimentation through a gradient may affect the ultrastructure of some subcellular components through changing osmotic pressure, shear forces, penetration of gradient materials, digestion due to gradient-contained enzymes, and possibly other factors. Ultrastructure can best be maintained by prior fixation of the starting sample with glutaraldehyde before the gradient fractionation.

For prefixation, substitute the homogenizing medium with a 2% solution of glutaraldehyde buffered with 0.1 M phosphate buffer, pH 7.3. Homogenize and prepare as described in *Sample Preparation* and allow the mixture to stand at least 6 hours in the refrigerator before fractionating in the zonal centrifuge. After the separation, choose the fraction(s) that contain the particles, dilute 1:1 with the same buffer used above and pellet the particles in an angle head rotor at 150,000 g for 1 hour. Carefully aspirate and discard the supernatant.

Prepare a 2% agar solution and maintain it at 45° for use. Put two drops of the warm agar (Pasteur pipette) on each pellet, mix well, and transfer the mixture to a glass microscope slide. Let the mixture solidify, cut it into 1 mm cubes with a razor blade, and prepare the cubes for embedding according to established techniques.[34]

Prefixation will preserve ultrastructure, but it can also alter enzyme activity so that assays may not work. In this case, make the zonal separation first and take only a portion of each fraction for electron microscope studies. There are two suggested methods for this approach, one quick way to show the size and numbers of particles present and the second a slow method for ultrastructure studies.

First is the free suspension technique wherein sucrose is removed from the particles by dialysis or by sedimentation. A drop of the sucrose-free suspension is put onto carbon-coated Formvar-covered grids and stained with 2% phosphotungstic acid (PTA) pH 7.0 for 15–45 seconds and then air-dried before examination in the microscope. The second

[33]N. G. Anderson, W. W. Harris, A. A. Barber, C. T. Rankin, and E. L. Candler, *Nat. Cancer Inst. Monogr.* 21, 253 (1966).
[34]J. H. Luft, *J. Biophys. Biochem. Cytol.* 9, 409 (1961).

method involves preparing pellets which are fixed and embedded for sectioning.[35]

Most Frequent System Malfunctions

1. Seal leaks. Can be due to multiple reasons. Most common is physical damage to Rulon seal by sucrose crystals, finger nails, or other abrasive items. Can also be due to undue seal wear caused by heating because of low seal coolant water flow. (Scratched seals can be easily repaired or replaced.) Can also be due to accidental pulling of the tygon tubing for seal coolant or both the center and/or edge lines, pulling the seal surfaces apart. The sealing surfaces are normally spring loaded to about 20 psi and are pivot mounted.

2. Poor vacuum. Most often occurs when the centrifuge refrigeration is kept on during the loading process and a protective cover is not used to limit air flow. Vacuum pump sublimates frost, and subsequent water in the vacuum oil reduces the amount of vacuum obtained. Rotor may thus heat up during high speed runs and thermal gradients may cause instability of the density gradient and cause zone widening.

3. Tight ball bearings in the seal assembly. The precision ball bearing race may become contaminated with sucrose from previous runs, crystallize, and "freeze". Thus when the seal assembly is put on the rotor at 2000 rpm, the bearing does not turn and so the "O" rings must slip. A burning odor may prevail. Usually the rotor is decelerated. The bearing can be "unfrozen" by putting the entire seal assembly in hot water. Grease can be removed by Freon treatment. Regrease before use. Tight bearings will get warm to the touch. If the housing gets too hot to touch, suspect that the grease is emulsified with sucrose and clean the bearings.

4. Noisy bearing. Can be due to lack of lubrication. (Use Corning Bearing lubricant No. 44 or manufacturer's suggested grease.) Can also be due to misalignment of seal assembly (not true for all types of assemblies).

5. Leaks from the rotor. The "O" ring under the lid may have been omitted, may not have been greased, or may have been cut during assembly of the rotor. Leaks due to this failure are noticeable during the loading of the gradient.

6. Rotor stops. Accidental hitting of the stop button, dialing in a low speed, or machine malfunction. Rotors that have been filled with gradient and/or sample can be restarted if they are accelerated very slowly to reorient the gradient. Some zone resolution can be lost.

---

[35]R. B. Ryel and G. B. Cline, Ala. Acad. Sci. Proc., in press.

7. Rotor wobble. Every rotor has vibrational frequencies (criticals) which it goes through during both acceleration and deceleration. At low speed (below 2000 rpm) the rotor wobbles noticeably at several ranges of rpm's. (Do not load rotor in one of these ranges.) By dialing in either a higher or lower speed the wobble will disappear. If rotors are partially filled they become highly unstable at higher speeds, i.e., greater than approximately 5000 rpm. This varies with the type of rotor. Decrease speed and fill rotor with additional cushion.

Common Operator Errors in Making Separations

1. Poor sample preparation: incomplete homogenization (large chunks of tissue), over homogenization, clumping (presence of ions), unfavorable pH, or the presence of other compounds.

2. Wrong sample density. When sample density is greater than the density of the top of the gradient, the sample submerges to its isopycnic level and sedimentation of particles begins from there.

3. Fast gradient loading. This can cause mixing of the gradient with subsequent loss in resolution. Best loading speeds range from 30 to 50 ml/minute.

4. Bubbles: Bubbles introduced to the periphery of the rotor rapidly float up through the gradient and mix it. Avoid solutions which degas on warming.

5. Seal leak: gradient or sample can be lost from the system by way of the seal. Two basic reasons for this: (a) line is still clamped on the opposite "end" of the system so that the fluid is pushed into a closed system. The seal is the safety valve in the system and yields to pressures about 18 psig; and (b) seal damage (see equipment malfunction).

6. Using large diameter tubing: Best resolution is retained by using smaller diameter tubing (e.g., ⅛ inch inside diameter).

7. Using a gradient which will not stop the particles in the time and at the speed the separation was made.

8. Loading the sample too fast: The best way is by hand at about 5–10 ml per minute. Loading speed must be increased when very large particles such as chloroplasts are to be separated on a rate basis.

9. Fast overlay addition: Jetting of overlay through rotor holes may mix and widen the starting sample zone.

10. Starting with solutions which are of unequal temperature: Thermal gradients can cause mixing.

11. Putting the gradient solutions in in the wrong order.

12. Pumping the gradient into the center of the rotor instead of into the edge line.

13. Misreading the density of the starting gradient solutions.

14. Leaving out the vanes of some early-type rotors during assembly.

Results

Figure 7 shows the results obtained by fractionating rat liver homogenate by protocol. Figure 8 shows human kidney; Fig. 9 shows liver from *Iguana;* and Fig. 10 shows how the protocol was modified to isolate virus inclusion bodies from infected *Heliothis zea*. Although the protocol was designed for rate separation, some large particulates, such as mitochondria, lysosomes, perixosomes, band together in a zone centered in 43.5% sucrose. All the fractions from the mammalian cells have not yet been identified.

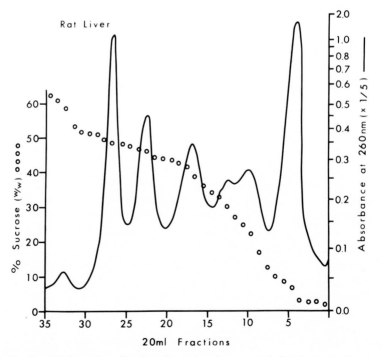

FIG. 7.   Result of rat liver fractionation using the protocol. Starting zone is centered in fraction 4 and sedimentation is from right to left.

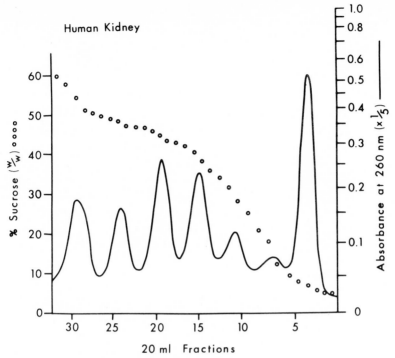

FIG. 8.   Fractionation of human kidney tissue using the protocol.

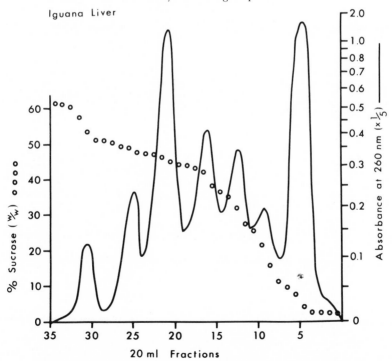

FIG. 9.   Fractionation of liver of *Iguana iguana*. Homogenized by 25 double strokes.

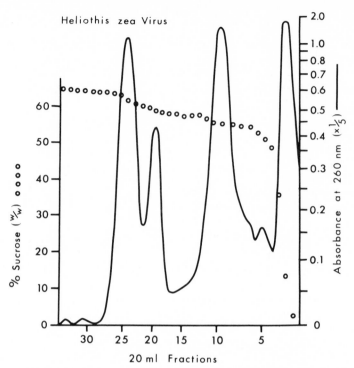

FIG. 10.   Fractionation of the nuclear polyhedra inclusion bodies (NPIB) from diseased cotton boll worm, *Heliothis zea*, showing how discontinuous gradients may be easily modified for fractionating many types of samples. NPIB's are centered in fraction 9. Separation was 3 hours at 30,000 rpm.

Appendix: Dictionary of Terms for Zonal Centrifugation

BAND. A zone.

BANDING. Centrifuging particles as zones. Generally relates to the stopping of the sedimenting particles in the gradient at their isopycnic levels.

BATCH-TYPE ZONAL ROTOR. A zonal rotor in which a fixed starting sample is layered over a gradient.

BUBBLE TRAP. Generally an inverted glass "Y" which is installed in the edge or sample lines to keep bubbles from getting into the rotor during loading or into the spectrophotometer during unloading and monitoring.

CLAMP. Generally a hemostat, used to stop flow in the tygon tubing.

COMBINATION GRADIENT. Gradient where factors in addition to density and viscosity are used to separate zones of material.

CONTINUOUS GRADIENT. Any gradient in which the major separating factor (i.e., density, viscosity, etc.) changes evenly down the gradient.

CONTINUOUS SAMPLE FLOW ROTOR. A zonal rotor which permits the constant replacement of depleted sample with fresh sample on a continuous and steady basis.

COOLANT LINES. Tygon tubing which carries cooling water to the seal assembly to decrease heat caused by friction between the static and dynamic seal parts.

CORE. The central portion of a zonal rotor which directs the flow of fluid, displaces part of the fluid volume of the rotor, and around which the septa fit (in some rotors).

CORE UNLOADING. Unloading a gradient from a dynamic loading rotor by displacement centripetally with piston solution.

DENSITY GRADIENT. A gradient where density changes as a function of volume of solution beginning with low density solution at the core of the rotor and gradually increasing to higher density at the periphery of the rotor.

DISCONTINUOUS GRADIENT. A gradient formed from 2 or more solutions, each differing in density so that a lighter density solution is always above a heavier density solution.

DYNAMIC LOADING OR UNLOADING. The process of putting in or taking out sample, gradient, or other solutions in a zonal rotor while the rotor is spinning. Requires a seal assembly.

EDGE AND SAMPLE LINES. Tygon tubing which lead to or from the seal assembly and which connect to paths in the rotor leading either to the periphery (edge) or to paths leading to the top of the gradient.

EDGE UNLOADING. Unloading a dynamic loading rotor by displacement of the gradient from the edge of the rotor. Can be done only in rotors designed specifically for this feature (B-XIV-A, B-XXIX, etc.). Such rotors can also be core unloaded.

EFFLUENT. The spent sample from a continuous sample flow rotor. This solution is the sample which has lost its particles to the gravitational force of the rotor.

FLO-BAND ROTOR. A continuous-sample-flow-with-particle-banding zonal rotor.

FRACTION. An aliquot of the gradient. The gradient is usually collected in sequential fractions either manually or with a fraction collector.

GRADIENT MATERIAL. A substance which can be used for the separation of particles or macromolecules as discrete zones.

GRADIENT PUMP OR ENGINE. A pumping machine which mixes two or more solutions of different density on a programmable basis for the

production of a gradient of some volume which can be loaded
into a zonal rotor.

GRADIENT REORIENTATION. The process of moving a density gradient
from vertical orientation in a rotor at rest to a horizontal orientation
at some rotating speed by slowly accelerating the rotor from rest
to permit the density gradient to climb the rotor wall and become
oriented in the horizontal plane. The process is reversed during
deceleration to let the gradient reorient back to the vertical position
for unloading.

HEAVY SOLUTION. A solution which has a density greater than other
solutions being used. A comparative term.

ISOPYCNIC OR ISODENSE BANDING. The stopping of sedimenting particles
in the gradient when the density of the gradient matches the
density of the particle in that gradient material.

LIGHT SOLUTION. A solution which has a density lighter than that of
other solutions being used.

LIQUID GRADIENT. A mixture of solute and solvent where the concentration of the solute changes across some given volume of mixture.

LOADING. The process of introducing a solution (gradient, sample,
underlay, overlay) into the zonal rotor to make a run.

MONITORING. The process of determining the location of banding zones
in the gradient stream during the unloading procedure.

OVERLAY. A solution of buffer or other material whose density is less
than the starting sample. It displaces the sample away from the
flat surface of the rotor core into a higher gravitational field.

PEAK. A zone of material detected by absorbance of light.

pH GRADIENT. Gradient where pH is varied down the sedimenting path
of materials.

PISTON. Solution of density equal to or greater than the underlay or
cushion. Piston is used to displace the gradient from the rotor
during core unloading.

RATE SEPARATION. Separation of various classes of particles on the
basis of differences in their size.

REFRIGERATION PLATE. A second piece of Plexiglas which fits loosely
over the rotor chamber to decrease wind movements in the chamber
and to decrease condensation on the cold wall of the centrifuge.

REOGRAD ROTOR. A rotor which is loaded statically and in which the
gradient is oriented during acceleration and reoriented during
deceleration.

ROTOR CAP. A small pressure-fitted cap which fits over the rotor spindle
and prevents the loss of the gradient during centrifugation in the
evacuated centrifuge chamber.

RUN. A separation in a zonal rotor.

SAFETY SHIELD. A piece of thick Plexiglas which fits tightly into the centrifuge chamber over the zonal rotor. Its purpose is to keep things from touching the spinning rotor during loading and unloading.

SAMPLE ZONE. The annular volume occupied in the gradient by the sample before acceleration of the rotor.

SEAL ASSEMBLY. The part of a zonal system which attaches to the rotor and permits loading and unloading while the rotor is spinning. This assembly contains water-cooled flat sealing surfaces of Rulon and stainless steel which face each other. One sealing surface spins with the rotor while the other is stationary. The dynamic portion spins inside a precision ball bearing which must be cleaned and greased occasionally.

SEPTA. Radial fins on the core which direct the rotor contents into sector-shaped components for more ideal sedimentation, carry solutions to the periphery, and make sure that the gradient in each sector accelerates or decelerates with the rotor. Generally 4 septa in the batch rotors and up to 6 in the flo-band rotors.

SOLUBILITY GRADIENT. A gradient which contains a substance or substances which affect the solubility of particles or macromolecules so that they either are precipitated and sediment faster or, if they are already precipitated, will be resolubilized and stop as zones.

STARTING SAMPLE. The sample introduced into a zonal rotor. Particles of interest are sedimented from the starting sample as zones.

STATIC LOADING OR UNLOADING. The process of putting in or taking out sample, gradient, or other solutions of certain zonal rotors while they are at rest (not spinning). May or may not require a seal assembly.

STEPS OR INTERFACES. The discrete areas between any two solutions which differ in density, viscosity, solubility, ionic concentration, etc.

THERMAL GRADIENT. Temperature gradient, generally undesirable in zonal rotors since temperature changes density and gradient mixing with a loss of resolution.

UNDERLAY OR CUSHION SOLUTION. A high density solution used to support the gradient away from the rotor wall and keep most particles from layering on the wall even though they may penetrate the gradient.

UNLOADING. Removal of the centrifuged gradient from the rotor. Can be done by displacement with piston solution for core unloading or by water for edge unloading.

VISCOSITY GRADIENT. A gradient of viscosity. For a single solution such as sucrose, viscosity changes follow density changes. This may be reversed for gradients where several different solutions are used.

ZONAL CENTRIFUGE. A centrifuge in which zonal rotors are used for separation of particulates and macromolecules.

ZONAL ROTOR. A hollow bowl rotor which spins about its verticle axis, has a central core, dividing septa, and holds a gradient in which useful separations can be made.

ZONE. A collection of particles of similar size or similar density which sediment together through a gradient. Since similar particles sediment together, zones of other sized particles may sediment faster or slower and thus be separated from each other.

[19] Extraction of Water-Soluble Enzymes and Proteins from Membranes

By HARVEY S. PENEFSKY and ALEXANDER TZAGOLOFF

Two classes of water-soluble proteins have been isolated from membranes. The first of these includes those proteins and enzymes which are present in the space which the membrane surrounds, but are not true components of the membrane itself. Examples of matrix enzymes are abundant in subcellular particles which have a high degree of internal organization (viz. mitochondria and chloroplasts). Matrix enzymes are characterized by solubility in water, by an absence of association with lipid, and by a lack of dependence on lipid for enzymatic activity. Matrix enzymes are usually isolated by procedures which disrupt the limiting membrane, thereby allowing the material entrapped by the membrane to be released.

A second class of proteins and enzymes are intrinsic components of the membrane which, when separated from their normal lipid environment, acquire the characteristics of classical water-soluble proteins. A large number of enzymes belonging to this class have been isolated from a wide spectrum of membranes. An important feature of such enzymes is that they are free of lipid and can be purified by standard procedures of protein fractionation.

The methods summarized below are illustrative of broad categories of approach to the problem of extracting proteins from membranes and of rendering these proteins soluble in dilute aqueous buffers. Of the examples of each type of approach listed in Table I, some were selected for detailed discussion. An exhaustive compilation, however,

TABLE I

Enzyme	Source	Procedure	Reference
Mitochondrial			
Coupling factor 1 (ATPase)	Beef heart	Nossal shaker	*a-c*
	Beef SMP*	Sonication	*d-f*
	Yeast	Nossal shaker	*g*
	Yeast SMP	Sonication	*h*
Coupling factor 2	Beef heart	Acetone powder	*i*
Coupling factor 3	Beef heart	Sonication	*j*
Oligomycin-sensitivity conferring protein	Beef SMP	Alkali	*k*
	ATPase particle	Alkali	*k*
Phosphoryl transferase	Beef heart	Sonication + EDTA	*l*
ATPase inhibitor	Beef heart	Alkali	*m*
Nonheme iron protein (hydroxylation)	Beef adrenal cortex	Sonication	*n*
Cytochrome c	Rat liver	Salt	*o*
	Rat liver	Cholate + deoxycholate + alcohol	*p*
	Beef heart SMP	Phospholipase A	*q*
Cytochrome c_1	Beef heart	Cholate, sodium lauryl sulfate	*r*
Cytochrome b	Beef heart	Succinylation	*s*
Cytochrome b_{555}	Mung bean	Lipase	*t*
Succinic dehydrogenase	Beef heart SMP	Acetone	*u*
	Beef heart SMP	Butanol + alkali	*v*
DPNH dehydrogenase	Beef heart	Acid ethanol	*w*
	Beef heart SMP	Phospholipase A	*x, y*
	Beef heart SMP	Thiourea	*y*
	DPNH-CoQ reductase particle	chaotropic reagents	*z*
	Yeast SMP	Ethanol	*aa*
α-Glycerphosphate dehydrogenase	Pig Brain	Acetone powder + phospholipase A	*bb*
Choline dehydrogenase	Rat liver	Acetone powder + phospholipase A	*cc*
Microsomal			
Phosphodiesterase	Hog kidney	*tert*-Amyl alcohol	*dd*
Alkaline phosphatase	Milk	*n*-Butanol	*ee*
Cytochrome b_5	Calf liver, rabbit liver	Lipase	*ff*
	Calf liver	Triton X-100 + deoxycholate	*gg*
	Rat liver	Trypsin	*hh, ii*
DPNH-cytochrome b_5 reductase	Calf liver	Phospholipase A	*jj*
P-420	Rabbit liver	Phospholipase A + deoxycholate	*kk*
TPNH-cytochrome c reductase	—	Trypsin	*hh, ii*
	Pork liver	Lipase	*ll*

5'-Nucleotidase	Rat liver	Sonication	*mm*
Mammalian plasma membrane			
Leucyl-β-naphthylamidase	Hepatoma	Papain	*nn*
Cholinesterase	Beef erythrocytes	*n*-Butanol	*oo*
	Human erythrocytes	Tween	*pp*
Bacterial plasma membrane			
ATPase	*Streptococcus faecalis*	Washing	*qq*
	Bacillus megaterium	dialysis + alkali	*rr*
Cytochrome *o*	*B. megaterium*	Lipase	*ss*
UGP-teichoic acid glucosyl transferase	*Bacillus subtilis*	Chaotropic	*tt*
Chloroplasts			
Coupling factor 1 (ATPase)	Spinach	EDTA	*uu*
Plastocyanin	*Chlorella*	Sonication	*vv, ww*

[a]M. E. Pullman, H. S. Penefsky, A. Datta, and E. Racker, *J. Biol. Chem.* 235, 3322 (1960).
[b]H. S. Penefsky, this series, Vol. 10 [82].
[c]M. E. Pullman and H. S. Penefsky, this series, Vol. 6 [34].
[d]D. H. MacLennan, J. M. Smoly, and A. Tzagoloff, *J. Biol. Chem.* 243, 1589 (1968).
[e]A. Datta and H. S. Penefsky, *J. Biol. Chem.* 245, 1537 (1970).
[f]L. L. Horstman and E. Racker, *J. Biol. Chem.* 245, 1336 (1970).
[g]G. Schatz, H. S. Penefsky, and E. Racker, *J. Biol. Chem.* 242, 2552 (1967).
[h]A. Tzagoloff, *J. Biol. Chem.* 244, 5020 (1969).
[i]J. Fessenden and E. Racker, *J. Biol. Chem.* 241, 2483 (1966).
[j]J. Fessenden and E. Racker, this series, Vol. 10 [84].
[k]D. H. MacLennan and A. Tzagoloff, *Biochemistry* 7, 1603 (1968).
[l]R. E. Beyer, this series, Vol. 10 [81].
[m]M. E. Pullman and G. C. Monroy, *J. Biol. Chem.* 238, 3762 (1963).
[n]T. Omura, E. Sanders, D. Y. Cooper, O. Rosenthal, and R. W. Estabrook, *in* "Non-Heme Iron Proteins" (A. San Pietro, ed.), p. 401. Antioch Press, Yellow Springs, Ohio, 1965.
[o]E. E. Jacobs and R. Sanadi, *J. Biol. Chem.* 235, 531 (1960).
[p]P. V. Blair, T. Oda, D. E. Green, and H. Fernández-Morán, *Biochemistry* 2, 756 (1963).
[q]K. S. Ambe and F. L. Crane, *Science* 129, 98 (1959).
[r]R. Bromstein, R. Goldberger, and H. Tisdale, *Biochim. Biophys. Acta* 50, 527 (1961).
[s]D. H. MacLennan, A. Tzagoloff, and J. S. Rieske, *Arch. Biochem. Biophys.* 109, 383 (1965).
[t]H. E. Kasinsky, H. Shichi, and D. P. Hackett, *Plant Physiol.* 41, 739 (1966).
[u]T. P. Singer, E. B. Kearney, and P. Bernath, *J. Biol. Chem.* 223, 599 (1956).
[v]D. Keilin and E. F. Hartree, *Biochem. J.* 41, 500 (1947).
[w]B. Mackler, *Biochim. Biophys. Acta* 50, 141 (1961).
[x]T. E. King and R. L. Howard, *J. Biol. Chem.* 237, 1686 (1962).
[y]T. E. King and R. L. Howard, this series, Vol. 10 [52].
[z]Y. Hatefi, personal communication; see also K. A. Davis and Y. Hatefi, *Biochemistry* 8, 3355 (1969).
[aa]B. Mackler, this series, Vol. 10 [53].
[bb]R. L. Ringler, S. Minakami, and T. P. Singer, *Biochem. Biophys. Res. Commun.* 3, 417 (1960).
[cc]T. Kimura and T. P. Singer, this series, Vol. 5 [76].
[dd]W. E. Razzell, this series, Vol. 6 [29].
[ee]R. K. Morton, *Nature (London)* 166, 1092 (1950).
[ff]P. Strittmatter and S. F. Velick, *J. Biol. Chem.* 221, 253 (1956).
[gg]A. Ito and R. Sato, *J. Biol. Chem.* 243, 4922 (1968).
[hh]A. H. Phillips and R. G. Langdon, *J. Biol. Chem.* 237, 2652 (1962).

[ii]T. Omura, P. Siekevitz, and G. E. Palade, *J. Biol. Chem.* **242**, 2389 (1967).
[jj]P. Strittmatter, this series, Vol. 10 [91].
[kk]T. Omura and R. Sato, *J. Biol. Chem.* **237**, PC1375 (1962).
[ll]B. S. S. Masters, C. H. Williams, Jr., and H. Kamen, this series, Vol. 10 [92].
[mm]C. S. Song, J. S. Nisselbaum, B. Tandler, and O. Bodansky, *Biochim. Biophys. Acta* **150**, 300 (1968).
[nn]P. Emmelot, A. Visser, and E. L. Benedetti, *Biochim. Biophys. Acta* **150**, 364 (1968).
[oo]J. A. Cohen and M. G. P. J. Warringa, *Biochim. Biophys. Acta* **10**, 195 (1953).
[pp]C. A. Zittle, E. S. Dellamonica, and J. H. Custer, *Arch. Biochem. Biophys.* **48**, 43 (1954).
[qq]A. Abrams, *J. Biol. Chem.* **240**, 3675 (1965).
[rr]M. Ishida and S. Mizushima, *J. Biochem. (Tokyo)* **66**, 133 (1969).
[ss]P. L. Broberg, and L. Smith, *Biochim. Biophys. Acta* **172**, 439 (1967).
[tt]Y. Hatefi and W. G. Hanstein, *Proc. Nat. Acad. Sci. U. S.* **62**, 1129 (1969).
[uu]R. E. McCarty and E. Racker, *Brookhaven Symp. Biol.* **19**, 202 (1966).
[vv]J. J. Lightbody and D. W. Krogman, *Biochim. Biophys. Acta* **131**, 508 (1967).
[ww]S. Katoh and A. Takamiya, *Biochim. Biophys. Acta* **99**, 156 (1955).

*Abbreviations: SMP, submitochondrial particles

was not attempted. Examples were selected only if it could be ascertained that the starting material was a well-washed membrane preparation containing the protein of interest. Methods of preparing and purifying plasma membranes of animal cells have been discussed by Maddy.[1]

Criteria of Solubility

Many workers consider their preparation "soluble" if it remains in the supernatant solution after centrifugation at 100,000 g for one or more hours. However, procedures such as sonic irradiation, mechanical disruption, or extraction at alkaline pH, may lead to dispersion of lipoprotein or to formation of small membrane fragments which do not sediment readily and thus may appear to be soluble. Acidification to pH 5 to 6 or addition of salts to a final concentration of 1–2% is a simple, though by no means infallible, step which may help to distinguish between proteins in true solution and suspended lipoprotein particles or lipoproteins. Lipoprotein material frequently will precipitate under such conditions and can be removed by low speed centrifugation. It should be borne in mind that some membrane proteins are soluble only in media of low ionic strength.[2]

Prolonged centrifugation at high force may also be employed to sediment particulate material. However, under such conditions water-soluble complexes of high molecular weight, i.e., on the order of several millions, may cosediment with membrane fragments. A useful procedure which permits separation of high molecular weight protein

[1]A. H. Maddy, *Int. Rev. Cytol.* **20**, 1 (1966).
[2]D. Mazia and A. Ruby, *Proc. Nat. Acad. Sci. U.S.* **61**, 1005 (1968).

from small membrane fragments is centrifugation through a column of sucrose. The mixture is applied to the top of a column of solution, the density of which is less than that of protein but greater than that of the lipoprotein particles. The proteins of interest move into the dense medium during centrifugation while the low density material remains near the top.

Protein Interactions with Membranes

While it is apparent that hydrophobic interactions make large contributions to the overall stability of membranes, the associations of individual proteins within the confines of larger molecular complexes may be due to electrostatic interactions as well. The concept of a protein as a partner in a highly integrated multiprotein complex suggests that many of the interactions are of a specific nature. It may well be that the type of surface fit postulated in antigen–antibody complexes will also participate in the organization of the macromolecular complexes of membranes. Antigen–antibody complexes can form spontaneously in aqueous solvents and are characterized by unusual stability. In view of these considerations, successful separation of a protein from the membrane may require conditions which not only weaken the forces that stabilize the complex, but also prevent reassociation due to highly specific interactions. The proteins of membranes may also be complexed with amphiphatic lipids, glycolipids, and the like.

These latter interactions are probably less specific than protein–protein interactions.

1. *Protein-Lipid Complexes.* Proteins may complex with amphiphatic lipid either through electrostatic or hydrophobic interactions.[3,4] Complexes stabilized only by electrostatic forces usually can be dissociated by exposure to salts.[3,4] Complexes stabilized by hydrophobic interactions can frequently be disrupted following extraction with lipophilic solvents,[3,4] or enzymatic degradation of the lipid.

2. *Protein-Metal Complexes.* Some proteins may be complexed to membrane proteins or lipids through metals such as Mg^{2+} or Ca^{2+}. Complexes of this type can be dissociated with the aid of appropriate chelating agents or by otherwise removing the metals involved.[5,6]

3. *Specific Protein-Protein Complexes.* Protein-protein complexes may be formed through interactions between complementary surfaces in which both electrostatic and hydrophobic forces singly or combined are

[3]D. E. Green and A. Tzagoloff, *J. Lipid Res.* 7, 587 (1966).
[4]D. E. Green and S. Fleischer, *in* "Metabolism and Physiological Significance of Lipids" (R. M. C. Dawson and N. Rhodes, eds.), p. 581. Wiley, New York, 1965.
[5]A. Abrams, *J. Biol. Chem.* 240, 3675 (1965).
[6]R. E. McCarty and E. Racker, *Brookhaven Symp. Biol.* 19, 202 (1966).

involved. Conditions which weaken these forces may lead only to a reversible dissociation of the complex and separation of the components by centrifugation may be poor. In many instances however, a protein is irreversibly released from the membrane by conditions which denature or otherwise modify one or more components of the complex. Thus, for example, a soluble ATPase is released from submitochondrial particles following incubation at 65°.[7]

Physical Methods

Homogenizers. Mechanical tissue and cell homogenizers have not been widely used to solubilize membrane proteins. Nevertheless, in a few instances efficient disruption and solubilization has been reported. The Nossal shaker[8] is a reciprocating device which shakes the contents of a small stainless steel cylinder at 6000 cycles per minute. The displacement is about 0.5 inch. The material to be processed is placed in the cylinder with a charge of glass beads which provide the abrasive action. An air-cooled machine with a cylinder capacity of 10 ml[9] and a water-cooled model with a capacity of 25 ml[10] are available commercially. The extent of disruption depends on several factors including the size of the glass beads used, the ratio of bead volume to liquid volume and the duration of shaking. Optimal conditions are best determined by trial and error. A systematic analysis of breakage conditions for a somewhat similar mechanical shaker has been presented.[11] The latter results may be useful in deciding on conditions for use in the Nossal shaker. The considerable heat generated during shaking may require operation with intermittent periods of cooling. The cylinder temperature of the air-cooled shaker may rise about 10° in 10 seconds during operation in a cold room. A similar rise in temperature occurs after about 1 minute when the water-cooled shaker is operated with circulating ice water. Nevertheless, deliberate operation of a shaker at high temperatures (above 45°) has led to improved solubilization of mitochondrial ATPase.[12] Formation of foam during shaking, which may be related to decreased yields of soluble proteins and of active enzyme, can be suppressed by partial evacuation of the shaking cylinder[13] or by addition of antifoam agents.[14]

---

[7] A. Tzagoloff, unpublished experiments.
[8] P. M. Nossal, *Aust. J. Exp. Biol. Sci.* **31**, 583 (1953).
[9] McDonald Engineering Company, Bay Village, Ohio.
[10] Lourdes Instrument Corp., Brooklyn, New York.
[11] A. Rodgers and D. E. Hughes, *J. Biochem. Microbiol. Tech. Eng.* **2**, 49 (1960).
[12] E. Racker, personal communication.
[13] M. E. Pullman, H. S. Penefsky, A. Datta, and E. Racker, *J. Biol. Chem.* **235**, 3322 (1960).
[14] H. S. Penefsky, unpublished experiments.

The French press[15] and the Gaulin homogenizer,[16] the latter operating on a principle similar to that of the French press and permitting large amounts of material to be processed continuously, are highly efficient when used to disrupt cells and may be adaptable to the solubilization of membrane proteins as well.

Sonic Oscillators. Exposure to sonic vibrations has proved to be an effective means of disrupting cells and subcellular membranes and also has been widely used as an aid in solubilizing membrane proteins. Of the two types of sonic oscillator available commercially, one contains a probe which may be immersed in the suspension to be treated[17] and the other consists of a chamber, the bottom of which is connected to the vibrating source.[18] Although both types of instrument are effective, the probe type is more versatile, since amounts as small as 0.5 ml, as well as large volumes of suspensions may be processed.

The efficiency of solubilization of membrane proteins by sonic oscillations is influenced by the power output of the instrument, the duration of exposure and the volume of material processed. Alterations in these parameters as well as alterations in the pH and the ionic strength of the medium make available a wide range of conditions of which the most suitable must usually be selected by trial and error. Sonic irradiation of membranes in media of very low ionic strength or at aklaline pH may give rise to greater solubilization. Since a considerable fraction of the energy transmitted by the sonic oscillator is dissipated into the medium as heat, efficient cooling may be essential. However, when the temperature was allowed to rise during sonic irradiation, greater solubilization of mitochondrial ATPase from submitochondrial particles[19] resulted. Jacketed vessels[17,18] as well as specially designed glass circulating cups[20] which provide efficient cooling are available commercially.

Chemical Methods

Extraction with Solutions of Low Ionic Strength. Proteins such as cytochrome *c* may be associated with membrane components such as protein or phospholipid through electrostatic interactions.[21] The results of a

[15]American Instrument Co., Silver Springs, Maryland.
[16]Manton-Gaulin Laboratory Homogenizer, Manton-Gaulin Manufacturing Co., Inc., Everett, Massachusetts.
[17]Heat Systems Company, Melville, New York. MSE Sonic Oscillator, Instrumentation Associates, Inc., New York, N.Y.
[18]Raytheon Company, Waltham, Massachusetts.
[19]A. Datta and H. S. Penefsky, *J. Biol. Chem.* 245, 1537 (1970).
[20]Rosett Cell, Heat Systems Company, Melville, New York.
[21]M. L. Das, E. D. Haak, and F. L. Crane, *Biochemistry* 4, 859 (1965).

number of studies[22,23] indicate that cytochrome c is readily extracted from swollen mitochondria with an isotonic solution of KCl.

Solubilization of membrane proteins has also been achieved by procedures involving prolonged dialysis against water. Solubilization of up to 50% of proteins from ghosts of *Bacillus subtilis* after dialysis for 48 hours against water at pH 8 has been reported.[24] A similar type of procedure was used by Harris[25] to solubilize proteins of the erythrocyte ghost. Considerable solubilization of proteins of the erythrocyte ghost was achieved[2] after dialysis for 18 hours against distilled water brought to pH 9.3 to 9.5 with NH₃. As much as 80% of the membrane protein remained in the supernatant solution after 1 hour at 100,000 g. Although 0.1% Triton X-100 was used during lysis of the cells, the function of this detergent was not clear, and substantial recovery of soluble membrane protein was observed (about 50%) without detergent. The authors emphasize the importance of low ionic strength for effective solubilization during dialysis. Addition of 100 mM KCl to the solubilized proteins led to sedimentation of about 70% of the material at 100,000 g. The preparation contained relatively less lipid than the original whole ghosts but more than a solubilized protein preparation obtained by *n*-butanol extraction.[26]

A soluble protein was extracted from lysed and washed red cell membranes by dialysis of the membranes for 24–48 hours against a solution containing 0.3 mM ATP and 50 mM mercaptoethanol, pH 7.5.[27] Approximately 20% of the membrane protein remained in the supernatant solution after centrifugation at 78,000 g for 2 hours. Dialysis against dilute solutions of EDTA in distilled water also was effective in solubilizing the protein.

Extraction with Concentrated Salts and with Urea. The effect of high concentrations of salt, urea, and compounds such as guanidinium chloride on the structure of biological macromolecules has been the subject of intensive investigation.

It was observed in a number of systems that a Hofmeister series resulted when anions which disrupted protein structure[28] or caused a loss of enzyme activity[29,30] were ranked in terms of relative effectiveness.

[22] E. E. Jacobs and R. Sanadi, *J. Biol. Chem.* **235**, 531 (1960).
[23] D. H. MacLennan, G. Lenaz, and L. Szarkowska, *J. Biol. Chem.* **241**, 5251 (1966).
[24] T. Yamaguchi, G. Tamura, and K. Arima, *J. Bacteriol.* **93**, 438 (1967).
[25] J. R. Harris, *Biochim. Biophys. Acta* **150**, 534 (1968).
[26] A. H. Maddy, *Biochim. Biophys. Acta* **117**, 193 (1966).
[27] V. T. Marchesi and E. Steers, *Science* **159**, 203 (1968).
[28] P. H. Von Hippel and K. Y. Wong, *Science* **145**, 577 (1964).
[29] I. Fridovich, *J. Biol. Chem.* **238**, 592 (1963).
[30] H. S. Penefsky and R. C. Warner, *J. Biol. Chem.* **240**, 4694 (1965).

This series is usually expressed as: $SCN^- > ClO_4^- > NO_3^- > I^- > BR^- > SO_4^- > CH_3COO^-$. Other types of anion and cation series,[31] including some which stabilize the structure of macromolecules also have been described.

The effectiveness of urea[27,32] and high concentrations of salts[33] as tools for the extractions of proteins from membranes has already been demonstrated. Recently a systematic examination was undertaken[34] of the effectiveness of the Na^+ salts of a Hofmeister series of anions, as well as urea and guanidinium chloride, as "extracting" agents for proteins from the plasma membrane of *Bacillus subtilis*, from erythrocyte ghosts, and from submitochondrial particles. It was found that treatment with $2 M$ $NaClO_4$ or with similar concentrations of urea or guanidine hydrochloride led to extraction of 27% or more of the total membrane protein. It should be noted that the authors considered the extracted proteins as "soluble" in the extraction medium if they did not sediment after 1.5–2 hours of centrifugation at 105,000 g. Rapid oxidation of lipids was observed in the presence of several of the extracting agents unless special precautions were taken.[34] Subsequently, the extraction of reduced diphosphopyridine nucleotide dehydrogenase and an iron-sulfur protein from submitochondrial particles and from complex I by low (approximately 0.5 M) concentrations of $NaClO_4$, NaSCN, or guanidine hydrochloride also was described.[35] The authors properly emphasized the *increased* water solubility of their preparations as a result of the solvents employed. The latter were called "chaotropic agents." Thus far, true water solubility has been achieved with DPNH dehydrogenase, an iron sulfur protein[35,36] and succinic dehydrogenase.[37]

A most serious limitation to the use of these agents is the well known observation, mentioned above, that these compounds are highly effective denaturants of proteins. In fact, the most effective extracting agents appear in general to be the most effective protein denaturants. Nevertheless, it may be assumed that some enzymes can refold with consequent resumption of enzymatic activity after exposure to conditions, e.g., 6 M urea, which lead to extensive unfolding. It may well be that enzymes extracted from membranes with the aid of these or other strong reagents will remain soluble and regain activity if permitted to refold in diluted aqueous buffers by the techniques described by Deal.[38]

[31] K. Hamaguchi and E. P. Geiduschek, *J. Amer. Chem. Soc.* **84**, 1329 (1962).
[32] Y. Kagawa and E. Racker, *J. Biol. Chem.* **241**, 2461 (1966).
[33] D. H. MacLennan, J. M. Smoly, and A. Tzagoloff, *J. Biol. Chem.* **243**, 1589 (1968).
[34] Y. Hatefi and W. G. Hanstein, *Proc. Nat. Acad. Sci. U. S.* **62**, 1129 (1969).
[35] K. A. Davis and Y. Hatefi, *Biochemistry* **8**, 3355 (1969).
[36] Y. Hatefi, personal communication.
[37] Y. Hatefi, K. A. Davis, and W. G. Hanstein, *Arch. Biochem. Biophys.* **137**, 286 (1970).
[38] W. C. Deal, *Biochemistry* **8**, 2795 (1969).

Alkali Extraction. The influence of pH on the solubilization of several membrane systems is illustrated in Fig. 1. It would seem that the titrations of charged groups in the membrane and the resulting repulsive forces may play an important role in the dissociation of membrane components. Though extensive solubilization occurs at a pH which causes loss of activity of many enzymes, components of the mitochondrial inner membrane have been successfully solubilized at pH 8–11. These include succinic dehydrogenase,[39,40] the oligomycin-sensitivity conferring protein,[41] and the ATPase inhibitor.[42]

Metal Chelators. Isolated cell membranes, particularly bacterial plasma membranes are stabilized by the inclusion of such divalent cations as Mg^{2+} in the suspending medium. Procedures that remove the bound metals can lead to release of proteins from the membrane. The amount of protein released varies with the source of membranes and depends on the extent of metal depletion. Effective procedures include washing

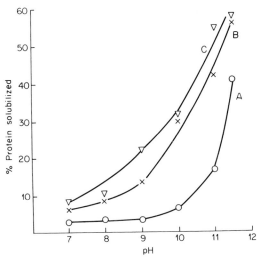

Fig. 1. The effect of pH on the solubility of membrane proteins. Membranes were first suspended in a buffer containing 0.25 M sucrose and 0.1 M Tris acetate, pH 7.5, at a protein concentration of 10 mg/ml. The pH of each mixture above was brought to the indicated pH with NaOH and then centrifuged at 105,000 g for 60 minutes. The percentage of protein solubilized represents the fraction of the total protein found in the clear supernatant solution. Curve A: Submitochondrial particles from beef heart. Curve B: Rat liver microsomes. Curve C: Plasma membranes from *Bacillus subtilis*.

[39]T. E. King, *Biochim. Biophys. Acta* **58**, 375 (1962).
[40]T. E. King, *J. Biol. Chem.* **238**, 4037 (1963).
[41]D. H. MacLennan and A. Tzagoloff, *Biochemistry* **7**, 1603 (1968).
[42]M. E. Pullman and G. C. Monroy, *J. Biol. Chem.* **238**, 3762 (1963).

by resuspension in the centrifuge, dialysis, and extraction with metal chelators. The Mg^{2+}-dependent ATPase of the ghosts of *Streptococcus faecalis* has been solubilized by repeated washing (as many as 7 times) with dilute buffers.[5] The success of the procedure appears to be related to the gradual removal of metals from the membrane. The ATPase of spinach chloroplats was released in soluble form when these organelles were diluted in buffers containing EDTA.[6] Dilution in the absence of the metal chelator was ineffective.

Organic Solvents. An informative review on the solubility of proteins in organic solvents has been presented by Singer.[43] The exposure of membranes to organic solvents under conditions which extract lipids or perturb the internal arrangements of lipids can lead to the extraction of proteins.

ACETONE. The extraction of soluble proteins from acetone powders is a classical procedure.[44] The extracted proteins are depleted of lipids since the latter usually remain in the acetone phase.

BUTANOL. Keilin and Hartree[45] first used *n*-butanol as a solubilizing agent in their studies on mitochondrial succinic dehydrogenase. Morton[46] emphasized the usefulness of this reagent in a study in which he solubilized a large number of particulate enzymes from intestinal mucosa. The low solubility of *n*-butanol in water combined with its lipophilicity appears to be related to the minimal denaturation of enzymes observed during extraction of membrane phospholipids.

In most procedures, *n*-butanol is added in a volume equal to that of the particle suspension which is maintained at about 5°. After mixing and centrifugation a two-phase system is obtained. The upper phase consists of butanol and membrane lipids while the lower phase contains solubilized protein. Lipid-rich material localizes at the interface whereas denatured proteins precipitate in the aqueous phase. The clear aqueous phase is collected by aspiration and dialyzed against a large volume of buffer to remove dissolved *n*-butanol.

An alternative procedure includes extraction with butanol of preparations of dried membranes. The lipid-depleted particles are then vacuum-dried and extracted with aqueous buffers. Cohen and Warringa[47] successfully isolated cholinesterase from ghost membranes of beef erythrocytes in this way.

[43]S. J. Singer, *Advan. Protein Chem.* 17, 1 (1962).
[44]R. K. Morton, this series, Vol. 1 [6].
[45]D. Keilin and E. F. Hartree, *Biochem. J.* 41, 500 (1947).
[46]R. K. Morton, *Nature (London)* 166, 1092 (1950).
[47]J. A. Cohen and M. G. P. J. Warringa, *Biochim. Biophys. Acta* 10, 195 (1953).

ETHANOL. DPNH dehydrogenase was solubilized by extraction with 9% ethanol of submitochondrial particles from beef heart[48] and yeast.[49] The incubations were carried out at 35° to 43° at pH 4.8 (beef) and pH 5.5 (yeast).

PYRIDINE. Blumenfeld[50] has reported the isolation of a water-soluble sialoprotein from pyridine extracts of red cell ghosts. One volume of the lysed and well-washed membranes were treated with one-half volumes of pyridine at 4°. The membranes dissolved quickly and were then dialyzed against cold distilled water for 18 hours to remove pyridine. The supernatant solution following centrifugation at 100,000 g for 30 minutes contained 30–40% of the membrane protein and all the sialic acid, most of which was present in the form of a soluble sialoprotein.

Surfactants. Surfactant reagents have been widely used to separate both lipoprotein and water-soluble proteins from membranes. Among the most frequently used surfactants are bile salts and synthetic detergents, such as Triton X-100, sodium lauryl sulfate, Lubrol,[51] and Tween. Cytochrome c_1 from beef heart mitochondria was obtained in water-soluble form and in good yield.[52] The enzyme was first extracted from the mitochondria with sodium deoxycholate and then purified via a number of steps which included treatment with sodium cholate, sodium lauryl sulfate, ammonium sulfate fractionation, and elution from calcium phosphate gels.[52] Cytochrome b_5 from rabbit liver microsomes was brought to a high degree of purity by Ito and Sato.[53] This protein was solubilized in a mixture of 1% each of Triton X-100 and sodium deoxycholate. Following ammonium sulfate fractionation, the preparation was exchanged with 1% Triton X-100 on a column of Sephadex G-50 and then eluted from a column of DEAE-Sephadex A-50, equilibrated with buffered 0.5% Triton X-100, with a linear gradient to 0.3 M KCl. Additional steps employed acetone to extract lipids from the preparation.

In order to distinguish between true solution and formation of surfactant-protein or lipoprotein complexes of low density which will not sediment in high centrifugal fields, it is necessary to remove the dispersing agent. Dialysis is frequently effective. Alternatively, bile salts as well as Triton and Tween can be separated from proteins by passage

[48]B. Mackler, *Biochim. Biophys. Acta* 50, 141 (1961).
[49]B. Mackler, this series, Vol. 10 [53].
[50]O. Blumenfeld, *Biochem. Biophys. Res. Commun.* 30, 200 (1968).
[51]General Biochemicals, Chagrin Falls, Ohio.
[52]R. Bomstein, R. Goldberger, and H. Tisdale, *Biochim. Biophys. Acta* 50, 527 (1961).
[53]A. Ito and R. Sato, *J. Biol. Chem.* 243, 4922 (1968).

through a column of Sephadex G-25. It should be noted that Sephadex column treatment may not be adequate if the detergent concentration is greater than 1%. Lubrol may be removed from extracts of microsomes by passage through a column of Sephadex LH-50.[54] Proteins which remain in solution following procedures designed to remove surfactants as described are usually considered soluble. However, tight binding of residual amounts of dispersant, sufficient to maintain solubility of protein preparations, is not ruled out. It should be possible to arrive at an accurate estimate of the amount of residual surfactant in a preparation by making use of surfactants labeled with radioisotopes when possible or by other types of sensitive assay such as the procedure of Mosbach et al.[55] for the estimation of deoxycholate and cholate. The latter method was employed to establish that the final purified preparation of cytochrome c_1 described above contained less than 0.005 mg cholate per milligram of protein.[52]

Chemical Modification. Changes in the net charge of membrane proteins may result in large changes in water solubility. In addition to alteration by pH and salts, the net charge on a protein can be modified by the introduction of new chemical groups. Succinylation of the ε-amino group of lysine residues with succinic anhydride has been shown to be a highly effective means of promoting the water solubility of membrane proteins.[56] The membrane preparation is suspended in 10 mM phosphate buffer, pH 7.5, so that the final membrane concentration is 5–20 mg/ml. An amount of solid succinic anhydride is used which represents a 50–100-fold molar excess over the expected lysine content of the preparation and is added slowly with rapid stirring at 4°. The pH of the reaction mixture is maintained between 7 and 8 by appropriate additions of alkali. The reaction is considered at an end when no further decrease in pH is observed. The extent of succinylation may be determined by assaying for ninhydrin-positive material before and after the reaction using the procedure of Moore and Stein.[57] Alternatively ^{14}C-labeled succinic anhydride may be used and the specific radioactivity of the modified protein determined. A serious disadvantage of this technique is that loss of enzyme activity may be irreversible since removal of covalently bound succinic acid requires strong hydrolytic conditions which may also result in peptide bond hydrolysis.

[54] J. L. Gaylor and C. V. Delwiche, *Anal. Biochem.* 28, 361 (1969).
[55] E. H. Mosbach, H. J. Kalinsky, E. Halpern, and F. E. Kendall, *Arch. Biochem. Biophys.* 51, 402 (1954).
[56] D. H. MacLennan, A. Tzagoloff, and J. S. Rieske, *Arch. Biochem. Biophys.* 109, 383 (1965).
[57] S. Moore and W. H. Stein, *J. Biol. Chem.* 176, 367 (1948).

A second procedure for introducing charged groups into a protein is maleylation.[58] The reaction of maleic anhydride with proteins appears to be similar to that of succinic anhydride except in that the former is more specific for the ε-amino groups of lysine whereas the latter is known to react as well with amino acids containing hydroxyl groups.[58] Maleylation is best carried out in borate buffer in the pH range of 8.5–9.5. Maleic anhydride, representing a 50- to 100-fold molar excess over the expected amino group content of the preparation, is added slowly with rapid stirring at 4°. The pH is maintained near 9 by appropriate addition of alkali and the reaction is considered completed when no further decrease in pH is observed. The extent of maleylation can be determined by titration of amino groups or with the aid of ^{14}C-labeled maleic anhydride. A spectrophotometric assay of the number of maleyl groups inserted into the protein may be based on the ratio of absorbance at 250 mμ and 290 mμ before and after the reaction.[58] In contrast to the succinyl-amino complex, the maleyl-amino complex is hydrolyzed under relatively mild conditions; the half-time for hydrolysis of maleyl-lysine at pH 3.5 is 11 hours at 37°.[58] Removal of covalently bound maleic acid may therefore be possible without major alterations in the primary structure of the protein molecule.

Enzymatic Methods

Treatment of membranes with enzymes which degrade lipids and proteins often results in the release of soluble protein components. This approach has been especially useful for the purification of microsomal enzymes and in the extraction of enzymes from the plasma membrane of mammalian tissues.

Lipase and Phospholipases. The most commonly used lipolytic enzymes are the lipases and phospholipases. Commercial preparations of pancreatic lipase (steapsin) may contain significant amounts of trypsin and amylase. The original method of Willstätter and Waldschmidt-Leitz,[59] or a modification of this procedure introduced by Crane,[60] may

[58]P. J. G. Butler, J. I. Harris, B. S. Hartley, and R. Leberman, *Biochemical J.* **103**, 78P (1967).
[59]N. Willstätter and E. Waldschmidt-Leitz, *in* "Die Methoden der Fermentforschung" (E. Bamann and K. Myrback, Eds.), Vol. II, p. 1560. Thieme, Leipzig 1941.
[60]B. S. S. Masters, C. H. Williams, Jr., and H. Kamen, this series, Vol. 10 [92].

be used to remove trypsin and amylase from the preparation. Some of the properties of pancreatic lipase are summarized in Table II. The optimal temperature for the enzyme with free glycerides as substrate, or when used to degrade membrane-bound lipids, is 37°. Although Ca^{2+} [61] is reported to activate the enzyme, monovalent cations such as Na^+ and NH_4^+ are also effective. The time required for adequate digestion must be checked with each sample. For specific examples of membrane-bound enzymes solubilized with pancreatic lipase, the reader is referred to the isolation of microsomal TPNH cytochrome c reductase[60] and cytochrome b_5.[62]

In addition to lipases, snake venom phospholipase preparations also have been widely used to solubilize membrane enzymes. The active component in snake venom is phospholipase A, which hydrolyzes most commonly occurring phospholipids to the corresponding lysophosphatides. Some of the properties of phospholipase A are listed in Table II. In addition to phospholipase A, crude preparations of snake venom also contain proteolytic enzymes. Since snake venom phospholipase A is extremely heat stable, proteolytic enzymes can be differentially inactivated by heat. A solution of snake venom dissolved in buffer at pH 5 is heated for 5 minutes at 100°. The insoluble material is removed by centrifugation and the supernatant solution containing the phospholipase A activity can be used without further purification. The digestion of membrane phospholipids is dependent upon Ca^{2+}. Although the enzyme is optimally active at $30° - 40°$, digestion can be carried out at lower temperatures when problems of enzyme instability arise. Examples of membrane-bound enzymes solubilized and purified with the aid of phospholipase A are listed in Table I.

Proteolytic Enzymes. Because of its relatively restricted specificity, trypsin is perhaps the most commonly used proteolytic enzyme for

TABLE II

PROPERTIES OF LIPASE AND PHOSPHOLIPASES

Preparation	Enzyme	Activity	pH optimum	Metal requirement
Snake venom (*Crotalus adamanteus, Naja naja*)	Phospholipase A	Hydrolysis of phosphatides to lysophosphatides	6–8	Ca^{2+} ($\sim 10^{-2}$ M)
Pancreatic lipase (steapsin)	Lipase	Hydrolysis of tri-, di-, and monoglycerides	7–8	Ca^{2+} ($\sim 10^{-2}$ M)

[61] E. E. Wills, *in* "Enzymes of Lipid Metabolism" (P. Desnuelle, ed.), p. 13. Pergmon Press, New York, 1961.

[62] P. Strittmatter and S. F. Velick, *J. Biol. Chem.* **221**, 253 (1956).

solubilizing membrane components. Inherent in the use of proteinases is the danger of enzymatic degradation during or following treatment. Although there are instances of membrane enzymes purified by this approach, it appears that the purified protein preparation in at least one case was heterogeneous with respect to molecular weight, suggesting partial hydrolysis during the isolation procedure. Thus cytochrome b_5 from microsomes purified with the aid of trypsin had a molecular weight of 12,000.[53] In contrast, when the same enzyme was purified by a procedure in which detergents were employed, the observed molecular weight of the enzyme in 4.5 M urea was 25,000. When trypsin is used, it is best to maintain a low ratio (about 1/100 to 1/200) of enzyme protein to membrane protein. The digestion is effectively terminated by the addition of soybean trypsin inhibitor to the reaction mixture.

[20] Extraction and Purification of Lipoprotein Complexes from Membranes

By ALEXANDER TZAGOLOFF and HARVEY S. PENEFSKY

Advances in our knowledge of the structure of membranes and the organization of constituent enzymes has made it abundantly clear that many membrane functions are carried out by macromolecular enzyme complexes. These complexes are composed of proteins and associated phospholipids. The latter frequently are necessary for enzymatic activity. Elsewhere in this volume,[1] procedures are described for the isolation, in water-soluble form, of membrane enzymes and proteins. Most of the materials isolated by these procedures are in fact components of large lipoprotein complexes. The methodology described below deals with the purification of such complexes. Although the term "soluble" is often used to describe lipoprotein complexes, it should be emphasized that the "solubility" depends on the presence of dispersing agents such as bile salts or detergents.

The purification of membrane lipoproteins presents a number of problems distinct from those encountered in the purification of classical water-soluble enzymes. For example, techniques of ion exchange chromatography and electrophoresis which often permit remarkable purifications of water-soluble proteins, generally are not applicable

[1] H. S. Penefsky and A. Tzagoloff, see this volume [19].

to membrane lipoproteins. However, drawbacks imposed by the limited methodology available for the fractionation of particulate enzymes are compensated in part by the fact that, in most cases, the isolation of a membrane in itself provides a substantial purification of the enzymes associated with it. Whereas, purification of several thousandfold may be needed to achieve homogeneity of soluble cytoplasmic enzymes, hundredfold purifications and often less, may be sufficient for membrane enzymes. This consideration applies particularly to those enzymes which are associated with highly specialized cellular membranes. For example, very pure preparations of cytochrome oxidase are obtained after only 8- to 10-fold purification from beef-heart mitochondria. The purification factor is even less when the enzyme is purified from submitochondrial membranes.

The emphasis here will be on fractionation with the aid of bile salts, since these reagents have been most useful for membrane lipoproteins. The use of synthetic detergents as well as organic solvents also will be described.

Fractionation with Bile Salts

Bile salts were first used to solubilize and purify electron transfer components of mitochondria.[2-4] The technology of bile salts fraction was later developed to a high degree of sophistication by Hatefi and his colleagues[5-8] and led to the separation of the electron transfer chain into four well defined lipoprotein complexes, some of which were isolated in a high state of purity.

Properties of Bile Salts

The most commonly used bile salts, deoxycholate and cholate, contain hydroxyl groups and a carboxyl group (cf. Fig. 1). The ampiphatic character of bile salts is a consequence of the fact that the polar groups stick out on one side of the molecule. Thus, bile salts have a polar and an apolar side. Such compounds are characterized by their ability to form micelles in aqueous solvents and to interact with other polar and apolar compounds. Although the properties of bile salts in aqueous

[2] K. Okunuki and E. Yakausizi, *Proc. Imp. Acad. (Tokyo)* **16**, 144 (1940).
[3] W. W. Wainio, P. Person, B. Eichel, and S. J. Cooperstein, *J. Biol. Chem.* **192**, 349 (1951).
[4] L. Smith and E. Stotz, *J. Biol. Chem.* **209**, 819 (1954).
[5] Y. Hatefi, A. G. Haavik, and D. E. Griffiths, *J. Biol. Chem.* **237**, 1676 (1962).
[6] Y. Hatefi, A. G. Haavik, and P. Jurtshuk, *Biochim, Biophys. Acta* **52**, 106 (1961).
[7] Y. Hatefi, A. G. Haavik, and D. E. Griffiths, *J. Biol. Chem.* **237**, 1681 (1962).
[8] L. R. Fowler, S. H. Richardson, and Y. Hatefi, *Biochim. Biophys. Acta* **64**, 170 (1962).

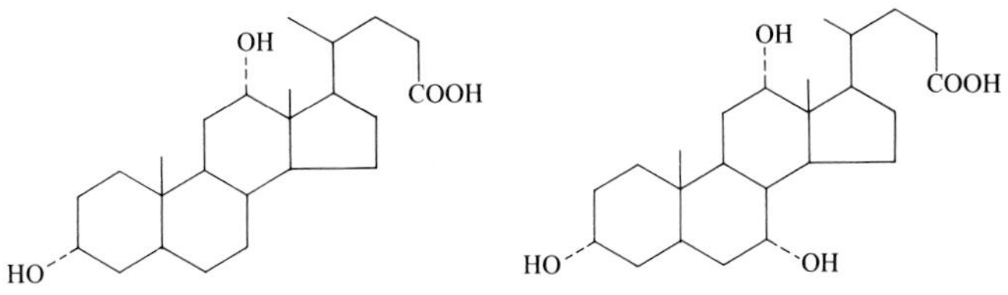

Deoxycholic acid   Cholic acid

FIG. 1.   Structural formulas of cholic and deoxycholic acid.

solutions have been studied, their micellar structure is not precisely known. Bile salt micelles differ from micelles of detergents in their size; the former are smaller. The aggregation number of bile salt micelles vary depending on the temperature and salt concentration and may range 4 to 8 molecules per micelle for cholate and 9 to 13 for deoxycholate.[9] The larger micelles are observed in the presence of $0.15-1.0\ M$ concentrations of monovalent cations.

Bile salt micelles may be precipitated from aqueous solutions by high concentrations of salt. Ammonium sulfate, for example, is effective over a fairly wide range of salt concentration. Moreover, the concentration of ammonium sulfate at which precipitation ensues is different for cholate and deoxycholate and depends to some extent on the concentration of detergent (Fig. 2). Figure 2 also illustrates the fact that precipitation of mixtures of cholate and deoxycholate by ammonium sulfate is affected by the ratios of the two detergents, suggesting formation of mixed micelles. This property is important from the standpoint of the purification of membrane lipoproteins, since most fractionation procedures employ ammonium sulfate precipitation. In fact, the concentration of ammonium sulfate at which a lipoprotein–bile salt complex precipitates, depends upon whether the bile salt is cholate or deoxycholate.

Cholate and deoxycholate can be purchased as the sodium or potassium salt or as the free acid. It is advisable to recrystallize these reagents before use, since they usually contain bile pigments and other contaminants. Both deoxycholic and cholic acid are recrystallized from 50% aqueous alcohol solutions and are freed of bile pigments by charcoal treatment. Aqueous solutions of deoxycholate (10% w/v, pH 7–8) and cholate 20% (w/v, pH 7–8) are prepared by conversion of the free acid to the sodium or potassium salt with alkali. Stock solutions should be stored at room temperature, since the bile salts will precipitate from solution if stored in the cold.

[9] A. F. Hoffman and D. M. Small, *Annu. Rev. Med.* **18**, 333 (1967).

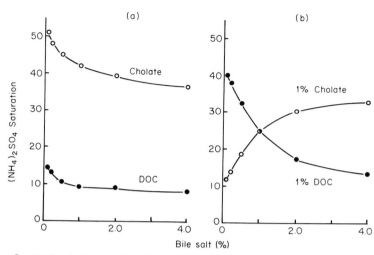

Fig. 2. (a) Precipitation of sodium cholate and sodium deoxycholate (DOC) with ammonium sulfate. (b) Precipitation with ammonium sulfate of sodium cholate in the presence of 1% sodium deoxycholate (O—O) and of sodium deoxycholate in the presence of 1% sodium cholate (●—●). Ammonium sulfate was added as a saturated solution of the salt to the bile salts dissolved in 0.25 M sucrose containing 10 mM Tris·acetate, pH 7.5. The data recorded indicate the saturation of ammonium sulfate at which the bile salts start to precipitate from solution.

With the exception of a few studies on the interaction between bile salts and proteins, there is virtually nothing concrete known about the interaction of bile salts with membranes or about the mechanism of solubilization of membrane lipoproteins. The ability of surfactants, either bile salts or other detergents, to disperse membranes is probably due to the effect of these compounds on the hydrophobic interactions between the lipoprotein components of the membrane. The lipoprotein-bile salt complexes may be expected to carry a net negative charge due to the carboxyl groups of the bile salt molecule. If these charges are distributed over the surface of the complex, forces of charge repulsion would cause a breakdown of the membrane and hence a solubilization of the lipoprotein complexes. In this model, hydrophobic interactions are vizualized as occurring between the apolar portion of the bile salt and the apolar region of the protein rather than that of the phospholipid. The model is supported by the fact that bile salts and other detergents solubilize membrane complexes even when the latter are not associated with phospholipid. Deoxycholate, for example, is an effective solubilizing agent for cytochrome oxidase in the presence and in the absence of associated lipids.[10]

[10]A. Tzagoloff, and D. H. MacLennan, Biochim. Biophys. Acta 99, 476 (1965).

Although low concentrations of bile salts do not appear to effect a separation of phospholipid and protein of membrane complexes, separation does occur at high concentrations.[10] The concentration of bile salts which causes a separation of the lipids and proteins will depend upon the stability of the interactions between lipids and protein and will vary for different lipoproteins.

Most lipoprotein complexes are solubilized in the range of 0.1 to 0.5 mg bile salt per milligram of protein at a protein concentration of 10–20 mg/ml. The activity of many enzymes is not affected at these levels of bile salts. This is probably a reflection of the fact that the primary site of interaction is at the surface of the enzyme complex. Loss of activity at higher concentrations is associated with dissolution of the complex. The concentration at which these reagents will cause dissociation depends again on the stability of the tertiary and quaternary structure of the complex. In contrast to bile salts, anionic detergents, such as sodium lauryl sulfate, are stronger denaturing agents.

Solubilization of Lipoprotein Complexes

The first objective in bile salt fractionation is to find minimal conditions for solubilization of the enzyme. For this purpose it is advantageous to carry out a preliminary experiment in which various conditions for solubilizing the enzymes are examined. The following tests can be performed on small samples of material

Bile Salts Alone. A series is prepared in which the concentration of either cholate or deoxycholate is varied. The bile salt is added to a suspension of the membranes at a fixed concentration of protein. After incubation at 0° for 10 minutes the samples are centrifuged at 105,000 *g* for 30 minutes. This distribution of enzyme activity is measured in the extract and in the pellet.

Bile Salts plus Salt. The solubilizing activity of bile salts is enhanced by monovalent salts, e.g., NaCl, KCl.[11] Another series can be prepared in which the effect of the concentration of bile salts is examined in the presence of a fixed concentration of salt. A useful titration is to vary the concentration of either deoxycholate or cholate from 0.1 to 1.0 mg per milligram of particle protein in the presence of 1 *M* KCl. The particles are suspended in 0.25 *M* sucrose–0.01 *M* Tris chloride, pH 8.0, at a protein concentration of 10–20 mg/ml. After incubation at 0° for 10 minutes, the suspensions are centrifuged at 105,000 *g* for 30 minutes and the distribution of activity is measured.

Bile Salt at Different pH. The solubilization of membrane lipoproteins with bile salts may be influenced by the pH of the suspending medium.

[11]R. K. Burkhard and G. Kropf, *Biochim. Biophys. Acta* **90**, 393 (1964).

As a rule, better solubilization is achieved at more alkaline pH values, although caution need be exercised in view of the danger of enzyme inactivation in the alkaline pH range.

Sequential Extraction with Bile Salts. It is often found that an enzyme is not extracted under conditions which solubilize a major portion of the membrane lipoproteins. This situation can be exploited by carrying out an initial extraction with a level of bile salt just below that which will extract the enzyme. Subsequent treatment of the particles at a higher level of bile salt results in extraction of the enzyme with fewer contaminants. Differential solubilization of membrane lipoprotein complexes was used by Hatefi *et al.*[6] in their studies on the purification of the electron transfer enzymes of beef heart mitochondria. The flavoenzymes and coenzyme Q-cytochrome *c* reductase were extracted from mitochondria with 0.3 mg of deoxycholate per milligram of protein. Under these conditions the cytochrome oxidase complex remained particulate, but was solubilized by subsequent extraction at a higher concentration of deoxycholate.[8]

Purification of Lipoprotein Complexes Solubilized with Bile Salts

Membrane lipoproteins solubilized by deoxycholate or cholate can be precipitated by ammonium sulfate, as well as other salts, over a fairly wide range of salt concentration, thus allowing the separation of different lipoprotein fractions. When used at fairly low concentration (0.1–1.0%) both deoxycholate and cholate can be removed by simple procedures although such a step may lead to reaggregation of the lipoproteins. The particulate material formed in this way can be treated with the same or different reagents under another set of conditions, thus permitting further purification. The following procedures have been used to purify enzymes solubilized with bile salts.

Salt Fractionation. The concentration of salt required to precipitate a particular lipoprotein will depend on the bile salt or combination of bile salts present in the extract. Lipoproteins solubilized with deoxycholate alone precipitate over a narrow range of ammonium sulfate concentrations (∼ 15–35% saturation). Extracts obtained with deoxycholate are therefore difficult to purify extensively with salt. The problem can usually be overcome by the addition of cholate to the extract, since this reagent increases the range of salt concentration over which lipoproteins will be precipitated. Extracts containing 0.5–1.0% cholate may be fractionated with ammonium sulfate over a range of 30–70% saturation.

As a rule, protein fractions which precipitate at low ammonium sulfate saturation have a low content of phospholipid and are poorly soluble in aqueous buffers. Fractions obtained at high saturations of ammonium sulfate are rich in phospholipid and are readily soluble. Since ammonium sulfate also precipitates both cholate and deoxycholate, it is not surprising that lipoprotein fractions precipitated with salts contain variable amounts of bile salts. Lipoproteins which precipitate at high salt saturations, generally contain sufficient endogeneous bile salts to form stable solutions.

Differential Precipitation by Removal of Bile Salts. Solubilized membrane lipoproteins may be induced to aggregate when bile salts are removed. The fact that lipoproteins exhibit widely varying solubility properties and that some require lower concentrations of bile salts than others for solubilization, can be used to advantage in fractionating bile salt extracts. Thus the less soluble lipoproteins will be the first to precipitate when the concentration of bile salt(s) in the extract is reduced. Differential precipitation of lipoproteins can be effected by several procedures.

Frequently the purity of an enzyme solubilized with deoxycholate or cholate can be improved by dialysis of the extract against dilute buffer. Controlled dialysis may lead to a lowering of the bile salt concentration to a level at which either the desired enzyme or contaminating lipoproteins will aggregate to form insoluble precipitates. If lowering the concentration of the bile salt causes the enzyme to precipitate, further purification can be achieved by resolubilization with bile salts and subsequent salt fractionation.

Since bile salts cause a redistribution of phospholipids, some enzymes may be enriched in phospholipids during the extraction. A density gradient centrifugation could be useful in separating the lipoproteins. Although density gradient centrifugation has not been used in conjunction with bile salt fractionation, Kaplan *et al.*[12] were able to purify the mitochondrial pyridine nucleotide transhydrogenase, solubilized with digitonin, with this technique.

Several generalized schemes for bile salt fractionation are shown in Fig. 3. The examples cited in Table I should provide the reader with more details on the principles and practice of bile salt fractionation.

Purification of Lipoproteins with Organic Solvents

Higher alcohols, particularly *tert*-amyl alcohol and *n*-butyl alcohol, have been used to fractionate membrane lipoproteins. Although these organic solvents do not solubilize lipoproteins, they appear to redistribute membrane phospholipids,[13] thereby causing, in some instances, a separa-

---

[12]N. O. Kaplan, this series, Vol. 10 [57].
[13]S. Fleischer and G. Brierley, *Biochim. Biophys. Acta* 53, 609 (1961).

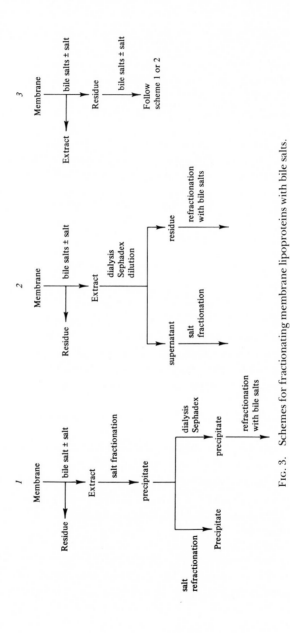

FIG. 3.   Schemes for fractionating membrane lipoproteins with bile salts.

tion of lipoproteins from other components of the membrane. Lipoprotein fractions which are separated by these procedures are usually in a particulate form.

Tert-Amyl Alcohol. Mitochondrial cytochrome oxidase,[14] succinate cytochrome *c* reductase[15] and components of the electron transfer system of liver microsomes[16] have been purified by initial extraction of the membranes with *tert*-amyl alcohol. Mitochondria treated with *tert*-amyl alcohol (10% by volume) in the presence of 0.25% KCl, separate into two distinct layers after centrifugation at 79,000 *g* for 20 minutes. The lighter, fluffy layer is enriched in succinic–cytochrome *c* reductase[15] while the heavier layer is enriched in cytochrome oxidase.[14] These enzymes have been purified after initial separation with *tert*-amyl alcohol by further fractionation with bile salts.

Beef liver microsomes, treated under the same conditions, also yield a light fluffy layer which is enriched 2- to 3-fold in most of the electron transfer components of the microsomes. This fraction is membranous and contains approximately 50% by weight of phospholipids.[16] DPNH–cytochrome *c* reductase and a lipoprotein complex containing cytochrome P-450 were purified from the electron transfer membranes of liver microsomes by extraction with bile salts and ammonium sulfate fractionation.[16]

n-Butyl Alcohol. The use of *n*-butyl alcohol to solubilize membrane proteins was described by Morton[17] (see also article [20][1]). In the resulting two-phase system, the butyl alcohol layer contains extracted lipids, whereas the water layer contains those membrane proteins which are converted to water-soluble species. However, in general, some lipoproteins gather at the interphase and can be further fractionated with bile salts or detergents. Penn and Mackler[18] have used this approach for the purification of microsomal DPNH-cytochrome *c* reductase. After initial solubilization with deoxycholate, the enzyme was further purified by treatment with *n*-butyl alcohol. A suspension of the particulate enzyme in 25% saturated ammonium sulfate was mixed with an equal volume of *n*-butyl alcohol. After centrifugation, DPNH–cytochrome *c* reductase was concentrated at the interphase while denatured proteins were in the lower aqueous phase.

[14]D. E. Griffiths and D. C. Wharton, *J. Biol. Chem.* **236**, 1850 (1961).
[15]D. E. Green and R. K. Burkhard, *Arch. Biochem. Biophys.* **92**, 312 (1961).
[16]D. H. MacLennan, A. Tzagoloff, and D. G. McConnell, *Biochim. Biophys. Acta* **131**, 59 (1967).
[17]R. K. Morton, *Nature (London)* **166**, 1092 (1950).
[18]N. Penn and B. Mackler, *Biochim. Biophys. Acta* **27**, 539 (1958).

Purification of Membrane Lipoproteins with Synthetic Detergents

Two general classes of synthetic detergents have been used to solubilize lipoproteins. These are the ionic detergents which carry an acidic group, (e.g., sodium lauryl sulfate, or a basic group, e.g., cetyldiethylammonium bromide, and the nonionic detergents, e.g., Tritons and Tweens). The ionic detergents are usually the more potent solubilizing agents. They are, however, effective protein denaturants, thereby limiting their usefulness in the purification of membrane enzymes. Nonionic detergents, on the other hand, are milder reagents which have been used successfully to solubilize a number of membrane lipoproteins without loss of enzymatic activity.

As is the case with bile salts, nonionic detergents are more effective solubilizing agents when used in conjunction with salts. In general, the same considerations which have been discussed for the use of bile salts apply. Triton X-100 is one of the most commonly used nonionic detergents. This reagent will solubilize most membrane lipoproteins at concentrations of 1 to 5% (w/v) and in the presence of 0.2 to 1.0 M KCl. On the other hand, fractionation with salts frequently cannot be applied to extracts containing nonionic detergents. Thus, Triton extracts are difficult to fractionate with ammonium sulfate because the detergent separates from solution as an oily layer. However, lipoproteins solubilized with nonionic detergents can be fractionated by ion-exchange chromatography, provided the eluting solvents contain a concentration of detergent sufficient to maintain the lipoprotein in solution.

Jacobs et al.[19] have purified cytochrome oxidase from beef heart mitochondria by sequential extraction with different Tritons followed by fractionation on DEAE-cellulose. Mitochrondia were extracted with 1.5% Triton X-114 in presence of 0.2 M phosphate. This treatment solubilized cytochromes b, c_1, and c. Cytochrome oxidase was then solubilized with 4% Triton X-100 in the presence of 0.2 M phosphate. The enzyme was further purified on DEAE-cellulose by elution with phosphate buffer containing 1% Triton X-100.

Since low concentrations of Triton (approximately 1%), will extract most membrane lipids, the enzyme obtained by Jacobs et el.[19] was depleted of phospholipid. This fact should be considered when sequential extractions are employed and enzyme activity is dependent on phospholipid.

Nonionic detergents may be removed from lipoproteins by the same tactics used in the removal of bile salts.

[19]E. E. Jacobs, E. C. Andrews, W. Cunningham, and F. L. Crane, *Biochem. Biophys. Res. Commun.* **25**, 87 (1966).

TABLE I

Enzyme	Source	Reagents	Reference
Adenosine triphosphatase (oligomycin sensitive)	Beef heart mitochondria (B.H.M.)	Deoxycholate (Doc.), cholate	a–c
Adenosine triphosphatase (oligomycin sensitive)	Yeast mitochondria (Y.M.)	Doc., cholate	d
Adenosine triphosphatase	Rabbit muscle microsomes	Doc.	e, f
Coenzyme Q–cytochrome c reductase	B.H.M.	Doc., cholate	g, h
Cytochrome c oxidase	B.H.M.	t-Amyl alcohol, cholate	i
	B.H.M.	Doc., cholate	j–o
	B.H.M.	Triton	p
	Y.M.	Cholate, Doc.	c, q, r
DPNH-coenzyme Q reductase	B.H.M.	Doc., cholate	s
DPNH-cytochrome c reductase	B.H.M.	Doc., cholate	s
	Beef liver microsomes	Doc., n-butanol	t
	Beef liver microsomes	t-Amyl alcohol, Doc., cholate	u
DPNH-oxidase	B.H.M.	Doc., cholate	v
Formate-dehydrogenase cytochrome b, complex	Escherichia coli	Doc.	w
β-hydroxybutyric dehydrogenase	B.H.M.	Cholate	x
Pyridine nucleotide transhydrogenase	B.H.M.	Digitonin	y
Succinate-coenzyme Q reductase	B.H.M.	Doc., organic solvents	z
Succinate-cytochrome c reductase	B.H.M.	t-Amyl alcohol, cholate	aa
Succinate oxidase	B.H.M.	Doc., cholate	v

[a] Y. Kagawa and E. Racker, *J. Biol. Chem.* **241**, 2467 (1966).

[b] A. Tzagoloff, K. H. Byington, and D. H. MacLennan, *J. Biol. Chem.* **243**, 2405 (1968).

[c] K. Kopaczyk, J. Asai, D. W. Allman, T. Oda, and D. E. Green, *Arch. Biochem. Biophys.* **123**, 602 (1968).

[d] A. Tzagoloff, *J. Biol. Chem.* **244**, 5020 (1969).

[e] A. Martonosi, *J. Biol. Chem.* **243**, 71 (1968).

[f] D. H. MacLennan, *J. Biol. Chem.*, **245**, 4508 (1970).

[g] Y. Hatefi, A. G. Haavik, and D. E. Griffiths, *J. Biol. Chem.* **237**, 1681 (1962).

[h] J. S. Rieske, W. S. Zaugg, and R. E. Hansen, *J. Biol. Chem.* **239**, 3023 (1964).

[i] D. E. Griffiths and D. C. Wharton, *J. Biol. Chem.* **236**, 1850 (1961).

[j] W. W. Wainio, P. Person, B. Eichel, and S. J. Cooperstein, *J. Biol. Chem.* **192**, 349 (1951).

[k] L. R. Fowler, S. H. Richardson, and Y. Hatefi, *Biochim. Biophys. Acta* **64**, 170 (1962).

[l] W. W. Wainio, *J. Biol. Chem.* **212**, 723 (1955).

[m]K. Okunuki, I. Sekuzu, T. Yonetani, and S. Takemori, *J. Biochem.* (*Tokyo*) **45**, 847 (1958).

[n]T. Yonetani, *J. Biol. Chem.* **236**, 1680 (1961).

[o]S. Horie and M. Morrison, *J. Biol. Chem* **238**, 1855 (1963).

[p]E. E. Jacobs, E. C. Andrews, W. Cunningham, and F. L. Crane, *Biochem. Biophys. Res. Commun.* **25**, 87 (1966).

[q]H. M. Duncan and B. Mackler, *J. Biol. Chem.* **241**, 1694 (1966).

[r]I. Sekuzu, H. Mizushima, and K. Okunuki, *Biochim. Biophys. Acta* **85**, 516 (1964).

[s]Y. Hatefi, A. G. Haavik, and D. E. Griffiths, *J. Biol. Chem.* **237**, 1676 (1962).

[t]N. Penn and B. Mackler, *Biochim. Biophys. Acta* **27**, 539 (1958).

[u]D. H. MacLennan, A. Tzagoloff, and D. G. McConnell, *Biochim. Biophys. Acta* **131**, 59 (1967).

[v]P. V. Blair, T. Oda, D. E. Green, and H. Fernández-Morán, *Biochemistry* **2**, 756 (1963).

[w]A. W. Linnane and C. W. Wrigley, *Biochim. Biophys. Acta* **77**, 408 (1963).

[x]I. Sekuzu, P. Jurtshuk, Jr., and D. E. Green, *J. Biol. Chem.* **238**, 975 (1963).

[y]N. O. Kaplan, this series, Vol. 10 [57].

[z]D. M. Ziegler and K. A. Doeg, *Arch. Biochem. Biophys.* **97**, 41 (1962).

[aa]D. E. Green and R. K. Burkhard, *Arch. Biochem. Biophys.* **92**, 312 (1961).

Section IV

Separation Based on Solubility

Articles Related to Section IV

[21] Fractionation of Protein Mixtures with Organic Solvents

By Seymour Kaufman

With the introduction of powerful new procedures for protein fractionation, such as are described elsewhere in this volume, some of the older methods, particularly organic solvent fractionation, appear to have fallen into disuse.

One of the reasons for this relative neglect of organic solvent fractionation is the fact that exposure to organic solvents can denature proteins. Proper technique is therefore more important with this procedure than with most; a badly poured column of DEAE-cellulose or Sephadex may give poor resolution; a badly performed organic solvent fractionation can destroy all the enzyme activity.

In spite of this potential disadvantage, organic solvent fractionation does offer certain advantages when compared to other procedures. It can be much faster than the column techniques, a run taking several hours rather than, as is often the case with column procedures, several days. Compared to ammonium sulfate fractionation, it offers an important practical advantage. Because of the lower density of the organic solvent–water mixtures, compared to concentrated salt solutions, lower speeds or shorter times of centrifugation are required to collect the precipitated protein. Balancing this advantage to some extent, is the fact that the addition of the organic solvent increases the volume of material to be centrifuged.

Since salt itself is not the precipitating agent in organic solvent fractionation, one can take advantage of very specific protein-electrolyte interactions. This is a particularly powerful variable because neutral salts, in general, increase the solubility of proteins in the presence of organic solvents. In a classic paper in this field,[1] the power of the technique has been succinctly summarized, "The balance between (1) the precipitating action of ethanol and (2) the interaction with salt permits attainment of a variety of conditions under which the protein to be separated may be brought to any desired solubility. This balance is different at constant ethanol and salt concentration for each pH and temperature, and for each protein component."

[1] E. J. Cohn, L. E. Strong, W. L. Hughes, Jr., D. L. Mulford, J. N. Ashworth, M. Melin, and H. L. Taylor, *J. Amer. Chem. Soc.* **68**, 459 (1946).

In the present article, technique rather than theory will be discussed. Some of the theoretical aspects of the subject have been reviewed.[2,3]

The discussion will also be limited to the use of ethanol as the organic solvent. The techniques described are, however, applicable to the other solvents that have been commonly used, such as methanol and acetone. Although there have been claims that methanol has less of a denaturing effect on plasma proteins than ethanol,[4] the available evidence appears to be insufficient for any generalization that would apply to most proteins.

General Procedure

As already mentioned, organic solvents can potentially denature proteins. Much of the technique involved in the use of organic solvents for protein fractionation, therefore, is aimed at avoiding protein denaturation. The most important factor in this respect is to work at temperatures below 0°.

Low temperature can be achieved in many different ways. One can carry out the whole procedure in a cold room maintained at a fixed temperature such as −5° or −10°. In the author's experience, such rooms are more effective in chilling the investigator than the protein solutions, and are inadequate for maintaining the protein mixture at a low and constant temperature during the ethanol addition. A preferable technique is to perform the fractionation in the laboratory and to cool the protein solution. This can be done either in a mechanically refrigerated bath, or with the use of a bath containing a dry-ice–ethanol–water mixture. In the latter case, it is convenient to prepare liter batches of ethanol–water mixtures that differ in their freezing points by 10°. Usually three mixtures with freezing points of −10°, −20°, and −30° will suffice. By varying the amount of dry-ice added to a given ethanol–water mixture, intermediate temperatures can be achieved and maintained.

Although temperature is a useful variable in organic solvent fractionation (see later section), it is advisable to carry out a trial experiment at a moderately low temperature, such as −5° or −10°. It is essential that the protein solution not be allowed to freeze during the fractionation procedure because this will lead to higher than desired ethanol concentrations in the liquid phase. In practice, therefore, the protein solution is cooled to about −1° before the ethanol addition is started.

[2] J. T. Edsall, *Advan. Protein Chem.* 3, 384 (1947).
[3] S. J. Singer, *Advan. Protein Chem.* 17, 1 (1962).
[4] L. Pillemer and M. C. Hutchinson, *J. Biol. Chem.* 158, 299 (1945).

As more ethanol is added, the temperature is gradually lowered until the desired temperature has been reached. It is most convenient to select an ethanol–water cooling bath that freezes at this temperature. In the procedure to be described, it has been assumed that $-10°$ is the temperature that has been selected for the fractionation.

At the start of the fractionation, a beaker containing the protein solution and large enough to contain the ethanol to be added, is placed in a pan that is filled with the ethanol–water mixture at a freezing point of $-10°$ (Fig. 1). The volume of ethanol–water mixture should be at least two or three times larger than that of the protein solution. The top of the beaker is covered with a sheet of *parafilm* or aluminum foil with holes cut out for thermometer, stirring rod, and inlet tube for the ethanol addition. The covering is important in order to minimize the possibility of splashing the contents of the cooling bath into the beaker. The protein solution is stirred mechanically. Good stirring is essential to avoid high local concentrations of ethanol. A magnetic stirrer can be used but is not reliable; occasionally the stirring bar behaves capriciously and moves away from the center of the beaker and vibrates in a corner rather than rotating in the center.

A measured amount of 95% ethanol (the volume of ethanol to be added should be measured at $20°$ rather than at the lower temperature) is added to the reservoir, which is cooled to $-30°$ to $-50°$ with a dry-ice–ethanol mixture held in a polyethylene jacket surrounding the reservoir. Since dissolving ethanol in water liberates heat, it has sometimes been

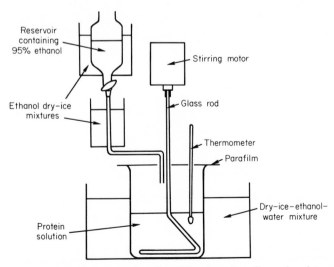

Reservoir containing 95% ethanol

Stirring motor

Ethanol dry–ice mixtures

Glass rod

Thermometer

Parafilm

Dry–ice–ethanol–water mixture

Protein solution

FIG. 1.   Bench-top arrangement for ethanol fractionation of protein mixtures.

recommended that the fractionation be carried out with more dilute ethanol–water mixtures, such as 50 or 70% ethanol, so as to avoid initial heating effects. In the author's experience, if all the other recommendations are followed, this precaution is unnecessary. The temperature of the ethanol-water mixture around the beaker is brought to about −0.1 by the addition of dry-ice. When this temperature is reached, the stopcock on the reservoir is opened and the ethanol is allowed to drop into the well-stirred protein solution. The rate of addition should be slow enough so that discrete drops of ethanol are being added. As addition of ethanol proceeds, more dry-ice chunks are added to the bath until the desired temperature is reached. After all the ethanol has been added, stirring is continued for another 15 to 20 minute equilibration period at the constant temperature. The mixture is then centrifuged in a refrigerated centrifuge at the same temperature as the one attained during the fractionation, −10° in the present case. It is important to calibrate the centrifuge before it is used to make certain that the selected temperature setting will maintain a solution at the desired temperature during a centrifugation at the same speed and time as will be used in the experiment. Obviously, some of the insoluble proteins may dissolve if the temperature is allowed to rise, and some of the soluble proteins may precipitate if the temperature falls, during the centrifugation. A 15-minute centrifugation at 3000 g is usually sufficient to pack the precipitated protein.

After centrifugation, the supernatant solution is poured into a beaker for further fractionation with ethanol. The tubes containing the precipitated protein are stored in an inverted position for a few minutes at a low temperature (−10° to −20°) to allow excess ethanol to drain: a deep-freeze is convenient for this purpose. Finally, the walls of the tubes are wiped with absorbent tissue paper. The precipitates are dissolved in a cold buffer solution with the help of a glass rod. It is desirable to use a sufficient volume of buffer so that the final concentration of residual ethanol will be low enough to avoid denaturation. Alternatively, ethanol can be removed from the protein precipitate by lyophilization.[1]

The procedure just outlined is designed to minimize protein denaturation. Many precautions are incorporated into the method. They are summarized below for emphasis.

1. Temperature below 0° at all times
2. Adequate rate of stirring during ethanol addition
3. Ethanol to be added is precooled to very low temperature
4. Cooling bath of sufficient volume to ensure large capacity for rapid heat transfer

5. Discrete drop addition of ethanol to minimize local high concentration of ethanol

6. Careful removal of the supernatant solution from the precipitated protein to minimize contamination with excess ethanol

7. Solution of the precipitated protein in sufficient volume of buffer to avoid high final ethanol concentration

8. Equilibration period after the ethanol has been added

9. Temperature during centrifugation is the same as that during equilibration period

The last two precautions are more critical for maximum fractionation than for maximum recovery of activity.

Variables

Temperature. Since low temperatures are essential during the fractionation, this is a limited variable. Within the limitation, however, it is still a useful one. The solubility of most proteins in ethanol–water mixtures decreases as the temperature is decreased. It is often possible to add ethanol to a given concentration, collect the insoluble proteins, and then, without any further addition of ethanol, lower the temperature to precipitate other proteins. This process can be repeated by a second lowering of the temperature.

Protein Concentration. It is desirable to keep the protein concentration low to minimize protein-protein interaction but high enough to minimize losses due to denaturation and to solubility of the protein in ethanol; proteins do have a finite solubility in ethanol so that at a starting concentration of 1–2 mg protein per milliliter, a significant fraction of the protein may be soluble in the final ethanol–water mixture. An initial concentration of protein between 5 and 20 mg/ml is a reasonable compromise between these opposing factors.

pH. The solubility of proteins in organic solvents varies markedly with pH, most of them being least soluble at their isoelectric points. The pH of the protein mixture to be fractionated must, therefore, be controlled. This is conveniently done by addition of a suitable buffer to the protein solution before the fractionation is started. A buffer concentration of 0.01–0.05 M is adequate. To minimize protein-protein interactions, it is advantageous to choose a pH wherein the majority of proteins will carry the same net charge.

Ionic Strength. As mentioned earlier, neutral salts increase the solubility of proteins in ethanol. High salt concentrations (above 0.1–0.2 M) are therefore a disadvantage because more ethanol will be required to precipitate a protein at the higher salt concentration. Furthermore, if the salt concentration is high enough, some of the salt will precipitate as the ethanol concentration is increased. For these reasons, it is not advisable to perform an ethanol fractionation on an undialyzed protein fraction obtained by ammonium sulfate precipitation. On the other hand, there is little advantage to dialyzing the ethanol-precipitated protein prior to carrying out an ammonium sulfate fractionation, if the precautions for avoiding excess ethanol in the protein precipitate have been followed.

Multivalent Cations. The complexes of proteins with cations such as Zn^{2+} and Ca^{2+} are less soluble in both water and organic solvents than the uncomplexed protein. This property is of particular value in the isolation of a protein that is extremely soluble in the organic solvent–water mixtures. The final volumes that must be centrifuged in the case of a very soluble protein may be so great that the entire procedure becomes cumbersome. The protein usually can be precipitated at one-half or one-third the concentration of organic solvent if a suitable multivalent cation is added before addition of the organic solvent.

The other use of multivalent cations in organic solvent fractionation of proteins is that they may permit a separation of proteins that cannot be achieved in their absence. Two proteins may have almost identical solubilities in an organic solvent–water mixture whereas the solubilities of their Zn^{2+} complexes may be markedly different.

In the author's experience with organic solvent fractionations of Zn^{2+}-protein complexes, an initial concentration of 0.02 M Zn^{2+} is adequate. It should be noted that if the protein solution is buffered with phosphate, the addition of the zinc salt will lead to a precipitate of zinc phosphate. While this may not interfere with the fractionation, it does introduce some uncertainty about the concentration of Zn^{2+} remaining in solution and adds the possible complication of adsorption of some of the protein to the zinc phosphate precipitate. It is preferable to avoid the use of phosphate buffers in this case.

[22] Water-Soluble Nonionic Polymers in Protein Purification[1]

By MELVIN FRIED and PAUL W. CHUN

Because of their operational simplicity and practical usefulness, fractionation techniques employing nonionic water-soluble high poly-

[1]Supported by Research Grant GB 6472 from the National Science Foundation.

mers are of increasing significance in studying biologically important macromolecules and cell particulates. However, these procedures have not been extensively applied to the purification of enzymes, possibly due to general lack of knowledge about their potential.

Bacteria and Viruses. A two-phase system of dextran and polyethylene glycol (PEG) has been employed in a countercurrent distribution procedure to separate two different strains of *Escherichia coli*.[2] Several types of viruses, notably bacteriophage, tobacco mosaic virus, vaccinia virus, and various polio and echo virus strains have been separated by distribution between two phases composed of buffers plus either dextran and PEG or dextran and methyl cellulose.[2-4]

PEG solutions have been used to precipitate plant viruses,[5,6] and infectious bacteriophage particles have also been separated by sedimentation with PEG.[7,8]

Nucleic Acids. An aqueous dextran and PEG two-phase system has been employed to separate single-stranded DNA and double-stranded DNA,[9] and the characteristics of the distribution of various RNA and DNA preparations in dextran and methyl cellulose systems have also been measured.[2]

Proteins. The isolation, separation, and purification of proteins in various aqueous high polymer systems has also been studied. A countercurrent purification of human ceruloplasmin in a dextran:PEG two-phase system has been described[2]; various classes of plasma proteins and individual molecular species have been isolated and purified by the use of systems involving solutions of dextran, PEG, or similar nonionic polymers.[10-13] A complete scheme using PEG for the separation of several plasma constituents and purification of immunoglobulins has been reported.[14] It has also been observed[15] that dextran decreases the

[2]P. -Å. Albertsson, *Methods Biochem. Anal.* 10, 229 (1962).

[3]P.-Å. Albertsson and G. Frick, *Biochim Biophys. Acta* 37, 230 (1960).

[4]P. -Å. Albertsson, "Methods in Virology" (K. Maramorosch and H. Koprowski, eds.), Vol. 2, p. 303. Academic Press, New York, 1967.

[5]T. T. Herbert, *Phytopathology* 53, 361 (1963).

[6]J. H. Venekamp and W. H. M. Mosch, *Virology* 22, 503 (1964).

[7]R. Leberman, *Virology* 30, 341 (1966).

[8]K. R. Yamamoto, B. M. Alberts, R. Benzinger, L. Lawhorne, and G. Treiber, *Virology* 40, 734 (1970).

[9]B. M. Alberts, *Biochemistry* 6, 2527 (1967).

[10]J. Krøll and R. Dybkaer, *Scand. J. Clin. Lab. Invest.* 1, 31 (1964).

[11]A. Polson, G. M. Potgieter, J. F. Largien, G. E. F. Mears, and F. J. Joubert, *Biochim. Biophys. Acta* 82, 463 (1964).

[12]P. Turini and M. R. Buzzesi, *Boll. Soc. Ital. Biol. Sper.* 40, 1985 (1966).

[13]P. H. Iverius and T. C. Laurent, *Biochim. Biophys. Acta* 133, 371 (1967).

[14]P. W. Chun, M. Fried, and E. F. Ellis, *Anal. Biochem.* 19, 481 (1967).

[15]M. Keler-Baĉoka and Z. Puĉar, *Clin. Chim. Acta* 14, 24 (1966).

gel electrophoretic mobility of β-lipoproteins and increases the amount of antigen-antibody complex precipitating in the region of antigen excess.[16]

Enzymes. There are very few examples of high polymer fractionation applied to enzyme purification in the literature. Janssen and Ruelius[17] have described the purification of a flavoprotein alcohol oxidase from several species of basidiomycete by fractional precipitation and crystallization with PEG, and, in unpublished work, Criss has used PEG fractionation in the purification of adrenal glucose-6-phosphate dehydrogenase.[18]

Principles

In several references cited above, the macromolecular separations were explained on the basis of either coprecipitation or of partition between various polymer:water phases.[2,3] Many investigators are of the opinion that the interaction between the particle of macromolecule and the polymer involves the formation of a complex.[2,11]

Relationships have been observed between this complex formation and the ionic strength of the solution, as well as with the molecular weights and/or molecular dimensions of both the precipitating polymer and the macromolecule or particle precipitated, and it was therefore postulated that the complex formation is dependent in some way on the charge on the protein.[11] Other workers[14] observed that low concentrations of PEG were required to precipitate proteins of high molecular weight and of corresponding high molecular volume, whereas higher concentrations of PEG were required to precipitate proteins of lower molecular weight and volume. Variations of pH and ionic strength affected the selectivity of precipitation. These authors, then suggested that it was the dielectric constant of the medium which influenced the precipitability of the protein and concluded that the precipitation of proteins by polyglycols is a result of the removal of water from the hydration envelope of the protein molecules, causing a corresponding alteration of the dielectric constant of the surrounding medium, which would alter the steric relationships of the hydrophilic groups of the protein molecules. This later proposal leads to the currently accepted hypothesis[13,19] which holds that precipitation of macromolecules from solution by polymers is due to the ability of the polymer to exclude the

[16]K. Hellsing and T. C. Laurent, *Acta Chem. Scand.* 18, 1303 (1964).
[17]F. W. Janssen and H. W. Ruelius, *Biochim. Biophys. Acta* 151, 330 (1968).
[18]W. E. Criss, Ph.D. Dissertation, University of Florida, Gainesville, 1968.
[19]T. C. Laurent, *Biochem. J* 89, 253 (1963).

protein, sterically, from part of the solvent. This brings the protein solution to its solubility limit. Laurent has adduced considerable data in support of this hypothesis,[19] including the observation that the degree of polymerization of the precipitating agent has little significance once it exceeds a certain minimum. Also, the relative decrease in solubility of proteins in the presence of polymer solutions is independent of the absolute solubility of the protein, of salt concentration, and of pH. These findings are all consistent with the assumption that the precipitation phenomenon is a result of an exclusion of protein from the volume occupied in solution by the polymer. The excluded volume should be a function solely of the concentration of the polymer and independent of protein concentration or of any other factor. The strong dependence of the solubility of a protein on its size[11,14] further substantiates this hypothesis. In general the larger the protein the less soluble it is in high polymer solutions, although one may suppose that the shape, i.e., the asymmetry of the molecule, which would affect its effective size, should also be taken into consideration.

The phenomenon of exclusion is one aspect of protein isolation and precipitation that is relatively unexplored but which offers many possibilities both for the practical separation of macromolecules and as a tool with which to study certain of their characteristics. Proteins, virus, red cells, subcellular particulates, nucleic acid, etc., are precipitated, as noted above, in the presence of high polymers. In a more physiological context, such precipitations in the presence of high polymers may play an important role in the formation of bone, lipid deposition in arterial walls, the formation of collagen fibers, and the deposition of amyloid in connective tissue, all cases in which the mechanism of molecular exclusion appears to be operating. Studies on the exclusion of macromolecules from high polymer media[20,21] have led to the hypothesis that such molecular exclusion depends on the steric factors operating in a multicomponent system which, in the cases described, is a three-component system. The three-component system reflects an interaction of solvent with two different molecular species of solute, in which one species of solute excludes the other in a continuing equilibrium interaction which regulates the distribution of the particles between two phases or between two kinds of solute:solvent complex. Such continuing interaction could also regulate various physiological processes. For example, the high molecular weight polysaccharides in the extracellular spaces of connective tissue are linked within a structural network

[20] A. G. Ogston and C. F. Phelps, *Biochem. J.* 78, 827 (1960).
[21] T. C. Laurent and J. Killander, *J. Chromatog.* 14, 317 (1964).

of protein fibers to the proteins themselves. Laurent[22] describes this as a polysaccharide phase separate from other tissue compartments. Globular proteins moving through such a system will interact with the polysaccharides, and the mechanism of steric exclusion will affect the properties of the proteins and also their distribution. Ogston[23] has considered the nature of the protein fiber network in a three-component system and the possibility of an object or a macromolecule penetrating such a suspension. The exclusion of protein from high polymer media is described by a fiber model based on Poisson's distribution function and the probability of penetration of a spherical protein particle into a fibrous network of high polymer molecules. The theoretical derivation of the probability distribution function has also been described.[24]

Operationally, it is not necessary to have visibly or otherwise physically distinct phases, since a simple high polymer:water system may be considered to have two phases even though the high polymer is, within limits, certainly soluble in the solvent. Thus it becomes possible to utilize systems involving single species of polymer for the effective fractionation and purification of proteins and enzymes.

Methods of Fractionation

Precipitating Agents. Among the water-soluble, nonionic polymers which may be used with this technique are the polyethylene glycols, PEG-2000, PEG-4000, and PEG-6000 (General Biochemicals); Aldosperse ML-14 (Glycol Chemicals, Inc.); 1,4 nonylphenylethylene oxide (NPEO NP-14, J. T. Baker Chemical Co.). PEG of higher molecular weight (e.g., 20,000) cannot be used because its high viscosity in water (approximately 96,000 centistokes/sec at 95°), renders centrifugations and other manipulations quite difficult.

Sodium dextran sulfate 2000 (Pharmacia Fine Chemicals) may be employed to precipitate lipoproteins in certain preliminary protein purifications.[25]

Purification of Polymer. PEG-6000 was found to be the most suitable for general purposes and was chosen for all the subsequent separations. PEG-6000 was further purified for these studies by dissolving in acetone and then precipitating with ether.[4] This procedure is effective in removing impurities that absorb light in the ultraviolet region at 290 mμ from the commercially available polymers.

---

[22]T. C. Laurent, "The Chemical Physiology of Mucopolysaccharides." Little, Brown, Boston, Massachusetts, 1967.
[23]A. G. Ogston, *Trans. Faraday Soc.* 54, 1754 (1958).
[24]P. W. Chun, J. I. Thornby, and J. G. Saw, *Biophys. J.* 9, 163 (1969).

The desired concentrations of PEG-6000 are achieved by using a stock solution prepared by dissolving 50% (w/v) of the PEG in distilled water at 37°.

Removal of Precipitants from Protein Fractions. After the fractional precipitation procedures, several examples of which are outlined below, each precipitate is redissolved in phosphate buffer, and twice reprecipitated with 35% (w/v) $(NH_4)_2SO_4$. These precipitates are redissolved in buffer and added to a suspension of DEAE-cellulose. Polyethyleneglycol remains unabsorbed to the anion exchanger under these conditions and is readily removed from the resin by washing with 10 mM phosphate buffer, pH 7.0. The absorbed proteins can be recovered by eluting with 0.1 M KCl. The resultant protein solutions are subsequently dialyzed against water and lyophilized. An alternate and more rapid method for removing PEG from immunoglobulins and other serum proteins is to precipitate the immunoglobulins with cold 20% ethanol, centrifuge, redissolve the precipitate in buffer, dialyze, and lyophilize. Pressure dialysis and ultrafiltration may also be used to remove the relatively low molecular weight precipitant from the proteins.

Examples of Applications

Precipitation of Human Serum Proteins

Purification of Immunoglobulins from Cohn Fraction II. One hundred milligrams of Cohn fraction II are dissolved in 10 ml of 55 mM phosphate buffer, pH 7.0, $\mu = 0.1$, and then precipitated with PEG-6000 at a final polymer concentration of 35% (w/v). The precipitate is dissolved in 10 ml of phosphate buffer and further purification steps are carried out according to the flow chart in Fig. 1. Recovery data, Table I, show the immunoglobulins in fractions III, IV, and V of Fig. 1 to comprise about 62% of the total protein of the starting material. The maximum yield of purified immunoglobulins is obtained at PEG concentration between 8.5 and 10.5%, as indicated also in Fig. 4.

Upon immunoelectrophoresis of fractions III, IV, and V in parallel with the original commercial materials at similar concentrations, single sharp lines are seen in the γ region as compared with multiple bands with unpurified preparations.

Isolation of Plasma Proteins with PEG-6000. A mixture of Cohn fractions I, II, IV, and V is prepared by dissolving 100 mg of each in 40 ml of potassium phosphate buffer, pH 7.0, $\mu = 0.1$. This solution is then subjected to stepwise precipitation with PEG-6000. Figure 2 shows the precipitation patterns and the separation of individual components obtained from such a mixture. Fibrinogen and γ-globulins precipitate at PEG concentrations of 1–2% and 8–10%, respectively, in buffers of low ionic

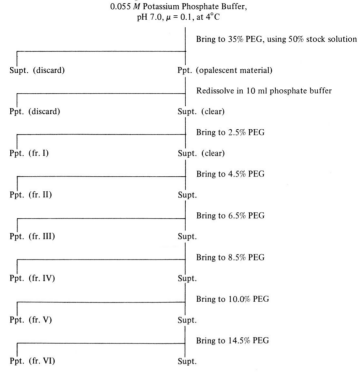

100 mg Cohn Fraction II Dissolved in
0.055 M Potassium Phosphate Buffer,
pH 7.0, μ = 0.1, at 4°C

Bring to 35% PEG, using 50% stock solution

Supt. (discard)   Ppt. (opalescent material)

Redissolve in 10 ml phosphate buffer

Ppt. (discard)   Supt. (clear)

Bring to 2.5% PEG

Ppt. (fr. I)   Supt. (clear)

Bring to 4.5% PEG

Ppt. (fr. II)   Supt.

Bring to 6.5% PEG

Ppt. (fr. III)   Supt.

Bring to 8.5% PEG

Ppt. (fr. IV)   Supt.

Bring to 10.0% PEG

Ppt. (fr. V)   Supt.

Bring to 14.5% PEG

Ppt. (fr. VI)   Supt.

FIG. 1.   Isolation of immunoglobulins from Cohn fraction II. All additions were carried out at 4°C with stirring, and the suspensions were allowed to stand for 5 minutes. They were then centrifuged at 35,000 g for 30 minutes.

TABLE I
RECOVERY OF IMMUNE GLOBULIN FRACTIONS
AFTER HIGH-POLYMER FRACTIONATION

Fraction precipitated (Fig. 1)	Percent PEG concentration	Percent recovery (dry weight basis[a])
I	2.50	4.20
II	4.50	12.36
III	6.50	12.50
IV	8.50	27.00
V	10.00	24.05
VI	14.50	9.12
VI supernatant	14.50	1.62
		Σ = 90.85

[a]Dry weight measurements were consistent with protein determinations by the procedure of Lowry *et al., J. Biol. Chem.* **193**, 265 (1951).

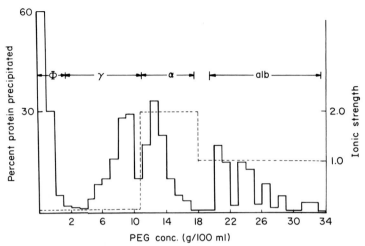

FIG. 2. Isolation of plasma proteins with PEG-6000; 100 mg each of Cohn fractions I, II, IV, and V were dissolved in 40 ml of 55 mM potassium phosphate buffer, pH 7.0, $\mu = 0.1$: $-$, % protein precipitated (% of each of the fractions added to the mixture); ---, ionic strength. The identities of the proteins were established electrophoretically.

strength. The maximum precipitation of γ-globulins occurs at 12–14% PEG, provided the ionic strength of the buffer is increased to 2.0 by the addition of KCl. Albumin in a buffer of $\mu = 1.0$ precipitates at 20–35% PEG in several discrete fractions, rather than as one large peak as is the case with the globulins. When each of the discrete precipitated albumin fractions is examined ultracentrifugally and by immunoelectrophoresis, no differences are observed.

It is not possible to isolate homogeneous β-globulin fractions by high-polymer precipitation; and since the commercial β-globulins (Cohn fraction III) available are insoluble in potassium phosphate buffers, even at high ionic strengths, conditions for their fractionation are still not developed. The β-globulins also appear to interfere with the separation of other serum components, and are therefore removed by precipitation with 2% sodium dextran sulfate[25] at the beginning of the fractionation procedure. Residual sodium dextran sulfate is removed by the succeeding steps of the fractionation procedure.

Isolation of Immunoglobulins from Whole Human Serum. The following procedure is based on the results obtained in preliminary experiments with individual proteins and with synthetic mixtures of Cohn fractions I, II, III, and V.

Five milliliters of whole serum, diluted with 5 ml of 55 mM phosphate buffer, is dialyzed against the same buffer for 20 hours prior to PEG

[25]D. G. Cornwell and F. S. Kruger, *J. Lipid Res.* **2**, 110 (1961).

precipitation. Sodium dextran sulfate is added to a final concentration of 2%. The precipitate that results, presumably a lipoprotein complex,[25] is removed by centrifugation at 35,000 g for 30 minutes. The supernatant liquid is redialyzed against phosphate buffer overnight to remove residual dextran sulfate before precipitating with PEG at a concentration of 35%. This precipitate is removed by centrifugation at 35,000 g for 30 minutes and is dissolved in 10 ml of phosphate buffer. Further precipitation steps are carried out according to the flow chart in Fig. 3. Each precipitated fraction that results is freed of PEG by the DEAE-cellulose method described above. The yield of immunoglobulins from 5 ml of whole serum is 45–50 mg.

Analysis of the immunoglobulin fractions III, IV, and V from Fig. 1 or fractions I, II, III of Fig. 3, by the Ouchterlony double-diffusion technique reveals a complete identity with monospecific anti-γ_G serum. However, when the fractions are run against monospecific anti-γ_M or anti-γ_A sera, trace amounts of γ_A and γ_M globulins may be detected. This would indicate that all immunoglobulins, γ_A, γ_M, and γ_G are present in the fractions obtained by PEG precipitation.

Fractionation of Adrenal Glucose-6-phosphate Dehydrogenase. The adrenal cortices of beef adrenal glands taken immediately after slaughter are minced and then homogenized in 10 mM, pH 6.7 phosphate buffer containing 5% glycerol, 2.7×10^{-3} M β-mercaptoethanol and 2×10^{-4} M, EDTA, in a ratio of 1 g tissue to 2 ml of buffer. This homogenate is centrifuged at 2000 g for 15 minutes to remove cell debris and unbroken cells. The supernatant fluid is centrifuged at 105,000 g for 1 hour in a Spinco preparative ultracentrifuge, and that high-speed supernatant is removed from the tubes with a syringe fitted with a long needle to avoid disturbing the thick compact lipid layer at the top of the tubes. The enzyme is precipitated from the high-speed supernatant by adding an equimolar amount of 0.1 M acetate buffer, pH 4.5, containing 15% ethanol while mixing rapidly. The suspension is centrifuged at 25,000 g for 15 minutes, and the precipitate is redissolved in 5 mM, pH 7.8, phosphate buffer containing glycerol, β-mercaptoethanol, and EDTA, as above. This solution which has a specific activity of 3.0 and represents a 10-fold purification over the whole homogenate,[26] is subjected to the procedure listed in Fig. 4, which outlines a stepwise precipitation with PEG-6000. Each precipitated fraction is freed of PEG by the DEAE-cellulose method described above, and assayed by the method of Criss and McKerns.[26] The majority of the enzyme activity precipitates between 41 and 50% PEG, resulting in a more than 2-fold purification with little loss of activity.

[26] W. E. Criss and K. W. McKerns, *Biochemistry* 7, 125 (1968).

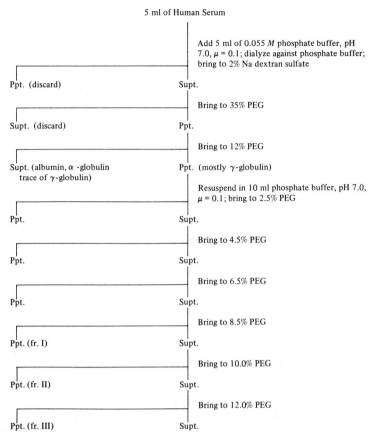

FIG. 3.   Isolation of immunoglobulins from whole human serum. All additions were carried out at 4°C with stirring, and the suspensions allowed to stand for 5 minutes. They were then centrifuged at 35,000 g for 30 minutes.

An analogous procedure has been used[17] for the purification and crystallization by fractional precipitation of alcohol oxidase. By careful addition of PEG to the partially purified enzyme solution to a point at which the PEG:enzyme mixture was just turbid, it was possible to precipitate the enzyme in crystalline form, resulting in a 3- to 5-fold purification with only minimal loss of activity.

Comments

The use of nonionic high polymers for the purification of proteins and enzymes deserves wider application. The method is simple and straightforward and results in products which appear to retain their native configuration as evidenced by measurements of physical prop-

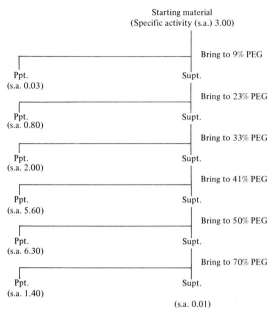

Fig. 4. PEG fractionation of adrenal glucose-6-phosphate dehydrogenase. Starting material was a partially purified enzyme preparation representing a 10-fold purification over whole homogenate. All additions were carried out at 4° with stirring; the suspensions were allowed to stand for 5 minutes then were centrifuged at 35,000 g for 30 minutes. s.a. = specific activity = 1 μmole of NADP reduced per minute per milligram of protein.

erties and biological activity. The technique may be applied in either batch or countercurrent distribution procedures and the precipitating agent is readily removed from the proteins isolated. Since the procedure is based on a principle of separation that is different from those involved in other techniques currently used, it provides an additional criterion for protein purity and protein characterization.

[23] Crystallization as a Purification Technique

By WILLIAM B. JAKOBY

Aside from considerations of esthetics, efforts designed to crystallize an enzyme are intended to achieve one of two aims: purification of the protein or preparation of a product suitable for structural studies. The former is the subject of this article. Methods for the latter are separately presented in detail.[1,2]

[1] M. Zeppezauer, this volume [24].
[2] D. R. Davies, this volume [25].

The purification method depends on the decrease in solubility of proteins in ammonium sulfate solution attendant to an increase in temperature. Protein, precipitated with salt, is extracted with ammonium sulfate solutions of *decreasing* concentration at or near 0°. Thus, contaminating proteins, soluble at higher concentrations of ammonium sulfate, are extracted first and thereby removed. Extracts obtained in this manner are allowed to warm to the temperature of the room during which period crystallization results.

In a series of over 100 proteins with which crystallization has been attempted in the author's laboratory, failure to crystallize has not yet been observed. Table I presents a diverse group of proteins and the conditions under which they have been crystallized. It is noteworthy that the composition of the extraction mixture, other than that of the salt concentration, was not found to affect the procedure. Although not listed in the table, extraction media have included sodium chloride solutions; imidazole, glycine, and diethylamine as buffers; a pH range from 6 to 9; and the presence of a wide variety of substrates, mercaptans, and other stabilizing agents.

Reagents

A saturated solution of ammonium sulfate is prepared by heating 1 liter of water to which has been added 800 g of a highly purified grade of the salt. The resultant solution is stored at 0° to 4° for several days before use. Crystals of ammonium sulfate should accumulate during this period. The clear, supernatant solution is considered to be 100% saturated in ammonium sulfate.

Solutions of the various reagents, including the buffer and such factors as are required for maximum stability of the enzyme, are prepared in concentrated form.

The desired enzyme, preferably at a stage of 50% or greater purity, is concentrated to a volume of 4 ml or less. A minimum of 4 mg of protein is suggested but 10 mg is recommended.

Procedure.[3] The protein solution is transferred to a 15-ml centrifuge tube[4] placed in an ice bath, and sufficient ammonium sulfate is added to salt-out the enzyme. The salt may be added either in the form of the powdered solid or as a saturated solution; the former is preferred so as to minimize solution volume. The suspension is allowed to stand in

---

[3]W. B. Jakoby, *Anal. Biochem.* **26,** 295 (1968).

[4]A Corex, glass tube (Corning No. 8441) has been suitable and has not resulted in breakage under the conditions specified here. Glass is used so that the extraction procedure can be followed visually.

CRYSTALLINE PROTEINS OBTAINED BY THE TECHNIQUE DESCRIBED[a]

No.	Protein	Molecular weight	Buffer (mM)	pH	Extracting solutions		Recovery in crystals (%)
					Other additions	Salt concentrations,[b] % of saturation	
1	Quinolinate phosphoribosyltransferase[c]	165,000	P_i, 50	7.1	None	50, 46, 42	82
2	Thyroglobulin[d]	660,000	–	–	1% NaCl	42, **39**, 37	70
3	Thyroid-stimulating hormone[e]	28,000	P_i, 50	7.0	None	50, 45, **40**, 36	76
4	Aldehyde dehydrogenase[f]	200,000	P_i, 50	7.0	25% glycerol +50 mM thioglycerol	**55**, 51, 47, 43	<30
5	Tartronic semialdehyde reductase[g]	unknown	Tris, 40	7.4	5 mM mercaptoethanol	50, 45, **40**, 35	50
6	Tartronic semialdehyde reductase[h]	104,000	P_i, 100	7.0	3 mM mercaptoethanol	55, **45**, 40, 35	75
7	Malate dehydrogenase[i]	43,000	P_i, 30	7.2	None	60, **55**	35
8	Tartrate dehydrogenase[j]	145,000	P_i, 100	7.0	4 mM mercaptoethanol	50, 45, **40**, 35	40
9	Dihydroxyfumarate reductive decarboxylase[k]	63,000	Tris, 40	7.4	5 mM mercaptoethanol	65, 50, **45**, **40**, 36	90
10	Hydroxypyruvate reductase[l]	76,000	P_i, 30	7.2	3 mM mercaptoethanol	50, **45**, **40**, 35	70

[a] Adapted from W. B. Jakoby, Anal. Biochem. **26**, 295 (1968).
[b] *Italicized* numbers indicate crystallization; **boldface** type denotes a large yield.
[c] P. M. Packman and W. B. Jakoby, J. Biol. Chem. **240**, 4107 (1965).
[d] W. B. Jakoby, L. L. Labaw, H. Edelhoch, I. Pastan, and J. E. Rall, Science **153**, 1671 (1966).
[e] P. Condliffe and W. B. Jakoby, Endocrinology **80**, 203 (1967).
[f] C. R. Steinman and W. B. Jakoby, J. Biol. Chem. **242**, 5019 (1967).
[g] L. D. Kohn and W. B. Jakoby, J. Biol. Chem. **243**, 2465 (1968).
[h] L. D. Kohn, J. Biol. Chem. **243**, 4426 (1968).
[i] L. D. Kohn and W. B. Jakoby, J. Biol. Chem. **243**, 2465 (1968).
[j] L. D. Kohn, P. M. Packman, R. H. Allen, and W. B. Jakoby, J. Biol. Chem. **243**, 2479 (1968).
[k] L. D. Kohn and W. B. Jakoby, J. Biol. Chem. **243**, 2486 (1968).
[l] L. D. Kohn and W. B. Jakoby, J. Biol. Chem. **243**, 2494 (1968).

the ice bath for 5 minutes and the protein is collected by centrifugation at 2° and 10,000 g for 15 minutes. The supernatant solution is discarded.

To the protein residue at the bottom of the centrifuge tube is added 1 ml of the highest concentration of a graded series of ammonium sulfate solution containing both the desired buffer and whatever supplements are appropriate to ensure stability. The protein is evenly and finely suspended in the added solution by trituration at 2° with a glass rod for 3 to 5 minutes. Care is taken to perform the procedure in the cold. After suspension of the protein, the tube remains in the ice bath for an additional 3 to 5 minutes and is then centrifuged for 5 minutes at 10,000 g and 2°. After centrifugation, the supernatant solution is immediately poured into a second centrifuge tube at room temperature and protected from dust by covering with a small inverted beaker. The residue from the first extraction is reextracted in an identical manner except for the use of an ammonium sulfate solution containing a slightly lower concentration of the salt.

The graded series of ammonium sulfate solutions will vary as a function of the solubility range of the protein. Thus, an enzyme which is known to be precipitated between 45 and 60% of saturation would be extracted with ammonium sulfate solutions of 65, 61, 58, 55, and 52% of salt saturation. A protein precipitating at 25 to 35% of saturation would be treated with ammonium sulfate solutions of 38, 36, 34, 32, and 30% of saturation. It is significant that only the upper range of ammonium sulfate concentration is required, i.e., those salt concentrations at which the solubility of the protein is relatively limited. If the range of salt is found to be too high, the extraction procedure is simply continued in succeeding steps, using the same increment of decreasing salt, until crystallization occurs.

Crystallization can take place within minutes of transfer of the extract to room temperature but may require an hour or overnight incubation. Crystallization is signaled by the appearance of turbidity. The silky sheen, often termed "schlieren," which is observed with several crystalline enzymes, is usually not realized and is never evident in the first few hours of crystallization.[5] The data in Table I demonstrate some of the patterns of crystallization in terms of the salt concentration used for extraction.

---

[5] The "schlieren" effect is a function of uniformity of particle size rather than of crystallinity and has been observed, under other conditions, with uniform but amorphous particles.

Establishing crystallinity. Crystals obtained by this method have been as large as 50 μ but, in the main, are produced in the range of 1 to 4 μ in their largest axis. This small size does not allow dependence on the more usual microscopic techniques in judging crystallinity. However, the substitution of a dark-field condenser and the insertion of a frontal stop into the usual oil-immersion objective, are sufficient.[6] The dark-field arrangement makes visible the crystalline nature of the product as indicated by the clear outlining of the crystal faces resulting from the refractile image. Frequently, particularly if turbidity develops quickly after transfer to room temperature, large refractile clusters of small crystals are evident. Focusing on any one plane of this mass at a time will reveal the regular structure of its individual components.

Comments

The invariant success of the method in yielding crystals led to questioning the nature of the products. Examination of several of the crystalline proteins by electron microscopy, revealed the presence of the ordered structure expected for crystals.[3,7] Because of their small size, none have been subjected to X-ray analysis for structure determination.

The ability to form crystals with ease should not mislead the investigator as to the value of the accomplishment. Subjecting an enzyme preparation of a low degree of purity to the procedure and finding crystals as a result, allows few conclusions as to the nature of the product. The abbreviated aphorism of the computor industry, GIGO,[8] is clearly applicable. Rather, it is expected that the procedure will be used in attempts to remove minor impurities as a final or penultimate stage of purification, using specific activity as a guide. Repetition of the procedure[9] until constant specific activity is attained may be worthwhile from the viewpoint of purification but should not be used as a criterion of purity.

---

[6]For the enzymologist intent on checking crystallinity, expensive equipment is unnecessary. A student microscope, equipped with an oil-immersion lens for viewing bacteria, can be inexpensively converted by addition of the two items noted in the text. One source for these is the American Optical Company (item 214F) which sold these items for 100 dollars in 1970.

[7]W. B. Jakoby, L. Labaw, H. Edelhoch, I. Pastan, and J. E. Rall, Science 153, 1671 (1966).

[8]Garbage in, garbage out.

[9]In the case of several enzymes, placing the crystalline suspension into an ice bath results in solution. However, after remaining in contact with the mother liquor at room temperature for several days, the crystals will no longer dissolve at 2° in the salt solution and may be stored at that temperature.

[24] Formation of Large Crystals

By Michael Zeppezauer

Routine methods for the preparation of micro crystals from proteins of different physical and chemical properties are well developed.[1] However, for X-ray and neutron diffraction work, for electron paramagnetic resonance spectroscopy and related methods it is necessary or desirable to use protein crystals of a certain minimal size (linear dimensions of at least 0.2 mm) and a high degree of order. The crystals should also possess a certain mechanical strength. These conditions are not always fulfilled simultaneously.

The formation of large crystals of a globular protein depends essentially on a cooperation of factors governing its solubility.[2-4] Temperature, pH, ionic strength, the nature of the counterions, e.g., buffer composition, and the presence of other ligands such as coenzymes, inhibitors, and heavy metals, are all important variables. The crystals obtained under different conditions may show different habits and even different crystallographic symmetry, i.e., polymorphism. Modifications obtained due to bound coenzymes, substrates, or inhibitors may indicate an important alteration of a tertiary structure. On the other hand, polymorphism caused by difference in the nature of added anions, cations, or of pH is in most cases probably related to the occurrence of other lattice types which are built up from protein particles with different net charge and varying kinds and numbers of counterions, the protein conformation is generally unaffected. That crystals may contain protein molecules with a distinct net charge is documented by the observation of several solubility minima in the determination of solubility as a function of pH.[5] It is also a general experience in the field of protein crystallography, that good crystals are obtained within a narrow range of pH for proteins precipitated by salts, and within a narrow range of pH, ionic strength, and protein concentration when crystallizing by means of organic solvents. This critical pH region does not always lie near the pH of the isoionic point. The development of a reproducible method for the growth of large protein crystals also

[1] W. B. Jakoby, *Anal. Biochem.* **26**, 295 (1968); This volume [23].
[2] A. A. Green and W. L. Hughes, this series, Vol. 1, p. 67.
[3] M. Dixon and E. C. Webb, *Advan. Protein Chem.* **16**, 197 (1961).
[4] J. A. Rupley, *J. Mol. Biol.* **35**, 455 (1968).
[5] R. Czok and T. Bücher, *Advan. Protein Chem.* **15**, 315 (1960).

includes the control of factors determining the kinetics of nucleation and growth. Important considerations are the rate of approach to the point of supersaturation where nucleation sets in, the absence of heterogeneous nuclei, and sometimes the nature of the transport of matter in the protein solution, i.e., whether by convection or mere diffusion.

The various techniques utilized in the past to meet these needs have been briefly reviewed. The most versatile and accurate method to test individually and systematically the variety of parameters influencing crystal growth is equilibrium dialysis against an agent leading to crystallization.[6a,6b]

Crystallization by Equilibrium Dialysis

A. Principle

Crystallization in diffusion cells was described early in the X-ray structural analysis of hemoglobin.[7] In the simplest form of the method[8] a dialysis bag containing a variable amount of protein solution is used, a minimum quantity of 0.5 to 1.0 ml being practicable. The bag is submerged in a suitable buffer solution. By small, daily additions of concentrated salt solution, organic solvent, or very dilute buffer the composition of the low molecular weight components in the system is gradually changed to the point where crystallization of the protein starts. The initiation of crystallization is often indicated by opalescence or cloudiness inside the cell. Any changes in the buffer-precipitant system are avoided until the first crystals have appeared and have reached the maximum obtainable size. The lower the degree of supersaturation, the smaller will be the number of single crystals formed. If subsequent additions of precipitant are made very cautiously, these few single crystals can grow larger without the formation of many new crystals. When the major fraction of the protein has been converted into crystals, a slight excess of precipitant is added to prevent damages during the mounting and soaking procedures.

B. Tools

1. Dialysis Membranes

The usual dialysis tubings and membranes are composed of cellophane and other cellulosa derivatives. Sometimes ultrafiltration devices,

[6a]M. Zeppezauer, H. Eklund, and E. Zeppezauer, *Arch. Biochem. Biophys.* 126, 564 (1968).
[6b]P. Dunnill, *J. Crystal Growth* 6, 1 (1969).
[7]I. Boyes-Watson, E. Davidson, and M. F. Perutz, *Proc. Roy. Soc. Ser. A* 191, 83 (1947).
[8]H. Theorell, *Biochem. Z.* 252, 1 (1932).

equipped with collodion bags have been used for crystallization.[9] Most membranes contain small amount of impurities, which may be removed by boiling in millimolar EDTA (sodium salt) solution or 1% sodium bicarbonate solution for 10 minutes. Alternatively, the following procedure is recommended:

One molar acetic acid is drawn through the tubing by applying reduced pressure and the acid is followed by washing with quartz distilled water. Thereafter a buffer at pH 9.5, usually 0.2 M Tris·chloride or Tris·sulfate is used for washing. The tubing is then rinsed with water and is immersed in the buffer to be used. Proteins with formula weight of less than about 10,000 diffuse through most of the commercial dialysis membranes. The reduction of pore size by treatment with acetic anhydride in pyridine has been described by Craig and co-workers.[10]

2. Diffusion Cells

In a large series of experiments, the use of dialysis bags calls for vast quantities of protein. The amount of protein can be greatly reduced by using microdiffusion cells made from capillaries. The method allows the use of identical concentrations of protein but very much smaller volumes, thereby conserving valuable starting material.

a. *Capillaries Closed by Dialysis Membranes.* The diffusion cells are made from 30 mm or 50 mm long capillary tubes of Pyrex glass or of Plexiglas (acrylonitrile polymer) which have a wall thickness of $\geqslant 3$ mm and $\geqslant 1.2$ mm, respectively. The ends of the tubes are carefully ground flat and the edges rounded in order to avoid damage to the membranes (Fig. 1.)

A piece of dialysis membrane is attached across the bottom of the cell by means of a ring of soft, transparent PVC tubing. The ring has two or three feet, each about 5 mm long, which serve to hold the membrane off the bottom of the vessel in which the dialysis cell will be placed. (The feet may be cut out conveniently by means of a wedge-shaped punch.) The protein solution, thoroughly centrifuged, is injected into the vertically directed cell by means of thin surgical tubing or a very thin capillary. A continuous column of liquid should result: air bubbles must be avoided. The upper end of the cell is closed with paraffin foil attached by means of a small ring of PVC-tubing (Fig. 2.) This cell is then placed vertically into a suitable vessel containing at least enough liquid to submerge the membrane. There should be no air bubbles on either side of the membrane. Crystallization inside such a capillary can then be attempted by changing the conditions in the outer liquid.

[9]A. Leibman and P. Aisen, *Arch. Biochem. Biophys.* **121**, 717 (1967).
[10]L. C. Craig and W. Konigsberg, *J. Phys. Chem.* **65**, 166 (1961). cf. Brownstone, *Anal. Biochem.* **27**, 25 (1969).

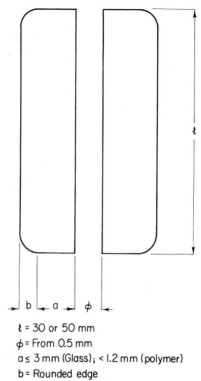

l = 30 or 50 mm

ϕ = From 0.5 mm

$a \leq$ 3 mm (Glass); < 1.2 mm (polymer)

b = Rounded edge

Fig. 1. Membrane-closed diffusion cell. Material: Glass or transparent polymer (Perspex, Plexiglas, Lucite, or similar).

Capillary diameters of 0.7–1.5 mm are suitable for screening crystallization conditions. It is more convenient to use somewhat larger capillary diameters when the crystals are to be mounted for X-ray work, i.e., $\phi \geq 1.5$ mm (Table I).

b. *Capillaries Closed by Gel Diaphragms.* If only minute amounts of protein are available, it is possible to perform the crystallization experiments directly in an X-ray capillary tube. Capillaries used for this purpose have a wall thickness of about 0.01 mm and cannot be closed by a dialysis membrane. However, they may be converted into microdiffusion cells in the following way:

i. *Cells for the crystallization of high molecular weight proteins ($M_w > 10^4$).* X-ray capillaries with a diameter of 0.7 mm are carefully cleaned with HNO_3 and double-distilled water. A solution[11] containing 475 mg of recrystallized acrylamide, 25 mg of N,N'-methylene-bis-acrylamide, and

[11] K. Mosbach and R. Mosbach, *Acta Chem. Scand.* **20**, 2807 (1966).

Fig. 2.   Micro dialysis cell ready for use (filled with a solution of blue-dextran).

TABLE I

APPROXIMATE VOLUME OF LIQUID (MICROLITERS) INSIDE A CAPILLARY AS FUNCTION
OF DIAMETER AND LENGTH OF THE LIQUID COLUMN

ϕ (mm)	Length (mm)							
	5	10	15	20	25	30	40	50
0.5	1	2	3	4	5	6	8	10
0.7	2	4	6	8	10	12	16	20
1.0	4	8	12	16	20	24	32	40
1.5	9	18	27	36	45	54	72	90
2.0	16	32	48	64	80	96	128	160
2.5	25	50	75	100	125	150	200	250
3.0	35	70	105	140	175	210	280	350
3.5	48	96	144	192	240	290	384	480
4.0	63	126	190	250	315	380	500	630

20 mg of ammonium persulfate in 9.5 ml of water or buffer is carefully deaerated *in vacuo*. If the capillary tube is dipped into this solution, a liquid column about 1 cm high is obtained at the narrow end of the capillary. This solution forms a gel upon a short exposure to 90° or after irradiation with a UV lamp. The capillary is then filled with buffer and is placed for 2–3 days in a vessel containing the same buffer to allow residual traces of the monomers and catalyst to diffuse out of the gel. Crystallization experiments are started by withdrawing the buffer with thin tubing and then injecting the protein solution into the cell. The upper end of the capillary is closed with paraffin. The capillary is placed vertically into the outer solution with the very fragile end standing on a plug of Pyrex-wool.

If crystallization occurs only at high concentrations of salt or organic solvent, a gel diaphragm polymerized in either water or dilute buffer may shrink excessively. In such cases, the diaphragm should be prepared from a solution of the monomer in water or buffer, with the solvent containing about half the concentration of the required precipitant. Organic solvents inhibitory to vinyl polymerization should be avoided. When the cell has been filled with protein solution, it is immediately put into an outer solution containing the same concentration of precipitant as the monomer solution.

ii. *Cells for low-molecular weight proteins* ($M_w < 10^4$). The diaphragm is prepared as above, but the monomer solution should contain a higher concentration of acrylamide, e.g., 1.9 g acrylamide, 0.1 g N,N'-methylene-bis-acrylamide, and 20 mg ammonium persulfate in 8 ml of water or buffer. After the diaphragm has been polymerized, two wax or picein balls may be applied around the capillary for use in the final mounting of the crystals for the X-ray exposure (Fig. 3).

c. *Cells for work at high concentrations of organic solvent.* Certain organic solvents may, at higher concentrations, extract additives (plasticizers) from the PVC tubings which are used to attach membranes to the capil-

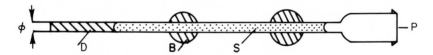

D = Acrylamide polymer diaphragm
S = Protein solution
B = Picein balls
P = Paraffin membrane
ϕ = Inner diameter 0.7 or 1.0 mm

FIG. 3.   Micro diffusion cell made from an X-ray capillary.

laries. The gel plugs in the capillaries described under b tend to detach from the inside walls at very high solvent concentrations. When an increase of the protein concentrations and/or a change of pH cannot circumvent the need for very high solvent concentrations, one may prefer an all-Teflon cell of the type shown in Fig. 4. This type of cell is machined from a Teflon rod. On both the cell, *A*, and the membrane holder, *B*, a simple ring, two edges are carved out. They serve to simplify the gentle removal of the ring when the experiment is completed. For storage of the contents, a lid, *C*, is used instead of *B*.[12]

Cleaning. Capillaries made from lithium beryllium borate glass (socalled Lindemann capillaries) are immersed in concentrated nitric acid overnight, resting on Pyrex-wool. After being rinsed with distilled water they are treated with an alkaline laboratory detergent and are then sequentially rinsed with quartz-distilled water and alcohol. Thereafter, they are dried in air. The capillaries with thick walls may be treated in the same way. Deposits resisting the above-mentioned agents are removed with a mixture consisting of 400 ml of water, 300 ml of concentrated nitric acid, 100 ml of hydrofluoric acid, and 15 g of dodecyl sulfate or other acid-resistant detergent. The mixture is mixed and stored in a resistant plastic container, care being taken to protect eyes, skin, and clothes against droplets of the solution. Exposure of a few minutes to this mixture is usually sufficient. The capillaries are to be picked up with plastic forcaps for rinsing in order to avoid metal contamination.

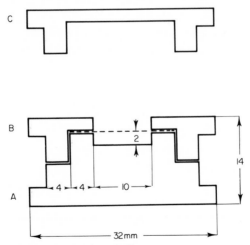

FIG. 4.   Micro dialysis cell made from Teflon.

[12]Similar cells made from optically treated glass are supplied on request by HELLMA, Optische Werkstätten, 7840 Müllheim, Germany.

3. Precipitating Agents

Different kinds of precipitants have to be explored systematically; they include inorganic salts, organic salts, and organic solvents. In each case, the highest grade of purity obtainable is just sufficient. It is better to explore the effect of systematically added substances, e.g., small amounts of metal salts, upon crystal growth using ultrapure buffers and precipitants than to trust to the beneficial action of impurities that are abundant in many analytical grade chemicals (see section C).

Inorganic Salts. Due to its high solubility, ammonium sulfate is the salt most frequently used. Proteins crystallizing at low or moderate ionic strength may also be precipitated by alkali sulfates, by chlorides, or by magnesium sulfate. It appears that salt-precipitated crystals, in general, possess good mechanical strength. However, at a high ammonium ion concentration several proteins fail to combine with certain ligands, which would otherwise form the desirable heavy atom derivatives for X-ray analysis. In such cases, mixtures of sodium and potassium phosphates are preferred over ammonium sulfate. They are soluble to equally high molarity, at least at room temperature. Substances not available in ultrapure form should be recrystallized from quartz-distilled water containing EDTA and, subsequently, from water. A number of inorganic substances may be precipitated with absolute alcohol from their concentrated aqueous solutions after extraction with dithizone (0.01% w/v) or 8-hydroxyquinoline (0.1% w/v) in chloroform, and then with chloroform alone in order to remove traces of metals.

Organic Salts. When crystallization is to be tried at a pH at which common ammonium salts liberate ammonia, quaternary ammonium salts sometimes prove useful. Tetramethylammonium chloride and (2-hydroxyethyl)trimethylammonium (choline) chloride are commercially available and can be recrystallized from alcohols or from mixtures of ethanol and ethyl acetate. Salts with other anions, unless available, may easily be prepared by neutralizing commercial tetramethylammonium hydroxide or choline hydrogen carbonate solutions with the appropriate acid and subsequent freeze-drying. Many of the quaternary salts are very hygroscopic and should be dried in a vacuum chamber after recrystallization. Quaternary ammonium ions with long aliphatic chains are moderately soluble. They usually act as detergents and precipitate proteins via the formation of insoluble complexes; the procedure often leads to denaturation.[13]

Organic Solvents. Crystals grown from salt solutions may prove unsatisfactory for certain types of problems. In such cases, organic solvents may

[13]H. E. Schultze and J. F. Heremans, "Molecular Biology of Human Plasma Proteins," Vol. 1, p. 261, Elsevier, Amsterdam, 1966.

provide an alternative means of crystallization. Although many common, spectroscopically pure alcohols, ketones, and cyclic ethers can be utilized in protein crystallization, their volatility makes the subsequent handling of the crystals quite difficult. Therefore, relatively nonvolatile organic solvents that are completely miscible with water should be used. 2-methyl-2,4-pentanediol (MPD)[14] has been used successfully to grow crystals for the X-ray analyses of ribonuclease[15] and horse liver alcohol dehydrogenase.[16] (In the latter case the enzyme is denatured by the salt concentrations needed to produce crystals from inorganic salt solutions.) MPD is commercially available and is freed from impurities (traces of amines and variable amounts of keto compounds) by treatment with mixed-bed ion exchangers and careful fractional distillation *in vacuo* over K_2HPO_4 or KBH_4. Alternatively, MPD is purified by stirring with charcoal (20% w/w, A grade, finely powdered) overnight and subsequently filtering through Celite. Prior to the addition of KBH_4, traces of water are removed by distillation at reduced pressure.

Even dimethyl sulfoxide, dimethyl formamide, and similar solvents, as well as nitriles, are possible precipitating agents.[17,18] Crystals of horse liver alcohol dehydrogenase, grown from acetonitrile, show cell dimensions identical to those of crystals grown from ethanol or MPD solutions. Despite its volatility acetonitrile does not distill from water mixtures as quickly as ethanol. Mounting of acetonitrile-grown crystals therefore appears to be rather convenient, at least at $+4°C$.[19]

The use of sulfoxides, amides, and nitriles is especially favorable in cases where one wishes to diffuse highly reactive or poorly water-soluble substances into the protein crystals. When adding uncharged ligands which attach noncovalently to the protein, one also must consider the possibility that dipolar solvents may compete for the ligand, and try solvents of different polarity.

C. Comments on Special Points of Interest

It is hoped that the general procedures outlined above will enable the reader to develop a proper crystallization method for every special case. Earlier crystallographic work has been summarized in work cited in foot-

[14]M. V. King, *Biochim. Biophys. Acta* 79, 388 (1964).

[15]G. Kartha, T. Bello, and D. Harker, *Nature (London)* 213, 862 (1967).

[16]C.-I. Brändén, E. Zeppezauer, T. Boiwe, G. Söderlund, B.-O. Söderberg, and B. Nordström, *in* "Pyridine Nucleotide-Dependent Dehydrogenases" (H. Sund, ed.), p. 129, Springer, Berlin, 1970.

[17]J. Drenth, W. G. J. Hol, J. W. E. Visser, and L. A. A. Sluyterman, *J. Mol. Biol.* 34, 369 (1968).

[18]B. Blombäck, M. Blombäck, and E. Holmberg, *Acta Chem. Scand.* 20, 2317 (1966).

[19]Drs. Å. Åkeson, C.-I. Brändén, and E. Zeppezauer, personal communication (1968).

notes 20-22. Table II summarizes information on the crystalline biopoly-mers studied very recently. Some of them were crystallized by simple addition of a precipitant ("batch crystallization"), some by equilibrium dialysis, and a few by other methods.

A great advantage of equilibrium dialysis is that the protein concen-tration is held constant until precipitation begins. If protein precipitates amorphously, the conditions in the outer liquid may easily be readjusted in order to dissolve the precipitate and to attempt another crystallization under changed conditions. Thus, the contents of one dialysis cell may be subjected to a number of different experiments provided, of course, that the protein is sufficiently stable over the time period.

Purity of the Proteins. It sometimes appears difficult to obtain completely monodisperse preparations or samples that are stable over the time re-quired to grow good crystals. Denatured material often precipitates first and introduces heterogeneous nuclei with concomitant precipitation of the bulk protein in microcrystalline form. It should be emphasized that operations such as heat denaturation, ethanol-chloroform treat-

TABLE II
BIOPOLYMERS RECENTLY CRYSTALLIZED FOR X-RAY DIFFRACTION WORK

Biopolymer	Method and technique used for crystallization
Aldolase[a] (rabbit muscle	High ionic strength, batch crystallization
Asparaginase[b] (*Erwinia carotovora*)	Alcohol addition at low ionic strength
Bacteriochlorophyll-protein[c] (*Chloropseudomonas ethylicum*)	High ionic strength reached by slow equilibration
Bence-Jones Protein[d] (human L-type)	Low ionic strength, equilibrium dialysis in micro cells
Human ceruloplasmin[e]	Low ionic strength; slow evaporation of a 7% solution near the isoionic point
Cytochrome *c* peroxidase[f] (yeast)	Low ionic strength; equilibrium dialys against water-MPD in micro cells
Cytochrome b_5 (calf liver[g], rabbit liver[h])	High ionic strength, batch crystal-lization[g] or equilibrium dialysis[h]
Fab fragment[i] (human myeloma protein I_gGI)	High ionic strength, equilibrium dialysis
Flavodoxin[j] (*Clostridium pasteurianum*)	Equilibrium dialysis at pH 6.4–6.8 at high ionic strength in micro cells
Glutamine synthetase[k] (*E. coli*)	High ionic strength, batch crystalliza-tion

[20] M. V. King, in "Crystal Data" (I. D. H. Donnay and G. Donnay, eds.), A.C.A. Monograph No. 5, p. 1263, American Crystallographic Association, Washington, 1963.
[21] B. W. Matthews, *J. Mol. Biol.* 33, 491 (1968).
[22] C. C. F. Blake, in "Amino Acids, Peptides, and Proteins", Vol. 1, Specialist Periodical Report, p. 154. The Chemical Society, London, 1969.

Metmyoglobin[l] (sperm whale), alkaline xenon complex	High pH and ionic strength, batch crystallization
Myoglobins[m] (tuna)	High ionic strength, batch crystallization
Protease[n] (*Penicillium janthinellum*)	High ionic strength, slow evaporation
Protease (*Sorangium*, sp. *Myxobacter* 405°, strain of *Arthrobacter[p]*)	High ionic strength, equilibrium dialysis in micro cells
Tropomyosin[q] (rabbit muscle)	Moderate high ionic strength, equilibrium dialysis
Tomato bushy stunt virus[r]	High ionic strength, batch crystallization
Thyroglobulin[s]	Temperature gradient applied at high ionic strength (Jakoby's method)
Transfer RNA[t,u,v]	Equilibration with solvent, either by isothermal distillation[t], or aided by a temperature gradient[u], batch crystallization[v].
Thioredoxin[w] (*E. coli*)	Equilibrium dialysis in micro cells against buffer-MPD near the isoionic point

[a]P. A. M. Eagles, L. N. Johnson, M. A. Joynson, C. H. McMurray, and H. Gutfreund, *J. Mol. Biol.* **45**, 533 (1969).

[b]A. C. T. North, H. E. Wade, and K. A. Cammack, *Nature (London)* **224**, 594 (1969).

[c]J. M. Olson, D. F. Koenig, and M. C. Ledbetter, *Arch. Biochem. Biophys.* **129**, 42 (1969).

[d]M. Schiffer, K. D. Hardman, M. K. Wood, A. B. Edmundson, M. E. Hook, K. R. Ely, and H. F. Deutsch, *J. Biol. Chem.* **245**, 728 (1970).

[e]A. G. Morell, C. J. A. Van Den Hamer, and I. Scheinberg, *J. Biol. Chem.* **244**, 3494 (1969).

[f] T. Yonetani, B. Chance, and S. Kajiwara, *J. Biol. Chem.* **241**, 2981 (1966). L. O. Hagman, L. O. Larsson, and P. Kierkegaard, *Int. J. Protein Res.* **1**, 283 (1969).

[g]F. S. Mathews and P. Strittmatter, *J. Mol. Biol.* **41**, 295 (1969).

[h]R. H. Kretsinger, B. Hagihara, and A. Tsugita, *Biochim. Biophys. Acta* **200**, 421 (1970).

[i]R. L. Humphrey, H. P. Avey, L. N. Becka, R. J. Poljak, G. Rossi, T. K. Choi, A. Nisonoff, *J. Mol. Biol.* **43**, 223 (1969).

[j]M. L. Ludwig, R. D. Andersen, S. G. Mayhew, and V. Massey, *J. Biol. Chem.* **244**, 6047 (1969).

[k]R. P. Bywater, C. H. Carlisle, and R. B. Jackson, *J. Mol. Biol.* **45**, 429 (1969).

[l]B. Schoenborn, *J. Mol. Biol.* **45**, 297 (1969).

[m]R. H. Kretsinger, *J. Mol. Biol.* **38**, 141 (1968).

[n]N. Camerman, T. Hofmann, S. Jones, and S. C. Nyburg, *J. Mol. Biol.* **44**, 569 (1969).

[o]M. N. G. James and L. B. Smillie, *Nature (London)* **224**, 694 (1969).

[p]H. Eklund, M. Zeppezauer, and C.-I. Brändén, *J. Mol. Biol.* **34**, 193 (1968).

[q]D. L. D. Caspar, C. Cohen, and W. Longley, *J. Mol. Biol.* **41**, 87 (1969).

[r]S. Harrison, *J. Mol. Biol.* **42**, 457 (1969).

[s]L. W. Labaw and T. E. Rall, *J. Mol. Biol.* **36**, 25 (1968).

[t]A. Hampel and R. Bock, *Biochemistry* **9**, 1873 (1970).

[u]H. H. Paradies, *Fed. Eur. Biol. Soc. Lett.* **2**, 112 (1968).

[v]F. Cramer, F. von der Haar, K. C. Holmes, W. Saenger, E. Schlimme, and G. E. Schulz, *J. Mol. Biol.* **51**, 523 (1970).

[w]A. Holmgren and B.-O. Söderberg, *J. Mol. Biol.* **54**, 387 (1970).

ment, and lyophilization during a purification procedure may be sufficient to introduce a conformational heterogeneity which can prevent the growth of large crystals. The omission of such treatment proved crucial in the work with human carbonic anhydrase C[23] and liver alcohol dehydrogenase.[19]

Screening. An already existing procedure for the crystallization of a protein may often form the basis for refinements leading to large crystals. Slowing down the addition of precipitant by successive dialysis steps rather than single-step addition may be sufficient. If not, change of pH or temperature could be effective. Otherwise, a broader screening program has to be devised in which most of the physical parameters are held nearly constant while a few are changed gradually and systematically. It is advisable, first to study the effect of variation of pH. Two series of dialysis experiments are performed at pH intervals of 0.3 or less within the tolerated pH range, using one salt and one solvent as precipitant. Thus, one will obtain a qualitative picture of the protein's solubility dependence on pH. Eventually, at some pH values the precipitates will contain more or less well developed crystals. One may then diminish the pH intervals, use different buffer substances, vary protein concentration, temperature, solvent, or salt, and use other parameters until the crystals are satisfactory. Sometimes, several solubility minima are observed. Extremely soluble proteins may perhaps precipitate and crystallize only near the isoionic point; it may be necessary to use solutions containing 5% protein or more. When satisfactory conditions have been found, the next step is to determine the solubility quantitatively.[24]

Solubility Curves. The handling of protein crystals and the performance of chemical reactions by diffusion of reagents into them are often made difficult by high concentrations of protein which remain dissolved in the mother liquor. Mother liquors saturated with protein usually result upon crystallization by evaporation of a protein solution or after dialysis at low ionic strength. They are very viscous, which may cause difficulty during the mounting of crystals for X-ray work. When a derivative is formed by reaction of such a crystal suspension with a heavy metal salt, voluminous amorphous precipitates form if the derivative is less soluble than the native protein. If it is more soluble, the crystals may dissolve again. Therefore, the development of a routine crystallization procedure should include the determination of solubility curves. To this end, a series of equilibrium dialysis experiments is performed with increasing concentrations of precipitant in the outer liquor. Using the outer liquor as

[23]B. Strandberg, B. Tilander, K. Fridborg, S. Lindskog, and P. O. Nyman, *J. Mol. Biol.* 5, 583 (1962).
[24]R. M. Herriott, this series, Vol. 4, p. 212.

reference, the mother liquor can be examined for its protein content after centrifugation of the crystals at the same temperature at which they were grown. From the solubility diagram obtained in this way, a final concentration of precipitant can be chosen which is adjusted after the growth is nearly completed and which ensures complete precipitation of the material. Proteins which crystallize on dialysis against buffers of low ionic strength will usually behave in this manner even when the starting buffer contains organic solvent, the concentration of which is finally raised to achieve complete precipitation. Unnecessarily high solvent concentrations may lead to denaturation and to destruction of the crystals.

The solubility diagram also yields information about the efficiency of a specific salt as precipitant for a given protein and could be especially important when the less common salts are tested.[3]

Undesirable and Unexpected Phenomena. It is very seldom that only a few experiments result in a definite crystallization method for a protein not previously crystallized.

The most common observation is the precipitation of amorphous or microcrystalline material, often at a high degree of supersaturation. The reason may be that pH, ionic strength and protein concentration, among other factors, are not optimal and should be varied in small increments. In case they are, one would first suspect that the critical point of supersaturation had been reached and passed too quickly. It is advisable then, to lower the concentration of precipitating agent slowly until the precipitate begins to dissolve. Often crystal growth sets in at this point and can be further promoted by alternate small and slow increase and decrease of the concentration of the precipitant. This method is also suitable to induce secondary growth from coagulated protein which deposits as droplets instead of irregular voluminous precipitates. On prolonged standing, the droplets convert to small crystals or clusters of crystals which continue to grow only after decrease in ionic strength. It may also happen that crystallization is retarded or does not occur at all in spite of high supersaturation. Seeding may then be helpful. Only a few seeds should be introduced into one cell or batch; microcrystalline seeds need be diluted with an excess of mother liquor.

Temperature and Convection. In general, protein crystals should be grown at the same temperature at which they are used. However, the success of Jakoby's method[1] indicates that a temperature gradient may be very useful in order to induce nucleation. A crystallization cell using a temperature gradient to effect slow equilibration of a *t*RNA solution with dioxane as the precipitating solvent, has been described by Paradies.[25]

[25]H. H. Paradies, *Fed. Eur. Biol. Soc. Lett.* **2**, 112 (1968).

Sometimes, it appears as though a sudden rise in the temperature of a supersaturated solution has the same effect as has the introduction of seeds. The thermoconvection caused by the temperature gradients is obviously not unfavorable in such cases. On the other hand the absence of convection, as maintained inside a capillary, favors crystal growth for certain proteins and diminishes the amount of amorphous precipitate.

Chemical Modification of the Protein. Many proteins seem to be extremely "unwilling" to form crystals suitable for physical studies, despite easy formation of microcrystals under varying conditions. One way to circumvent this difficulty is to test the analogous protein from another species. Another is to modify the protein chemically. Whenever possible, such a modification should be a reversible one. For example, a mercurial derivative of a protein may yield better crystals. By diffusing cysteine into the crystals, one can hope to remove the organomercury residue and thus obtain isomorphous crystals consisting of native protein.[26] In an analogous manner, specific enzyme inhibitors can be used, as well as metal ions and metal complexes. The solubility of the native protein may differ greatly from the solubility of one or more of its isomorphous derivatives (both heavy and light atom derivatives). In view of this one should try to crystallize the protein from solutions containing metal ions, both simple and complexed with organic and inorganic ligands.[2,14] The problems concerned with the preparation of derivatives—or native crystals respectively—have been dealt with in a review by Blake.[27]

[26]B. Strandberg, *Ark. Kemi* **28**, 1 (1967).
[27]C. C. F. Blake, *Advan. Protein Chem.* **23**, 59 (1968).

[25] Protein Crystallization: Micro Techniques Involving Vapor Diffusion

By David R. Davies and D. M. Segal

The vapor diffusion technique has been a conventional weapon in the armory of small molecule crystallographers for many years. It has also been used by some protein crystallographers as a method for very gradually approaching the conditions of crystallization. Recently, adapted for micro techniques, it has received widespread use in the crystallization of tRNA's, where only extremely small amounts of material were available.[1-7] This general method, described here, is equally applicable to the crystallization of proteins, particularly where the quantity of material is limited.

Methods

Materials

> Hamilton syringe, 10 μl micropipettes
> Micro culture slides with 3 mm-deep wells
> Desiccators
> Melting point capillaries
> Small circular dishes: we have these manufactured. They are approximately 3.5 inches in diameter and 1.25 inches deep. They have a ground glass lip and may be sealed with a plate glass lid, thus permitting observation of the sample without disturbing it. An improved version has a Lucite lid with ports for easy introduction and removal of samples and solvent.[8]

Procedure

The protein to be crystallized is brought to the desired concentration (usually about 1%) in dilute buffer, and a 10 μl droplet is placed in the well of a siliconed microculture slide. (Siliconing the slide helps to prevent spreading of the droplet.)

Crystallizations with Aqueous Salt Solutions. At any pH a preliminary test should be made to establish the concentration of salt, e.g., ammonium sulfate, at which the protein precipitates. This can be accomplished by adding strongly buffered saturated salt solutions to the 10 μl droplet with a 10-μl Hamilton syringe until the precipitation point is reached. Saturated salt solution is added to a second 10 μl droplet to give a concentration of salt in the drop about 10% below that of the precipitating concentration. The slide is then placed in a desiccator or a

[1] B. F. C. Clark, B. P. Doctor, K. C. Holmes, A. Klug, K. A. Marcker, S. J. Morris, and H. H. Paradies, *Nature (London)* 219, 1222 (1968).

[2] S. H. Kim, and A. Rich, *Science* 162, 1381 (1968).

[3] A. Hampel, M. Labanauskas, P. G. Connors, L. Kirkegard, U. L. Rajbhandary, P. B. Sigler, and R. M. Bock, *Science* 162, 1384 (1968).

[4] F. Cramer, F. Van Den Haar, W. Saenger, and E. Schlimme, *Angew. Chem.* 80, 969 (1968).

[5] J. R. Fresco, R. D. Blake, and R. Langridge, *Nature (London)* 220, 5174 (1968).

[6] C. D. Johnson, K. Adolph, J. J. Rosa, M. D. Hall, and P. B. Sigler, *Nature (London)* 226, 1246 (1970).

[7] D. R. Davies and B. P. Doctor, "Procedures in Nucleic Acid Research," Vol. 2, Harper and Row, New York 1971.

[8] A. Hampel and R. Bock, *Biochemistry* 9, 1873 (1970).

tightly covered dish together with a beaker containing about 2 ml of salt solution, buffered at the required pH, which is approximately 5% below the precipitating concentration. The desiccator is sealed, and the contents are allowed to stand for some days, during which the concentration of the precipitant in the sample is increased by equilibration with the contents of the reservoir. Periodic microscopic examination will reveal whether crystals appear. If, after a few days, neither a precipitate nor crystals are observed, the beaker may be removed from the desiccator and replaced with a beaker containing a slightly higher concentration of salt. In this manner, the salt concentration may be gradually increased until the protein in the drop is clearly all precipitated or until crystallization has occurred.

The same drop may be used many times providing the stability of the protein permits. A premature precipitate can frequently be redissolved simply by exposing the drop to a water atmosphere until the precipitate disappears.

When searching for appropriate crystallization conditions, it is necessary in general to cover a broad range of conditions, and the above procedure should be carried out simultaneously at different pH's and temperatures (at least 4° and room temperature). Protein concentration may also be varied to provide a 4-dimensional matrix of variables. The small amounts of material used and the ability to change salt concentration at will by vapor exchange permit a variety of conditions to be investigated.

Once the optimal conditions for crystallization have been established, crystals sufficiently large for X-ray analysis, i.e., minimum dimension >0.1 mm, may frequently be grown by means of a slower approach to equilibrium. This can be achieved by placing a cover slip over the well of the slide so that there is only a small aperture at the top of the well. Alternatively, after addition of salt, the solution to be crystallized may be sucked up into an unsealed melting point capillary tube, and allowed to equilibrate in a desiccator with salt solution at the desired concentration. The narrow bore of the capillary will reduce the rate of diffusion in both the vapor and the liquid phase.

Crystallizations from aqueous-organic solvents. The above procedure may be easily adapted to precipitation by organic solvents which are miscible with water. If a highly volatile precipitant is used, e.g., acetone or methanol, an accurate predetermination of precipitation conditions may be difficult to carry out due to rapid evaporation in air of the precipitant. In this case, the precipitation point may be determined by increasing the concentration of the volatile solvent by fairly large increments, using one of the vapor diffusion techniques. An approximate estimate of the precipitating concentration can thus be obtained, and crystallization can then be attempted by approaching this concentration more gradually. This method is not as time consuming as it may seem, since equilibrium is achieved much more rapidly with highly volatile solvents than with aqueous solutions of nonvolatile solutes. With volatile solvents, there is of course no need to add precipitant to the droplet prior to equilibration, although buffer must be added separately in order to obtain the required pH. With less volatile solvents, e.g., 2-methyl-2, 4-pentanediol, some solvent must be added initially to the protein solution in order to avoid taking an inordinately long time to reach equilibrium.

The Use of Seeding Solutions. Large crystals may sometimes be encouraged to grow by the use of a seeding solution.[6] Such a solution is prepared by crushing a single crystal or group of crystals in a "stabilizing" solvent in which the crystal fragments will not redissolve. The suspension may then be centrifuged with a low speed clinical centrifuge to remove large fragments and diluted in the stabilizing solvent to minimize the concentration of nucleation sites. The sample to be crystallized is brought to a precipitant concentration slightly lower than that required to yield crystals without seeding, and is then touched with a glass fiber which has been wetted with the seeding solution. This procedure frequently leads to the production of a few large crystals.

Section V

Chromatographic Procedures

Articles Related to Section V

Vol. I [11]. Separation of proteins by use of adsorbents. Sidney P. Colowick.

Vol. I [12]. The partition chromatography of enzymes. R. R. Porter.

Vol. I [18]. Chromatography of enzymes on ion exchange resins. C. H. W. Hirs.

Vol. V [1]. Column chromatography of proteins: Substituted celluloses. Elbert A. Peterson and Herbert A. Sober.

Vol. V [2]. Column chromatography of proteins: Calcium phosphate. Osten Levin.

Vol. XIX [70]. Water-insoluble derivatives of protealytic enzymes. Leon Goldstein.

Vol. XIX [71]. Cellulose-insolubilized enzymes. E. M. Crook, K. Brocklehurst, and C. W. Wharton.

Vol. XIX [72]. Water-insoluble thrombin. Benjamin Alexander and Araceli M. Engel.

Vol. XXI [3]. Chromatography of nucleic acids on hydroxyapetite. G. Bernardi.

[26] Chromatography of Proteins on Ion-Exchange Adsorbents

By S. Ralph Himmelhoch

Since the last review of ion-exchange chromatography of proteins in this series,[1] the use of adsorbents prepared in the laboratory has been largely supplanted by the use of commercially prepared products. This change has been necessary for the routine and convenient use of chromatographic procedures, particularly when separations requiring large quantities of adsorbent are being carried out. On the other hand, the commercial preparation of ion-exchange celluloses has introduced problems of its own, the most significant being both the lack of comparability of adsorbents designated similarly by different manufacturers, and, even more troublesome, lack of reproducibility in different lots of ostensibly identical adsorbents from the same manufacturer. The protein binding capacity of DEAE-cellulose from different manufacturers varies greatly.[2] The same adsorbents vary in physical properties from extremely coarse materials, incapable of forming very compact beds but permitting essentially any flow rate desired, to very fine materials that form quite compact beds but frequently impose severe restrictions on the range of flow rates obtainable. Thus the designation of the exact source and type of adsorbent under consideration is essential in the published description of any chromatographic experiment. It is no exaggeration to state that any analytical or preparative method for which this information is not provided is of little value to others who wish to reproduce it.

We will attempt to outline those general considerations which govern the use of ion-exchange chromatographic experiments, placing special emphasis on the precautions necessitated by the use of commercial adsorbents.

General Principles of Ion-Exchange Chromatography of Proteins

Adsorption of proteins to ion-exchange celluloses involves primarily the formation of multiple ionic bonds between charged groups on the protein and available groups of opposite charge on the adsorbent. Chromatographic separation then depends on the differential elution

[1] E. A. Peterson and H. A. Sober, Vol. 5 [1].
[2] S. R. Himmelhoch and E. A. Peterson, *Anal. Biochem.* 17, 383 (1966).

of the adsorbed proteins by a variety of techniques based either upon alteration of the charge state of the protein (pH), or upon the use of agents capable of "competing" with the adsorbed protein for the charged sites on the adsorbent. The affinity and capacity of a particular ion-exchange adsorbent for a specific protein depend upon both the pH and salt concentration, and these parameters must be specified accurately if data are to be significant. The capacity for protein-binding of an adsorbent depends on the number of charged groups incorporated, but not in direct proportion, since the distribution of charged groups on the adsorbent surface as well as their total number is important. When commercial adsorbents are used, great care must be employed in interpreting the data on capacity provided by the manufacturer. Lots of DEAE-celluloses specified as having very similar nitrogen content by one manufacturer have been found to vary more than 10-fold in their capacity to bind identical protein mixtures under identical conditions. Recent titration data in our laboratory have shown that, although the nitrogen content specified by the manufacturer was accurate, a large proportion of the nitrogen had been incorporated as a group of much weaker basicity in the case of adsorbents of low capacity.[3] Since these weaker groups would be largely suppressed under the conditions used to initiate the test chromatograms, they would not contribute to the capacity of the adsorbent bed. If the tests had been conducted at a lower pH the discrepancy might largely have disappeared.

Thus, the safest way to establish the useful capacity of these adsorbents is by empirical test, using a well-studied system of proteins, such as serum. The performance of such tests will be described later.

Types of Adsorbent Available–Matrix

In the table are listed the commercially available ion-exchange adsorbents in general use for the separation of protein mixtures. These adsorbents may first be classified into several large categories depending upon the matrix to which the charged group is attached. Although some use has been made of ion-exchange resins in protein separation, such adsorbents are not of general usefulness and have not been included in the list. The main categories of matrix available are cellulose (either untreated or as "purified" microcrystalline regions of the cellulose fiber), "Sephadex" (either G-25 or G-50).[4] and poly-acrylamide gel (Biogel series).[4] In general, cellulose provides an essentially open, hydrophilic supporting matrix whereas the gels, although

---

[3] E. A. Peterson, personal communication.
[4] See this Volume [27].

hydrophilic, have regions of the matrix available only to molecules within a certain size range. Although systematic studies are lacking, definite effects related to molecular size have been observed with the ion-exchange DEAE-Sephadexes. It has been found that the exchangers based on G-25 Sephadex are of adequate capacity for molecules below 10,000 MW, whereas exchangers based on G-50 Sephadex are of adequate capacity for molecules having molecular weight as high as 200,000. For molecules of MW greater than 200,000, exchangers based on G-25 Sephadex are again preferred, because of the better flow rates achieved. Such differences in capacity based on "exclusion" phenomena are also observed with ion-exchangers based on polyacrylamide gel, but not with those based on ordinary cellulose fibers. An instance of this type of effect is seen when equal aliquots of a properly equilibrated (e.g., with 5 mM succinic acid: 40 mM Tris, pH 8.6) sample of human serum are applied on the one hand to a bed of fibrous DEAE-cellulose or on the other to beds of DEAE-Sephadex A-25 and DEAE-Sephadex A-50. In the first case a series of colored bands quickly forms as the column is washed with starting buffer, a greenish or bluish band of ceruloplasmin being adsorbed near the top of the bed and successive yellow (albumin), red (hemoglobin-haptoglobin and siderophilin), and buff (β-lipoprotein) bands beneath it. In the case of DEAE-Sephadex A-25, a faint yellowish band extends over the whole column, and almost all the serum proteins are found in the starting buffer effluent. With DEAE-Sephadex A-50, a diffuse yellow band is formed almost twice as wide as the corresponding band on DEAE-cellulose, without differentiation into bands of different color. However, in spite of this nondescript initial banding, gradient chromatograms of serum on DEAE-Sephadex A-50, although differing from those on DEAE-cellulose in detail, provide comparable overall resolution.

It is of interest that the nondescript banding of serum proteins observed with DEAE-Sephadex is also seen when "microgranular" DEAE-cellulose made from the purified microcrystalline regions of the cellulose fiber is employed. This observation reflects the fact that this material is in some ways intermediate in properties between the two types of adsorbent matrix because the most open and accommodating portions of the cellulose fiber have been removed and covalent cross-linking added. Again, the resolution obtained in gradient chromatography with this adsorbent type is not, apparently, adversely affected by the effect. In fact, the finer particle size and higher density of the microgranular cellulose result in a more compact adsorbent bed, which provides higher resolution than the usually employed range of fibrous materials, although at some expense in flow rate.

ION-EXCHANGE ADSORBENTS IN COMMON USE FOR PROTEIN
CHROMATOGRAPHY

Designation	Ionizable group	Matrix available
Anion exchangers		
Aminoethyl-	$-O-CH_2-CH_2-NH_2$	Fibrous cellulose
Diethylaminoethyl-	$-O-CH_2-CH_2-N(CH_2CH_5)_2$	Fibrous cellulose, microgranular cellulose, Sephadex, polyacrylamide gel
Triethylaminoethyl-	$-O-CH_2-CH_2-\overset{X}{N(C_2H_5)_3}$	
Triethylaminoethyl-	$-O-CH_2-CH_2-N\ (C_2H_5)_3$	Fibrous cellulose
Guanidoethyl-	$-O-CH_2-CH_2-N-\overset{NH}{\overset{\|}{C}}-NH_2$	Fibrous cellulose, Sephadex
ECTEOLA-	Mix	Fibrous cellulose
Cation exchangers		
Carboxymethyl-	$O-CH_2-COOH$	Fibrous cellulose, microgranular cellulose, Sephadex, polyacrylamide gel
Phospho-	$-O-\overset{O}{\overset{\|}{P}}-OH$ $\overset{\|}{OH}$	Fibrous cellulose, Sephadex
Sulfoethyl-	$-O-CH_2-CH_2-O\overset{O}{\overset{\|}{S}}-OH$ $\underset{O}{\overset{\|}{}}$	Fibrous cellulose, Sephadex

The nature of the matrix to which charged groups are attached also affects other features of adsorbent behavior. The fibrous and microgranular cellulosic ion-exchangers, and the ion-exchange Sephadexes are reasonably inert to molar sodium hydroxide and molar HCl. However, unfavorable changes in the physical properties of the ion-exchange polyacrylamide gels occur on prolonged exposure to 1 M NaOH. It should be mentioned that certain types of ion-exchange cellulose, e.g., Selectacel Type 40, also break down into fine material on exposure to alkali, but these represent an exception rather than the rule among the ion-exchange celluloses.

Another significant difference between those adsorbents produced from hydrophilic gels and those from cellulose lies in the relative stability in volume of beds of these adsorbents upon changing the influent salt concentration and pH. Beds made from cellulosic adsorbents do not change appreciably in volume over a wide range of pH and salt concentration. Beds made from adsorbents based either on polyacrylamide gel or Sephadex, shrink to two-thirds or less of their initial volume when ordinary salt gradients are used and undergo changes of similar magnitude when changing from the charged to the uncharged state. Although these changes can be minimized by appropriate selection of experimental conditions, they constitute a significant disadvantage in the use of these adsorbents. Another disadvantage is the very large volume occupied by beds of these adsorbents for an equivalent ion-exchange capacity. The large effective "dead space" of such beds have a deleterious effect on the sharpness of the eluted peaks.

In general, the several adsorbent matrices are characterized by certain differences with respect to the range of flow rate obtainable. Fibrous celluloses, particularly those of the coarser, more flocculent, variety such as Whatman DE-1, can provide beds capable of yielding almost any desired flow rate, even with samples containing lipoproteins or other substances of unfavorable physical properties. Thus, this grade of adsorbent is extremely useful in carrying out operations of large scale on crude starting materials when a column operation is desired. On the other hand, the large interstitial spaces characteristic of beds made from such materials render them incapable of the resolution obtainable with other adsorbent types. Contrariwise, the ion-exchange gels, microgranular celluloses, and finer meshes of fibrous celluloses, can provide beds whose flow properties, though quite adequate for usual purposes, are in some instances inadequate. In addition, they are much more sensitive to the physical properties of applied samples, frequently occluding or allowing channeling when very crude, lipid-containing materials or solutions containing insoluble material are applied.

The ion-exchange Sephadexes and microgranular cellulosic adsorbents have the useful property of settling from suspension quite rapidly and forming compact and rather stable sediments. This property can be exploited for large-scale batch operations, making these adsorbents useful for many crude operations for which their resistance to flow when packed into columns would seem to render them unsuitable.

The physical structure of cellulosic matrices places a limit on the concentration of charged groups that can be attached without producing drastic changes in the physical properties of the resulting adsorbent. This is less true in the case of the gel matrices because of covalent crosslinks which stabilize the matrix. Although much of the additional charge that can be incorporated is probably internal to the gel beads and thus not available to large molecules, this feature can prove of great utility in chromatographing materials of lower molecular weight.

Types of Adsorbent Available–Charged Group

Each type of adsorbent matrix is available with a variety of charged groups. They are differentiated, in the first instance, by the sign of their charge (positively charged group incorporated = anion exchanger; negatively charged group = cation exchanger); and in the second instance by their strength as bases or acids, respectively.

Since proteins are amphoteric substances, they usually can be chromatographed on either anion or cation exchangers by an appropriate selection of conditions. This general principle is limited by the fact that some proteins are not stable in pH ranges on one or the other side of their isoelectric point. Nevertheless, unless problems of stability are shown to be prohibitive, the usefulness of both anion and cation exchange adsorbents should be empirically tested.

In general, in appropriate salt concentration, proteins will be adsorbed to anion exchangers at pH's above their isoelectric points and to cation exchangers at pH's below their isoelectric points. This principle can serve as a rule of thumb in choosing conditions for initial tests if information concerning the isoelectric properties of the proteins to be separated is available. It should be remembered, however, that in particular instances proteins can be successfully chromatographed on an adsorbent possessing ionized groups of the same charge as the net charge of the protein, presumably because distribution of charge on the protein is such that the protein molecule can present a surface with an effective charge of sign opposite to that of the adsorbent.

Once a decision has been made concerning the sign of the charged group desired, the choice as to its nature is made largely on the basis of the strength of the group required. The usual choice for a cellulosic anion exchanger in the pH range up to about 9 is DEAE-cellulose. The corresponding adsorbents based on a gel matrix are DEAE-Sephadex A-25 or A-50 (see above) or Biogel DM-2 or DM-20. Exceptions to this choice occur when exchangers of lower or of adjustable capacity are desired, as in chromatography of ribosomes and nucleic acids where special anion-exchange adsorbents are available (ECTEOLA-cellulose).

Although the instability of macromolecules of biological origin in the pH range above 9 makes chromatographic experiments in the high alkaline pH range rarely desirable, such adsorbents are available both on a cellulosic [GE-cellulose (guanidoethyl cellulose)] and gel [GE-Sephadex (guanidoethyl Sephadex)] matrix. These adsorbents have an additional advantage when quick regeneration of columns *in situ* is desired for experiments run below pH 11, because they remain completely in the charged form under these conditions and therefore do not require careful adjustment to the pH at which the adsorbent is to be equilibrated. Since interactions of the eluting buffers with the adsorbent are minimal, pH-fronting is a smaller problem than is the case with the weaker ion-exchangers.

The most usual choice of cation exchanger for the pH range above 3.5 is CM-cellulose or CM-Sephadex A-25 or A-50. Experience with the corresponding adsorbents based on a polyacrylamide matrix is very limited (Biogel CM-2, CM-30). For operations with cation exchangers in the low pH range two cellulose-based exchangers, P-cellulose (phosphocellulose) and SE-cellulose (sulfoethyl cellulose), and one gel based exchanger, SE-Sephadex (sulfoethyl Sephadex), are available. Although the P-cellulose has probably been more widely applied, the sulfoethyl-adsorbents have the at least theoretical advantage of a more stable ether bond between the charged group and the adsorbent. These adsorbents can of course be used in the higher pH range as well.

Preparation of Adsorbents for Use: Sieving and Washing

Although manufacturers frequently, and perhaps justifiably, claim to provide their adsorbents in a state requiring only equilibration as preparation for use, it is safer for the individual investigator to assume that all adsorbent types require exposure to certain preconditioning and general cleansing steps (sometimes with particular modification to fit his own special experimental requirements) before use.

In the case of cellulosic exchangers, the flow properties and, to a lesser extent, the resolution obtained with a given adsorbent are markedly affected by the particle size of the adsorbent. Although it is not possible to obtain true "mesh ranges" of cellulose particles, fractions with reproducible packing properties, flow rates, and resolution, can be obtained from the bulk adsorbent supplied by the manufacturer by subjecting it to a schedule of shaking over a set of U. S. Standard, 8-inch sieves. In our laboratory a small batch, about 100 g of adsorbent, is shaken over a set of 5 sieves (20, 40, 100, 230, and 325 mesh) for 1 hour on a motor-driven shaker. Fractions of 40–100 mesh and 100–230 mesh are employed for the general run of chromatographic use. Coarser fractions have special use when very high flow rates or filtrations of samples containing significant amounts of lipid or even particles are desired. Finer meshes find special use when the increased resolution which they can provide is more important than the slower flow rates at which they must be used.

Several commercially available adsorbents, in particular, the Whatman so-called "advanced" adsorbents, are provided in a narrower than usual range of particle size. Use of adsorbents of this type without preliminary sieving has proved entirely satisfactory for the general run of chromatographic work. Adsorbents based on polyacrylamide gel or Sephadex are not generally sieved before use. Although these adsorbents are convenient in this respect, they are, for the same reason, incapable of providing those extreme mesh fractions which are invaluable for some special applications.

Hydrogen bonds are formed within and among cellulose fibers and also prevent complete and immediate hydration of Sephadex and polyacrylamide particles. For this reason, a preliminary swelling step is required if these adsorbents are to be prepared from the dried form. In the case of the ion-exchange celluloses this can be accomplished by allowing the dry adsorbent powder to sink into 1 M sodium hydroxide (or in the case of CM-cellulose, into 0.5 M NaOH–0.5 M NaCl) and allowing the slurry to stand for 30 minutes at room temperature. Whatman provides the microgranular ion-exchange celluloses in a wet form which eliminates the requirement for standing in alkali, but it has been our practice nevertheless to begin preparing these adsorbents for use by allowing them to sink into 1 M NaOH (or 0.5 M NaOH–0.5 M NaCl for CM-cellulose). The ion-exchange Sephadexes may be swollen in distilled water and then washed just as described below for the ion-exchange celluloses. The manufacturer recommends that the polyacrylamide gel based adsorbents (Biogel DM series, Biogel CM series) be swollen directly in buffer without a preliminary washing procedure. It is probably unwise to treat these adsorbents with alkali.

In spite of the much improved quality control in the production of ion-exchange celluloses, it remains prudent for the investigator to wash even those adsorbents ostensibly provided by the manufacturer ready for use. A standard cycle of washes includes the following: 1.0 M NaOH (0.5 M NaOH–0.5 M NaCl in the case of CM-celluloses and Whatman "advanced microgranular adsorbents"), water, 0.5 M HCl, water, 1.0 M NaOH (0.5 M NaOH–0.5 M NaCl in the case of CM-celluloses and Whatman "advanced microgranular adsorbents"), and water until excess alkali has been removed from the filtrate (pHydrion paper test). The entire process can be carried out in a few minutes on a Büchner funnel fitted with a coarse fritted glass dish. The slow filtration of highly swollen cellulose in NaOH can sometimes be facilitated by addition of 1 M NaCl. If this maneuver is not effective, then it is necessary to transfer the cellulose to a large container, dilute it with about 10 volumes of water, and allow it to sediment. The length of time required for adequate sedimentation varies with the adsorbent and with the geometry of the vessel in which the procedure is carried out. In general, the fine materials which interfere markedly with the flow properties of the adsorbent will not sediment in any reasonable length of time although some of them will be occluded in the sediment. Since substantial losses of adsorbent can occur if decantation is carried out too soon, it is better to err on the side of overlong standing. In satisfactory cases a clear-cut line will separate sedimented adsorbent from a hazy supernatant suspension. The supernatant suspension can then be decanted and the procedure repeated until a clear supernatant liquid is obtained. Many adsorbents, e.g., the Whatman series, can be prepared with little or no need for removal of fines in this manner.

Equilibration of the Adsorbents for Use.

Each of the washed ion-exchangers under consideration contains substantial quantities of an acidic or basic group, frequently one with a pK' such that the adsorbent interacts with the starting buffer. Thus, it is mandatory before beginning a chromatographic experiment that the adsorbent bed be adjusted to such a state that it is in equilibrium with the buffer to be used.

The simplest case is encountered when an ion-exchange adsorbent is to be used at a pH far removed, i.e., greater than 1.5 unit, from the pK' of its charged group. Under these conditions, placing the adsorbent in the fully charged state (by washing it with an appropriate alkali in the case of a cation exchanger, or an appropriate acid in the case of an anion exchanger) achieves a condition close enough to equilibrium to permit easy equilibration with the starting buffer. Thus, when the washing procedure described above is completed, SE-cellulose or SE-Sephadex are ready for equilibration if the operation is to be performed above pH 3; CM-cellulose and CM-Sephadex, if the operation is to be performed above pH 6. GE-cellulose or GE-Sephadex used below pH 11 or DEAE-cellulose below pH 6 can similarly be prepared for easy equilibration with starting buffer by washing them with an excess of the acidic component of the buffer system to be employed. However, when the pH at which chromatography is to be performed is well within the buffering range of the charged group on the adsorbent, its adjustment to pH equilibration by washing with the starting buffer may require prohibitive quantities of buffer, of time, or of both, particularly if the starting buffer is very dilute. One frequently used expedient is to wash the adsorbent on the Büchner funnel with a more concentrated buffer of the same composition as starting buffer. This does not solve the problem entirely, however, since the pK' of the charged groups on the adsorbent shifts dramatically with changes in salt concentration. Thus, when such a column is washed later with the unconcentrated starting buffer and the ionic strength falls, the shift in pK' results in a new disequilibration. A more effective method is to suspend the adsorbent cake in sufficient starting buffer to produce a thin slurry and adjust its pH as measured by the glass electrode to the desired point by the addition of a solution containing the acidic or basic component of the starting buffer. The remaining minor equilibration can be easily accomplished by passing a reasonable volume of starting buffer through the packed column. The final equilibration of the column should never be assumed but always checked with accurate pH and conductivity meters.

Packing the Columns

Recommendations concerning the packing of columns have varied with respect to the use of pressure to achieve flow rates necessary to establish a stable adsorbent bed that will not "run dry" under gravity flow. The details of this dispute have been discussed elsewhere.[5] Here we recommend procedures which have proved useful in our laboratory.

Beds of cellulosic ion-exchangers up to about 200 ml in volume and of all but the finest mesh ranges are packed under a schedule of graduated air or nitrogen pressure in the following manner: The column to be used is fitted with a thick-walled conical glass reservoir by means of a liquid tight, pressure-fast seal which is conveniently achieved with clamped ball-socket joints and a silicone-rubber gasket. Marks are placed on the side of the column dividing the column height into ten equal segments. The washed and adjusted adsorbent is suspended in a volume of starting buffer equal to about half the volume of the adsorbent cake on the Büchner funnel, and the slurry is poured into the conical reservoir. The reservoir is sealed and attached to a source of regulated nitrogen or air pressure. When the rising bed has reached the first mark on the column, 5 psi of gas pressure is applied. The apparatus is gently swirled from time to time to maintain the adsorbent in the reservoir in even suspension, and the applied pressure is increased 1 psi as the packed adsorbent bed reaches each new division on the column to a final pressure of 15 psi when the top division has been reached.

This procedure is varied for very coarse adsorbents by a marked increase in the dilution of the initial adsorbent slurry. In extreme cases, as with coarse meshes of Whatman adsorbents designated "Floc," 10 volumes of buffer per volume of washed adsorbent cake are required With

[5] E. A. Peterson *in* "Laboratory Techniques in Biochemistry and Molecular Biology." T. S. Work and E. Work (eds.), Wiley, New York, in press.

very fine adsorbents, including the Whatman adsorbents designated "microgranular," the column is allowed to pack to full height under gravity flow and then exposed to 5 psi of pressure until a bed of stable height is reached. It has been our practice to use a similar procedure with the ion-exchange Sephadexes, but perfectly satisfactory results can be obtained by following the manufacturer's recommendation to pack entirely under gravity-induced flow.

A special problem is presented in the packing of very large beds, of the order of 600 ml and more of ion-exchange celluloses. One practical solution is to pack columns by pumping a suitable slurry of the adsorbent (usually about 5 volumes of buffer for each volume of washed adsorbent cake) into a column initially filled with the starting buffer. Caution must be taken to ensure that the diameter of the pump tubing and the entry port of the column are large enough to prevent clogging by the adsorbent slurry, and the horizontal portions of the tubing should be minimized. The slurry from which the adsorbent is pumped should, of course, be stirred constantly at a rate adequate to keep the adsorbent effectively suspended. The flow rate employed must be judiciously chosen for the particular application; as a general guide, a flow rate of 5 l/hour provided by a "Roll-flex" pump with 0.5 cm i. d. tubing was found sa ˙isfactory for packing fibrous DEAE-cellulose into 4.5 cm i. d. columns. The packed column should be washed with starting buffer until the pH and conductivity of the influent and effluent solutions are equal.

Testing of Commercial Adsorbents for Capacity

In spite of recent improvement in quality control by manufacturers, the prudent investigator should test each batch of ion-exchange adsorbent for adequate protein binding capacity, particularly since many members of a laboratory may depend on a central supply of adsorbent for a variety of work. This is easily accomplished and can prevent a tremendous amount of wasted labor. A convenient test for DEAE-celluloses has been in use for several years now. It depends on the fact that human serum contains several colored proteins whose position on a column can be estimated visually. A stock of human serum obtained from a unit of out of date blood is stored frozen in 1-ml aliquots. When a test is to be performed a single aliquot is thawed and dialyzed against two changes of 24 ml each of 40 mM Tris–5mM phosphoric acid, pH 8.6. A standard quantity (200 mg) of each adsorbent to be tested, washed as described above, is suspended in 1 ml of 40 mM Tris–5 mM phosphoric acid, pH 8.6, and packed under gravity flow into a tuberculin syringe fitted with a porous polyethylene disk cut to size with a cork borer. The column thus formed is washed with 5 ml of starting buffer under gravity flow. Dialyzed serum (0.2 ml) is applied to the column and washed in with 1 ml of starting buffer. After this wash the distance between the bottom of the visible protein band and the top of the column is measured. For satisfactory adsorbents, the distance should be less than one-third of the total column height.

Similar tests can be devised as needed for other adsorbents, but are probably less mandatory, since the uniformity of commercial batches of CM-cellulose and the Sephadex ion-exchangers has appeared to be better than is the case with DEAE-cellulose.

Preparation of Sample for Application

The means of preparing a sample for application to ion-exchange columns depends on the type of experiment contemplated. For high resolution analytical chromatograms, particularly when the position of components emerging near the starting buffer concentration is important, careful equilibration of the sample with starting buffer is required. This is generally accomplished either by dialysis or by using columns of Sephadex G-25 or G-50. A convenient and effective schedule using 1/4 inch "Visking" tubing is to dialyze the sample in a 4° room against two 20-fold volumes of the starting buffer in an apparatus that assures stirring of both the contents of the sack and of the dialyzate. The first volume of dialyzate is allowed to equilibrate overnight, and the second for about 6 hours. This permits completion of dialysis at a time of day convenient for the performance of chromatographic experiments. The use of Sephadex columns for this purpose is adequately described elsewhere.[4]

In enzyme purification, it is frequently necessary to prepare large volumes of relatively crude material for application to ion-exchange columns. It is usually possible to accomplish this with sufficient precision by dilution of the preparation, monitored with a conductivity meter and pH meter, and thus avoid the cumbersome problems inherent in dialyzing very large volumes.

Choice of Column Size

No general rule can be given for choice of column size. This depends not only on the particular adsorbent type and quality used and the nature of the proteins to be adsorbed and separated, but also on the purpose of the experiment and the mode of elution chosen. In general, for high resolution gradient chromatography, the adsorbed protein bands should occupy no more than about 10% of the adsorbent bed under starting conditions. On the other hand, many applications can tolerate much higher ratios of bound sample to adsorbent. A ratio of 10 g of adsorbant per gram of protein can be tried, and this ratio modified as found necessary by the results of the first experiment. For initial steps in enzyme purification, the choice of the smallest column capable of binding all of the activity of interest is frequently an advantage when rapid concentration and gross purification is to be achieved by adsorption and step elution. Under these circumstances, the maximum amount of more loosely bound contaminating material will be displaced from the column by the enzyme and other more tightly bound proteins. The use of such an experimental system, although simple in principle, requires care in checking the capacity of the adsorbents used, since small errors with regard to estimation of capacity will result in displacement of the enzyme itself from the adsorbent bed.

Choice of Mode of Elution

The two most frequently utilized methods of elution have special areas of utility. For high resolving power and relative freedom from artifacts, gradient elution is the clear method of choice. A wide variety of apparatus for producing gradients of complex shape is available. Perhaps most flexible is the "Varigrad,"[6] which is commercially available and can produce essentially any gradient shape required for practical chromatographic problems. Apparatus of this type is necessary when the elution of several components of widely different affinities for the adsorbent is desired in a reasonable volume or when strictly linear pH gradients are required. Simpler apparatus is suitable for enzyme purifications because linear gradients in salt concentration are usually quite adequate.[7]

Step elution is most useful when performing purification procedures in which the high effluent concentration of the component being isolated and the rapid operation achieved are more important than the sacrifice in resolution which is usually encountered. In fact, the adsorption of proteins to minimum sized columns and subsequent step elution of the adsorbed protein, constitutes one of the most effective available methods for protein concentration. Methods employing elution with multiple steps of buffer of successively higher eluting power have several shortcomings. These include the frequent artifactitious subdivision of single substances into multiple "components" and the requirement for constant attention by the investigator, who must be present to apply each new eluting agent to the column.

---

[6]E. A. Peterson and H. A. Sober, *Anal. Chem.* 31, 857 (1959).
[7]C. W. Parr, *Biochem. J.* 56, (1954).

Use of Preliminary Probe Experiments

Since a full-scale, completely analyzed chromatogram can be a very laborious affair, and since a variety of conditions must frequently be explored to assure the investigator that he is utilizing favorable conditions, it is highly advisable to apply small-scale, easily performed experiments prior to full-dress studies on columns of the usual type. With small beds made in "syringe" columns such as those described above for testing batches of adsorbents, and assuming the availability of a reasonably accurate and rapid assay procedure, it is possible in a single working day or less to study the general chromatographic properties of a molecule under study with respect to binding to several adsorbents throughout a range of pH, and with a variety of buffers. The recovery and stability of the molecule under the various conditions can also be ascertained. The gathering of similar information through full-scale formal chromatographic experiments might require several months of work.

Choice of Pumping Methods and Collection Equipment

A wide variety of good pumps and fraction collection equipment are commercially available for performance of chromatographic experiments. The use of a pump, rather than of gravity-induced flow, is in most instances a distinct convenience and, in some, a necessity. Frequently the resistance of beds of ion-exchange adsorbents is sufficiently high to prevent attainment of suitable flow rates with reasonable hydrostatic heads. Variable flow rates will sometimes be obtained even under a constant hydrostatic head because of changes in column resistance produced by the gradient or by changes in the effluent viscosity as proteins are eluted from the column. Thus, as a general recommendation, a pump should be used.

With a dependable pump the flow rates obtained will be reasonably constant, permitting use of the most dependable mode of fraction collector operation, namely, time-indexing. Experience has shown that time-indexed collection will result in fewer experimental mishaps than are obtained with drop-counting or volume-indexing devices. However, the latter are conveniently applied whenever gravity flow is used since they prevent wide variation in fraction volume.

Acknowledgment

The author would like to thank Dr. Elbert A. Peterson for his guidance and assistance in preparing this chapter.

[27] Gel Filtration

By JOHN REILAND

Introduction

Gel filtration is a liquid column chromatographic method of separating solute molecules according to differences in molecular size. The separation is achieved by percolating the sample through a bed of porous, uncharged gel particles. Several other names, similar in meaning, have been given to the method. Among these are: gel chromatography, gel permeation chromatography (GPC), exclusion chromatography, and restricted-diffusion chromatography.

As with other types of chromatography, it is rarely, if ever, possible to attribute the separation achieved by this method completely to the action of a single phenomenon. Thus gel filtration separations, while principally due to molecular sieve effects, may also involve ion exchange, sorptive, and, in the presence of immiscible solvent systems, liquid–liquid partition effects. All these influences must be considered when interpreting the data obtained in gel filtration experiments.

In recent years, gel filtration techniques have been given an increasingly important role in both the preparative and analytical separation of biomolecules. Reviews of the theory, techniques, and applications of gel filtration are given in the works of Fischer[1] and Determann.[2] Useful information is further provided in the technical literature and in lists of references supplied by the manufacturers of gel filtration media.

[1]L. Fischer, *in* "Laboratory Techniques in Biochemistry and Molecular Biology" (T. S. Work and E. Work, ed.). North Holland Publ., Amsterdam, 1969.
[2]H. Determann, "Gel Chromatography 2nd Edition." Springer, New York, 1969.

Basic Principle

The molecular sieve principle upon which the gel filtration method is based can be briefly described as follows:

Particles of a gel material possessing a spongelike, porous matrix structure of controlled dimensions are equilibrated in an appropriate solvent and then packed in a chromatographic column. The sample, consisting of a mixture of substances which differ in molecular size, is applied to the bed surface; this is followed by more of the same solvent in order to cause the sample to percolate through the bed. As this occurs, molecules too large to diffuse into the porous gel structure move very rapidly through the bed within the space between the gel particles, and are thus quickly eluted. Molecules small enough to penetrate the matrix structure of the gel, on the other hand, are temporarily but repeatedly delayed in their migration through the bed. These smaller molecules are sequentially eluted in order of decreasing molecular size (Fig. 1). Immediately after elution of all solutes, another sample can be applied to the bed surface and the process repeated. Unlike some other types of column chromatography, no gradients are used for elution and no regeneration of the bed is required. While gel filtration methods that per-

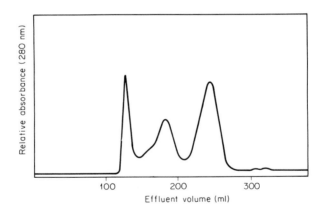

Fig. 1.   Elution profile of normal human serum as obtained by gel filtration on Sephadex G-200. The macroglobulins (excluded from gel matrix) are found in the first peak; 7 S immunoglobulins and ceruloplasmin in the second peak; albumin and transferrin in the third peak. Dry particle size: 40–120 μ; sample volume: 2.0 ml; bed dimensions: 2.5×87 cm. Eluent: 50 mM NaPO$_4$, 0.1 M NaCl, 0.02% NaN$_3$, pH 6.9; eluent flow rate: 3.2 ml cm^{-2} hr.$^{-1}$ Data were supplied by I. M. Easterday, Pharmacia Fine Chemicals, Inc.

mit the use of organic solvents have been developed,[3] only aqueous systems will be discussed in this chapter.

Advantages

The wide acceptance of gel filtration is due to the simplicity, rapidity, and economy of the method. It can be used whenever sufficient molecular size differences exist among sample substances, yielding highly reproducible results. Solute recoveries approach 100% and scale-up to large sample volumes is easily accomplished. Gel filtration is an extremely gentle method which rarely causes denaturation of labile substances.

Limitations

Few limitations to the use of gel filtration exist among biochemical applications. The viscosity of both the sample and eluent must be kept low, since the method depends upon efficient migration and diffusion of solute molecules within the gel bed. Because the maximum pore sizes of gel filtration media are necessarily limited, the range of dimensions of substances which can be purified is restricted. While most proteins, large nucleic acid molecules, viruses, and small organelles are readily fractionated, gels suitable for the fractionation of relatively large organelles and intact cells have not yet been discovered. Sorption of solutes to the gel matrix can occur, for example with aromatic substances, resulting in delayed elution or poor recoveries; often, however, this same phenomenon can be used to advantage in many purification procedures, permitting separations that would not be feasible solely on the basis of differences in molecular dimensions.

Physical Characterization of Gel Filtration Media

Water Regain (W_r)

The W_r is defined as the grams of water absorbed upon hydration of 1 g of the xerogel (from ξηρός, dry). This value refers only to the water contained *within* the gel particles; it does not refer to the water trapped between particles in a packed bed of the materials. W_r values are not given for agarose gels, which are available only in the hydrated state; instead, the approximate concentration of the agarose solution from which the gel is made is supplied. Both W_r and agarose concentration are related to the size of the pores in the matrix structure; in a given gel series the lower the matrix content of the hydrated product, the larger

[3] J. Sjövall, E. Nystrom, and E. Haati, *Advan. Chromotogr.* 6, 119 (1968).

the average pore size. In this chapter, "low W_r" gels are defined as possessing W_r values less than 7.5. "High W_r" gels have W_r values equal to or greater than this figure. Agarose gels are categorically considered as "high W_r" gels.

Exclusion Limit

The molecular weight of the smallest molecule incapable of penetrating the pores of the gel matrix structure is defined as the exclusion limit, and is thus an index of maximum effective pore size. Pore dimensions could probably be stated most accurately in terms of the average pore radius, but this is not practical since molecular radii of gyration (Stoke's radii) of sample substances are often unknown. Molecular weight exclusion limits, therefore, remain the most generally useful manner of evaluating this important characteristic.

It is, of course, always necessary to make note of the configuration of the molecule used to determine the exclusion limit. The values obtained will reflect the degree of coiling and folding of the molecule. Thus exclusion limits are relatively low for linear, random coil molecules, such as dextrans, and relatively high for molecules possessing more compact structures, such as globular proteins.

Fractionation Range

The fractionation range of a gel represents the molecular weight spectrum over which separation can be expected to occur. The upper limit of the fractionation range is equal to the exclusion limit of the gel; the lower limit is somewhat arbitrary. The fractionation range corresponds to the more or less linear region of a sigmoid curve which shows the relationship between the logarithm of the molecular weight and one of several functions of the elution volume (Fig. 2). Interpolation of such "*selectivity curves*" is used to advantage in the estimation of polymer molecular weights from gel filtration data.[4-6] This method, while technically facile and relatively inexpensive, yields data which must be interpreted in light of molecular shape as well as molecular weight.

Particle Shape

The ideal shape for particles of any chromatographic medium is spherical, because the most uniform column packing and highest eluent flow rates can be obtained when this is the case. In the case of the agarose gels the spherical product has been shown to provide superior resolu-

[4]P. Andrews, *Biochem. J.* **91**, 222 (1964).
[5]H. Determann, *J. Chromatogr.* **25** 303 (1966).
[6]G. K. Ackers, *J. Biol. Chem.* **243**, 2056 (1968).

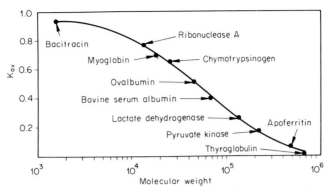

FIG. 2.   Selectivity curve for Sephadex G-200, showing K_{av} values as a function of the molecular weights of several proteins. Data were supplied by I. M. Easterday, Pharmacia Fine Chemicals, Inc.

tion as well.[7] Practically all gel filtration media now marketed are in the form of minute spherical beads. For best results, the older granular materials should not be used. If doubt exists concerning the particle shape, a low-power microscope can be used to examine the product in either the dry or hydrated state.

Particle Size

As with other chromatographic media, a reduction in gel particle size will result in an increase in resolution (Fig. 3). Unfortunately, decreasing the gel particle size also causes a reduction in eluent flow rate. Most operations will therefore reflect a compromise between these two considerations, where particles averaging approximately 50–300 μ in diameter (hydrated) are most often used for typical laboratory operations. For large-scale preparative and industrial processes, larger particle sizes may be indicated. This is particularly true in desalting operations, where great resolution is not required but a high flow rate is essential. Particles less than 50 μ in diameter (hydrated) are indicated when maximum resolution is needed, and for thin-layer gel filtration techniques.[8–10] Some manufacturers indicate gel particle dimensions by mesh sizes in lieu of stating particle diameters.

The gel particle size range is also an important criterion of the quality of the medium. Preparations in which the diameters of the hydrated

[7]M. K. Joustra, *in* "Modern Separation Methods of Macromolecules and Particles" (T. Gerritsen, ed.), p. 183. Wiley (Interscience), New York, 1969.
[8]B. G. Johansson and L. Rymo, *Acta Chem. Scand.* **18**, 217 (1964).
[9]P. Andrews, *Biochem. J.* **91**, 222 (1964).
[10]C. J. O. R. Morris, *J. Chromatogr.* **16**, 167 (1964).

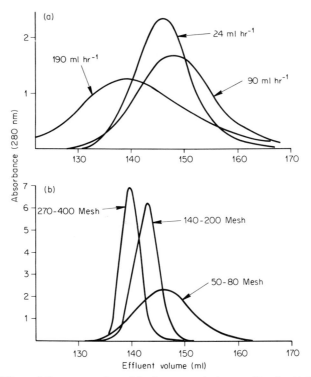

Fig. 3. Effects of flow rate and particle size upon elution profile of uridylic acid chromatographed on Sephadex G-25 beds. Bed dimensions: 2×65 cm. (A) Column packed with 50–80 mesh fraction and eluted at several flow rates. (B) Columns packed with gels of several particle sizes and eluted at a constant flow rate of 24 ml hr.$^{-1}$ [P. Flodin, *J. Chromatogr.* **5**, 103 (1961).]

gel particles vary by greater than a factor of three should be sized prior to use in order to obtain a narrower particle size distribution.

Chemical Characterization of Gel Filtration Media

Cross-linked Dextran Gels (Sephadex) (Table I)[11]

The starting material used for the production of these gels is a linear dextran (poly-α-1,6-D-anhydroglucopyranose). After purification, this polysaccharide is allowed to react with epichlorohydrin, an epoxide which introduces glyceryl side chains and cross-linkages within and between the dextran molecules. Thus, a three-dimensional porous matrix, which swells but does not dissolve in water, is formed. By regulating the molecular weight of the dextran and the fraction of epichlorohydrin in

[11]The tables are grouped in the Appendix.

the reaction mixture, it is possible to regulate the extent of cross-linking, thereby controlling the average pore size within the matrix.[12]

At present, eight dextran gels are commercially available, designated Sephadex G-10 through G-200. The nomenclature of Sephadex gels is derived from their W_r values.

All dextran gels are now produced in the form of spherical particles, although many were previously available in granular form. The gels are marketed as the xerogel and must be hydrated prior to use.

Dextran gels have generally high resistance to chemical attack by reagents commonly used in biochemical work. They are insoluble, unless chemically degraded, in all solvents tested. They show excellent stability in organic solvents, weakly alkaline and acidic solutions, and concentrated urea and guanidine solutions. With concentrated eluent solutions or at extremes of pH, some change in W_r values and exlusion limits may be encountered. Below pH 2, cleavage of the glycosidic linkages occurs while above pH 12, alkaline degradation of the polysaccharide takes place. Both these processes are accelerated at elevated temperatures. Since the time of exposure is also an important factor, however, it is often possible to use dextran gels at extremes of pH, if the exposure is brief.

Sorption of aromatic and heterocyclic solutes to the gels is a commonly encountered phenomenon which results in retarded solute migration. Cyclodextrins may also be sorbed to the gel matrix. Borate ion forms a complex with dextran and may alter the chromatographic behavior of the gels.

Destruction of dextran gels by dextranase-producing microorganisms is rare, but has been known to occur in improperly maintained columns.

All dextran gels contain approximately 10–20 microequivalents of carboxyl groups per gram of the xerogel. The eluent ionic strength should be ≥ 0.02 in order to preclude ionic interaction between charged solutes and these carboxyl groups, which otherwise would cause retardation or acceleration of such solutes in their migration through the bed. The use of eluents containing oxidizing agents can increase the carboxyl content of the gels, and should therefore be avoided.

Small amounts of free dextran may be found in the effluent from beds of dextran gels; this is most pronounced with the high W_r gels. Most of this is extracted during initial processing of the slurry and stabilization of the packed bed, but persistent extraction at very low levels can be expected to occur for an indefinite period. Therefore, if polysaccharides are being chromatographed, the use of polyacrylamide gels may be

[12] P. Flodin, "Dextran Gels and Their Applications in Gel Filtration." Dissertation, Univ. of Uppsala, Sweden, 1962.

preferable, particularly if the sample mass is small or if contamination with dextran must absolutely be avoided.

Polyacrylamide Gels (Bio-Gel P) (Table II)[11]

These gels are produced by copolymerizing acrylamide ($H_2C =$ $CHCONH_2$) with a cross-linking agent, N,N'-methylenebisacrylamide ($H_2C = CHCONHCH_2NHCOCH = CH_2$).[13] By varying the concentrations of these monomers, products differing in W_r are produced. The ten commercially available gels are designated Bio-gel P-2 through P-300; the nomenclature is derived from the approximate molecular weight exclusion limits.

All polyacrylamide gels are made in the form of spherical particles and are marketed as the xerogel; they must be hydrated prior to use.

Polyacrylamide gels are insoluble in water and common organic solvents. Concentrated salt, urea, and guanidine solutions are well tolerated. The materials are stable from pH 1 to pH 10, but will adequately tolerate exposure at pH values outside that range for short periods of time. Lengthy exposure of the gels to strongly acidic or alkaline conditions should be avoided, especially at high temperatures, because extensive hydrolysis of the amide groups may occur. The carboxyl groups thus produced increase the ion exchange capacity of the material, which is initially less than 50 nanoequivalents per gram of the xerogel. Sorption of very acidic, very basic, and aromatic compounds may occur, especially when eluents of low ionic strength are employed; this effect is less pronounced than with dextran gels in most cases.

Polyacrylamide gels are not subject to microbial degradation, although organisms can grow in the presence of the inert material, utilizing substrates present in solutions in which the gels are suspended.

Agarose Gels (Sepharose; Bio-Gel A) (Table III)[11]

Agarose is an essentially linear polysaccharide obtained from agar and composed of alternating residues of D-galactose and 3, 6-anhydro-L-galactose.[14] The quality of the agarose used for gel production is a critical factor in the elimination of ionic groups from the finished product; to this end, negatively charged agaropectin molecules should be removed as completely as possible. Gels formed at agarose concentrations as low as 1% are commercially available. Although the exact nature of the forces responsible for gelling are unknown, hydrogen bonding is thought to be at least partly responsible.[7]

---

[13]S. Hjertén and R. Mosbach, *Anal. Biochem.* **3**, 109 (1962).
[14.]C. Araki, *Bull. Chem. Soc. Jap.* **29**, 543 (1956).

Agarose gels are stable from pH 4.0 to 9.0, a somewhat more restrictive range than in the case of other gel filtration media. Concentrated salt solutions are well tolerated, but concentrated urea and guanidine solutions may shorten the useful life of gels of low agarose concentration.

The physical rigidity of agarose gels surpasses that of the polyacrylamide and dextran gels at equivalent polymer concentrations, permitting relatively high eluent flow rates.

The fractionation ranges of available agarose gels partially overlap those of the dextran and polyacrylamide gels. Since the separation power of the latter exceeds that of the agarose gels within the region of overlap,[7] it is suggested that dextran gels be used in lieu of agarose gels whenever possible, unless the higher flow rates obtainable with agarose gels are required. No comparison of the relative separation powers of agarose and polyacrylamide gels has yet appeared.

Agarose gel particles are damaged upon drying, and for that reason the gels are marketed as slurries in water. Prolonged and vigorous stirring of the slurries should be avoided so that destruction of the rather friable particles does not occur. Freezing of agarose gels may be deleterious to the gel structure. Heating the gels above 40° causes them to dissolve.

Elementary Mathematical Concepts

While an extensive knowledge of the mathematical basis of gel filtration is not required for most applications, a knowledge of fundamental principles is helpful. More detailed discussions of the theoretical aspects of gel filtration can be found elsewhere.[15-18]

Definitions of Terms

V_t = Total bed volume, the volume occupied by the gel bed. This is best determined by prior water calibration of the column or, alternatively, by geometry.

V_e = Elution volume, the volume of effluent which precedes elution of a specific solute in the sample. This is measured from start of sample application to the half-height point of the rising edge of the effluent peak.

[15] M. Joustra, in "Protides of the Biological Fluids" (H. Peeters, ed.), Vol. 14, p. 533. Elsevier, Amsterdam, 1967.
[16] T. C. Laurent and J. Killander, J. Chromatogr. 14, 317 (1964).
[17] G. K. Ackers, Biochemistry 3, 723 (1964).
[18] K. H. Altgelt, Advan. Chromatogr. 7, 3 (1968).

V_o = Void volume, the volume of the space between the gel particles. This can be determined by measuring the V_e of a solute incapable of penetrating the gel matrix. In a bed of rigid spheres of uniform diameter the V_o is approximately $0.35 \, V_t$.[19]

V_i = Inner volume, the volume of the liquid contained within the gel particles. This can be approximated by

$$V_i = m \, W_r \tag{1}$$

where m = mass of xerogel used in preparation of the bed. The V_i can also be estimated by chromatographing a sample of tritium oxide. The elution volume of this substance is equal to $V_o + V_i$; the V_i can then be determined by subtracting the value obtained for the V_o.

V_p = Polymer volume, the volume occupied by the matrix substance.

Mathematical Characterization of the Gel Bed and Solute Behavior

A packed bed of any gel filtration medium can be mathematically described as

$$V_t = V_o + V_i + V_p \tag{2}$$

In view of this model, and in the absence of gel-solute interactions, gel filtration can be considered as a type of liquid–liquid partition chromatography, where solutes are partitioned between the V_o and V_i spaces. The extent to which a solute is able to penetrate the V_i space is determined by its molecular dimensions with respect to the dimensions of the pores of the matrix material. The behavior of a solute within a given gel is represented ideally by the distribution coefficient

$$K_d = \frac{V_e - V_o}{V_i} \tag{3}$$

Accordingly, solutes possessing molecular dimensions which preclude their access to the V_i of the bed will be simultaneously eluted in a single peak at the V_o with $K_d = 0$, whereas smaller molecules will follow sequentially in order of decreasing molecular size with $0 < K_d \leq 1$.

By rearranging Eq. (3) to give

$$V_e = V_o + K_d \, V_i \tag{4}$$

it is readily seen that the K_d value for a specific solute when chromatographed on a specific gel is an index of the fraction of the V_i space to which the solute has access.

Because of the difficulty of accurately determining V_i values, the K_d formula is often not practical for characterizing the solute behavior;

[19]H. Susskind and W. Becker, *Nature (London)* **212,** 1564 (1966).

consequently a revised formula, which eliminates the need for the V_i term, is frequently used. By replacing V_i with $V_i + V_p$, Eq. (3) is transformed to:

$$K_{av} = \frac{V_e - V_o}{V_i + V_p} = \frac{V_e - V_o}{V_t - V_o} \tag{5}$$

Note that to find the values for K_{av} only the readily obtainable values for V_t, V_o, and V_e are required.

Sorption of solutes to the gel matrix has the effect of elevating both K_{av} and K_d values. Ionic interaction between a solute and the gel matrix, on the other hand, can result in either an increase or decrease in these values, depending upon whether the solute is retarded or accelerated in its passage through the bed. Both K_d and K_{av} values are independent of column geometry and are thus of value in calculating bed dimensions and sample volumes when scale-up is contemplated.

Selection of Gels

Separation by gel filtration can be qualitatively divided into two categories: *group separation* and *fractionation*. The gel, the bed dimensions, and the sample volume will depend upon the category into which the separation falls.

Group Separation

Group separation involves separation of the sample into two groups that differ widely in molecular size, e.g., desalting of proteins, or elimination of phenol from nucleic acid preparations. In such cases, the best separation will be obtained when the large molecules are excluded from the gel matrix ($K_{av} = 0$), while the small molecules are able to penetrate the matrix to a considerable extent ($K_{av} \sim 1$). For example, if a protein mixture is to be desalted, a gel should be chosen that would exclude all protein molecules in the mixture. Since most proteins have molecular weights in excess of 5 or 6×10^3 either Sephadex G-25 or Bio-gel P-6 would be a suitable medium. When one of these products is used, all protein molecules will appear in the void peak of the elution pattern. If the desired molecules in the excluded group are known to be somewhat higher in molecular weight than those given in the example, the use of a gel with a higher exclusion limit, such as Sephadex G-50 or Bio-Gel P-30, will give a more efficient separation. Substances of low molecular weight can be effectively desalted with the highly cross-linked Bio-Gel P-2 or Sephadex G-10 gels.

A special case of group separation is the exchange of buffer ions in solutions of macromolecules. This technique is useful for sample preparation prior to ion exchange chromatography or other techniques, where contamination of the sample with ions other than those present in the starting buffer solution is undesirable. To remove these contaminants, the sample may be applied to a Sephadex G-25 or Bio-Gel P-6 column which has been equilibrated with the starting buffer solution. The sample is then eluted with more of the same solvent. The macromolecules in the sample will be separated from the contaminating ions, and they will appear in the void volume dissolved in the starting buffer solution.

Fractionation

Fractionation is used here to designate the separation of substances from complex mixtures when molecular size differences are much smaller than in group separation. An example is the purification of a specific enzyme from a mixture of proteins. In fractionation experiments, K_{av} differences are usually on the order of 0.1–0.2. If the approximate molecular weights of desired components are known, it is possible to select the proper gel for a given separation by consulting fractionation range charts (Tables II–IV). The gel with the lowest fractionation range capable of effecting the separation should be chosen if maximum resolution is to be achieved. As an example, Bio-Gel P-300 or Sephadex G-150 can be used for the purification of protein molecules ranging up to about 400,000 in molecular weight; molecules larger than this will appear unseparated as a single peak at the void volume. Currently available gel filtration materials have fractionation ranges which extend up to 150 million daltons for globular molecules.

When the purification of a solute of unknown molecular weight is contemplated, the selection of an appropriate gel medium for preliminary experiments must be made on a presumptive basis. Careful evaluation of the elution patterns from such pilot studies will suggest adjustments for improving the separation in subsequent experiments.

Selection of Equipment for Gel Filtration

Chromatographic Column

Design. Much of the success in any chromatographic procedure rests upon the equipment used; in many instances failure to attain a given separation can be traced to poorly designed or faulty apparatus.

The most important item is the *chromatographic column* used to contain the gel medium. A well designed column should incorporate the following features: (a) *Sturdy plastic and glass construction.* Metals may interfere with biological activity of samples or corrode. (b) *A properly designed bed support.* This should not become clogged, sorb solutes, nor permit the passage of chromatographic media. The best materials currently used for fabrication of bed supports are very fine (400 mesh) polyamide (nylon) or polytetrafluoroethylene (PTFE) cloth. Porous plastic bed supports may become partially or completely clogged, resulting in either slow or nonuniform eluent flow. Sintered glass has these same defects, and it also possesses undesirable sorptive characteristics. Glass bead and glass wool bed supports should not be used for critical chromatographic experiments: they possess undesirable sorptive characteristics; they may pass particles of media; they have large dead volumes and do not offer uniform resistance to eluent flow. Bed supports should be of the same diameter as the column tube. (c) *Minimal dead volume beneath bed support.* A volume about 1/1000 that of the total column volume is acceptable for most purposes. Larger dead volumes are unnecessary and may result in loss of resolution and dilution of effluent fractions. (d) *Narrow-bore outlet tubing.* For most laboratory-scale applications 1.0 mm i.d. tubing is recommended. Unnecessarily wide-bore tubing has the effect of increasing the dead volume as noted above; for the same reason tubing length should be minimized. PTFE and polyethylene capillary tubings are often preferred for their chemical inertness, but are easily kinked and difficult to attach to apparatus. Soft polyvinyl chloride capillary tubing is convenient for most purposes, but may release substances which absorb ultraviolet light. (e) *Flexible design to permit use with accessories.* The availability of plunger-type flow adaptors is an important consideration if operation in ascending eluent flow mode is contemplated. Flow adaptors permit automatic sample application and are required for recycling chromatography or connection of columns in series. (f) *Water jacket.* If experiments are to be conducted at other than the ambient temperature, or if temperature controlled operation is needed, a jacketed column will be required. Even if not filled with water, a jacket helps to maintain uniform temperature distribution throughout the bed by virtue of its insulating qualities.

Dimensions. The dimensions of the gel bed are important since too small a bed can result in incomplete separation, whereas one that is too large may result in excessive dilution of the effluent fractions. Unfortunately, it is not always possible to accurately predict the optimum bed dimensions for a given separation, and adjustments in either sample size or bed dimensions may be found desirable after a trial experiment. A few basic rules can be given, however, for preliminary selection of bed dimensions.

For *desalting* applications, the total bed volume should be 4 to 10 times that of the sample volume, and the ratio of column length to diameter on the order of 5:1 to 15:1. *Fractionation* of complex mixtures is accomplished on bed volumes 25–100 times as great as the sample volume, with length-to-diameter ratios of 20:1 to 100:1.

Best results will be obtained when the desired effluent peaks are separated at the baseline, with very little effluent volume between peaks. In this manner, minimal dilution of effluent fractions is caused.

Special Techniques to Increase Effective Bed Length. Occasionally gel beds of extraordinary length are required to effect a separation. If beds in excess of 1 meter in length are needed, it is suggested that two or more short columns, connected in series by means of flow adaptors, be used in lieu of a single, long, cumbersome column.[15]

Recycling techniques[20] can also be used to increase effective bed length, and they are recommended if the sample contains only a few desired components with nearly equal K_{av} values. Such systems are, however, difficult to operate; this disadvantage is partially offset by the necessity of packing only one column. Required equipment for recycling chromatography includes an eluent pump and effluent monitor, in addition to flow adaptors and a 4-way valve.

Constant Pressure Eluent Reservoir (Mariotte Flask)

The most common manner of supplying eluent to the column is by gravity flow. With low W_r gels the hydrostatic pressure drop across the bed rarely limits the flow rate; and relatively high, unregulated eluent pressures may be used. With high W_r gels, however, the relatively low hydrostatic pressures employed (Table IV) must be carefully controlled in order to prevent the gel from becoming excessively compacted, with an attendant lowering of the flow rate. For this reason, and to stabilize the flow rate, the use of a constant pressure eluent reservoir is recommended. If the design illustrated in Fig. 4 is used, the pressure drop across the gel bed is measured from the bottom of the air vent tube to the point where the effluent issues from the system. The pressure will remain constant as long as the liquid level remains above the bottom of the air vent tube. The vertical lengths of any air spaces in the system must be subtracted from the pressure measured in this manner.

Preparation of Eluents and Gel Slurries

Eluent Composition

Since gel filtration materials are unaffected by most buffer ions commonly used for biochemical work, selection of these ions will usually be dictated by the stability requirements of desired sample components. Elution patterns obtained under different ionic conditions may not

[20]Killander, J., *Biochim. Biophys. Acta* 93, 1 (1964).

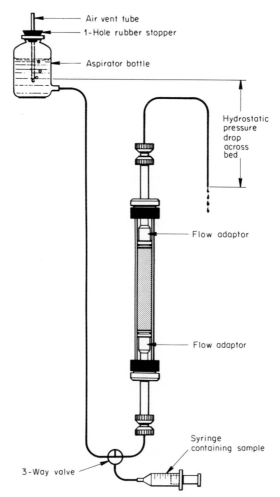

Air vent tube

1-Hole rubber stopper

Aspirator bottle

Hydrostatic pressure drop across bed

Flow adaptor

Flow adaptor

Syringe containing sample

3-Way valve

FIG. 4. Arrangement for ascending eluent flow mode operation using constant pressure eluent reservoir and flow adaptors. Gel beds cannot run dry if the tip of the outlet tubing is positioned above the bed surface.

agree; such differences are most often due to alteration of sample materials, and are unlikely to occur as a result of alteration of gel characteristics. Certain organic solvents and highly concentrated solutions of salts, urea, or guanidine, however, can cause shrinkage or expansion of the gel matrix, with an attendant change in exclusion limits.[12]

All gel filtration media currently in use contain small amounts of ionizable groups. In order to rule out the possibility of interaction between these groups and charged groups on sample solutes, it is suggested that the eluent ionic strength be ≥ 0.02.

When isolation of materials unadulterated by the presence of buffer salts is desired, the use of a volatile buffers should be considered. Suitable buffer substances for this purpose include ammonium formate, bicarbonate, and acetate; pyridine, ethylenediamine, and aminoethyl alcohol. These substances are readily removed from the fractions by lyophilization.

In order to prevent bubble formation within the gel bed, eluant solutions should be degassed by boiling or by aspiration. Degassed solutions should be stored in tightly stoppered bottles. The temperature of the eluent solution should be the same as that of the gel bed.

Preparation of Dextran and Polyacrylamide Gel Slurries

Both the dextran and polyacrylamide gels are hydrated by simply adding the required mass of the xerogel (Tables I and II) to an excess of the eluent employed, stirring only to disperse aggregates. Doubling the W_r value yields an approximation of the bed volume which can be expected from 1 g of the xerogel. *Under no circumstances should vigorous or prolonged stirring be employed, since damage to the gel structure may result.* This is especially true for high W_r gels such as Sephadex G-200, Bio-Gel P-300 and the agarose gels, all of which have delicate, easily damaged structures. Both dextran and polyacrylamide xerogels may be hydrated at room temperature or at 100° on a boiling water bath; no detectable differences in the gels are consequently produced, although the times required for hydration are different.

After hydration, fine particles, if present, should be removed by elutriation with a 200% excess volume of the eluent. This operation is most conveniently carried out in a large glass cylinder. After inverting the cylinder, the slurry is allowed to stand undisturbed until $\sim 95\%$ of the larger particles have settled, with the smaller particles remaining in suspension. At this time the supernatant layer should be removed with suction. A small amount of the material is thus discarded with each repetition of the process. With beaded gels, one or two such steps are usually all that is required.

Finally, the slurry should be degassed as described for eluents. If the slurry is to be stored, action should be taken to prevent microbial growth.

Preparation of Agarose Gel Slurries

Since these products are supplied in the form of concentrated suspensions in water, the only preparation that is required is elutriation with eluent in order to remove any fine particles that may be present.

Degassing of the diluted slurry is advised, but the boiling method cannot be employed because agarose gels begin to dissolve above 40°.

Packing Gel Filtration Columns

The procedure used for column packing depends upon the W_r of the gel employed.

Packing Low W_r Gels ($W_r < 7.5$)

1. The hydrated and degassed slurry should be brought to the temperature at which the column is to be operated. Gel beds should not be subjected to temperature changes during or after packing because contraction or expansion of water may cause packing irregularities.

2. Slurry concentration should be adjusted to permit convenient pouring. Relatively concentrated slurries can be used with success if packing is properly executed; they should, however, be dilute enough to permit air bubbles to escape to the surface. Eluent is used to dilute slurry prior to packing.

3. The column should be mounted in a sturdy, vibration-free support, using at least two clamps; a plumb line or small level may be used to assure that it is in a vertical position.

4. The bottom flow adaptor, if required, should be installed. The plunger should be withdrawn maximally in order to permit subsequent adjustment.

5. The column should be filled with eluent and checked for leaks. The outlet should be opened to permit flushing of all air from beneath bed support and outlet tubing. If necessary, suction can be applied from above the bed support in order to extract recalcitrant bubbles. The outlet should be closed, and the volume of eluent in the column adjusted to approximately 15% of total column volume.

6. If available, a gel reservoir may be attached to the top of the column.

7. The gel slurry is then added, filling the column.

8. The gel particles are allowed to settle for 10 minutes. The column outlet is then opened, permitting the excess eluent to drain from above the bed.

9. When the eluent is within 2–5 cm of the surface of the densely packed region of the bed, the outlet is closed. If needed, additional slurry may be added following gentle stirring of the top 2 cm of the bed surface in order to resuspend the gel particles. Failure to stir the bed surface before adding the additional slurry will result in formation of visible zones, caused by the slower settling of fine gel particles. The packing procedure may be repeated as required in order to build the desired bed length.

10. If required, the upper flow adaptor is installed while 2–3 cm of

eluent remains above the surface of bed. Care must be taken to avoid entrapment of air between bed surface and flow adaptor.

11. The eluent reservoir is connected to the column. The bed is stabilized by allowing at least 2 bed volumes of eluent to pass through bed in descending eluent flow mode. Following stabilization, conversion to ascending eluent flow mode may be made if dictated by experimental design (Fig. 4). The positions of flow adaptors, if used, are readjusted as required to maintain contact of the plungers with the gel bed.

12. The hydrostatic pressure is adjusted by raising or lowering the eluent reservoir to achieve the desired flow rate. Since low W_r gels are relatively rigid, rather high pressures can be used to achieve increased flow rates. An eluant pump may be installed, if desired.

13. Column packing efficiency should be checked as described below.

Packing High W_r Gels ($W_r \geq 7.5$)

These gels possess relatively nonrigid structures which are easily deformed under pressure. Thus, during column packing and elution care must be exercised in order to avoid the possibility of the bed becoming excessively compacted. At no time during the packing or elution of high W_r gel columns should the hydrostatic pressures given in Table IV[11] be exceeded; these represent the pressures at which maximum eluent flow occurs.

1. Steps 1 through 7 of the procedure described for low W_r gels should be followed.

2. The gel is then allowed to settle for 20 minutes. The outlet tubing should be positioned (Fig. 5) to provide the recommended hydrostatic pressure as indicated in Table IV. The hydrostatic pressure is measured from the liquid level in the column or gel reservoir to the tip of the outlet tubing. *The use of excessive hydrostatic pressures will result in lowered flow rates, and will necessitate repacking the column.* The column outlet should then be opened to allow the eluent to drain from above the bed.

3. If additional gel is required, the column outlet should be shut off and the bed surface stirred as described above; the tip of the outlet tubing should then be positioned in order to readjust the hydrostatic pressure. The packing procedure should be repeated as indicated in order to achieve the desired bed length.

4. The upper flow adaptor, if required, should be installed while 2-3 cm of eluent remains above the surface of the bed. Air bubble entrapment must be avoided.

5. A constant pressure eluent reservoir should then be connected to the column inlet. The bed is then stabilized by allowing at least 3 bed

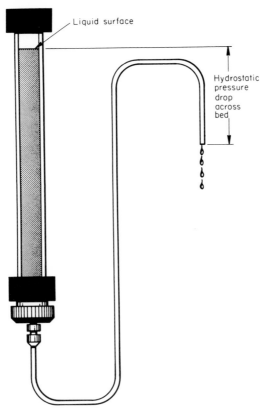

Liquid surface

Hydrostatic
pressure
drop
across
bed

FIG. 5.   When packing high W_r gels, the hydrostatic pressure drop across the bed must
not at any time exceed recommended limits. Refer to Table IV.

volumes of eluent to pass through the bed at the proper hydrostatic
pressure in descending eluent flow mode. Following stabilization of the
bed, conversion to ascending eluent flow mode may be made if dictated
by experimental design (Fig. 4). The flow adaptors should be readjusted
as required to maintain contact of plungers with the bed surface.

6. If an eluent pump is used, care should be taken not to exceed
the flow rate established with a gravity system at the proper hydrostatic
pressure. To allow for possible bed resistance change, the pumping
rate should be decreased 10% from the flow rate obtained with gravity
elution.

Checking for Column Packing Irregularities; V_0 Determination

Checking for column packing irregularities and V_0 determination are
carried out simultaneously by chromatographing a sample consisting
of a solution of an excluded, colored molecule. By observing the

migration of the colored zone as it passes through the bed, defects in the bed which result in nonuniform eluent flow are readily detected. The elution volume (V_e) of an excluded molecule is equal to the bed void volume (V_0).

Blue dextran 2000,[21] consisting of a high molecular weight ($M_w = 2 \times 10^6$) dextran fraction covalently coupled to a blue dye, is often used for these purposes.

The recommended concentration of blue dextran 2000 is 0.2% when used with all dextran gels, all polyacyrlamide gels and agarose gels of 6% concentration and above. With agarose gels below 6% concentration, a 0.5% solution of blue dextran 2000 should be employed. Higher concentrations than those recommended may produce artifacts due to viscous streaming of the sample within the gel bed. The required sample volume is related to the bed dimensions; a volume sufficient to allow visualization of the blue zone as it migrates throughout the entire length of the bed should be used. Relative absorption maxima for blue dextran 2000 in water exist at 265 and 630 nm. Solutions should be prepared the day they are used, since the material slowly hydrolyzes in solution, liberating free dye molecules which may become sorbed to gel filtration media. Should this occur, the sorbed material can usually be removed by passing a zone of 5% serum albumin solution through the bed.

Blue dextran 2000 can be successfully used with agarose beds since it contains a fraction of very high molecular weight molecules which are excluded from the gel matrix such that a small peak corresponding to the V_0 will be seen at the beginning of the elution pattern.

The presence of air bubbles, channeling and incompletely packed regions within the gel bed can be detected by observing the transmission of light from a small lamp held behind the column.

Properly packed beds can be expected to function satisfactorily for extended periods of time, provided microbial contamination and other adversities are avoided. Even with repeated use, column survival times in excess of one year are not uncommon.

Sample Preparation and Application

Sample Preparation

In order to prevent contamination of the bed surface and to assure continued high flow rates, samples should be completely free of undissolved substances. If particulate matter is present, filtration or high speed centrifugation can be used to clarify the solution.

Lipids and lipoproteins may be sorbed to gels and should therefore

[21]Supplied by Pharmacia Fine Chemicals, Inc., Piscataway, New Jersey 08854.

be removed prior to sample application. This can best be achieved by preliminary filtering of the sample through a short bed of a low W_r gel, by high speed centrifugation, or by precipitation with dextran sulfate.[1]

If the sample viscosity is too great, broadening of zones will occur, resulting in decreased resolution (Fig. 6). This effect is apparently due to rheological instability of the sample zone in addition to decreased solute diffusion, since excluded solutes also show broadened zones when the viscosity is elevated. To avoid such difficulties it is important that the viscosity of the sample relative to that of the eluent (η_{rel}) not exceed 2. The η_{rel} of the sample can be estimated by using a viscometer, or by observing the time required to empty a 1-ml volumetric pipette filled with the sample as compared to the time required to empty the

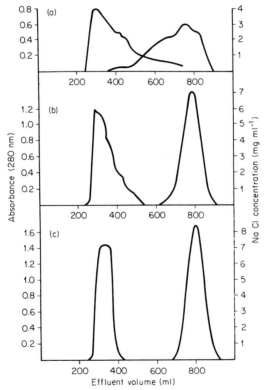

FIG. 6.   Influence of sample viscosity upon elution pattern of a mixture of hemoglobin 0.1% and sodium chloride 1.0% chromatographed on a 4 × 85 cm bed of Sephadex G-25 at a constant flow rate of 180 ml hr.$^{-1}$ Dextran 2000 was added to sample in (A) at 5% final concentration ($\eta_{rel} = 11.8$), and to sample in (B) at 2.5% final concentration ($\eta_{rel} = 4.2$), in order to increase viscosity. Sample in (C) was not altered by addition of dextran. [P. Flodin, *J. Chromatogr.* 5, 103 (1961).]

same pipette filled with the eluent. For dilute aqueous eluents, a sample η_{rel} of 2 corresponds approximately to the viscosity of blood serum, or to a protein concentration of about 70 mg/ml. If the sample viscosity is higher than recommended and dilution of the sample is impractical, the eluent viscosity can be increased by the addition of solutes, in order to satisfy the viscosity requirements. Additives useful for this purpose include dextrose, sucrose, or dextran.

The volume of the sample must be adjusted according to the type of separation and the bed volume. For group separations, the sample volume should be 10–25% of the bed volume, whereas for fractionation the optimal sample volume is usually about 1–4% of the bed volume. In most instances sample volumes can be adjusted to provide a more satisfactory separation when experiments are repeated. The use of excessively large samples will prevent baseline separation of sample solutes, whereas suboptimal sample volumes will result in excessive dilution of fractions (Fig. 7). Thin-layer gel filtration techniques[8–10] have been devised for use with samples 1–5 μl in volume.

Sample Application

Sample application is a critical step in liquid chromatography because unnecessary sample dilution and nonuniform penetration into the bed can result in zone broadening, thereby adversely affecting resolution.

In most instances the sample solution can be applied directly to the flat, drained bed surface. If a large volume of eluent is present above the bed surface, preliminary removal of the majority of the liquid with suction is a useful time-saving step, particularly if the eluent flow rate is low. If the bed is not perfectly flat, the gel grains on the surface should be carefully stirred, and then allowed to settle. The column outlet is then opened to drain off the residual eluent until the surface of the bed is just barely dry, without permitting the balance of the bed to run dry. At this point the column outlet should be shut off, and the sample applied directly to the surface of the bed. For this purpose a pipette with a relatively wide tip (e.g., a plain glass tube or a pipette with a broken tip) is recommended; relatively low pressures will be required to discharge large quantities of solution within a short time, so that sample application occurs rapidly, with minimal disturbance of the bed surface. The tip of the pipette should initially be held within 1–2 mm of the bed surface, and raised gradually as the sample is applied. After placing the sample on the bed surface, the column outlet is opened to allow the sample to penetrate the bed until the surface is again just barely dry. The sample should be followed by one or two

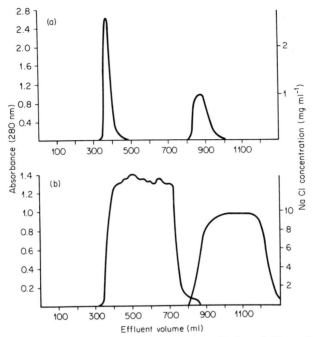

Fig. 7. Influence of sample size upon resolution and dilution of effluent fractions on a 4 × 85 cm bed of Sephadex G-25 (100–200 mesh). (A) Sample: 10 ml containing 0.1% hemoglobin and 0.1% sodium chloride. (B) Sample: 400 ml containing 0.001% hemoglobin and 1% sodium chloride. The dilution factor for the combined hemoglobin fractions was 10 in (A) and 1.25 in (B). [P. Flodin, *J. Chromatogr.* 5, 103 (1961).]

applications of a small volume of the eluent, using a technique similar to that employed for sample application. In this manner dilution of the sample by the eluant which follows is minimized. As soon as the sample has penetrated the bed sufficiently beneath the surface, several centimeters of eluent are placed above the bed, the eluent reservoir is connected to the column inlet, and elution is allowed to proceed. If desired, a sample applicator device consisting of a short cylinder of acrylic plastic with fine polyamide cloth stretched across the bottom may be placed on top of the bed in order to prevent disturbance of the bed surface during sample application and elution. If this device is used, sample application must be carried out rapidly to prevent penetration of the sample into the space between the column wall and the sample applicator tube. Alternatively, to facilitate sample application on high W_r gel beds, a layer of a low W_r gel such as Sephadex G-10 or Bio-Gel P-2 may be placed on top of the bed surface. The use of glass wool or filter paper for the purpose of protecting the bed surface is discouraged;

such materials often possess undesirable sorptive qualitites and can cause decreased or nonuniform eluant flow through the bed.

A second useful sample application technique involves layering a relatively dense sample under the eluent at the bed surface, in the manner of a *pousse-cafe*. If the sample density is not sufficiently high to permit use of this technique, glucose, sucrose, or another suitable solute may be added for that purpose. The sample is then slowly applied at a point several millimeters above the bed surface, and allowed to fall through the eluent above the bed surface, forming a layer at the gel-eluent interface. This is most easily accomplished with a hypodermic syringe fitted with narrow bore (*ca.* 1.0 mm i.d.) flexible plastic tubing; care should be exercised to avoid contamination of the buffer layer above the sample when the tubing is manipulated.

The sample may also be applied through a plunger-type flow adaptor. This method has the advantages of being highly reproducible, rapid, and readily adaptable to automatic operation; in addition, it permits the sample to be applied either at the top of the column, or at the bottom as required for ascending eluent flow chromatography. The bed must be thoroughly stabilized and it is essential that the flow adaptors be in contact with the bed surface, in order to avoid excessive dilution of the sample and subsequent tailing of effluent peaks. The sample can then be applied through the flow adaptor tubing by gravity flow, with a hypodermic syringe, or with a small peristaltic pump. The point where the sample is introduced into the system must be located as close as possible to the bed surface. Unnecessarily long tubing between the sample reservoir or syringe and the bed will result in significant dilution of the sample. For this same reason, the use of a sample loop made of coiled capillary tubing (usually PTFE) is not recommended when maximum resolution is required. If a three-way valve has been included in the system, it is a simple matter to switch to eluent flow once the sample has been applied. Considerable care must be taken to avoid the introduction of air into the system. If this does occur, temporarily reversing the flow through the system will usually discharge the air.

Investigators unskilled in sample application should practice these techniques prior to chromatographing critical samples. The use of an aqueous solution of blue dextran 2000,[21] 0.2%, alone or in conjunction with another colored molecule, provides a sample which permits easy detection of errors in technique.

Elution of Beds

Elution of the packed bed is accomplished either by gravity flow, or with an eluent pump.

Gravity Elution

In order to help maintain a constant eluent flow rate, a constant pressure eluent reservoir is recommended, especially with high W_r gel beds (Fig. 4). The pressure drop across the bed is measured from the bottom of the air vent tube to the point where the effluent issues from the system; this pressure will remain constant as long as the liquid level remains above the tip of the air vent tube. The vertical length of the air space above the bed, if any, must be deducted from the pressure so measured. For low W_r gels, flow rates are proportional to the hydrostatic pressure drop across the bed. For high W_r gels, however, the maximum pressures recommended in Table IV must not be exceeded; to do so will result in a decrease in the maximum flow rate obtainable (Fig. 8). Needle valve control of eluent pressure is not recommended, since the output pressure is dependent upon the input pressure. Gel beds cannot run dry if the tip of the outlet tubing is positioned above the bed surface.

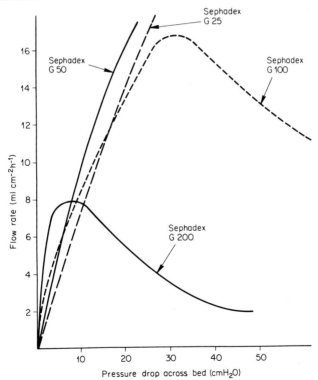

FIG. 8.   Influence of hydrostatic pressure drop across bed upon flow rate through Sephadex beds. Bed dimensions: 2.5 × 50 cm. [M. Joustra, *in* "Protides of the Biological Fluids" (H. Peeters, ed.), Vol. 14, p. 533. Elsevier, Amsterdam, 1967.

Pump Elution

The use of a pump for stabilizing eluent flow rates or for elution of large beds is occasionally desirable. While no problems are likely to be encountered with low W_r gel beds, avoidance of excessive pressures with high W_r gel is necessary. Thus, to allow for variations in pumping rate and bed resistance, it is recommended that pump flow rates not exceed 90% of those obtainable by gravity elution at the proper hydrostatic pressure drop across the bed.

Remarks Applicable to Both Elution Systems

The use of excessive pressures with high W_r gel beds with either a gravity or pumped system will occasionally cause gel particles to penetrate the column bed support. Should this occur, the column must be repacked.

The presence of small bubbles in the system is usually due to the release of dissolved gases from eluents; to preclude this occurrence, all eluents, especially those of low ionic strength or high pH, should be degassed prior to use. Degassing techniques are described under *Preparation of Eluents and Gel Slurries.*

The maximum eluent flow rate which provides a satisfactory separation can only be determined empirically. In laboratory columns, however, eluent flow rates of approximately 2–5 ml per hour per cm² of cross-sectional bed area are used with low W_r gels, whereas high W_r gel beds can be operated at eluant flow rates above 10 ml per hour per cm² of cross-sectional bed area. When high W_r gels are used, flow rates will often be limited by the physical resistance of the packed gel bed.

Resolution can usually be improved by lowering flow rates (Fig. 3); at very low flow rates, longitudinal diffusion or convection may limit the resolution which can be obtained.[6]

In general, the maximum flow rate obtainable is inversely proportional to the bed length. Although the total maximum eluant flow rate through a larger diameter bed is greater than that seen with a smaller diameter bed, the maximum eluent flow rate per unit of cross-sectional area is inversely proportional to the bed diameter. This latter effect is assumed to be due to the diminishing role played by the column wall in supporting the gel bed, as the column diameter is increased.

Some reduction in the maximum attainable eluent flow rate is occasionally seen with high W_r gel beds after prolonged use, apparently due in part to gradual compaction of the bed. Inversion of the bed, without changing the eluent flow mode, is often a successful means of increasing the flow rate, should a reduction in flow rate occur.

The belief that ascending eluant flow mode elution of gel beds results in substantially increased eluant flow rates by achieving a "fluidized bed" state has not been supported. The deposition of fine gel particles, if present, on the outlet bed support may, however, be prevented or delayed. If properly prepared slurries of spherical particles are employed, no appreciable differences in eluant flow rates are likely to be observed. Notwithstanding, ascending flow mode elution may be desirable from a standpoint of placement of apparatus, or for other reasons of convenience.

Maintenance of Gel Beds

Prevention of Microbial Growth

While the chemical nature of the gels is such that they are unlikely to serve as microbial substrates, substrates are often included in eluents and samples. If left in the presence of the gels, these substances may encourage microbial growth, especially during storage of packed columns or slurries. Fortunately, microbial growth is seldom encountered while columns are in use.

Accordingly, if microbial growth is to be prevented, phosphate ions and all substrates should be eluted from gel beds prior to storage. Microbial growth can be further inhibited by storing packed beds or slurries in the presence of an antimicrobial agent, if required. Substances commonly employed for this purpose are sodium azide 0.02%, trichlorobutanol 0.02%, ethylmercuric thiosalicylate 0.01%, and phenylmercuric acetate 0.01%. A detailed discussion of these agents and their applications is given by Fischer.[1]

Sterilization of Gel Beds

If sterile effluent fractions are required, two approaches can be taken: sterile beds can be prepared and sterile techniques used throughout the experiment, or effluent fractions can be sterilized after collection from an unsterile bed. The latter method is usually preferred due to the technical difficulties involved in maintaining sterility during chromatographic operations.

Hydrated dextran and polyacrylamide gels can be sterilized in the autoclave, if suspended in distilled water. Dextran gels may become discolored if heated in the presence of phosphate ions. Agarose gels cannot be autoclaved since they begin to dissolve at temperatures above 40°.

Cleanup of Beds Contaminated with Sorbed Substances

Lipids and lipoproteins most often appear as a faint yellow zone at the surface of the bed. If deposits are extensive, they can be the cause of a lowered eluent flow rate. The quickest solution to the problem is to remove the contaminated region with a spatula, replacing with an equivalent amount of gel slurry, if necessary. A nonionic detergent, such as Tween 80 (5% aqueous solution), is also successful for lipid removal and can be used with all three gel types.

Both carbocyclic and heterocyclic aromatic compounds tend to be sorbed to the gel matrix. Such compounds should be eluted as soon as possible because often they become more difficult to remove the longer they are left in the presence of the gel; this is notoriously the case with fluorescein isothiocyanate residues. Aromatic compounds can usually be removed from gel beds by elution with aqueous buffers, but in some instances prolonged washing may be required. Passage of a zone of serum albumin through the bed has been used with success; this method is often useful in the removal of sorbed blue dextran 2000 dye residues. Organic solvents may be effective, but will cause dehydration of the gels, necessitating rehydration and repacking of columns. An increase in ionic strength or a change in the pH of the eluent may be helpful in accelerating release of sorbed aromatic compounds. While cleanup of contaminated beds is usually successful, in some instances it may be most economical to discard contaminated gels, and repack the column with fresh material.

Acknowledgment

The assistance of I. M. Easterday, R. L. Easterday, C. Siebert, A. Friedman, and R. C. Leaf in the preparation of the manuscript is gratefully acknowledged.

Appendix
TABLE I
PHYSICAL CHARACTERISTICS OF DEXTRAN GELS (SEPHADEX)[a]

Designation	Dry particle diameter (μ)	Water regain (W_r)	Expected bed volume (ml g^{-1})	Approximate fractionation ranges[b] Globular proteins (daltons)	Linear dextrans (daltons)	Time required for hydration (hours) 20°	100°[c]
Sephadex G-10	40–120	1.0 ± 0.1	2 – 3	700	700	3	1
Sephadex G-15	40–120	1.5 ± 0.2	2.5– 3.5	1500	1500	3	1
Sephadex G-25 Coarse	100–300						
Medium	50–150	2.5 ± 0.2	4 – 6	1000–5000	100–5000	6	2
Fine	20– 80						
Superfine	10– 40						
Sephadex G-50 Coarse	100–300						
Medium	50–150	5.0 ± 0.3	9 –11	1500–30000	500–10000	6	2
Fine	20– 80						
Superfine	10– 40						
Sephadex G-75	40–120						
Superfine	10– 40	7.5 ± 0.5	12 –15	3000–70000	1000–50000	24	3
Sephadex G-100	40–120						
Superfine	10– 40	10.0 ± 1.0	15 –20	4000–150000	1000–100000	48	5
Sephadex G-150	40–120						
Superfine	10– 40	15.0 ± 1.5	20 –30	5000–400000	1000–150000	72	5
Sephadex G-200	40–120						
Superfine	10– 40	20.0 ± 2.0	30 –40	5000–800000	1000–200000	72	5

[a] Pharmacia Fine Chemicals Inc., 800 Centennial Avenue, Piscataway, New Jersey 08854. Data in table was supplied by manufacturer.
[b] Exclusion limits are equal to upper limits of fractionation ranges. Values for superfine gels may be somewhat less than shown.
[c] In boiling water bath.

TABLE II
PHYSICAL CHARACTERISTICS OF POLYACRYLAMIDE GELS (BIO-GEL P)[a]

Designation	Wet particle size (mesh)	Water regain W_r (± 10%)	Expected bed volume (ml g^{-1}) (± 10%)	Approximate fractionation ranges for globular proteins[b] (daltons)	Time required for hydration (hours) 20°	100°[c]
Bio-Gel P-2	50–100 100–200 200–400 –400	1.5	3.0	200–1800	4	2
Bio-Gel P-4	50–100 100–200 200–400 –400	2.4	4.8	800–4000	4	2
Bio-Gel P-6	50–100 100–200 200–400 –400	3.7	7.4	1000–6000	4	2
Bio-Gel P-10	50–100 100–200 200–400 –400	4.5	9.0	1500–20,000	4	2
Bio-Gel P-30	50–100 100–200 –400	5.7	11.4	2500–40,000	12	3
Bio-Gel P-60	50–100 100–200 –400	7.2	14.4	3000–60,000	12	3
Bio-Gel P-100	50–100 100–200 –400	7.5	15.0	5000–100,000	24	5
Bio-Gel P-150	50–100 100–200 –400	9.2	18.4	15000–150,000	24	5
Bio-Gel P-200	50–100 100–200 –400	14.7	29.4	30000–200,000	48	5

TABLE II (continued)

Designation	Wet particle size (mesh)	Water regain W_r (± 10%)	Expected bed volume (ml g^{-1}) (± 10%)	Approximate fractionation ranges for globular proteins[b] (daltons)	Time required for hydration (hours)	
					20°	100'[c]
Bio-Gel P-300	50-100 100-200 -400	18.0	36.0	60000-400,000	48	5

[a]Bio-Rad Laboratories, 32nd and Griffin Avenue, Richmond, California 94804. Data in table were supplied by manufacturer.
[b]Exclusion limits are equal to upper limit of fractionation range. Values for linear molecules, where applicable, are somewhat lower; manufacturer does not supply exact values.
[c]In boiling water bath.

TABLE III

PHYSICAL CHARACTERISTICS OF BEADED AGAROSE GELS

Product[a]	Agarose content of particles (approximate %)	Particle diameter (wet)	Approximate fractionation ranges[b]	
			Globular proteins and viruses (daltons)	Linear dextrans (daltons)
Bio-Gel A−0.5 M	10	50-100 mesh 100-200 mesh 200-400 mesh	<10^4 to 5 × 10^5	c
Bio-Gel A−1.5 M	8	50-100 mesh 100-200 mesh 200-400 mesh	<10^4 to 1.5 × 10^6	c
Bio-Gel A−5 M	6	50-100 mesh 100-200 mesh 200-400 mesh	10^4 to 5 × 10^6	c
Bio-Gel A−15 M	4	50-100 mesh 100-200 mesh 200-400 mesh	4 × 10^4 to 1.5 × 10^7	c
Bio-Gel A−50 M	2	50-100 mesh 100-200 mesh	10^5 to 5 × 10^7	c
Bio-Gel A−150 M	1	50-100 mesh 100-200 mesh	10^6 to 1.5 × 10^8	c
Sepharose 6B	6	40-210 μ	10^4 to 4 × 10^6	10^4 to 10^6

TABLE III (continued)

Product[a]	Agarose content of particles (approximate %)	Particle diameter (wet)	Approximate fractionation ranges[b]	
			Globular proteins and viruses (daltons)	Linear dextrans (daltons)
Sepharose 4B	4	40–190 μ	10^4 to 2×10^7	10^4 to 5×10^6
Sepharose 2B	2	60–250 μ	10^4 to 4×10^7	10^4 to 2×10^7

[a]Bio-Rad Laboratories, 32nd and Griffin Ave., Richmond, California 94804 (Bio-Gel A); Pharmacia Fine Chemicals, Inc., 800 Centennial Avenue, Piscataway, New Jersey 08854 (Sepharose). Both products are preserved with 0.02% sodium azide; Bio-Gel A preparations contain, in addition, 0.001 M tris + 0.001 M EDTA. Data were supplied by the manufacturers.

[b]Exclusion limits are the same as upper limits of fractionation ranges.

[c]Values for linear molecules are approximately 30–50% of those shown for globular molecules.

TABLE IV

RECOMMENDED MAXIMUM HYDROSTATIC PRESSURE DROP ACROSS GEL BEDS

Gel Product	Limit (cm H_2O)
Sephadex G-10 through G-50	>100
Sephadex G-75	50
Sephadex G-100	35
Sephadex G-150	15
Sephadex G-200	10
Bio-Gel P-2 through P-60	>100
Bio-Gel P-100	50
Bio-Gel P-150	35
Bio-Gel P-200	25
Bio-Gel P-300	15
Agarose Gels- 10%	>50
Agarose Gels- 8	>50
Agarose Gels- 6	50
Agarose Gels- 4	40
Agarose Gels- 2	20
Agarose Gels- 1	10

TABLE V

TROUBLESHOOTING CHART—POOR FLOW RATE

Cause	Remarks
1. Column bed support clogged with gel particles	Use bed supports made from fine polyamide or PTFE cloth. Sintered glass and porous plastic bed supports are easily clogged, especially by high W_r gels.

TABLE V (*continued*)

Cause	Remarks
2. Excessively high packing pressure and/or operating pressure	Do not exceed recommended hydrostatic pressure drop across bed (Table IV). High W_r gels are most sensitive to excessive pressures. Column should be repacked.
3. Presence of fine particles, broken beads	Do not stir slurry during hydration. Remove fine particles by elutriation of slurry. Check condition of particles with low-power microscope.
4. Use of granular rather than spherical beaded products	Use only beaded gels. Check particle shape using low-power microscope.
5. Small particle size gel used	Maximum flow rate on gels is directly related to square of mean diameter of particles.
6. Bed surface contaminated with dust, undissolved sample materials, precipitates, lipids, etc.	Keep column covered to protect bed surface; centrifuge or filter samples to remove all undissolved materials. Do not attempt to chromatograph lipemic sera. Filter eluent solutions prior to use. Remove and/or replace contaminated gel at surface of bed.
7. Clogged filter atop bed	Top filters are neither required nor desirable if proper techniques of sample application are used.
8. Incorrect use of eluent pump with low W_r gels	Operate column at lower flow rate than expected with gravity system. Column should be repacked.
9. Microbial growth within bed	Remove substrates and phosphates from bed at completion of experiment. Store gel beds and eluents under conditions which prevent microbial growth. Discard contaminated gels and eluents.
10. Gel incompletely hydrated	Allow sufficient time for hydration of gel before packing column. Column should be repacked.
11. Air present in capillary tubing, orifices	Clear air from system prior to packing column. Use suction to start flow if necessary. Use only degassed gel slurries and eluents.
12. Temperature change.	Use bed only at temperature at which packed. Column should be repacked.
13. High eluent viscosity	Maximum eluent flow rate is inversely related to eluent viscosity.

TABLE VI
TROUBLESHOOTING CHART — POOR RESOLUTION

Cause	Remarks
1. Incorrect gel type.	Double-check exclusion limit, fractionation range.
2. Sample not applied properly.	Use care in sample application to assure even application with minimal dilution.
3. Poor column design	Use properly designed equipment with acceptable bed support material, small dead space volume, and narrow-bore outlet tubing.
4. Sample inappropriate for gel filtration	Molecules of similar dimensions cannot usually be separated by gel filtration.
5. High sample relative viscosity	Reduce relative viscosity to recommended level.
6. Bed packing irregularities; bubbles; channeling; separation of gel bed. Incomplete bed stabilization	Check bed packing uniformity with Blue Dextran 2000. Do not let bed run dry. Stabilize bed before applying sample. Use only degassed gel slurries and eluents. Column should be repacked.
7. Insufficient bed length	Increase effective bed length. Consider connection of columns in series and recycling techniques.
8. Incorrect gel particle size	Resolution varies inversely with gel particle diameter.
9. Partially clogged filter on top of bed	Top filters are neither required nor desirable if proper techniques of sample application are used.
10. Contaminated bed	Check for bacteria, sorbed substances, dirt, and air bubbles. Use only filtered, degassed eluents. Removal of salts during chromatography will occasionally cause proteins to precipitate on bed.
11. Temperature fluctuation	Use bed only at temperature at which packed. Avoid drafts and sunlight. Column should be repacked.
12. Flow rate too high	Decrease flow rate to recommended level.
13. Column not properly mounted	Check vertical position with plumb line or level. Do not permit vibrations from motors, etc., to be transmitted to column.
14. Excessively high packing and/or operating pressure	Do not exceed recommended hydrostatic pressure drop across bed. High W_r gels are most sensitive to excessive pressures. Column should be repacked.

TABLE VI (*continued*)

Cause	Remarks
15. Surface of bed contaminated with lipoproteins, lipids, etc.	Remove contaminated region of bed and replace with fresh gel if indicated.

TABLE VII

TROUBLESHOOTING CHART — POOR SAMPLE RECOVERY

Cause	Remarks
1. Contaminated bed	Check for microorganisms, other contaminants
2. Removal of cofactors	Replenish cofactors. Try combining fractions to restore activity.
3. Eluent inadequately buffered; pH, ionic strength incorrect	Adjust conditions accordingly.
4. Ion exchange binding	Use buffer with ionic strength above 0.02. This is especially important when small masses of positively charged substances are chromatographed.
5. Antimicrobial agent reacting with sample	Use alternate method of preventing microbial growth.
6. Sample substances precipitating on bed	Removal of salts will occasionally result in precipitation of proteins on the bed. Adjust conditions to prevent precipitation.
7. Dilution of protein	Some proteins show increased lability at low concentrations.

[28] A Method for Anaerobic Column Chromatography

By R. REPASKE

Column chromatography of an oxygen-sensitive enzyme (hydrogenase) was accomplished with more than 90% recovery using the anaerobic technique described. Previously, very low recovery was obtained by the usual method of incorporating reducing agent in the elution buffer. The anaerobic system consisted of a column and an elution buffer vessel, both modified to facilitate anaerobic assembly and operation. During operation the volume of eluted fractions removed from the system was balanced by an equivalent input of an oxygen-free gas.

The system was assembled from individual units (Fig. 1; units A, B, and C) from which air was flushed with O_2-free gas. These units could be opened subsequently for making additions or for making connections between units if a sufficiently large outflow of gas was maintained to create an anaerobic atmosphere at the opening. The anaerobic system could be assembled, operated, and maintained with low levels of contaminating oxygen if attention was given to small details such as keeping a slight positive pressure in the system and flushing capillary tubing with gas before making the terminal connection.

Multiple gas lines were required for flushing several parts of the system simultaneously. For most purposes large volumes of gas were needed and the usual cylinder gas regulator gave adequate control. However, while collecting fractions, small volumes of gas at very low pressure were required. It was essential to insert a reducing valve capable of delivering small volumes of gas at a few ounces positive pressure. Connections between units were made with polyethylene capillary tubing having female Kel-F Luer fittings which connected with male Luer ends on the glass tubing.

Anaerobic Buffers and Sorbent

Anaerobic buffer solutions needed for elution and for packing the column were prepared first. Eluent buffers were placed in unit B (Fig. 1), and additional buffer to be used to pack the column bed was placed in a reagent bottle. (Reagent bottles with bottom tubulation could be used in place of unit B.) The tops for unit B had a long-stemmed glass tube for bubbling O_2-free gas through the liquid and a short-stemmed vent tube. The same tube arrangement in a rubber stopper was used with reagent bottles. For convenience the tubes were provided with stopcocks for quick closure. Reducing agent was introduced shortly before

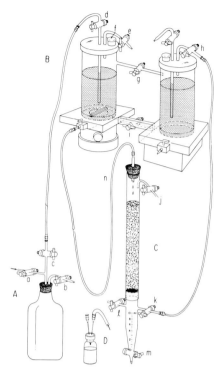

Fig. 1.   Schematic drawing of the assembled anaerobic column system. *A*, a plastic bottle which receives low-pressure O_2-free gas through stopcock *a*;   *B*, a Plexiglas two-chamber eluent buffer vessel for either step or for gradient elutions;   *C*, the column; and *D*, a serum bottle containing a collected fraction being vented with O_2-free gas.

gassing was completed (20 minutes) by injecting the solution through the serum bottle stopper (*f*) in the lid of *B* or by raising the reagent bottle stoppers; escaping gas formed an adequate seal against air. Buffer equilibrated sorbent slurry was also placed in a reagent bottle and was made anaerobic as described above.

The Plexiglas eluent buffer vessel (*B*) consisted of two identical compartments which were used separately for step elutions or together for gradient elutions. With gradients, the connection above the liquid was needed to maintain equal pressure in both compartments. The long tube used to add O_2-free gas to the buffer also functioned as a Mariotte tube during operation of the column. The screw cap covers had an 'O' ring seal to provide an air-tight closure.

Anaerobic Column

The column (*C*) was a modified commercially available column. Immediately below the rubber stopper a small-diameter glass tubing enters the column, its open end directed downward. O_2-free gas admitted to the column through this tube was used for flushing the column and for providing a protective baffle of gas at the top through which anaerobic buffer, sorbent slurry or enzyme could be added without exposure to air. Once the column had been made anaerobic and stoppered, it could be safely reopened if gas was simultaneously admitted through stopcock *j*. The bottom member of the column, a male standard taper joint with a sintered-glass disk support, was modified by replacing the drip tube with a 12-ml conical centrifuge tube. Stopcocks were located as shown (*k, l,* and *m*).

The column was made anaerobic by flushing gas for 15–20 minutes through both *j* (with stopper removed) and *k*; gas exhaust in the bottom section was alternated between *l* and *m*. Stopcock *k* and the exhaust stopcock were closed while *j* remained open and anaerobic buffer and sorbent slurry were poured through the gas flowing out of the top of the column. Because the bottom portion of the column was closed, buffer did not pass through to the collecting tube. Buffer flow required for packing the sorbent to form the column bed was controlled by releasing pressure through *l*. As a precaution against air entering the open stopcock, gas flow was maintained by simultaneously opening stopcock *k*. With subsequent additions of buffer or sorbent, liquid flow was controlled by opening or closing these stopcocks. (It should be noted that each time bottles containing buffer or sorbent were opened, it was necessary to flush them again with O_2-free gas.) The top of the column was sealed with a solid rubber stopper as stopcock *j* was closed.

Assembly of Components

Unit *B* was next connected to the low-pressure O_2-free gas source. The gas was connected through *a* to the plastic bottle *A*, which acted as a safety buffer against sudden pressure changes. *A*, each of the stopcocks (*b, c*), and the capillary tubing were flushed before the tubing was connected to *d*. Small amounts of air in the glass tubing before stopcock *d* were vented through *B* via *h* (or *e* if one elution buffer was used) and through the capillary tubing to be connected to the bottom part of the column. After about 5 minutes the tubing was connected to stopcock *k* and venting was continued for a time through *l*. Stopcocks *k* and *l* and the gas source were then closed. Air in tubing *n* was displaced with anaerobic buffer from *B*.

Operation of the Column

Enzyme was now added to the column after removal of the stopper with a protective flow of gas from *j*; enzyme was washed in with anaerobic buffer by controlling column flow with k and l as described above. The stopper with tubing n was placed on the column as j was closed. The buffer flow rate was established and the gas pressure was adjusted so that each drop of buffer caused a gas bubble to rise from the Mariotte tube. When the system was completely assembled, stopcock k had to be opened to provide pressure equalization between B and the collecting tube if the column was to flow. When a gradient elution was used, stopcocks g and i were opened and the magnetic stirrer was started.

Fractions of the desired volume were collected and dispensed (m) into preflushed serum bottles while O_2-free gas was directed into the bottle. A serum bottle stopper sealed the bottle and gas was flushed through the head space and vented by hypodermic needles (D).

[29] Chromatography of Proteins on Hydroxyapatite

By Giorgio Bernardi

Chromatography of proteins on hydroxyapatite[1] (HA[2]) columns is a separation technique developed in Tiselius' laboratory.[3-7] The use of HA as prepared by Tiselius *et al.*[4-6] has superseded that of other calcium phosphates previously employed in protein purification.[8]

The relatively slow acceptance of chromatography on HA columns by protein chemists seems to be due largely to three main factors: the rather laborious preparation procedure; the unknown mechanism of interaction of proteins with HA; the introduction of cellulose ion-exchangers

---

[1]Hydroxyapatite, not hydroxylapatite, is the name recommended by Wyckoff[34] since "hydroxyl" implies the derivatives being named after the subtituted ion, a usage which is not observed in the corresponding fluorine and chlorine derivatives (e.g., fluorapatite, chlorapatite, not fluoridapatite, chloridapatite).

[2]Abbreviations: HA, hydroxyapatite; NAP, KP, equimolar mixtures of NaH_2PO_4, Na_2HPO_4, and KH_2PO_4, K_2HPO_4, respectively. The pH is close to 6.8, and the ionic strength is equal to about twice the molarity. The abbreviation PB (phosphate buffer) used by some authors does not indicate the cation. Because the eluting power of phosphates is different for different salts, the abbreviation PB is discouraged.

[3]S. M. Swingle and A. Tiselius, *Biochem. J.* **48**, 171 (1951).

[4]A. Tiselius, *Ark. Kemi* **7**, 445 (1954).

[5]A. Tiselius, S. Hjertén, and O. Levin, *Arch. Biochem. Biophys.* **65**, 132 (1956).

[6]S. Hjertén, *Biochim. Biophys. Acta* **31**, 216 (1956).

[7]O. Levin, this series, 5, 27.

[8]C. A. Zittle, *Advan. Enzymol.* **14**, 319 (1953).

by Peterson and Sober in 1956[9] and of Sephadex by Porath and Flodin in 1959.[10]

The first difficulty seems to have been overcome with the advent of commercially available HA preparations (see following section). As far as the second point is concerned, recent work has led to a better understanding of the interaction between proteins and HA,[11,12] the influence of the secondary and tertiary structure of proteins on their chromatographic behavior,[11] the parameters which determine the resolving power of the columns,[13, 14] and the theoretical basis of chromatography of macromolecules endowed with a rigid structure on HA columns.[15, 16] These investigations have contributed to establishing HA chromatography as a major protein separation technique. Since the basic mechanism of fractionation on HA is quite different from those underlying ion exchange and gel filtration, the three methods usefully complement each other.

The present article will deal only with some basic problems in protein chromatography on HA columns, without attempting to cover the theoretical aspects of this technique or its applications to specific proteins. It may be worthwhile to recall that HA chromatography has become in recent years a standard technique also in the field of nucleic acids.[17–21]

Methods

Preparation of Hydroxyapatite

Preparation Procedure of Tiselius et al.[5]

The following is a description of this procedure as used in the author's laboratory.

Materials. The following analytical grade reagents (Merck, Darmstadt, Germany) are used routinely: $CaCl_2 \cdot 2H_2O$, $Na_2HPO_4 \cdot 2H_2O$, $NaH_2PO_4 \cdot H_2O$, $K_2HPO_4 \cdot 3H_2O$, KH_2PO_4.

[9]E. A. Peterson and H. A. Sober, *J. Amer. Chem. Soc.* **78**, 751 (1956).
[10]J. Porath and P. Flodin, *Nature (London)* **183**, 1657 (1959).
[11]G. Bernardi and T. Kawasaki, *Biochim. Biophys. Acta* **160**, 301 (1968).
[12]G. Bernardi, *Biochim. Biophys. Acta,* submitted for publication.
[13]T. Kawasaki and G. Bernardi, *Biopolymers,* **9**, 257 (1970).
[14]T. Kawasaki and G. Bernardi, *Biopolymers,* **9**, 269 (1970).
[15]T. Kawasaki, *Biopolymers,* **9**, 277 (1970).
[16]T. Kawasaki, *Biopolymers,* **9**, 291 (1970).
[17]G. Bernardi, *Nature (London)* **206**, 779 (1965).
[18]G. Bernardi, *Biochim. Biophys. Acta* **174**, 423 (1969).
[19]G. Bernardi, *Biochim. Biophys. Acta* **174**, 435 (1969).
[20]G. Bernardi, *Biochim. Biophys. Acta* **174**, 449 (1969).
[21]G. Bernardi, this series, Vol. 21 [3].

Preparation of Brushite, CaHPO₄·2H₂O. Two liters each of 0.5 M CaCl₂ and 0.5 M Na₂HPO₄ are fed at a flow rate of 250 ml/hour, using a multi-channel peristaltic pump, into a 5-liter beaker containing 200 ml of 1M NaCl. The addition is done while stirring just enough to avoid sedimentation of the brushite precipitate. After addition, brushite is allowed to settle. The supernatant liquid is decanted, and the precipitate is washed with two 4-liter volumes of distilled water.

Conversion of Brushite into Hydroxyapatite, Ca₁₀ (PO₄)₆ (OH)₂. Brushite is suspended in 4 liters of distilled water and stirred. 100 ml of 40% (w/w) NaOH is added. The mixture is heated to boiling in 40–50 minutes, and then boiled for 1 hour with simultaneous, gentle stirring. The precipitate is allowed to settle completely, and the supernatant fluid is siphoned off. The precipitate is washed with 4 liters of water and the supernatant fluid is siphoned off when a 2 cm-layer of precipitate is formed on the bottom of the beaker. This is the only time during the procedure when a complete settling of the precipitate is not allowed in order to eliminate the "fines." The precipitate is then washed once more, allowing a complete settling. At this point, the precipitates from two preparations are pooled and suspended in 4 liters of 10 mM sodium phosphate buffer, pH 6.8 (NaP²), and just brought to boiling; frank boiling at this point is to be avoided. After suspension in 4 liters of 10 mM NaP, the precipitate is boiled for 5 minutes. This operation is repeated twice more using 10 mM NaP and, again, using 10 mM NaP; in both cases boiling is carried out for 15 minutes. A yield of 400–500 ml of packed precipitate is obtained from two pooled preparations.

Storage of Hydroxyapatite. The final precipitate, composed of bladelike crystals, may be stored in 1 mM NaP for several months at 4° without any change in its chromatographic behavior. The addition of chloroform as a preservative is not necessary. During resuspension of HA crystals, strong agitation should be avoided, since this fractures the crystals and their aggregates, thereby rendering them unsuitable for column chromatography.

Alternative Preparation Procedures

Other methods for preparing hydroxyapatite have been described by Main, Wilkins, and Cole,[22] Anacker and Stoy,[23] Jenkins,[24] and Siegelman *et al.*[25, 26] The results reported with these preparations are limited so that it is difficult to judge their relative merits.

[22]R. K. Main, M. J. Wilkins, and L. Cole, *J. Amer. Chem. Soc.* **81**, 6490 (1959).
[23]W. F. Anacker and V. Stoy, *Biochem. Z.* **33**, 141 (1958).
[24]W. T. Jenkins, *Biochem. Prep.* **9**, 83 (1962).
[25]H. W. Siegelman, G. A. Wieczorek, and B. C. Turner, *Anal. Biochem.* **13**, 402 (1965).
[26]H. W. Siegelman and E. F. Firer, *Biochemistry* **3**, 418 (1964).

Commercial Hydroxyapatite Preparations

A preparation obtained according to the procedure of Tiselius *et al.*[5] is sold by Bio-Rad laboratories (Richmond, California), either as a suspension in 1 m*M* NaP or as a dry powder. Another preparation is sold by Clarkson Chemical Co. (Williamsport, Pennsylvania). Commercial HA preparations met with criticisms from several laboratories when they were first made available. Comments on the preparations sold during the past two years have been generally favorable.

Experimental Techniques with Columns

For general instructions on column chromatography the reader is referred elsewhere.[27,28] Some features that are more specific to HA columns are briefly noted here.

Packing of the Columns. This is done by adding a suspension of HA crystals in Na or K phosphate buffers, pH 6.8, (NaP or KP[2]), to columns partially filled with the same buffer; the column outlet is progressively opened only after a 1-cm layer of HA has settled. Further additions of the HA suspensions are made to fill the column. The filling operation may be facilitated by the extension of the column with a glass tube of the same diameter. Alternatively, columns may be prepared by adding the HA suspension to a funnel mounted on the top of the column, the whole system being full of starting solvent; the HA suspension in the funnel is maintained under gentle stirring during the procedure. This procedure, suggested for Sephadex,[29] allows homogeneous packing.

Adsorption and Elution. As a rule, the sample is loaded in the solvent with which the column was previously equilibrated; generally this is a low-molarity NaP or KP solution.

NaP or KP of increasing concentration are generally used to elute proteins. NaP cannot be used at 4° at molarities higher than 0.5 *M*, because of the limited solubility of Na_2HPO_4. Columns are normally operated under a slight pressure (30–50 cm of water). If controlled by a pump, the flow rate should not be kept higher than that of a column flowing under a slight hydrostatic pressure. The phosphate concentration in the column effluent is usually checked by refractive index measurement, phosphorus analysis, or conductimetry.

Column Regeneration. If elution of adsorbed material is complete, the column may be reequilibrated with the starting buffer and used again

[27]This volume, [26] and [27].
[28]L. Fisher, *in* "Laboratory Techniques in Biochemistry and Molecular Biology" (T. S. Work and E. Work, eds.), Vol. I, p. 151. North Holland Publ., Amsterdam, 1969.
[29]P. J. Flodin, *J. Chromatogr.* 5, 103 (1961).

although, preferably, after removal of the top layer. The same column can be re-used 3 or 4 times.

Recovery of Irreversibly Adsorbed Materials. The HA bed may be extruded from the column and treated in one of the following ways: (a) placed in dialysis bags and dissolved by dialysis against 1 M EDTA, pH 8.0; (b) eluted with 0.1 M NaOH; (c) dissolved in 1 N HCl.

The Adsorption-Elution Process

A systematic exploration of the parameters involved in the chromatography of proteins on HA columns has just begun.[13, 14] Therefore, it may be useful to review briefly the basic features of the adsorption-elution process and to present the limited information available so far.

Adsorption

Adsorption may be done in batch or on a column. Five sets of parameters should be considered: (a) the HA bed, (b) the material to be absorbed, (c) the solvent, (d) the temperature at which adsorption takes place, and (e) the time of contact of the protein solution with HA, respectively.

HA Bed. The total volume of packed HA crystals, V_t (total volume), is equal to the sum of three terms—the volume of the "dry crystals," V_c (crystal volume), the volume of the solvent bound to the HA crystals and inaccessible to the material to be adsorbed, V_i (inner volume), and the volume of the solvent between the HA crystals and accessible to the material to be adsorbed V_o (outer volume):

$$V_t = V_c + V_i + V_o \tag{1}$$

(i) The total volume of the packed HA bed, V_t, can be determined by measuring its dimensions. (ii) The outer volume, V_o, can be determined by measuring the elution volume of a nonadsorbed substance, such as methyl orange, eosin, fuchsin, or methyl red,[7] i.e., the volume of the solvent which leaves an HA column between loading and appearance of the dyestuffs in the effluent. (iii) The inner volume, V_i, can be calculated from the difference, $(V_o + V_i) - V_o$, the term $(V_o + V_i)$ being determined by measuring the loss in weight, at 110°, of a known amount of packed HA crystals. (iv) The crystal volume, V_c, may be calculated from the difference, $V_t - (V_o + V_i)$.

HA preparations, obtained as described in the preparation procedure of Tiselius, packed under stirring and equilibrated with 1 mM KP exhibit linear flow rate *vs.* pressure drop diagrams. A pressure drop (hydrostatic pressure divided by the length of the column) of 10 results in a flow rate of ~ 100 ml/cm²/hour. For these preparations, $V_o = 0.82$, $V_i = 0.10$, and $V_c = 0.08$ ml/ml HA bed. The density of packed HA crystals (wet) is equal to 1.17 g/ml. The value found for V_0 is quite reproducible for preparations obtained according to the method described above and is definitely higher than that (0.60–0.75) reported by Levin.[7] Obviously, HA preparations obtained by different procedures, or preparations in which crystals were fractured, will have different properties. Since HA crystals are in the form of lamellae, it is likely that mechanical breakdown does not cause a very large increase in the surface available for adsorption.

The Material To Be Adsorbed. Two parameters are of interest: (1) The amount of material to be adsorbed (this should be established from the known capacity of HA), (2) the concentration of the protein in the solution to be adsorbed. The coexistence of different materials to be adsorbed need also be considered, since their presence will lead to competition for the adsorbing sites and cause displacement effects.

The Solvent. The concentration of eluting ions at the adsorption step is obviously a critical parameter in determining the capacity of HA for a given material to be adsorbed. The presence in the solvent of substances having a stronger affinity for calcium than phosphate, e.g., EDTA and citrate, can decrease the capacity of HA to zero.

Temperature. Temperature will affect the adsorption phenomenon itself (the adsorption isotherm), the ionization of phosphate ions, and the secondary structure of the proteins to be adsorbed. The effect of temperature on adsorption and on phosphate ionization is not important, yet deserves to be investigated in detail. The effect on the protein structure may cause serious changes in their affinity for HA; denatured proteins in 8 M urea have a lower affinity for HA than native proteins.[11]

Time of Contact between Proteins and HA Necessary to Reach Adsorption Equilibrium. If adsorption is done on a column rather than in batch, one should consider the flow rate while loading the protein solution.

Elution

This is generally performed by increasing the concentration of eluting ions, usually phosphate, either stepwise or continuously. Stepwise elution may be used in both batch and column operation; elution with a concentration gradient can be used only with columns. In both cases, the flow rate of the eluent should be maintained within certain limits to avoid a deformation of the chromatographic peaks.

Stepwise Elution. This procedure is very useful when separating two or more adsorbed substances which have known and different elution molarities. Its two main disadvantages, when used with columns, are the following: (i) tailing of the peaks: substances with strongly curved adsorption isotherms and therefore extended elution ranges cannot be eluted by a solvent of constant composition without tailing, unless elution is so strong that the R_f is close to 1.0[5]; (ii) "false peaks": single substances with strongly curved isotherms, may give rise to several peaks, each new concentration step releasing an additional amount of substance.[5]

Gradient Elution. Two parameters are very important in determining resolving power: (i) the length of the column, L; (ii) the slope of the gradient of the column, *grad.* In the usual case of linear concentration gradients, *grad* may be calculated from Eq. 2

$$grad = \frac{\Delta M}{V} \cdot \frac{S}{V_o / V_t} \qquad (2)$$

where ΔM is the difference in the molarity of phosphate between the initial and the final buffer; V, the total volume of the buffer; S, the cross-sectional area of the column; and V_o and V_t, the outer and the total volume of the column as already defined. If S and V are expressed in cm^2 and cm^3, respectively, *grad* represents the increase in phosphate molarity per centimeter of column.

When elution is carried out with a linear phosphate gradient, the chromatographic behavior of a protein is characterized by two parameters[13]: (i) the elution molarity, m_{elu}, defined as the phosphate molarity at which the center of the protein peak is eluted; the center of gravity of the peak is given by Eq. 3,

$$\overline{V} = \int Vf \cdot dV \ / \ \int f \cdot dV \qquad (3)$$

where f is the distribution function of the peak, and V the volume of the solvent; (ii) the width of the peak. This can be calculated as its standard deviation and should be normalized by dividing it by S (Eq. 4).

$$\sigma = [\int (V\text{-}V)^2 f \ dV \ / \ \int f \ dV]^{1/2} \cdot \frac{1}{S} \qquad (4)$$

Both chromatographic parameters, m_{elu} and σ, depend upon several factors including column length, slope of the gradient and the presence of other chromatographic components. These relationships have been studied in some detail[13, 14] for five proteins endowed with rigid structure and of different sizes: cytochrome c, lysozyme, β-lactoglobulin A, calf skin tropocollagen, and T2 phage.

The main conclusion of this work may be summarized as follows:

In the case of small protein molecules (cytochrome c, lysozyme, β-lactoglobulin A) m_{elu} markedly increased with increasing column length and slope of the gradient (Fig. 1 a–c; Fig. 2 a). In the case of tropocollagen, m_{elu} increases with increasing column length, but is not dependent upon the slope of the gradient (Fig. 1 d; Fig. 2 b). T2 phage particles show an m_{elu} which is independent of both column length and slope of the gradient (Fig. 1 e). The behavior of m_{elu} appears therefore to be different for proteins having different molecular weights.

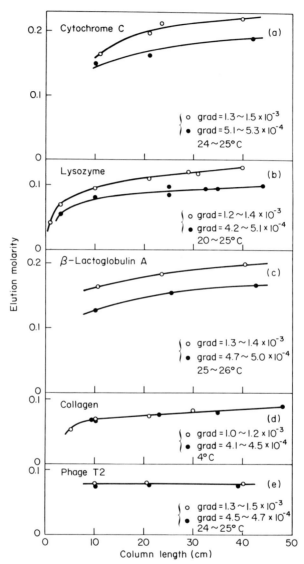

Fɪɢ. 1. Elution molarity of (a) cytochrome c, (b) lysozyme, (c) β-lactoglobulin A, (d) collagen, and (e) T2 phage, as a function of column length and slope of the gradient Load was 2 mg of protein, except in the case of T2 phage, where 1 A_{260} unit was used. Elution was carried out at room temperature with a linear molarity gradient of potassium phosphate, except in the case of collagen, where elution was done at 4° with sodium phosphate containing 0.15 M NaCl and 1 M urea. Columns of 1 cm or 0.5 cm diameter were used with flow rates of 30–60 ml/hour or 7–14 ml/hour, respectively. Yields were close to 100% in all cases. The slope of the gradient, the load and the width of the peak have been normalized for a column diameter of 1 cm. [T. Kawasaki and G. Bernardi, *Biopolymers*, **9**, 257 (1970)].

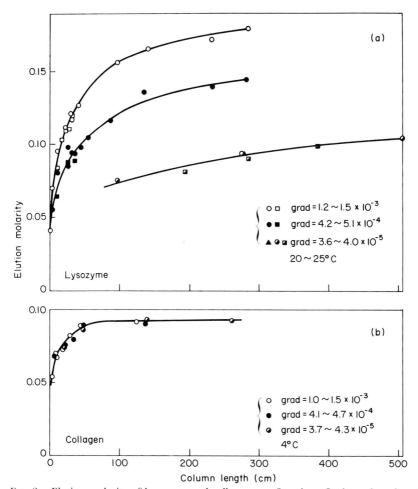

FIG. 2.   Elution molarity of lysozyme and collagen as a function of column length and of the slope of the gradient over a wider range of experimental conditions. Loads were (▲) 0.6 mg, (○, ●,◑) 2 mg (□, ■,◪), and 10 mg for lysozyme; the load was 2 mg for collagen. Certain values for collagen (◑) were obtained by extrapolation to zero load. In the case of columns longer than 150 cm, several columns connected by capillary tubing were used. Other parameters of the procedure are the same as for Fig. 1. [T. Kasawaki and G. Bernardi, *Biopolymers*, 9, 257 (1970)].

The width of the peak, σ, increases while m_{elu} decreases with increasing load, m_{elu} showing only slight dependence upon load. If two proteins are cochromatographed, the m_{elu} of the lower-eluting one is decreased, whereas that of the higher-eluting one remains the same; this displacement effect decreases with decreasing load (Fig. 3). How-

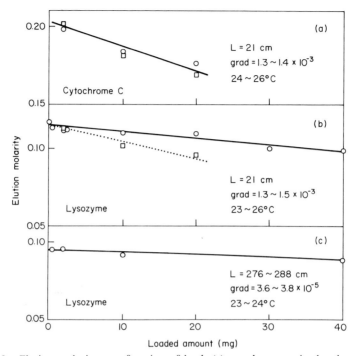

Fig. 3. Elution molarity as a function of load: (a) cytochrome *c*, in the absence (○) and in the presence (□) of the same amount of lysozyme; (b) lysozyme, in the absence (○) and in the presence (□) of the same amount of cytochrome *c*; (c) lysozyme alone, under experimental conditions different from those used in (a) and (b). Other parameters of the procedure are the same as in Fig. 1. [T. Kawasaki and G. Bernardi, *Biopolymers*, **9**, 257 (1970)].

ever, the improved separations that one would expect as a consequence of increased load are offset by the concomitant increase in σ (Fig. 4). Within experimental error, σ is not affected by the presence of another chromatographic component (Fig. 4). If the load is small enough, both m_{elu} and σ become essentially independent of both load and presence of other chromatographic components (Fig. 4). Similar observations have been made with collagen.

A lower slope of the gradient leads to a better resolution for the separation of proteins of similar size. In the case of the separation of small proteins among themselves, the best column length increases with the decrease of the slope of the gradient. If proteins are larger, the column length may be shorter and good resolution can still be obtained. In fact, in the case of collagen, a column length of about 80 cm is sufficient, whereas 10 cm is adequate for phage T2.

In the separation of small (10^4 daltons) and large ($>10^5$ daltons) proteins, the slope of the gradient must be low in order to obtain a good

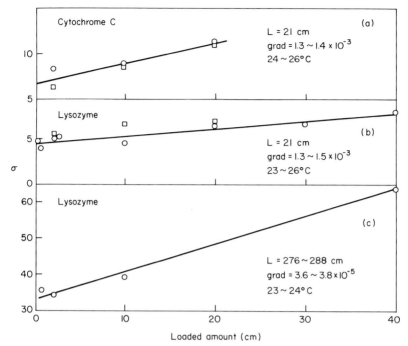

FIG. 4. Standard deviation of the peaks obtained in the experiments of Fig. 3. [T. Kawasaki and G. Bernardi, *Biopolymers*, **9**, 257 (1970)].

resolution if the former have a lower elution molarity. No general conclusion can be reached if the latter have a lower elution molarity.

Mechanism of Adsorption and Elution

In order to understand the mechanism of adsorption and elution of proteins, it is important to know which chemical groups are involved in the protein–HA interaction. As far as proteins are concerned, both negative groups, carboxyl and phosphate, and positive groups can interact with HA. The binding sites for the negatively charged groups of proteins (as well as of nucleic acids) are, in all likelihood, the calcium ions at the surface of the HA crystals. The importance of calcium ions in the adsorption process was first recognized by Tiselius et al.,[5] who noticed that treatment of HA with compounds having a very strong affinity for calcium, e.g., citrate, decreases its adsorption capacity for proteins. Dyes forming sparingly soluble salts with calcium ions, such as the crystal violet and alizarin dyes, have a strong affinity for HA. In contrast with the negative groups, binding sites for the positive groups of proteins are not known. These may be the same calcium ions or, more likely, negatively charged groups. In this connection it is important to stress that HA crystals are amphoteric and that isoelectric points of different HA preparations have been found to range from 6.5 to 10.2.[30] HA prepared according to Tiselius *et al.*[5] has a net positive charge in 1 mM KP[20] and is, therefore, a basic HA. A very important difference in the interaction of positive and negative groups of proteins with HA is that the adsorption of positive groups is strongly reduced by salts such as KCl or NaCl whereas that of negative groups is practically not changed.

Chromatography of Amino Acids

Tiselius *et al.*[5] reported that neutral and dicarboxylic amino acids show very weak or no adsorption on the columns. Basic amino acids do have slight affinity; arginine and lysine have an R_f of about 0.4 in mM NaP but display considerable tailing. In contrast, Hofman[31] reported that aspartic acid had by far the lowest R_f of 20 amino acids chromatographed in thin layers of HA.

Chromatography of Synthetic Polypeptides[11,12]

Chromatography of Polypeptides Containing Carboxylic Groups. Poly-L-glutamate and poly-L-aspartate show a rather conspicuous affinity for HA, being eluted at about 0.25 M and 0.35 M KP, respectively. Several observations deserve to be mentioned here in connection with the chromatographic behavior of polypeptides containing carboxylic groups. (1) Statistical copolymers of poly-L-glutamate with phenylalanine, lysine and serine, are eluted at a slightly lower molarity than poly-L-glutamate, the concentration necessary for elution decreasing with decreasing glutamate content. (2) If carboxyl groups are esterified, the polymer is not retained by a column equilibrated with 1 mM KP, as shown by a copolymer of DL-histidine and benzyl-L-glutamate (1:1 molar ratio). (3) Chromatography of poly-L-glutamate and poly-L-aspartate in the presence of 8 M urea caused no change in the eluting molarity. This result, at variance with what is found in the case of proteins endowed with a rigid structure (in which case denaturation causes a drastic drop in the elution molarity of proteins[11]), is not surprising since poly-L-glutamate already is in a random coil configuration at neutral pH. (4) Carrying out the elution with a linear gradient in which the limiting buffers were formed by 1 M KCl–0.001 M KP and 0.5 M KP, respectively (therefore at a practically constant ionic strength, since KP has a ionic strength which is equal to twice its molarity when its dissociation is complete) did not change the eluting molarity of poly-L-glutamate. More recent experiments[12] in the presence of 3 M KCl, also did not

[30]S. Mattson, E. Kontler-Andersson, R. B. Miller, and K. Vantras, *Kgl. Lantbruks-Hoegsk. Ann.* **18**, 493 (1951), quoted by S. Larsen, *Nature (London)* **212**, 212 (1966).
[31]A. F. Hofman, *Biochim. Biophys. Acta* **60**, 458 (1962).

alter significantly the phosphate eluting molarity of poly-L-glutamate. These results are identical to those obtained with native DNA, where the phosphate eluting molarity is practically unaffected by the presence of KCl.

Chromatography of Basic Polypeptides. The chromatographic patterns of poly-L-lysine (MW = 75,000) poly-L-arginine (MW = 9500), and poly-L-ornithine (MW = 15,800) are characterized by their strong affinity for HA, except for a fraction not retained by columns equilibrated with 1 mM KP.[32] In fact, these basic polypeptides were so strongly retained that they could not be eluted with gradients reaching a concentration of 1 M KP, and only very poorly with even higher concentrations (see legend of Fig. 5). An exception to this rule is found with a low-molecular-weight poly-L-lysine (MW = 7,000), in which case a large aliquot of the retained material is eluted with the 1 mM–1 M KP gradient as a series of sharp peaks (Fig. 5A); this pattern is not modified by the presence of 7 M urea in the eluting buffer (Fig. 5B), a result to be expected in view of the random-coiled configuration of poly-L-lysine at neutral pH (see above).

The adsorption of the basic polypeptides by HA is strikingly different from that just described for the acid polypeptides, in that adsorbed basic polypeptides can be easily and completely eluted by 3 M NaCl or 3 M KCl.

Chromatography of Neutral Polypeptides. Poly-L-tyrosine, poly-L-proline, and a poly-L-serine of very low molecular weight (MW = 1600) were not retained by HA columns equilibrated with 1 mM KP. Poly-L-histidine also showed this behavior.

Chromatography of Proteins

Only two particular cases will be considered here.

Basic Proteins. The chromatographic behavior of highly basic proteins is similar to that of basic polypeptides. Phosphate concentrations required for elution are high, and NaCl or KCl have strong effects on adsorption.[12] For instance, a lysine-rich histone fraction in which 30% of the amino acid residues consist of lysine is eluted at a high phosphate molarity, 0.55 M NaP. However, if the histone solution is loaded in 0.01 M KP–3 M KCl, the protein is not retained. Another basic protein, lysozyme, normally eluted by about 0.12 M KP, is not retained by HA columns equilibrated with 0.01 M KP–2 M KCl. In fact, it is possible to elute lysozyme with a KCl gradient at a concentration of about 0.25 M. The lowering of the elution molarity by added NaCl had been ob-

---

[32] This fraction, variable in amount in different polypeptide preparations, consists of ultraviolet-absorbing impurities.

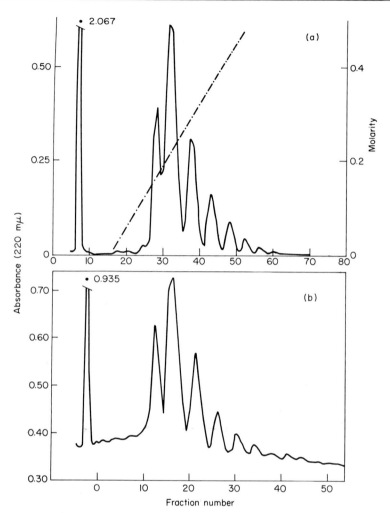

Fig. 5. (a) Chromatography of poly-L-lysine (Yeda-Miles, Rehovoth, Israel; MW = 7000). In 4 ml of 0.01 M KP, 59 A_{220} units were loaded onto a 1 × 22 cm column. The unretained sharp peak represents 23% of the load. A 0.01 M to 1 M KP gradient eluted a series of peaks representing 47% of the load. A similar run with poly-L-lysine (MW = 75,000) showed a first sharp peak representing 20% of the load; a 0.01 to 1 M KP gradient did not elute any material; direct elution with 2 M KP yielded only 11% of the load. (b) Chromatography of poly-L-lysine (Yeda-Miles; MW = 7000). In 4 ml of 0.01 M KP-7 M urea, 57.6 A_{200} units were loaded on to a 1 × 23.5 cm column. The unretained sharp peak represents 10% of the load; a 0.01 M to 1 M KP gradient in 7 M urea eluted a series of peaks representing 56% of the load. [G. Bernardi, unpublished results].

served previously for γ-globulin by Hjertén,[6] who remarked that only γ-globulin, a protein with a relatively high isoelectric point, showed this effect among the serum proteins. Tiselius *et al.*[5] observed that protamine is adsorbed more strongly at high pH than at neutrality, contrary to the behavior of proteins with acid or neutral isoelectric points, and that both protamine and lysozyme are more easily eluted by cacodylate than by phosphate buffer.

Phosphoproteins. Phosphoproteins represent another special case in that they also have a very high affinity for HA. It has been shown[33] that two egg-yolk phosphoproteins, α- and β-lipovitellin, identical in amino acid and lipid composition but different in their phosphorus content, could be separated easily on HA. When elution was performed with a molarity gradient, instead of the stepwise technique originally used, it could be shown[11] that β-lipovitellin, the electrophoretically slow component, was eluted by 0.4 *M* KP, whereas α-lipovitellin, the fast component, was eluted by 0.75 *M* KP. The third egg-yolk phosphoprotein, phosvitin, a protein in which almost 50% of the amino acid residues are phosphoryl serines, was eluted by an exceptionally high phosphate molarity, 1.2 *M* KP.

The very high eluting molarities required by phosphoproteins in comparision with nucleic acids may be due, in part, to the monoesterified phosphate groups in contrast to the diesterified groups of nucleic acid. Another consideration is that phosphoproteins have runs of phosphoryl-serine which form areas of a very high density of groups able to interact with HA.

[33]G. Bernardi and W. H. Cook, *Biochim. Biophys. Acta* **44**, 96 (1960).
[34]R. W. G. Wyckoff, "Crystal Structures," Wiley (Interscience), New York, 1951.

[30] Preparation of Calcium Phosphate Gel Deposited on Cellulose

By MASAHIKO KOIKE and MINORU HAMADA

Calcium phosphate gel[1] is a common reagent for the purification of enzymes by batchwise procedures. However, it is impossible to use the gel in columns because of poor flow characteristics unless mixed with a suitable filter aid, such as Super-Cel[2,3] or cellulose.[4] Some of the

[1]D. Keilin and E. F. Hartree, *Proc. Roy. Soc. Ser.* B **124**, 297 (1938). See also S. P. Colowick, this series, Vol. 1, [11].
[2]S. M. Swingle and A. Tiselius, *Biochem. J.* **48**, 171 (1951).
[3]A. Tiselius, S. Hjertén, and Ö. Levin, *Arch. Biochem. Biophys.* **65**, 132 (1956).
[4]V. Massey, *Biochim. Biophys. Acta*, **37**, 310 (1960). See also this series, Vol. 9, [52].

diluents themselves adsorb protein and lead to irregular elution patterns. Our experience is that it is the most important to mix the gel with cellulose homogeneously. If the column is not homogeneously packed, protein bands may tilt or streak during elution. These disadvantages are overcome by depositing the gel directly on cellulose, much like frost flowers. The procedure was introduced by Price and Greenfield,[5] and a modified, detailed procedure for its preparation is described here.

Preparation of Calcium Phosphate Gel Deposited on Cellulose

Reagents

Cellulose, nonionic, powder [Whatman Chromedia CF 1 (coarse) or CF 11 (fine)]

$CaCl_2 \cdot 2H_2O$, analytical grade

KH_2PO_4, analytical grade

HCl, 0.1 N

NH_4OH, 4 N

Nessler reagent[6] (may be obtained commercially from Fisher Scientific Co.). Dissolve 22.5 g of iodine in 20 ml of water containing 30 g of potassium iodide. After solution is complete, add 30 g of pure metallic mercury and shake the mixture well, keeping it from becoming hot by immersing the flask in tap water from time to time. Continue until the supernatant liquid has lost all the yellow color due to iodine. Remove an aliquot of supernatant solution and test a portion by adding a few drops of a 1% soluble starch solution. Unless the starch test for iodine is obtained a few drops of an iodine solution, of the same concentration as employed above, is added until a faint excess of free iodine can be detected. Dilute to 200 ml and mix well. To 975 ml of an accurately prepared, 10% sodium hydroxide solution, add the entire solution of potassium mercuric iodide solution. Mix thoroughly and allow to clear by standing.

Procedure. To 9.1 g of KH_2PO_4 is added 33 ml of 1 N HCl; the mixture is warmed until it dissolves. After cooling to room temperature, 14.7 g of $CaCl_2 \cdot 2H_2O$ is added and diluted to a final volume of 50 ml with deionized and redistilled water. The solution is added to 41 g of cellulose powder suspended in about 130 ml of water; Whatman CF 1 is the most desirable, although CF11 is also useful. The mixture is stirred rapidly for no more than 2 minutes, and 55 ml of 4 N NH_4OH is added.

[5]V. E. Price and R. E. Greenfield, *J. Biol. Chem.* **209**, 365 (1954). See also L. J. Reed and C. R. Willms, this series, Vol. 9, [50].
[6]F. C. Koch and T. L. McMeekin, *J. Amer. Chem. Soc.* **46**, 2066 (1924).

Stirring is continued for about 10 minutes. The pH of the mixture should be above 9.0. The slurry becomes quite thick and is allowed to stand overnight in a refrigerator. The supernatant fluid is decanted; two such lots of gel-cellulose are combined. The combined gel-cellulose is washed by decantation with 3-liter volumes of water until the supernatant fluid is negative to Nessler reagent. During the washing procedure care is taken to remove fine particles. The gel-cellulose is collected by low-speed centrifugation, resuspended in 1 liter of appropriate buffer (0.02 M to 0.1 M phosphate buffer, pH 6.0 to pH 7.0) or water and is packed into the columns. The gel-cellulose may be stored at either 5° or room temperature.

To increase the flow rate, 40 g of cellulose powder (Whatman CF 1 or CF 11) are suspended in the same phosphate buffer as that used for the gel-cellulose and is added with vigorous stirring to the washed gel-cellulose. The addition of bulk cellulose is performed immediately before packing of the column.

Preparation of Calcium Phosphate Gel-Cellulose Column

Prior to packing columns of the gel-cellulose, it is very important to remove dissolved gases by evacuating the suspension with a water aspirator for about half an hour after equilibration to the chosen temperature. To a suitable glass chromatographic tube, fitted with a detachable sintered glass plate covered with a circle of filter paper, is added the appropriate buffer. A slurry of the gel-cellulose is added in a few portions to the tube. The gel-cellulose is allowed to pack by gravity and is equilibrated with appropriate buffer by washing it with several volumes of the buffer at the selected temperature. Generally, by using about 100 ml of gel-cellulose in a column, 4.2 cm in diameter, a flow rate of about 20–30 ml per hour is obtained. By addition of untreated cellulose, as noted above, the flow rate is increased to between 30 and 50 ml per hour.

Method of Massey[4]

Another method for preparation of calcium phosphate gel-cellulose is that of Massey: To a suspension of 200 ml of 10% (w/v) Whatman CF 11 cellulose powder, 100 ml of calcium phosphate gel (30 mg/ml), prepared by the method of Swingle and Tiselius,[2] is added with stirring. The mixture is stirred for 20 minutes, then allowed to stand until the gel-cellulose settles. After decantation of the supernatant fluid, the gel-cellulose is washed by decantation with several batches of water of 1-liter each. The gel-cellulose is collected and suspended in an appropriate volume of water or buffer, as mentioned above.

Comments

For the purification of some enzymes, it is required that the gel-cellulose be aged. On the other hand, purification of the α-ketoglutarate dehydrogenase complex[7] from extracts of pig heart, requires that the gel-cellulose should be less than 1 month old.

In this laboratory, Whatman CF 1 is used most frequently and has been successful in the purifications of the pyruvate dehydrogenase complex,[8] lipoamide dehydrogenase,[9,10] and D-amino acid oxidase,[11] among others.

[7] M. Hirashima, T. Hayakawa, and M. Koike, *J. Biol. Chem.* 242, 902 (1967).

[8] T. Hayakawa, M. Hirashima, S. Ide, M. Hamada, K. Okabe, and M. Koike, *J. Biol. Chem.* 241, 4694 (1966).

[9] S. Ide, T. Hayakawa, K. Okabe, and M. Koike, *J. Biol. Chem.* 242, 54 (1967).

[10] T. Hayakawa, T. Kanzaki, T. Kitamura, Y. Fukuyoshi, Y. Sakurai, K. Koike, T. Suematsu, and M. Koike, *J. Biol. Chem.* 244, 3660 (1969). See also M. Koike and T. Hayakawa, Vol. 18, [51].

[11] K. Yagi and T. Ozawa, *Biochim. Biophys. Acta* 56, 413 (1962).

Section VI

Separation Based on Specific Affinity

Articles Related to Section VI

[31] Affinity Chromatography

By Pedro Cuatrecasas and Christian B. Anfinsen

General Considerations

Conventional procedures of protein purification are generally based on relatively small differences in the physicochemical properties of the proteins in the mixture. They are frequently laborious and incomplete, and the yields are often low. The selective isolation and purification of enzymes and other biologically important macromolecules by "affinity chromatography" exploits the unique biological property of these proteins or polypeptides to bind ligands specifically and reversibly.[1-7] The method is related in principle to the use of "immunoadsorbents" introduced as chemically defined materials for chromatographic separation of antibodies by Campbell and his colleagues in 1951[8] and subsequently employed widely as a standard immunological procedure.[9] Affinity chromatography, as summarized in the present article, exploits the phenomenon of specific biological interaction in a large variety of protein-ligand systems. A solution containing the macromolecule to be purified is passed through a column containing an insoluble polymer or gel to which a specific competitive inhibitor or other ligand has been covalently attached. Proteins not exhibiting appreciable affinity for the ligand will pass unretarded through the column, whereas those which recognize the inhibitor will be retarded in proportion to the affinity existing under the experimental conditions. The specifically adsorbed protein can be eluted by altering the composition of the solvent so that dissociation occurs.

In principle, specific adsorbents can be used to purify enzymes, antibodies, nucleic acids, cofactor or vitamin-binding proteins, repressor proteins, transport proteins, drug or hormone receptor structures, sulfhydryl group containing proteins, and peptides formed by organic

[1]P. Cuatrecasas, M. Wilchek, and C. B. Anfinsen, *Proc. Nat. Acad. Sci. U. S.* 61, 636 (1968).
[2]P. Cuatrecasas, *J. Biol. Chem.* 245, 3059 (1970).
[3]P. Cuatrecasas, in *Biochemical Aspects of Solid State Chemistry*, ed. G. R. Stark. New York, Academic Press. In press.
[4]P. Cuatrecasas, *J. Agr. Food Chem.*, In press (1971).
[5]I. Kato, and C. B. Anfinsen, *J. Biol. Chem.* 244, 1004 (1969).
[6]P. Cuatrecasas and C. B. Anfinsen, *Annu. Rev. Biochem.*, 40, In press (1971).
[7]P. Cuatrecasas, *Nature*, 228, 1327 (1970).
[8]D. H. Campbell, E. Leuscher, and L. S. Lermann, *Proc. Nat. Acad. Sci. U. S.* 37, 575 (1951).
[9]I. Silman and E. Katchalski, *Annu. Rev. Biochem.* 35, 873 (1966).

synthesis. Affinity chromatography may also be useful in concentrating dilute solutions of proteins, in removing denatured forms of a purified protein, and in the separation and resolution of protein components resulting from specific chemical modifications of purified proteins. Inherent advantages of this method of purification are the rapidity and ease of a potentially single-step procedure, the rapid separation of the protein to be purified from inhibitors and destructive contaminants, e.g., proteases, and protection from denaturation during purification by active site ligand-stabilization of protein tertiary structure.

Successful application of the methods will depend in large part on how closely the particular experimental conditions chosen permit the ligand interaction to simulate that observed when the components are free in solution. Careful consideration must therefore be given to the nature of the solid matrix, the dependence of the interaction on the structure of the ligand, the means of covalent attachment, and the conditions selected for adsorption and elution. The specific conditions for purification of a particular protein must be highly individualized since they will reflect the uniquely specific biological property of selective interaction with a given ligand. The following general considerations should serve as guidelines in the preparation of adsorbents for affinity chromatography.

Solid Matrix Support. An ideal insoluble support should possess the following properties: (1) It must interact very weakly with proteins in general, to minimize the nonspecific adsorption of proteins. (2) It should exhibit good flow properties which are retained after coupling. (3) It must possess chemical groups which can be activated or modified, under conditions innocuous to the structure of the matrix, to allow the chemical linkage of a variety of ligands. (4) These chemical groups should be abundant in order to allow attainment of a high effective concentration of coupled inhibitor (capacity), so that satisfactory adsorption can be obtained even with protein-inhibitor systems of low affinity. (5) It must be mechanically and chemically stable to the conditions of coupling and to the varying conditions of pH, ionic strength, temperature, and presence of denaturants (e.g., urea, guanidine hydrochloride) which may be needed for adsorption or elution. Such properties also permit repeated use of the specific adsorbent. (6) It should form a very loose, porous network which permits uniform and unimpaired entry and exit of large macromolecules throughout the entire matrix; the gel particles should preferably be uniform, spherical, and rigid. A high degree of porosity is an important consideration for ligand–protein systems of relatively weak affinity (dissociation constant of 10^{-5} M or greater), since the concentration of ligand *freely available* to the protein must be quite high to permit interactions strong enough to physically retard the downward migration of the protein through the column. This probably explains why certain solid supports, such as cellulose, have been used so success-fully as immunoadsorbents but have been of only minimal utility in the purification of enzymes.

The use of hydrophilic cellulose derivatives in the purification of anti-bodies is described in detail elsewhere in this volume and need not be discussed further here. In some cases cellulose derivatives have been used in the purification of enzymes: (1) aminophenol diazotized to cellu-lose containing resorcinol residues in ether linkage has been used to purify tyrosinase[10]; (2) flavokinase has been specifically adsorbed to car-boxymethyl cellulose and to cellulose containing flavin derivatives[11,12]; (3) avidin has been partially purified with the use of cellulose reacted with biotinyl chloride.[13] Derivatives of cellulose also have been used to purify nucleotides,[14] complementary strands of nucleic acids,[15] and cer-tain species of transfer RNA.[16] Although cellulose derivatives may be advantageous in specific instances, they are generally much less useful than the agarose derivatives because their fibrous and nonuniform character impedes proper penetration of large protein molecules. Highly hydrophobic polymers, such as polystyrene, display poor com-munication between the aqueous and solid phases. A technique for the immobilization of organic substances on glass surfaces, described recent-ly by Weetall and Hersh,[17] may be a promising tool for the preparation of specific adsorbents.

Various polysaccharide hydrophilic polymers are very useful solid supports. Commercially available, cross-linked dextran derivatives (Sephadex, Pharmacia) possess most of the desirable features listed above except for their low degree of porosity. For this reason they are relatively ineffective as adsorbents for purification of enzymes of even low molecular weight. However, beaded derivatives of agarose,[18] another polysaccharide polymer, have nearly all the properties of an ideal ad-sorbent[1-6] and are commercially available (Pharmacia; Bio-Rad Labora-tories). The beaded agarose derivatives have a very loose structure which allows molecules with a molecular weight in the millions to diffuse

[10]L. S. Lerman, *Proc. Nat. Acad. Sci. U. S.* **39**, 232 (1953).
[11]C. Arsenis and D. B. McCormick, *J. Biol. Chem.* **239**, 3093 (1964).
[12]C. Arsenis and D. B. McCormick, *J. Biol. Chem.* **241**, 330 (1966).
[13]D. B. McCormick, *Anal. Biochem.* **13**, 194 (1965).
[14]E. G. Sander, D. B. McCormick, and L. D. Wright, *J. Chromatogr.* **21**, 419 (1966).
[15]E. K. F. Bautz and B. D. Holt, *Proc. Nat. Acad. Sci. U. S.* **48**, 400 (1962).
[16]S. L. Erhan, L. G. Northrup, and F. R. Leach, *Proc. Nat. Acad. Sci. U. S.* **53**, 646 (1965).
[17]H. H. Weetall and L. S. Hersh, *Biochim. Biophys. Acta* **185**, 464 (1969).
[18]S. Hjertén, *Arch. Biochem. Biophys.* **99**, 446 (1962).

readily through the matrix. These polysaccharides can readily undergo substitution reactions by activation with cyanogen halides,[19,20] are very stable, and have a moderately high capacity for substitution.[2,3,7] Synthetic polyacrylamide gels also possess many desirable features and are available commercially in beaded, spherical form, in various pregraded sizes and porosities (Bio-Gel, Bio-Rad Laboratories). Recently described derivatization procedures permit attachment of a variety of ligands and proteins to polyacrylamide beads.[2,7,21] These beads exhibit uniform physical properties and porosity, and the polyethylene backbone endows them with physical and chemical stability. Preformed beads are available which permit penetration of proteins with molecular weights of about one half million (Bio-Gel P-300). Their porosity, however, is diminished during the chemical modifications required for attachment of ligands, and in this respect the polyacrylamide beads are inferior to those of agarose.[2,22] The principal advantage of polyacrylamide is that it possesses a very large number of modifiable groups (carboxamide). Thus, highly substituted derivatives may be prepared for use in the purification of enzymes, which exhibit poor affinity for the attached ligand.[2,22]

Considerations in Selecting a Ligand. The small molecule to be covalently linked to the solid support must be one that displays special and unique affinity for the macromolecule to be purified. It can be a substrate analog inhibitor, effector, cofactor, and, in special cases, substrate. Enzymes requiring two substrates for reaction may be approached by immobilizing one of the substrates, provided sufficiently strong binding is displayed toward that substrate in the absence of the other. Also, a substrate may be used if it binds to the enzyme under some conditions that do not favor catalysis, i.e., in the absence of metal ion, if the pH dependence of K_m and of k_{cat} are different, or at lower temperatures.

The small molecule to be insolubilized must possess chemical groups that can be modified for linkage to the solid support without abolishing or seriously impairing interaction with the complimentary protein. If the strength of interaction of the free complex in solution is very strong, i.e., a K_i of about 1 mM, a decrease in affinity of 3 orders of magnitude upon preparation of the insoluble derivative may still leave an effective and selective adsorbent. The important parameter is the effective experimental affinity—that displayed between the protein in solution and the insolubilized ligand under the experimental conditions chosen. In practice it has been very difficult to prepare such adsorbents for enzyme-

[19]J. Porath, R. Axén, and S. Ernbäck, *Nature (London)* **215,** 1491 (1967).

[20]R. Axén, J. Porath, and S. Ernbäck, *Nature (London)* **214,** 1302 (1967).

[21]J. K. Inman and H. M. Dintzis, *Biochemistry* **8,** 4074 (1969).

[22]E. Steers, P. Cuatrecasas, and H. Pollard, *J. Biol. Chem.* **246,** 196 (1971).

inhibitor systems whose dissociation constants under optimal conditions in solution are greater than 5 mM.[22] It is possible in theory, however, to prepare adequate adsorbents for such systems if a sufficiently large amount of inhibitor can be coupled to the solid support.

It is becoming increasingly clear that, for successful purification by affinity chromatography, the inhibitor groups critical in the interaction with the macromolecule to be purified must be sufficiently distant from the solid matrix to minimize steric interference with the binding process.[2,4,22] Steric considerations appear to be most important with proteins of high molecular weight. The problem may be approached by preparing an inhibitor with a long hydrocarbon chain, an "arm," attached to it, which can in turn be coupled to the insoluble support. Alternatively, such a hydrocarbon extension "arm" can first be attached to the solid support.[2]

General Considerations in the Preparation of Specific Adsorbents. The ligand must be coupled to the solid support under mild conditions that are tolerated well by both ligand and matrix. The coupled gel must be washed exhaustively to ensure total removal of the material not covalently bound. In some cases, such as with highly aromatic compounds, e.g., estradiol, complete removal of adsorbed material is very difficult and may require many days of continuous washing. Organic solvents may occasionally be required for this purpose. In rare instances, i.e., Congo red, effective washing can be achieved only by washing the column with large quantities of the protein to be purified, i.e., amyloid, followed by elution to regenerate the column. Whenever possible, radioactivity or other sensitive indicators should be present on the ligand to assist in monitoring the washing procedure. Aqueous dimethyl formamide (50%, v:v) has been useful in accelerating the washing of agarose and polyacrylamide derivatives; both tolerate this solvent well.

An accurate method for determining the amount of material attached to the solid support should be available. This is preferably done by determining the amount of ligand (by radioactivity, absorbance, amino acid analysis, etc.) released from the substituted matrix by acid or alkaline hydrolysis. Exhaustive digestion with pronase or carboxypeptidase, followed by amino acid analysis, has been used in some cases in which oligopeptides are attached to agarose. It is also possible to determine the radioactive content of the unhydrolyzed solid support by assay in suspension with Cab-O-Sil (Beckman); estimates made in this way are generally low even if appropriate internal standards are used for correcting efficiencies of counting. An alternative means of quantitation is to estimate the amount of ligand not recovered in the final washings. This is less accurate when a very large excess of ligand is added during the coupling procedure or when appreciable noncovalent adsorption to the solid matrix occurs which demands large volumes of solvent for thorough washing. It is operationally more useful to express the degree of ligand substitution on the solid matrix in terms of concentration, such as micromoles of ligand per milliliter of packed gel, rather than on the basis of dry weight.

Conditions for Adsorption and Elution of Proteins. The specific conditions for adsorption are dictated by the specific properties of the protein to be purified. Affinity purification need not be restricted to column procedures. In certain circumstances it may be advisable to use batch purification methods. This is the case, for example, when small amounts of protein are to be extracted from very crude or particulate protein mixtures with an adsorbent of very high affinity. Column flow rates in such cases may be severely compromised, and purification can be more expeditiously carried out by adding a slurry of the specific adsorbent to the crude mixture, followed by batchwise washing and elution. In some cases involving very high affinity complexes, such as observed with certain antibody–antigen systems, it is preferable to adsorb the protein to the solid support and to wash extensively while the gel is packed in a column, elute the protein by removing the matrix and incubating in suspension in an appropriate buffer. For reasons of thoroughness of mixing, higher dilution of the insoluble ligand, and easier control of time and temperature, this means of elution requires less drastic conditions and the yields may be higher.

Protein specifically adsorbed to a solid carrier matrix will eventually emerge from the column without altering the properties of the buffer if the affinity for the ligand is not too great. With this type of elution, however, the protein is generally obtained in dilute form. In most cases elution of the protein will require changing the pH, ionic strength, or temperature of the buffer. Dissociation of proteins from very high-affinity adsorbents may require protein denaturants, such as guanidine hydrochloride or urea. Ideal elution of a tightly bound protein should utilize a solvent which causes sufficient alteration in the conformation of the protein to decrease appreciably the affinity of the protein for the ligand but which is not sufficiently severe to completely denature or unfold the protein. The eluted protein should be neutralized, diluted, or dialyzed at once to permit prompt reconstitution of the native protein structure. This may be tested by rechromatographing the purified protein to determine changes in adsorption to the affinity column. Elution of the specifically adsorbed protein can also be achieved with a solution containing a specific inhibitor or substrate. The inhibitor can either be the one that is covalently linked to the matrix (and must be used at higher concentrations), or preferably another, stronger competitive inhibitor.

An alternative method of eluting the specifically bound protein is to selectively cleave the matrix–ligand bond, thus removing the intact ligand–protein complex. Excess ligand can then be removed by dialysis or by gel filtration on Sephadex. This approach can be applied to ligands attached to Sepharose by azo linkage, and by thiol or alcohol ester bonds. These will be discussed under the appropriate sections.

Preparation and Use of Agarose Bead Adsorbents

Of the beaded agarose derivatives commercially available, Sepharose 4B (Pharmacia) is the most useful for affinity chromatography. It is more porous than the 6B derivative, and has considerably greater capacity than the 2B gels. Chemical compounds containing primary aliphatic or aromatic amines can be coupled directly to agarose beads after activation of the latter with cyanogen bromide at alkaline pH (Scheme I). The chemical nature of the intermediate formed by cyanogen halide treatment of polysaccharide derivatives is not known, but the products formed upon coupling with amino compounds appear to be principally derivatives of amino carbonic acid and isourea.[23] Notably, both of these postulated linkage groups retain the basicity of the amino group of the coupled ligand. It must be borne in mind, therefore, that even if the charge on the ligand amino group is known to contribute to the binding interaction, the resultant agarose ligand gel may nevertheless demonstrate considerable binding effectiveness.

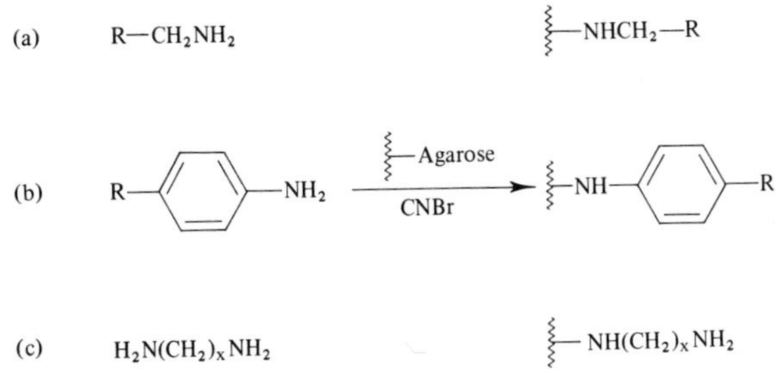

(a) $R{-}CH_2NH_2$ $\xi{-}NHCH_2{-}R$

(b) $R{-}\langle\bigcirc\rangle{-}NH_2$ $\xrightarrow[\text{CNBr}]{\xi{-}\text{Agarose}}$ $\xi{-}NH{-}\langle\bigcirc\rangle{-}R$

(c) $H_2N(CH_2)_x NH_2$ $\xi{-} NH(CH_2)_x NH_2$

SCHEME I

Beaded agarose gels, unlike the cross-linked dextrans, cannot be dried or frozen, since they will shrink severely and essentially irreversibly. Similarly, they will not tolerate many organic solvents. Dimethyl formamide (50%, v:v) and ethylene glycol (50%, v:v) do not adversely affect the structure of these beads. These solvents are quite useful in situations where the compound to be coupled is relatively insoluble in water (e.g., steroids, thyroxine, tryptophan derivatives) since the coupling step

[23]J. Porath, *Nature (London)* **218,** 834 (1968).

can be carried out in these solvents. Similarly, the final, coupled derivative can be washed with these solvents to remove strongly adsorbed or relatively water-insoluble material.

Agarose beads, both before and after activation and coupling, exhibit very little nonspecific adsorption of proteins provided the ionic strength of the buffer is 0.05 M or greater. The coupled, substituted adsorbents of agarose can be stored at 4° in aqueous suspensions with an antibacterial preservative (sodium azide, toluene), for periods of time limited only by the stability of the bound inhibitor or ligand; they should, as mentioned above, not be frozen or dried for storage. Agarose beads tolerate 0.1 M NaOH and 1 N HCl for at least 2–3 hours at room temperature without detectable adverse alteration of their physical properties, and without cleavage of the covalently linked ligand, if the latter itself does not possess labile bonds. The specific adsorbents formed from these beads can therefore be used repeatedly even after exposure to relatively extreme conditions; the limitation in this respect is more likely determined by the stability of the ligand than of the matrix. Agarose beads are not destroyed by exposure to temperatures as high as 45° for many minutes. They tolerate quite well 6 M guanidine·HCl or 7 M urea solutions for prolonged periods; a very small amount of dissolution may occur after exposure to 6 M guanidine·HCl for 2–3 days at room temperature. These protein denaturants may therefore be used to aid in the elution of specifically bound proteins, for thorough washing of columns in preparation for re-use, and in washing off protein that is tightly adsorbed during the coupling procedure.

Procedure.[2] A given volume of decanted agarose, containing a magnetic stirring bar, is mixed with an equal volume of water and the electrodes of a pH meter are placed in this suspension. The procedure is performed in a well ventilated hood. Finely divided solid cyanogen bromide (50–300 mg per milliliter of packed agarose) is added at once to the stirred suspension, and the pH is immediately raised to 11 with NaOH. The molarity of the NaOH solution will depend on the amounts of agarose and cyanogen bromide added; it should vary between 2 M (for 5–10 ml agarose and 1–3 g of cyanogen bromide) and 8 M (for 100–200 ml of agarose and 20–30 g of cyanogen bromide). The pH is maintained at 11 by constant manual titration. Temperature is maintained at about 20° by adding pieces of ice as needed. The reaction is complete in 8–12 minutes, as indicated by the cessation of base uptake; no solid CNBr should remain. A large amount of ice is then rapidly added to the suspension, which is transferred quickly to a Büchner funnel (coarse disk) and washed under suction with cold buffer. The buffer should be the same as that which is to be used in the coupling stage, and the volume of wash should be 10–15 times that of the packed agarose, larger volumes (20–30 times) should be used if a protein is to be coupled. The Büchner funnel, containing the moist, washed agarose is removed from the filtering flask and its outlet is covered tightly with Parafilm. The compound to be coupled, in a volume of cold buffer equal to the volume of packed agarose, is added to the agarose and the suspension is immediately mixed (in the Büchner funnel) with a glass stirring rod. The entire procedure of washing, adding the ligand solution, and mixing should consume less than 90 seconds. It is important that these procedures be performed rapidly and that the temperature be lowered, since the "activated" agarose is unstable. The suspension is transferred from the Büchner funnel to a beaker containing a magnetic mixing bar and is *gently* stirred at 4° for 16–20 hours. Care must be taken to avoid vigorous stirring since Sepharose can be physically disrupted, resulting in material with poor flow rates. The substituted agarose is then washed with large volumes of water and appropriate buffers until it is established with certainty that ligand is no longer being removed.

The quantity of ligand coupled to agarose can be controlled by varying several parameters.[2] Most important is the amount of ligand added to the activated agarose (Table I). When highly substituted derivatives are desired, the amount of inhibitor added should, if possible, be 20–30 times greater than that which is desired in the final product. For ordinary procedures, 100–150 mg of cyanogen bromide is used per milliliter of packed agarose, but much higher coupling yields can be obtained if this amount is increased to 250–300 mg (Table II). The larger amounts of CNBr should be used only if highly substituted derivatives are desired since greater heat is evolved and longer reaction times

TABLE I

EFFICIENCY OF COUPLING OF 3'-(4-AMINOPHENYLPHOSPHORYL)DEOXYTHYMIDINE 5'-PHOSPHATE TO AGAROSE, AND CAPACITY OF THE RESULTING ADSORBENT FOR STAPHYLOCOCCAL NUCLEASE[a]

Expt.	μMoles of inhibitor/ml agarose		Mg of nucelase/ml agarose	
	Added	Coupled	Theoretical[b]	Found
A	4.1	2.3	44	–
B	2.5	1.5	28	8
C	1.5	1.0	19	–
D	0.5	0.3	6	1.2

[a]One hundred milligrams of cyanogen bromide was added per milliliter of packed agarose, and coupling was performed in 0.1 *M* NaHCO$_3$, pH 9.0. Data from P. Cuatrecasas, M. Wilchek, and C. B. Anfinsen, *Proc. Nat. Acad. Sci. U.S.* **61**, 636 (1968).
[b]Assuming equimolar binding.

TABLE II

EFFECT OF pH ON THE COUPLING OF [^{14}C]-ALANINE TO ACTIVATED AGAROSE[a]

Conditions for coupling reaction		Alanine coupled
Buffer	pH	(μmoles per ml agarose)
Sodium Citrate, 0.1 M	6.0	4.2
Sodium phosphate, 0.1 M	7.5	8.0
Sodium borate, 0.1 M	8.5	11.0
Sodium borate, 0.1 M	9.5	12.5
Sodium carbonate, 0.1 M	10.5	10.5
Sodium carbonate, 0.1 M	11.5	0.2

[a]Sixty milliliters of packed Sepharose 4B was mixed with 60 ml of water and treated with 15 mg of cyanogen bromide as described in the text. To the cold, washed activated Sepharose was added 2.2 mmoles of [^{14}C]alanine (0.1 μCi/μmole) in 60 ml of cold distilled water, and 20-ml aliquots of the mixed suspension were added rapidly to beakers containing 5 ml of cold 0.5 M buffer of the composition described in the table. The final concentration of alanine was 15 mM. After 24 hours, the suspensions were thoroughly washed, aliquots were hydrolyzed by heating at 110° in 6 N HCl for 24 hours, and the amount of released radioactivity was determined. Data from P. Cuatrecasas, *J. Biol. Chem.* 245, 3059 (1970).

are required. The pH at which the coupling stage is performed will also determine the degree of coupling since it is the unprotonated form of the amino group which is reactive. Compounds containing an α-amino group (pK about 8) will couple optimally at a pH of about 9.5–10.0 (Table II). Higher pH values should not be used, since the stability of the activated Sepharose decreases sharply above pH 10.5 (Table II). Compounds with primary aliphatic aminoalkyl groups, such as the ϵ-amino group of lysine or ethylamine (pK about 10), should be coupled at pH values of about 10, and a large excess of ligand should be added. The most facile coupling occurs with compounds bearing aromatic amines, due to the low pK (about 5) of the amino group; very high coupling efficiencies are obtained at pH values between 8 and 9 (Table I).

It is possible, in some cases, to increase the amount of ligand coupled to agarose by repeating the activation and coupling procedures of the already substituted material, provided the inhibitor is stable for about 10 minutes at pH 11.

Quantitation of the amount of amino acid or protein coupled to agarose is best determined by hydrolyzing a lyophilized aliquot of material in 6 N HCl, 110°, for 24 hours. After hydrolysis, the brown-black suspension is filtered twice by passing through a Pasteur pipette containing glass wool, or through a Millipore filter; the solution is dried and its composition is determined with an amino acid analyzer.

The operational capacity for specific adsorption of the agarose adsorbents[1] (Table I) can be tested by passing slowly an amount of enzyme or protein in excess of the theoretical capacity through a sample of adsorbent packed in a column, washing with buffer until negligible protein emerges in the effluent, and eluting (Fig. 1). Alternatively, small samples of preferably pure protein can be added successively to the column until detectable protein or enzymatic activity emerges in the effluent.[1] The total amount added, or that eluted, is considered the operational capacity.

Examples 1: Purification of Staphylococcal Nuclease with a Competitive Inhibitor.[1] The inhibitor of staphylococcal nuclease, 3'-(4-aminophenylphosphoryl)deoxythymidine 5'-phosphate (pdTp-aminophenyl) is an ideal ligand for preparation of specific agarose adsorbents because it is a strong competitive inhibitor (K_i, 10^{-6} M), its 3'-phosphodiester bond is not cleaved by the enzyme, it is stable at pH values of 5–10, the

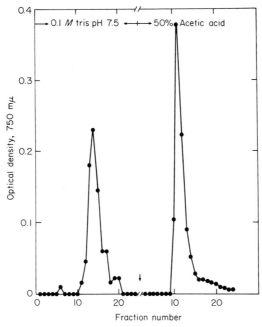

Fig. 1. Determination of the operational capacity of a column of agarose-RNase-S-protein (0.8 × 19 cm) for RNase-S-peptide. The column having a bed volume of 9.6 ml and containing 27 mg of bound S-protein on the basis of amino acid analysis, was washed with 200 ml of 0.1 M Tris·HCl buffer, pH 7.5. It was then perfused with 3.8 ml of an aqueous solution containing 4.2 mg of S-peptide. After thorough washing with buffer (0.1 M Tris·HCl, pH 7.5) bound S-peptide (2.3 mg) was eluted with 50% acetic acid. Since 27 mg of S-protein should combine, theoretically, with approximately 5 mg of S-peptide, it may be calculated that the binding efficiency of the immobilized S-protein was about 45%. Data from I. Kato and C. B. Anfinsen, *J. Biol. Chem.* 244, 1004 (1969).

pK of the aromatic amino group is low, and the amino group is relatively distant from the basic structural unit (pTp-X) recognized by the enzymatic binding site.[24] This inhibitor couples to agarose with high efficiency, and the resulting adsorbent has a high capacity for the enzyme (Table I). Columns containing this specific adsorbent completely and strongly adsorb samples of pure and crude nuclease (Fig. 2). Virtually no enzyme escapes from such columns with washes exceeding 50 times the column bed volume if the amount of protein applied does not exceed 30% of the operational capacity. Elution is accomplished by washing with solutions of low pH (acetic acid, pH 3) or high pH (NH$_4$OH, pH 11). The eluted protein emerges in a very small volume, and the material can be lyophilized directly.

If very crude enzyme solutions with total protein concentration greater than 20 mg/ml are passed through such columns, some of the enzyme escapes, probably because of interaction with the major protein fraction. Rechromatography usually will allow complete removal of the residual

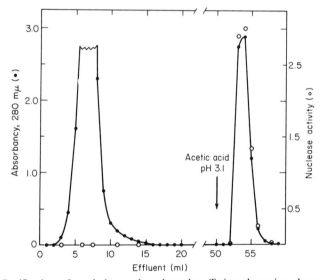

FIG. 2. Purification of staphylococcal nuclease by affinity adsorption chromatography on a nuclease-specific agarose column (0.8 × 5 cm) (sample B, Table 1.). The column was equilibrated with 50 mM borate buffer, pH 8.0, containing 10 mM CaCl$_2$. Approximately 40 mg of partially purified material containing about 8 mg of nuclease was applied in 3.2 ml of the same buffer. After 50 ml of buffer had passed through the column, 0.1 M acetic acid was added to elute the enzyme. Nuclease, 8.2 mg, and all the original activity was recovered. The flow rate was about 70 ml per hour. Data from P. Cuatrecasas, M. Wilchek, and C. B. Anfinsen, *Proc. Nat. Acad. Sci. U.S.* **61**, 636 (1968).

[24]P. Cuatrecasas, M. Wilchek, and C. B. Anfinsen, *Biochemistry* **8**, 2277 (1969).

enzyme. Very dilute solutions of enzyme should be passed through these columns at moderately slow flow rates. The columns can be used repeatedly, and over protracted periods, without detectable loss of effectiveness. These affinity adsorbents can also be used in the separation of active and inactive nuclease derivatives from samples subjected to chemical modification[25,27] (as will be described), and they are effective in stopping enzymatic reactions.[1]

Example 2: Purification of α-Chymotrypsin with an Inhibitor Containing an "Arm."[1] An agarose adsorbent containing a relatively weak competitive inhibitor, D-tryptophan methyl ester, $(K_i, 10^{-4} M)$ is relatively ineffective in adsorbing α-chymotrypsin even though 10 μmoles of inhibitor are present per milliliter of agarose (Fig. 3B). Only a slight retardation of the enzyme is observed. However, by interposing a 6-carbon chain between the same inhibitor and the agarose matrix, dramatically stronger adsorption of the enzyme occurs (Fig. 3C). This illustrates the marked steric interference that results when the ligand is attached too closely to the supporting gel. DFP-treated α-chymotrypsin (Fig. 3D), chymotrypsinogen, pancreatic ribonuclease, subtilisin, and trypsin are not adsorbed by this specific affinity adsorbent. Greater than 90% recovery is obtained on elution with distilled water titrated to pH 3 with acetic acid. It is also possible to elute the protein with a 0.018 M solution of the competitive inhibitor, β-phenylpropionamide $(K_i, 7$ m$M)$.

Example 3: Purification of DAHP Synthetase with an Effector.[28] An agarose adsorbent containing L-tyrosine (3.2 μmoles/ml), an effector of 3-deoxy-D-arabinoheptulosonate 7-phosphate (DAHP) synthetase $(K_i, 5 \times 10^{-5} M)$, was capable of separating this enzyme from a crude protein mixture (Fig. 4). As with α-chymotrypsin, the enzyme is merely retarded, and it is not necessary to change the nature of the buffer to obtain elution. Nevertheless, a very useful derivative is obtained which separates the weakly adsorbed enzyme from the bulk of the protein and from a related enzyme.

Example 4: Batch Purification of a Vitamin-Binding Protein.[29] An early example of affinity chromatography was the purification of avidin on biotin-cellulose.[13] A much more effective adsorbent for avidin can be prepared by coupling a derivative of biotin containing a 6-carbon chain, biocytin (ε-N-biotinyl-L-lysine), to Sepharose. The lysyl portion serves as an "arm" to hold the essential structural portion of the vitamin, the

[25]P. Cuatrecasas, M. Wilchek, and C. B. Anfinsen, *J. Biol. Chem.* **244**, 4316 (1969).
[26]P. Cuatrecasas, *J. Biol. Chem.* **245**, 574 (1970).
[27]H. Taniuchi and C. B. Anfinsen, *J. Biol. Chem.* **244**, 3864 (1969).
[28]W. C. Chan and M. Takahashi, *Biochem. Biophys. Res. Commun.* **37**, 272 (1969).
[29]P. Cuatrecasas and M. Wilchek, *Biochem. Biophys. Res. Commun.* **33**, 235 (1968).

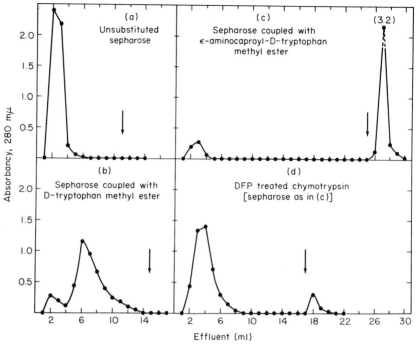

FIG. 3. Affinity chromatography of α-chymotrypsin on inhibitor-agarose columns. The columns (0.5 × 5 cm) were equilibrated and run with 50 mM Tris·HCl buffer, pH 8.0. Each sample (2.5 mg) was applied in 0.5 ml of the same buffer. Fractions of 1 ml were collected at a flow rate of about 40 ml per hour at room temperature. α-Chymotrypsin was eluted with 0.1 M acetic acid, pH 3.0 (arrows). Peaks preceding the arrows in B, C, and D were devoid of enzyme activity. In preparing the agarose adsorbent, 65 μmoles of inhibitor was added per milliliter of Sepharose, 10 μmoles of which was found to be covalently attached, yielding an effective concentration of inhibitor, in the column, of about 10 mM. Data from P. Cuatrecasas, M. Wilchek, and C. B. Anfinsen, *Proc. Nat. Acad. Sci. U.S.* 61, 636 (1968).

ureido and thiophan rings, at a good distance from the matrix. Adsorbents containing very small amounts (0.02 μmole per milliliter of agarose) of this compound which binds very tightly to avidin (K_i, 10^{-15} M) can be used to purify, in a batchwise manner, avidin from crude egg white in a single step (Table III). The binding of avidin is so strong that elution must be performed with 6 M guanidine·HCl, pH 1.5.

Example 5: Purification of Thyroxine-Binding Protein from Serum. Columns containing agarose with N-(6-aminocaproyl)thyroxine,[30] or thyroxine[30,31] itself, remove most of the thyroxine-binding globulin from serum. This

[30] R. A. Pages, H. J. Cahnmann, and J. Robbins, personal communication (1969).
[31] J. Pensky and J. S. Marshall, *Arch. Biochem. Biophys.* 135, 304 (1969).

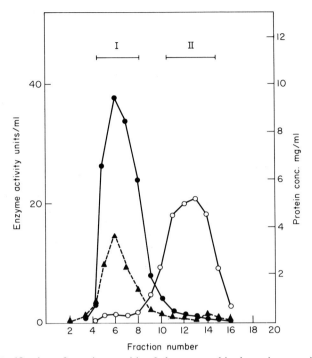

Fig. 4.   Purification of tyrosine-sensitive 3-deoxy-D-arabinoheptulosonate 7-phosphate synthetase by affinity chromatography. Chromatography of crude yeast extract on tyrosine-Sepharose column (0.6 × 15.5 cm), equilibrated with 0.2 M sodium phosphate buffer, pH 6.5 containing $CuSO_4$ (0.1 mM) and PMSF (0.1 mM). The sample consisted of 2 ml of crude extract, and 1-ml fractions were collected. ○—○, Tyrosine-sensitive DAHP synthetase; ●—●, phenylalanine-sensitive DAHP synthetase;   —   , protein concentration. Pooled fractions are indicated (I and II). Data from W. W. C. Chan and M. Takahashi, *Biochem. Biophys. Res. Commun.* **37**, 272 (1969).

protein, which has a dissociation constant of about $10^{-10}\,M$ for thyroxine and which is present at very low concentrations in serum (10–20 μg/ml), remains strongly adsorbed to the agarose after the column is washed with 0.1 M $NaHCO_3$ buffer, pH 9.0, it is eluted with 4 mM NaOH, pH 11.4.[30] Although not pure, the eluted protein is enriched considerably by this single-step procedure. Sepharose-4B, activated with 100 mg of cyanogen bromide per milliliter of Sepharose by the previously described procedures, is coupled with thyroxine or aminocaproyl-thyroxine (3 μmoles per milliliter of Sepharose) in 50% (v:v) ethylene glycol (to enhance thyroxine solubility), 50 mM $NaHCO_3$, pH 9.6; about 1.8 μmoles of thyroxine derivative is coupled per milliliter of agarose. Adsorbed [^{125}I]thyroxine is completely removed by washing with 20 volumes of the ethylene glycol buffer followed by 20 volumes of 0.1 M $NaHCO_3$, pH

TABLE III

PURIFICATION OF AVIDIN FROM EGG WHITE BY BIOCYTIN-AGAROSE CHROMATOGRAPHY

Sample	Protein (mg)	Avidin (mg)	Specific activity (μg biotin per mg protein)	Purification factor
Egg white	8000	2.2	0.0037	—
Washings of Sepharose	8000	0.2	0.0003	—
Elution from Sepharose	1.5[a]	1.6	14.6	4000

[a]Based on $E_{1\,cm}^{0.1\%}$ of 1.57 (M. D. Melamed and N. M. Green, *Biochem. J.* **89**, 591 (1963). Two milliliters of biocytin-Sepharose were added to 25 ml of 0.2 M NaHCO$_3$, pH 8.7, containing 8 g of crude, dried egg white. After a few minutes the solution was centrifuged at 20,000 g for 10 minutes. The Sepharose was suspended in bicarbonate buffer and poured onto a 1 × 5 cm column. The column was washed with about 25 ml of 3 M guanidine·HCl, pH 4.5, and the avidin was eluted with 6 M guanidine·HCl at pH 1.5. Avidin concentration was determined by addition of biotin ^{14}C, followed by dialysis for 24 hours at room temperature against large volumes of 0.1 M NaHCO$_3$ at pH 8.2; it was assumed that 13.8 μg of biotin are bound per milligram of avidin. Data from P. Cuatrecasas and M. Wilchek, *Biochem. Biophys. Res. Commun.* **33**, 235 (1968).

9.0. These studies indicated that ethylene glycol (50%, v:v), but not dimethyl sulfoxide, which was also investigated, is well tolerated by agarose.

Example 6: Resolution of Chemically Modified Enzyme Mixtures.[25-27] Studies of the functional effects of chemical modifications of purified enzymes frequently reveal incomplete loss of enzymatic activity. It is often difficult to determine whether this results from an altered protein possessing diminished catalytic properties or to residual native enzyme. In the latter case, separation of the active native and the catalytically inert proteins may not be possible. In certain cases, both problems may be resolved by using affinity adsorbents. Modification of staphylococcal nuclease by attachment of a single molecule of an affinity labeling reagent through an azo linkage to an active site tyrosyl residue, resulted in loss of 83% of the enzyme activity.[26] Chromatography of this enzyme solution on a nuclease-specific agarose column revealed that the residual activity was due entirely to a 20% contamination of native enzyme; complete resolution of the two components was possible by affinity chromatography (Fig. 5).[26] Similar separation was obtained with partially active preparations of staphylococcal nuclease modified with bromoacetamidophenyl affinity labeling reagents,[25] and residual native nuclease could be separated from an inactive peptide fragment obtained by specific tryptic cleavage.[27]

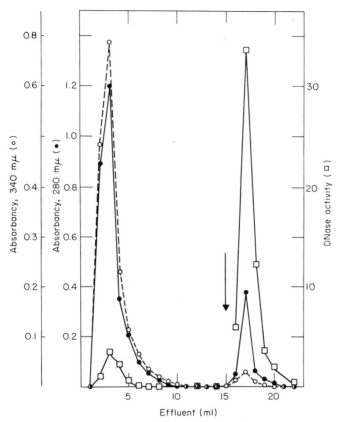

FIG. 5. Affinity chromatography on specific agarose column (Table I, Fig. 2) of staphylococcal nuclease treated with a 1.7-fold molar excess a diazonium labeling reagent derived from pdTp-aminophenyl. About 3 mg of chemically modified nuclease, containing 17% of the DNase activity of the native enzyme, were applied to a 0.5 × 7 cm column which contained Sepharose 4B conjugated with the inhibitor, pdTp-aminophenyl (0.8 μmole per milliliter of Sepharose). The column was equilibrated and developed with 50 mM borate buffer, pH 8.0, and 10 mM CaCl₂. The bound enzyme was eluted with NH₄OH, pH 11 (arrow). The small amount of enzymatic activity present in the early, unretarded peak could be removed by rechromatography of this peak through the same column. The specific activity of the small amount of protein adsorbing strongly to the column was identical with that of native nuclease. Data from P. Cuatrecasas, *J. Biol. Chem.* **245**, 574 (1970).

Example 7: Other Enzymes. Papain has been purified with an agarose adsorbent to which glycylglycyl-(o-benzyl)-L-tyrosine-L-arginine had been coupled by the cyanogen bromide procedure described earlier.[32] Bovine pancreatic ribonuclease can be purified with an agarose adsorbent

---

[32] S. Blumberg, I. Schechter, and A. Berger, *Israel J. Chem.* **7**, 125 p (1969).

containing 5'-(4-aminophenylphosphoryl)uridine-2'(3')-phosphate, prepared by the cyanogen bromide method.[33]

Preparation of and Coupling to Modified Agarose Beads

A number of stable chemical derivatives of agarose beads, which can be rapidly and easily prepared under mild aqueous conditions,[2,3,4,7] will be presented in order to provide greater versatility to the general method. Since selective adsorbents must be tailored to the special characteristics of the individual protein, and to the equally stringent ligand requirements of that protein, it is useful to have alternative methods of attaching ligands to agarose. In many cases it is much easier to modify the matrix support than the ligand. More specifically, these derivatizations provide procedures especially applicable to cases in which (1) an amino group on the ligand is not available or its synthesis is difficult; (2) hydrocarbon chains of varying length need to be interposed between the matrix and the ligand; and (3) only the mildest eluting conditions, i.e., neutral pH, absence of protein denaturants, are tolerated by the adsorbed enzyme. In the last case, the protein ligand complex is removed intact by specific cleavage of the ligand matrix bond. Some of the derivatives described below are now available commercially (Affitron Corp.).

ω-Aminoalkyl Derivatives of Agarose Beads.[2,7]

Aliphatic diamine compounds, such as ethylenediamine, may be substituted directly to agarose by the cyanogen bromide procedure described earlier (Scheme I, C). Although on theoretical grounds cross-linking might be expected to occur, in practice very little, if any, can be detected. This is probably due to the large excess of diamine added during the coupling stage, so that the amino groups on the agarose must compete with a very large excess of amino groups free in solution. It is possible, therefore, to prepare agarose derivatives having free amino groups which extend a good distance from the solid matrix, depending on the nature of the group, $-(CH_2)_x$. The ω-aminoalkyl groups can be used, in turn, for attachment of other functional groups, ligands, or proteins.

Procedure — Preparation of Aminoethyl Agarose. To an equal volume of a Sepharose-4B water suspension is added 250 mg of cyanogen bromide per milliliter of packed gel, and the reaction is performed as described above. The washed, activated Sepharose is suspended in an equal volume of cold distilled water containing 2 mmoles of ethylenediamine per

[33]M. Wilchek and M. Gorecki, *Eur. J. Biochem.* 11, 491 (1969).

milliliter of Sepharose; the pH of this solution is adjusted to 10.0 with 6 N HCl before it is added to the Sepharose. After reaction for 16 hours at 4°, the gel is washed with large volumes of distilled water. This treatment results in a derivative having about 12 μmoles of aminoethyl group per milliliter of Sepharose. Different ω-aminoalkyl derivatives can be prepared in a similar way by using other diamines. 3,3'-Diamino-dipropyl amine has been one of the most useful for a variety of purposes.[2,3,22]

Color Test with Sodium 2, 4, 6-Trinitrobenzenesulfonate (TNBS)[2]

This simple color test, modified[2] from that described by Inman and Dintzis,[21] is very useful in following the course of the various agarose and polyacrylamide derivatizations to be described. One milliliter of saturated sodium borate is added to a slurry (0.2–0.5 ml, in distilled water) of the agarose or polyacrylamide gel. Three drops of a 3% aqueous solution of sodium 2, 4, 6-trinitrobenzenesulfonate are added; the stock solution may be stored, in a brown glass bottle, for at least 2 weeks. At room temperature the color reaction of the gel beads is complete within 2 hours. The following color products are formed with various derivatives: unsubstituted agarose or polyacrylamide beads, yellow; derivatives containing primary aliphatic amines, orange; derivatives containing primary aromatic amines, red-orange; unsubstituted hydrazide derivatives, deep red; carboxylic acid and bromo-acetyl derivatives, yellow. The degree of substitution of amino gel derivatives by carboxylic acid ligands, and of hydrazide gels by amino group containing ligands, can be conveniently estimated from the relative color intensity of the washed, coupled gel.

Coupling Carboxylic Acid Ligands to Aminoethyl Agarose.[2,7]

Ligands containing free carboxyl groups may be coupled directly to the ω-aminoalkyl-Sepharose derivatives described above with a water-soluble carbodiimide (Scheme II, A). Two examples of such reactions are presented below.

Procedure a: Preparation of Estradiol-Sepharose.[34] Three hundred milligrams of 3-O-succinyl-[³H]estradiol (prepared by reacting [³H]-estradiol with succinic anhydride in 50% dimethyl formamide-pyridine; specific activity, 0.3 μCi/mole) is added in 40 ml of dimethyl formamide to 40 ml of packed aminoethyl Sepharose-4B (2 μmoles of aminoethyl groups per milliliter). The dimethyl formamide is required to solubilize estradiol. The pH of this suspension is brought to 4.7 with 1 N HCl. Five hundred milligrams (2.6 mmoles) of 1-ethyl-3-(3-dimethylamino-

---

[34]P. Cuatrecasas and G. A. Puca, unpublished observations (1970).

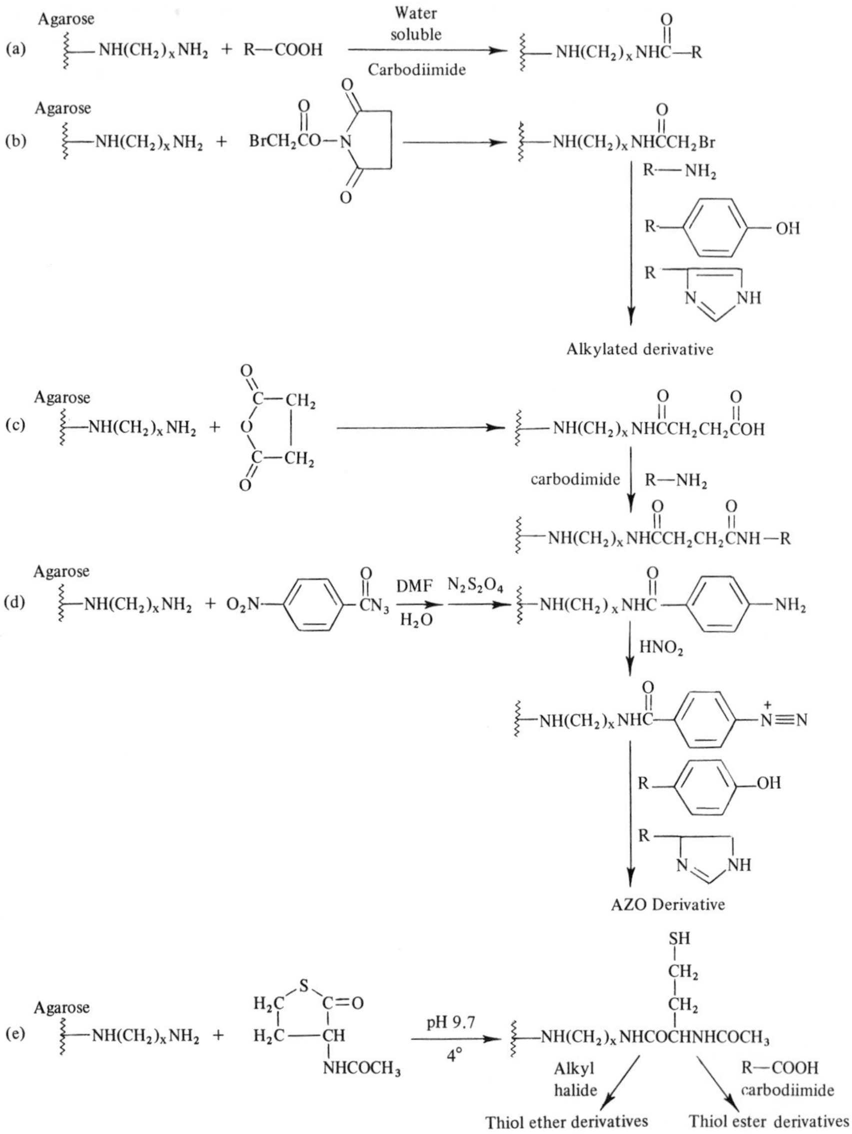

(a) Agarose $\{$—NH(CH$_2$)$_x$NH$_2$ + R—COOH $\xrightarrow[\text{Carbodiimide}]{\text{Water soluble}}$ $\{$—NH(CH$_2$)$_x$NHC(=O)—R

(b) Agarose $\{$—NH(CH$_2$)$_x$NH$_2$ + BrCH$_2$CO—N(succinimide) $\longrightarrow$ $\{$—NH(CH$_2$)$_x$NHCCH$_2$Br

    R—NH$_2$

    R—⟨⟩—OH

    R—(imidazole)N⟨⟩NH

    Alkylated derivative

(c) Agarose $\{$—NH(CH$_2$)$_x$NH$_2$ + (succinic anhydride) $\longrightarrow$ $\{$—NH(CH$_2$)$_x$NHCCH$_2$CH$_2$COH

carbodimide | R—NH$_2$

$\{$—NH(CH$_2$)$_x$NHCCH$_2$CH$_2$CNH—R

(d) Agarose $\{$—NH(CH$_2$)$_x$NH$_2$ + O$_2$N—⟨⟩—C(=O)CN$_3$ $\xrightarrow[\text{H}_2\text{O}]{\text{DMF} \ \text{N}_2\text{S}_2\text{O}_4}$ $\{$—NH(CH$_2$)$_x$NHC—⟨⟩—NH$_2$

HNO$_2$

$\{$—NH(CH$_2$)$_x$NHC—⟨⟩—N$^+$≡N

R—⟨⟩—OH

R—(imidazole)N⟨⟩NH

AZO Derivative

(e) Agarose $\{$—NH(CH$_2$)$_x$NH$_2$ + (thiazolidine H$_2$C—S C=O / H$_2$C—CH / NHCOCH$_3$) $\xrightarrow[\text{4°}]{\text{pH 9.7}}$ $\{$—NH(CH$_2$)$_x$NHCOCHNHCOCH$_3$ (with SH—CH$_2$—CH$_2$ side chain)

Alkyl halide ↙   ↘ R—COOH carbodiimide

Thiol ether derivatives   Thiol ester derivatives

SCHEME II

propyl) carbodiimide, dissolved in 3 ml of water, is added over a 5-minute period and the reaction is allowed to proceed at room temperature for 20 hours. The substituted agarose is then washed continuously, while packed in a column or on a Büchler funnel with 50% aqueous dimethyl formamide until radioactivity is absent from the effluent. The completeness of washing is checked by collecting 200 ml of effluent wash, lyophilizing, adding 2 ml of dimethyl formamide, and determining radioactivity. Before use, it is essential to incubate the agarose in an albumin buffer (3%, v/v) to determine if there is release of substances into the medium which are capable of inhibiting the binding of ^3H estradiol to the uterine binding protein, this is an excellent way of detecting non-covalently bound estradiol. About 0.5 μmole of [^3H]-estradiol are covalently bound to the agarose matrix. This is determined by counting the Sepharose in Cab-O-Sil, or by hydrolyzing the ester bond by exposure to 0.1 N NaOH for 60 minutes at room temperature, followed by centrifugation and estimating radioactivity in the super-natant fluid. With such derivatives it is possible to elute strongly ad-sorbed serum estradiol binding protein by cleaving the estradiol agarose bond with mild base, thus avoiding the use of protein denaturants. This estradiol-agarose derivative is very effective in extracting estradiol binding proteins from serum and from calf uterine extracts.[34] Estradiol has been attached to polyvinyl and cellulose polymers by different pro-cedures.[35]

Procedure b: Preparation of Organomercurial-agarose for Separation of SH-proteins.[2,7] Packed aminoethyl-Sepharose, 35 ml (capacity, 10 μmoles/ml) is suspended in a total volume of 60 ml of 40% dimethyl formamide; 2 mmoles of sodium p-chloromercuribenzoate is added; the pH is adjusted to 4.8; and 5 mmoles of 1-ethyl-3-(3-dimethyl-aminopropyl) carbodiimide are added. The pH is maintained at 4.8 by continuous titration with 0.1 N NaOH for one hour. After reaction at room temperature for another 18 hours, the substituted agarose is washed with 4 liters of 0.1 M NaHCO$_3$, pH 8.8, over an 8-hour period. Complete substitution of the agarose amino groups is confirmed by the TNBS color test, which yields a yellow gel. The derivative can bind 40–50 mg of horse hemoglobin per milliliter of packed gel. The effectiveness of the binding, compared to that of a derivative prepared by attaching a bifunctional organomercurial to SH-Sephadex G25 by much more complicated procedures,[36] is reflected in the fact that solutions of low pH (2.7, acetic acid) or complexing agents, such as EDTA (0.1 M), remove the protein only very slowly. The most effective elution is achieved by passing a 50 mM cysteine or dithiothreitol solution into the column and stopping the flow for 1 or 2 hours before collecting the effluent. Advantages of the present procedures include the ease of

[35] B. Vonderhaar and G. C. Mueller, *Biochim. Biophys. Acta* 176, 626 (1969).
[36] L. Eldjarn and E. Jellum, *Acta Chem. Scand.* 17, 2610 (1963).

preparation, the use of agarose with its inherently large capacity to bind large proteins by permitting their diffusion throughout the mesh, the circumvention of conditions which irreversibly destroy the agarose beads, i.e., drying and dioxane, and the stability of the final derivative.

Preparation of Bromoacetyl Agarose Derivatives.[2,7]

Bromoacetamidoethyl-Agarose can be prepared under mild aqueous conditions by treating aminoethyl-Sepharose-4B with O-bromoacetyl-N-hydroxysuccinimide (Scheme II, B). This derivative of agarose can react with primary aliphatic or aromatic amines as well as with imidazole and phenolic compounds. Additionally, proteins readily couple to bromoacetamidoethyl-Sepharose, forming insoluble derivatives in which the protein is located at some distance from the solid support.

Procedure. In 8 ml of dioxane are dissolved 1.0 mmole of bromoacetic acid and 1.2 mmole of N-hydroxysuccinimide. To this solution, 1.1 mmole of dicyclohexylcarbodiimide is added. After 70 minutes, dicyclohexylurea is removed by filtration, and the entire filtrate, or crystalline bromoacetyl-N-hydroxysuccinimide ester, is added, without further purification, to a suspension, at 4°, which contains 20 ml of packed aminoethyl-Sepharose (2 μmoles of amino groups per milliliter) in a total volume of 50 ml at pH 7.5 in 0.1 M sodium phosphate. After 30 minutes, the product is washed with 2 liters of cold 0.1 N NaCl. Quantitative reaction of the amino groups occurs as shown by the loss of orange color with the TNBS test. The bromoacetamidoethyl-Sepharose gel can be stored at 4° as a suspension in distilled water (pH about 5.5), or can be treated with a primary amine, R-NH$_2$ (Scheme II, B). Reaction with 10 mM pdTp-aminophenyl as a 50% (v:v) suspension in 0.1 M NaHCO$_3$, pH 8.5, for 3 days at room temperature, followed by reaction for 24 hours at room temperature with 0.2 M 2-aminoethanol to mask unreacted bromoacetyl groups, results in attachment of 0.8 μmole of inhibitor per milliliter of packed agarose. A similar reaction with [³H]5'-GMP resulted in the covalent attachment of 0.2 μmoles/ml. Such bromoacetyl-Sepharose derivatives have also been used to insolubilize [³H]estradiol for use in the purification of serum and uterine estradiol binding proteins.[34] Proteins can be similarly attached by reacting in 0.1 M NaHCO$_3$, pH 9.0, for 2 days at room temperature, or for longer periods at lower temperatures.

Preparation of Succinylaminoethyl-Agarose[2,7]

This derivative is prepared by treating aminoethyl-Sepharose with succinic anhydride in aqueous media at pH 6.0 (Scheme II, C). One

millimole of succinic anhydride is added per milliliter of packed aminoethyl-Sepharose (capacity, 8 μmoles/ml), in an equal volume of distilled
water at 4°, and the pH is raised to and maintained at 6.0 by titrating
with 20% NaOH. When no further change in pH occurs, the suspension
is left for 5 more hours at 4°. Complete reaction of the amino group
occurs as shown by the TNBS color reaction. Compounds containing
primary amino groups can be coupled at pH 5 to such carboxyl containing Sepharose derivatives in the presence of the water-soluble
carbodiimide, 1-ethyl-3-(3-dimethylaminopropyl) carbodiimide, by
the same procedures described above.

 Example: Purification of Bacterial β-Galactosidase.[22] The very "long"
diamine derivative, 3, 3′ -diaminodipropylamine, was coupled to
Sepharose (about 10 μmoles/ml of gel) by the procedure described for
the preparation of aminoethylagarose. The succinyl derivative was then
prepared as described above. One millimole of *p*-aminophenyl-β-D-
thiogalactopyranoside is added to 30 ml of the succinyl gel suspended
in 50 ml of distilled water. The pH is adjusted to 4.7, and 7.5 mmoles
of 1-ethyl-3-(3-dimethylaminopropyl) carbodiimide in 3 ml of water,
are added dropwise over a 5-minute period. The pH is maintained at
4.7 for 1 hour by titrating with 0.1 N NaOH. The suspension is stirred
at room temperature for another 16 hours, then washed with about
12 liters 0.1 N NaCl over an 8-hour period. The adsorbent, prepared
in this way, binds β-galactosidase from several bacterial sources very
strongly; elution of the protein from such a column is achieved by
passage of a buffer having a pH of 10, or with a substrate-containing
buffer. It is notable that gel derivatives prepared by attaching the same
ligand directly to agarose, or by the diazotization procedure (described
below), were ineffective in binding the enzyme. This compound, which
is a very weak competitive inhibitor with a K_i of about 5 mM, must be
attached to the solid matrix backbone by a rather long "arm."

*Preparation of p-Aminobenzamidoethyl-Agarose for Coupling Compounds
via Azo Linkage.*[2,7]

 Diazonium derivatives of agarose, which are capable of reacting with
phenolic and histidyl compounds, can be prepared under mild aqueous
conditions from *p*-aminobenzamidoethyl-Sepharose (Scheme II, D).
The azo-substituted agarose gels retain the properties of good flow
rate, porosity, capacity, and stability, all of which are necessary for
effective affinity chromatography.

 Procedure. Aminoethyl-Sepharose, in 0.2 M sodium borate, pH 9.3,
and 40% dimethyl formamide (v:v), is treated for 4 hours at room
temperature with 70 mM *p*-nitrobenzoyl azide (Eastman). The sub-

stitution is complete, as judged by the loss of color reaction with TNBS. The p-nitrobenzamidoethyl-Sepharose is washed extensively with 50% dimethyl formamide and is reduced by reaction for 60 minutes at 40° with 0.2 M sodium dithionite in 0.5 M NaHCO₃ at pH 8.5. The p-aminobenzamidoethyl-Sepharose derivative, in 0.5 N HCl, can be diazotized by treating for 7 minutes at 4° with 0.1 M sodium nitrite. This diazonium-Sepharose derivative may be utilized at once, without further washing, by adding the phenolic, histidyl, or protein component to be coupled in a strong buffer such as saturated sodium borate, and adjusting the pH with NaOH to 8 (histidyl) or 9.2 (phenolic); the reaction is allowed to proceed for about 16 hours at 4°. Strong tertiary amines, such as triethylamine, should not be used as buffers during the azotization reaction since they can react with the diazonium intermediate. Agarose derivatives containing estradiol in azo linkage at C-2 and C-4 of the A ring remove estradiol binding proteins from human serum and from calf uterine extracts.[34]

An important advantage of the azo-linked ligand derivatives of agarose or of polyacrylamide beads is that passage of a solution of 0.1 M sodium dithionite in 0.2 M sodium borate at pH 9, through such columns causes rapid and complete release of the bound inhibitor by reducing the azo bond. This allows easy and accurate estimation of the quantity of inhibitor bound to the gel. Of greater importance, the procedure permits elution of the intact protein-inhibitor complex under relatively mild conditions. For example, serum estradiol binding protein is denatured irreversibly by exposure to pH 3 or 11.5, and by low concentrations of guanidine hydrochloride (3 M) or urea (4 M). The protein, which binds estradiol very tightly (K_i about 10^{-9} M), can be removed in active form from the agarose-estradiol gel by reductive cleavage of the azo link with dithionite, but not with buffers of low pH or with guanidine·HCl.[34]

Preparation of Tyrosyl-Agarose for Coupling Diazotized Compounds[2,7]

A tripeptide containing a COOH-terminal tyrosine residue can be attached to agarose by the cyanogen bromide procedure described earlier. Ligands containing an aromatic amine, which can be diazotized, can then be coupled in high yield through azo linkages to the tyrosyl moiety (Scheme III). The ligand thus extends considerably from the matrix backbone. An example of such a coupling procedure which has been used to insolubilize deoxythymidine 5′-phosphate 3′-p-aminophenylphosphate (for staphylococcal nuclease) and p-aminophenyl-β-D-thiogalactoside (for β-galactosidase), and which is applicable to many other compounds is the following: 100 μmoles of inhibitor is

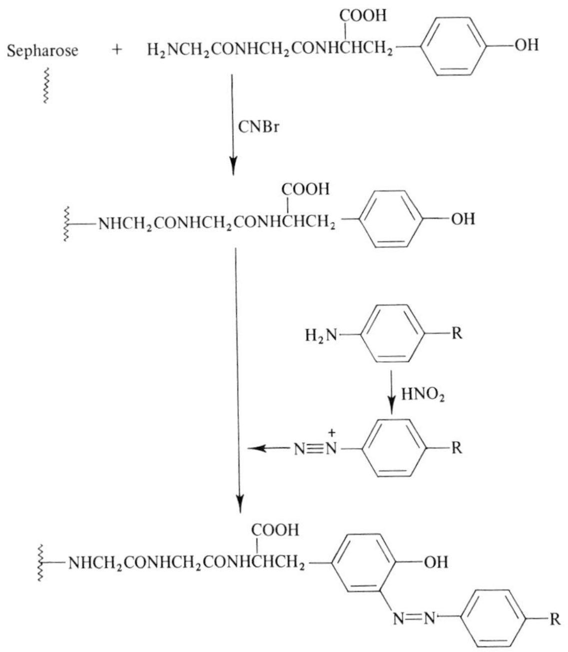

SCHEME III

dissolved in 1.5 ml of 1.5 N HCl, and placed in an ice bath; 700 μmoles of $NaNO_2$, in 0.5 ml of water, is added over a 1-minute period to the stirred ligand solution. After 7–8 minutes the entire mixture, without further purification, is added to a rapidly mixing suspension, in an ice bath, of tyrosyl-Sepharose containing 0.2 M Na_2CO_3. The pH is rapidly adjusted to 9.4. After 3 hours the agarose suspension is transferred to a Büchner funnel and washed. If dimethyl formamide is to be used in the reaction mixture, the buffer should be sodium borate in order to prevent precipitation of the salt. Staphylococcal nuclease is readily purified in one step by passage through a column of such a derivative containing the competitive inhibitor, pdTp-aminophenyl, in azo linkage. Congo red dye has also been attached to Sepharose in this manner; small amounts of amyloid protein are agarose adsorbed to this matrix after passage of crude material dissolved in 4 M guanidine· HCl at pH 6; elution is achieved with 6 M guanidine·HCl at pH 3. These ligand-protein complexes can be removed by treating the gel with sodium dithionite, as described earlier for other gels containing azo-linked ligands.

Preparation of Sulfhydryl Agarose Derivatives[2]

Thiol groups can be introduced into ω-aminoalkyl agarose derivatives by reaction with homocysteine thiolactones by using procedures similar to those used for thiolation of proteins (Scheme II, E).[37,38] Five grams of N-acetylhomocysteine thiolactone is added to a cold (4°) suspension consisting of an equal volume (50 ml) of aminoethyl Sepharose 4-B and of 1.0 M NaHCO$_3$, pH 9.7. The suspension is stirred gently for 24 hours at 4°, then washed with 8 liters of 0.1 N NaCl. The agarose derivative is then incubated in 0.05 M dthiothreitol, 0.5 M Tris·HCl buffer, pH 8.0, for 30 minutes at room temperature. The suspension is washed with 2 liters of 0.1 M sodium acetate buffer, pH 5.0. At this stage a strong red-brown color is obtained with the sodium trinitrobenzenesulfonate color test. The color produced with this reagent is completely lost after treating a sample of the thiol agarose with 50 mM iodoacetamide in 0.5 M NaHCO$_3$, pH 8.0, for 15 minutes at room temperature, followed by washing with distilled water. This confirms that complete substitution of the amino-agarose groups has occurred. The derivative can be stored for many months as an aqueous suspension at 4° provided that it is treated, before use, with dithiothreitol as described above.

The degree of sulfhydryl substitution can be conveniently determined by reacting the thiol-agarose with 5,5′-dithio-bis-(2-nitrobenzoic acid)[39] at pH 8, followed by centrifugation and determination of the absorbancy at 412 mμ in the supernatant. By this procedure, the derivative described above contains 0.58 μmole of sulfhydryl per milliliter of packed agarose. Reaction with 10 mM [^{14}C]iodoacetamide in 0.1 M NaHCO$_3$, pH 8.0, for 15 minutes at room temperature, results in uptake of 0.65 μmole of radioactivity per milliliter of agarose.

Ligands or proteins that contain alkyl halides can be readily coupled to sulfhydryl agarose derivatives through stable thioether bonds. The reactivity of the thiol gels with heavy metals may be of special value in affinity chromatography. Furthermore, agarose derivatives containing free sulfhydryl groups can be used to couple ligands by thiol ester linkage, as described below.

Coupling Ligands to Agarose by Thiol Ester Bonds.[2] Ligands containing a free carboxyl group can be coupled to sulfhydryl-agarose with water soluble carbodiimides. Although the thiol ester bonds thus formed are stable at neutral pH, cleavage is readily achieved by exposure for a few minutes to pH 11.5, or by treatment with 1 N hydroxylamine for about 30 minutes. Thus, it is possible by specific chemical cleavage to remove the intact ligand–protein complex from an adsorbent containing the specifically bound protein.

[37] R. Benesch and R. E. Benesch, *J. Amer. Chem. Soc.* **78**, 1597 (1956).
[38] R. Benesch and R. E. Benesch, *Proc. Nat. Acad. Sci. U. S.* **44**, 848 (1958).
[39] G. L. Ellman, *Arch. Biochem. Biophys.* **82**, 70 (1959).

N-Acetylglycine (0.2 *M*) is added to 40 ml of packed sulfhydryl-agarose (containing 0.7 μmole of thiol per milliliter of packed agarose) adjusted to 80 ml with distilled water. The pH is adjusted to 4.7, and 1.5 g of 1-ethyl-3-(3-dimethylaminopropyl) carbodiimide, dissolved in 3 ml of water, is added. The pH is maintained at 4.7 for 1 hour by continuous titration. The suspension is allowed to stir gently for 8–12 hours at room temperature. The agarose is then washed with 2 liters of 0.1 *N* NaCl. At this stage virtually no color is obtained with the sodium trinitrobenzene sulfonate color test, and the sulfhydryl content, determined with 5,5'-dithiobis-(2-nitrobenzoic acid),[39] is about 0.03 μmole per milliliter of Sepharose. All the sulfhydryl groups remaining free on the Sepharose are masked by treatment with 0.02 *M* iodoacetamide for 20 minutes at room temperature in 0.2 *M* NaHCO$_3$, pH 8.0. 3-*O*-Succinylestradiol has been coupled to sulfhydryl-Sepharose by these procedures. The resulting adsorbent is very effective in binding estradiol binding proteins from serum and uterine extracts.[34]

Preparation of Derivatives of Polyacrylamide Beads

As discussed earlier, the principal advantage of polyacrylamide beads over agarose is that a much higher degree of substitution is possible with the former. However, the considerably lower porosity of currently available beads may limit their use in the affinity chromatography of very large proteins. Another consideration is that some shrinkage of the gels occurs during formation of the acyl azide intermediate[2]; shrinkage would be expected to lead to decreased porosity. Thus, it may be possible to prepare an adsorbent containing a very high concentration of ligand, much of which is inaccessible to the protein to be purified.[22] Only the most porous beads, Bio-Gel P-300 (Bio-Rad Labs), have been found effective for the purification of staphylococcal nuclease, a protein with a molecular weight of 17,000. The procedures, based on those described by Inman and Dintzis,[21] described below, involve conversion of the carboxamide side groups to hydrazide groups which in turn are converted with nitrous acid into azyl azide derivatives, thereby allowing ready reaction with aliphatic and aromatic primary amines.

Preparation of Hydrazide Derivatives (Scheme IV, *1*).[2,21] Swollen polyacrylamide beads, in a volume of water equal to 1.5 times the value of the hydrated gel and containing 3–6 *M* hydrazine hydrate are heated in a constant temperature bath to between 45° and 50°. The procedure is carried out under constant stirring in a stoppered vessel in a hood. The time required for reaction is a function of the degree of derivatization desired; the amount of substitution varies linearly for at least an 8-hour period (Fig. 6). Thereafter, the hydrazide-acrylamide

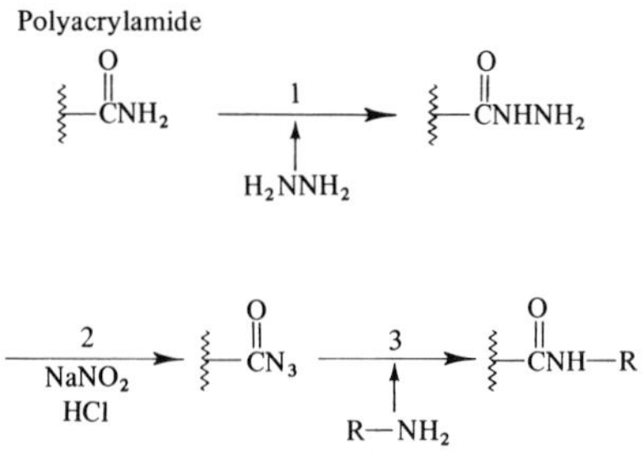

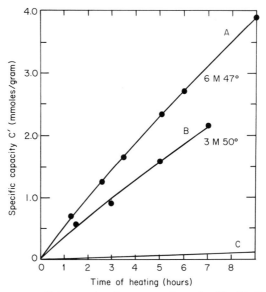

SCHEME IV

derivative is washed in a hood with large volumes of 0.1 M NaCl until the washes are free of hydrazine as determined by the TNBS color reaction. Bio-Gel P-300 derivatives are conveniently washed by repeated decantation using large volumes of saline, rather than on a Büchner funnel with its slower flow rate. Approximately 30 μmoles of hydrazide per milliliter of packed resin (Bio-Gel P-300) are formed after treatment for 3 hours at 47° in 3 M hydrazine. Greater substi-

FIG. 6. Time course of the reaction between polyacrylamide Bio-Gel P-60, 100–200 Mesh) and aqueous hydrazine. Molar concentrations of hydrazine and temperature are indicated. Curve C is the average time course of carboxyl group formation during the runs for curves A and B. Specific capacity, C', refers to mmoles of hydrazide group found on the amount of derivative produced from 1 g of the original dry polyacrylamide. Similar curves are to be expected for Bio-Gel beads other than P-60. Data from J. K. Inman and H. M. Dintzis, *Biochemistry* 8, 4074 (1969).

tution on Bio-Gel P-300 is not recommended because the physical properties of the beads are adversely affected. The polyacrylamide-hydrazide derivatives may be stored in aqueous media in the presence of a preservative, such as sodium azide or toluene, for at least 5 months.

Coupling of Primary Amines via the Azyl Azide Procedure (Scheme IV, 2, 3).[3,18] Primary aliphatic or aromatic amines can be coupled via the azyl azide derivative without intermediate washings or transfers. The hydrazide polymer (packed volume, 50 ml) is washed, and suspended in twice its volume of 0.3N HCl and the entire suspension is placed in an ice bath. Ten milliliter of 1.0 M NaNO$_2$ are added rapidly under vigorous magnetic stirring. After 90 seconds, the amine is added rapidly in about 20 ml of cold 0.2 M Na$_2$CO$_3$, and the pH is raised to 9.4 with 4 M NaOH. After reaction for 2 hours at 4°, the suspension is washed with distilled water and suspended for 5 hours in 2 M NH$_4$Cl–1 M NH$_4$OH, pH 8.8, in order to convert unreacted hydrazide to the carboxamide form. If a high degree of substitution is desired, the amine should be added in amounts 10- to 20-fold greater than that of the hydrazide content of the gel.

Other Polyacrylamide Derivatizations.[2] In general, the specific polyacrylamide adsorbents are more useful when the ligand is attached at some distance from the matrix. For this purpose ω-aminoalkyl, bromoacetamidoethyl, Gly-Gly-Tyr, and p-aminobenzamido ethyl derivatives can be prepared by procedures identical to those described earlier. Very satisfactory affinity adsorbents have been prepared for the purification of staphylococcal nuclease by many of these procedures. A derivative obtained by coupling pdTp-aminophenyl directly to Bio-Gel P-300 via the acyl azide step, which contained 12 μmoles of inhibitor per milliliter of gel, effectively adsorbed 22 mg of nuclease per milliliter of resin from a crude enzyme preparation. Enzyme was recovered by elution with NH$_4$OH at pH 11. Other ligands which have been coupled to polyacrylamide beads include Congo red, p-aminophenyl-β-D-thiogalactopyranoside, 5'-GMP, estradiol, adenosine 3', 5'-cyclic monophosphate, and several amino acids.

Coupling of Proteins and Peptides to Agarose for Affinity Chromatography

The previously described desirable features of agarose, particularly its porosity, permit the preparation of insoluble protein derivatives which can be used to exploit reversible macromolecular interactions. In addition to attaching proteins or peptides to agarose directly by the cyanogen bromide procedure, it is possible to use the bromoacetyl or the diazonium derivatives described earlier. Proteins coupled to such gels extend some distance from the matrix backbone by an "arm." This may be important in decreasing steric difficulties when interactions with other macromolecules are being studied. For example, purification or isolation of intact cells or particulate cell structures by insolubilized proteins may best be achieved with such derivatives.[3]

An important consideration in the covalent attachment of a biologically active protein to an insoluble support is that the protein should be attached to the matrix by the fewest possible bonds. This will enhance the probability that the attached macromolecule will retain its native tertiary structure. Proteins react with cyanogen bromide-activated agarose through the unprotonated form of their free amino groups. Since most proteins are richly endowed with lysyl residues, which are exposed to the solvent, it is likely that such molecules will have multiple points of attachment to the resin when the coupling reaction is performed at pH 9 or higher. The number of linkage points and the resultant biologic activity of the insolubilized protein can be controlled by manipulating the pH of the coupling reaction. Proteins having few amino groups, e.g., insulin, can be preferentially attached to agarose through the α-amino group by coupling at a relatively low pH (5.0) and high ionic strength.[40] Proteins with a large number of lysyl residues, and any protein of high molecular weight, yield much more efficient adsorbents when coupled to agarose at pH values in the neighborhood of 6.0–6.5 (see example 2, Fig. 7).[2,3] Only a small fraction of the lysyl residues, and the α-amino groups, will be significantly nonionized and thus capable of reaction with the cyanogen bromide-activated agarose.

In spite of the low pH, adsorbents containing a large amount of protein can still be obtained by the procedures described above by (1) using agarose activated with relatively large amounts of cyanogen bromide (250 mg per milliliter of packed Sepharose), (2) performing the coupling procedure at pH 6, and (3) using high protein concentrations during the coupling reaction.[2,3]

Because of its greater porosity, Sepharose 2B should be used for studies involving the isolation of particulate structures, such as cells, cell membranes, or viruses. Protein-agarose derivatives are quite stable when stored in suspension at 4°, and it is generally possible to use them repeatedly.

It should be stressed that when the biological activity of insoluble ligands or proteins is to be studied, special care must be taken to ensure *complete* removal of adsorbed, noncovalently bound material. This may require long and tedious washing procedures. It is recommended that, whenever possible, insolubilized protein derivatives be washed ex-

[40] P. Cuatrecasas, *Proc. Nat. Acad. Sci. U. S.* **63**, 450 (1969).

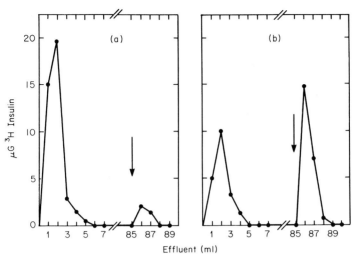

FIG. 7. Affinity chromatography patterns of [³H]insulin on columns containing purified antiporcine insulin sheep immunoglobulin which had been coupled by the cyanogen bromide procedure to agarose at pH 9 (A) and at pH 6.5 (B). Forty milligrams of purified γ-globulin, in 10 ml of 0.2 M sodium citrate, pH 6.5, or 8 ml of 0.2 M sodium bicarbonate, pH 9.0, were added to 5 ml of Sepharose 4B activated with 300 mg of cyanogen bromide per milliliter of gel. Based on the recovery of absorbancy (280 mμ) in the washes, about 90–95% of the protein was coupled in both cases. The specific adsorbent was washed with 20 times its volume of 6 M guanidine·HCl before use. [³H]Acetyl insulin, 40 μg, in 1 ml of 50 mM borate buffer at pH 8.5 containing 0.1 N NaCl, and 0.5% bovine albumin, was applied to a 12 × 0.6 cm column. The column was then washed with 85 ml of the above buffer, and elution of bound insulin was accomplished with 6 M guanidine·HCl (arrow). The theoretical capacity for insulin binding, based on the amount of antibody covalently linked and on the capacity of the antibody to bind insulin in solution, was about 30 μg for both adsorbents. Column A adsorbed 2 μg of insulin, or 7% of its theoretical capacity, whereas column B adsorbed 23 μg, or 77% its theoretical capacity.

tensively with 6 M guanidine·HCl. To further ensure that the biological function being measured is due to material covalently attached, the effect of varying the concentration of the insoluble derivative should be studied, and the possible release of the material must be examined after incubation in buffers containing high concentrations of albumin or other substances that simulate the biological environment.[40]

Example 1: Purification of Antibodies. Protein-agarose and haptene-agarose derivatives appear to be nearly ideally suited for use as immuno-sorbents.[41–43] For example, porcine insulin linked to agarose 2B through its single lysyl residue (B-29) can completely remove insulin antibodies from crude mixtures.[41] The very strong binding of antibody by this

[41] P. Cuatrecasas, *Biochem. Biophys. Res. Commun.* **35**, 531 (1969).
[42] L. Wofsy and B. Burr, *J. Immunol.* **103**, 380 (1969).
[43] G. Omenn, D. Ontjes, and C. B. Anfinsen, *Nature (London)* **225**, 189 (1970).

derivative is reflected by the need to use strong acid conditions (1 N HCl) to achieve elution of the major fraction of the antibody from the column. Recovery of antibody applied to such a column approaches 80%. Large-scale purification of sheep antiporcine insulin is readily achieved with these adsorbents. A column containing 20 ml of packed insulin-Sepharose described in Fig. 7 removed all the insulin antibody present in 200 ml of crude immune sheep serum. The combined washes of equilibrating buffer (500 ml) and of 0.1 M sodium acetate, pH 5.5 (1 liter), contained only 5% of the total antibody applied to the column. About 65% (150 mg) of the total insulin antibody applied was eluted sharply with acetic acid at pH 2.8 in a volume of 29 ml. Subsequent elution with 6 M guanidine·HCl removed about 15% of the antibody applied.

Example 2: Immunoglobulin-Sepharose for Purification of Antigens.[44] Purified insulin antibodies, obtained by affinity chromatography as described above, can in turn be covalently attached to agarose to prepare derivatives capable of isolating the corresponding antigen and other immunologically cross-reacting proteins. The studies on anti-insulin shown in Fig. 7 illustrate the importance of attaching large proteins by few linkage points. The immunoglobulin coupled to activated agarose at pH 9 could adsorb only 7% of its theoretical capacity for insulin, whereas that coupled at pH 6.5 could bind 77% of its theoretical capacity for insulin. Since the total protein content of both derivatives is the same, the former derivative must contain immunoglobulin which is incapable of effectively binding antigen.

Example 3: Polypeptide Hormones and Isolation of Cellular Receptor Structures. Insolubilized ligands or proteins can also be used to study interactions of various molecules with intact cells, and to explore membrane phenomena. Although the macromolecules involved were not attached *covalently* to the supporting matrix, the recent studies of Wigzell and Andersson[45] on the fractionation of immunologically active cells on columns of glass or plastic beads to which antigen coatings were tightly adsorbed indicate the power of the approach. Similar studies on the immunoadsorption of cells on reticulated polyester polyurethane foam-coated with antibodies have been reported by Evans, Wage, and Peterson.[46]

Insulin-agarose derivatives are capable of stimulating several different biological properties of intact isolated fat cells nearly as effectively as native insulin (Fig. 8).[40] Columns prepared with these insulin-agarose derivatives, in contrast to those with unsubstituted

[44]P. Buatrecasas; unpublished observations (1970).
[45]H. Wigzell and B. Andersson, *J. Exp. Med.* **129,** 23 (1969).
[46]W. H. Evans, M. G. Wage, and E. A. Peterson, *J. Immunol.* **102,** 899 (1969).

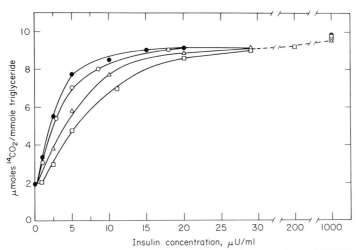

FIG. 8. Effect of insulin-agarose concentration on the oxidation of [U^{14}C] glucose by isolated adipose cells. Fat cells (3–8 μmoles of triglyceride) were incubated for 2 hours at 37° in 2 ml of medium containing 0.5 mM [U^{14}C] (0.1 μCi/μmole). The data presented are for native insulin (●), insulin coupled to Sepharose 2B (at pH 9) through lysine B29 (○), insulin coupled (at pH 5) through phenylalanine B1 (△), and insulin acetylated at positions A1 and B1 and coupled to agarose by lysine B29. Data from P. Cuatrecasas, *Proc. Nat. Acad. Sci. U. S.* **63**, 450 (1969).

agarose, bind intact fat cell "ghosts" very strongly.[44] The latter can be specifically eluted intact from the insulin-agarose columns with concentrated insulin solutions. These studies indicate that such insoluble hormone derivatives interact very effectively with membrane structures, and thus are promising tools for the isolation of "receptor" proteins (hormone, drug, etc.).

Example 4: Purification of Histidyl-tRNA with Enzyme-Sepharose.[47] Five milliliters of packed Sepharose 4B, activated with 1.3 g of cyanogen bromide, is reacted with 1.8 mg of purified histidyl-tRNA-synthetase (lacking other acyl tRNA-synthetase activities) in 50 mM phosphate buffer, pH 6.5. Virtually all the protein is coupled to Sepharose, since the washes contain essentially no absorbance at 280 mμ and only 1–3% of the enzyme activity is present. Columns containing this adsorbent strongly adsorb histidyl-tRNA, which can be eluted with 1 M NaCl (Fig. 9).

Example 5: Purification of Peptides Prepared by Organic Synthesis. A polypeptide corresponding to the amino acid sequence from residue 6 through 47 in staphylococcal nuclease (149 amino acids), obtained by organic synthesis, can be purified by passing the crude synthetic poly-

[47]F. Blasi and R. Goldberger, personal communication (1969).

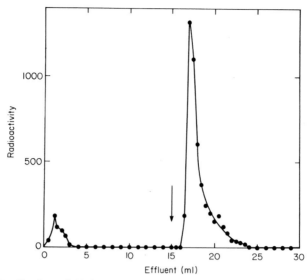

Fig. 9.   Purification of histidyl-tRNA on a 0.6 × 8 cm column containing agarose coupled with purified histidyl-tRNA-synthetase (0.36 mg/ml Sepharose). The column was equilibrated with 66 mM potassium cacodylate, pH 7.0, containing 8 mM reduced gluta-thione and 10 mM MgCl₂. About 0.4 optical density unit of [³H]histidyl-tRNA was applied to the column in 0.2 ml of the above buffer. Elution was obtained with 3.0 M NaCl, acetic acid pH 3; it is also possible to elute with 1 N NaCl. Data from F. Blasi and R. Goldberger, unpublished.

peptide mixture through a column containing a Sepharose-linked native peptide which comprises residues 49–149.[48] In solution these two peptides separately have no discernible tertiary structure or en-zymatic activity, but upon mixing there is regeneration of enzymatic activity. Advantage of the affinity of these two peptides is taken to ob-tain a "functional" purification of the synthetic peptide. In a similar manner, it is possible to purify synthetic RNase-S-peptide derivatives on agarose columns containing RNase-S-protein (Fig. 1).[5] The specific activity of the purified peptide is increased by this procedure, but the chemical and catalytic properties of the purified materials suggest the presence of closely related synthetic side products which bind tightly to the S-protein conjugate but which yield enzymatically in-active complexes with RNase-S-protein.

[48]D. Ontjes and C. B. Anfinsen, J. Biol. Chem. 244, 6316 (1969).

[32] Specific Elution with Substrate

By Burton M. Pogell and M. G. Sarngadharan

There are now numerous reports in the literature of the advantageous use of specific enzyme-substrate interaction for the purification of enzymes.[1] The present article summarizes recent successful applications of this principle and describes in detail two modified procedures that have been developed for the large-scale purification of rabbit liver fructose-1,6-diphosphatase (FDPase).[2] One of these methods utilizes elution of the enzyme from CM-cellulose by a concave salt gradient containing dilute substrate. This approach may be of general applicability in enzyme purification wherein the combined effects of increasing ionic strength and the affinity of a substrate, activator, or inhibitor for the enzyme bring about selective enzyme elution. The second procedure is more specific and serves to illustrate a case of ideal adaptation of the principle. Homogeneous FDPase was obtained in large quantities by a single adsorption and elution from CM-cellulose.

General Examples of Enzyme Purification

Noteworthy recent examples of application of the principle of specific enzyme-ligand affinity for enzyme purifications include the following:

(a) The purification of chicken pancreatic ribonuclease by elution from phosphocellulose with RNA.[3] A 60- to 70-fold increase in specific activity was achieved with a recovery of more than half of the applied activity. A very marked specificity was observed in these studies, since no elution was obtained with 5′-nucleotides, 3′- and 2′-nucleotides, alkali-digested RNA, or chicken pancreatic ribonuclease-digested RNA.

(b) The purification of rat liver pyruvate kinase by elution with fructose-1,6-diphosphate (FDP).[4] This case is particularly interesting, since it represents specific purification by elution with an allosteric modifier. A 30-fold purification with 60% recovery was achieved by elution from CM-cellulose at pH 6 with 0.5 mM FDP. The high specificity for elution of the kinase by either a positive effector (FDP) or a negative effector and

---

[1]B. M. Pogell, this series, Vol. IX [2].

[2]M. G. Sarngadharan, A. Watanabe, and B. M. Pogell, *J. Biol. Chem.* 245, 1926 (1970).

[3]J. Eley, *Biochemistry* 8, 1502 (1969).

[4]H. Carminatti, E. Rozengurt, and L. Jimenez de Asua, *Fed. Eur. Biol. Soc. Lett.* 4, 307 (1969).

product (ATP) was clearly established. A complete separation of pyruvate kinase from FDPase was also accomplished by prior elution of the phosphatase with a combination of AMP and a low concentration of FDP before elution of the kinase with 1 mM FDP.

(c) The purification of glucose-6-phosphate dehydrogenase from rat liver and human erythrocytes employing, as a major step, elution with glucose-6-phosphate from CM-cellulose[5] and CM-Sephadex,[6] respectively. In the latter case, 91% of the applied activity was recovered with a 51-fold increase in specific activity by elution with 2 mM substrate at pH 6. It is now possible to obtain pure enzyme from relatively small amounts of blood, thus making practicable structural studies on genetic variants of the enzyme in humans.

(d) The isolation of lysozyme from *Escherichia coli* by specific binding to a polysaccharide.[7] The enzyme was tightly bound to a column of chitin, a substrate, and completely separated from an acid polysaccharide inhibitor present in the bacteria. Also, the enzyme was markedly concentrated in this process, and the authors were able to demonstrate the presence of lysozyme in *E. coli* cells within 15 seconds after infection with T2 bacteriophage and to establish that, indeed, lysozyme from the phage entered the infected cells.

Purification of Rabbit Liver Fructose-1,6-Diphosphatase

Assay Method

See Sarngadharan *et al.*[2] and Pontremoli.[8] All reagents for enzyme assay and purification are prepared in glass-distilled water.

Definition of Unit and Specific Activity. A unit of enzyme is defined as the amount that will hydrolyze 1 μmole of fructose-1,6-diphosphate (FDP) per minute at pH 9.3 and 22°. Protein concentrations were determined assuming an extinction coefficient of 0.890 at 280 mμ for a 0.1% solution.[8] Specific activity is expressed as units of enzyme per milligram of protein.

Generalized Procedure

Principle. CM-cellulose is saturated with enzyme by the batchwise addition of dialyzed liver supernatant at pH 6. The column is eluted at pH 6 with a concave salt gradient containing substrate. As the salt concentration increases in the gradient, a point will be reached where the interac-

[5]T. Matsuda and Y. Yugari, *J. Biochem. (Tokyo)* **61**, 535 (1967).
[6]M. C. Rattazzi, *Biochim. Biophys. Acta* **181**, 1 (1969).
[7]I. F. Pryme, P. E. Joner, and H. B. Jensen, *Biochem. Biophys. Res. Commun.* **36**, 676 (1969).
[8]S. Pontremoli, this series, Vol. IX [112a].

tion of substrate and enzyme becomes sufficient to produce specific elution of enzyme from the column.

Step 1. Preparation of pH 6 Supernatant. Normal rabbits, fasted overnight, are killed by blows on the head and exsanguinated. The livers (ca. 500 g) are quickly removed and placed in ice-cold 154 mM KCl. All further operations are carried out at 0–4°. The livers are homogenized for 2 minutes at medium speed in a Waring Blendor (1 gallon capacity) in 4 volumes of 154 mM KCl per gram of tissue. The homogenate is centrifuged for 60 minutes at 19,000 rpm in a Spinco type 19 rotor (53,700 g at tip). The supernatant is removed by means of a large syringe, passed through glass wool to trap lipid particles, and thoroughly dialyzed against distilled water. Dialysis of 1–1.5 liters of supernatant against five changes of 25–30 liters of water is usually sufficient to lower the Cl⁻ content to a level where no precipitate is obtained upon addition of $AgNO_3$ to a solution cleared of protein by addition of perchloric acid.

One-hundredth volume of 0.5 M sodium malonate (pH 6.0 at 1:100 dilution) is added to the dialyzed solution, and the pH is adjusted to 6.0 with glacial acetic acid. After standing for 30 minutes with occasional stirring, the suspension is centrifuged for 30 minutes at 9500 rpm in a Sorvall GSA rotor (14,600 g at tip), and the clear supernatant is collected.

Step 2. CM-Cellulose Column Chromatography. Fifty grams, wet weight, of CM-cellulose (Whatman, CM-52), equilibrated with 5 mM sodium malonate buffer, pH 6.0, is stirred mechancially in 244 ml of the same buffer, while 426 ml of freshly centrifuged pH 6 supernatant is added dropwise from a separatory funnel. The stirring is continued overnight.

A sufficient amount of enzyme solution should be added so that the CM-cellulose is fully saturated with enzyme. The exact quantity necessary varies slightly with different preparations and must be determined empirically. In general, saturation is assumed when 4–5% of residual activity remains in the supernatant from the slurry after removal of the cellulose fines by centrifugation. If larger quantities of enzyme are unadsorbed, further addition of equilibrated CM-cellulose is necessary.

The slurry is poured into a large, coarse sintered-glass funnel, and most of the liquid is removed by gentle suction. The deep red cake is then stirred in the funnel with 5 mM sodium malonate, pH 6.0, and the washings are allowed to drain without suction. Most of the remaining liquid is removed by a short application of suction. Washing is repeated until no more protein is removed ($A_{280} < 0.01$ in filtrate). (In the experiment reported here, the CM-cellulose and washings were separated by centrifugation. The slurry was suspended in 500 ml of buffer, centrifuged, and the supernatant was removed by decantation. Washing was repeated 10 times.)

A second portion of CM-cellulose (25 g wet weight), equilibrated with 5 mM sodium malonate, pH 6.0, is packed over a thin pad of glass wool in a 2.8 cm diameter column. The washed CM-cellulose-enzyme slurry is layered over this bed of cellulose and allowed to settle. A thin pad of glass wool is placed on top to stabilize the surface of the slurry. The column is eluted with substrate in a concave NaCl gradient at a flow rate of 30 ml/hour. A 9-chamber rectangular Varigrad (Buchler Instruments, Inc., Fort Lee, New Jersey; 4-liter capacity) containing a total volume of 2.25 liters is used to form the gradient. The first 6 chambers contain 0.2 mM FDP in 5 mM sodium malonate, pH 6.0, and the last 3 chambers contain 0.2 mM FDP plus 0.5 M NaCl in 5 mM malonate buffer. Typical results obtained are shown in Fig. 1. FDPase appeared in the effluent as a sharp symmetrical peak at a low salt concentration (0.03 M NaCl). Fractions 48–50 contained 42% of the added FDPase which had been purified 194-fold to an average specific activity of 7.5. Total enzyme recovery was 86%. In general, the specific activity of over half of the enzyme was increased more than 200-fold in such columns. Pooled fractions were lyophilized and could be further purified by Sephadex G-200 gel filtration (see Step 3 of specific procedure, below).

A major problem experienced with this approach was the simultaneous elution of some nonspecific protein with the enzyme. This difficulty was

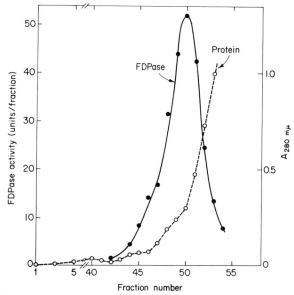

FIG. 1. Purification of FDPase on CM-cellulose by substrate elution in a concave NaCl gradient. Fraction volumes were 21 ml. ●—●, enzyme activity; ○ - - - ○, $A_{280m\mu}$.

overcome in the specific procedure described below, which yielded homogeneous FDPase by elution with dilute substrate at pH 6.8.

Specific Procedure

Principle. CM-cellulose is saturated with enzyme as in the generalized procedure. Increasing the pH of the slurry to 6.8[8] serves to weaken the binding of enzyme to the ion exchanger. A dilute solution of substrate will now directly elute the FDPase in pure form.

Step 1. Same as in generalized procedure.

Step 2. CM-Cellulose Column Chromatography. Two hundred and ten grams (wet weight) of CM-cellulose is saturated with pH 6 supernatant (1660 ml) and washed as described under the general procedure, step 2. Malonate buffer (5 mM, pH 6.8) is substituted for the buffer of lower pH. Washing is continued over a period of 3 days until no more protein is removed ($A_{280} < 0.01$ in filtrate). It is apparently difficult to obtain complete equilibrium between proteins adsorbed on CM-cellulose and in the water layer at this stage. Therefore, standing overnight in buffer by stoppering the sintered-glass funnel effects a further elution of protein, which requires several additional washes for complete removal. There is no elution or loss of enzyme activity in this process. About 23 liters of buffer is necessary for this step.

The washed slurry is packed over a layer of CM-cellulose (100 g wet weight), equilibrated with 5 mM sodium malonate, pH 6.8, in a 3.8 cm diameter column. The column is eluted with a solution of 0.06 mM FDP in 5 mM sodium malonate, pH 6.8, at a flow rate of 40–45 ml/hour. Fractions of 20 ml are collected. The enzyme activity appears as a sharp peak after about 160 ml, and recovery is quantitative (see the table). Fractions are pooled according to their specific activity, adjusted to pH 6.0 with 1 M malonic acid, and lyophilized. (Freeze-drying at pH 6.8 results in a significant loss of enzyme activity.)

If it is assumed that homogeneous enzyme has a specific activity of 22.5 units/mg of protein,[8] then 72% of the enzyme is eluted in pure form with a specific activity 300-fold greater than the pH 6 supernatant.

Step 3. Chromatography on Sephadex G-200. Pooled fractions with specific activity above 25 units/mg of protein are not treated further. Fractions with slightly lower specific activity can be further purified by gel filtration on Sephadex G-200. Pool II from the previous step and another sample of similar specific activity are combined and dissolved in 2 ml of water and applied to a Sephadex G-200 column (2.5 × 37 cm), equilibrated with 5 mM sodium malonate buffer, pH 6.0. The column is eluted with the same buffer, and fractions of 5 ml are collected at a flow rate of 7.5 ml/hour. The enzyme activity appears in a sharp symmetrical peak, and fractions containing the bulk of activity are pooled and lyophilized (see the table).

Specific Activity and Purity of Preparations

It was repeatedly found that some of the fractions eluted from CM-cellulose initially had slightly higher specific activities (24–30 units/mg protein) than those reported for homogeneous enzyme (20–22.5 units/mg protein).[8] However, after several freezings and thawings, the specific activities generally dropped to 22–23.

The purified enzyme was homogeneous upon polyacrylamide gel electrophoresis at pH 6.8 and 9.3.

General Considerations

A major modification in the present procedure for purification of FDPase is the batchwise saturation of CM-cellulose with enzyme. Twenty times more enzyme can be put on the CM-cellulose by this technique than by addition to a prepacked column. This method of saturation also permits a more efficient washing of the slurry before the column is packed. Since the enzyme appears in the early fractions upon substrate elution (specific procedure), complete removal of washable protein is crucial to obtain a high specific activity of the enzyme.

Specific elution with substrate or other ligand appears to be the result of the formation of an enzyme-ligand complex which produces a

PURIFICATION OF RABBIT LIVER FDPASE[a]

Step	Units	Protein (mg)	Recovery (%)	Specific activity (units/mg)
1. Dialyzed liver supernatant	1960	36,360[b]	–	0.05
pH 6 supernatant	2160	27,110[b]	100	0.08
2. CM-cellulose column				
Pool I	947	32.0	–	29.6
Pool II	618	26.2	–	23.6
Pool III	574	113.5	–	5.1
Total	2139	–	99	–
3. Sephadex G-200 column (pool II + a fraction of step 3 from another batch; total units 1006; specific activity 24.2)	831	32.0	82.5	26.0

[a] CM-cellulose column: fractions of 20 ml were collected; Pool I, fractions 7 and 8; Pool II, fraction 9; Pool III, fractions 10–14. Sephadex G-200 column: fractions of 5 ml were collected.

[b] Protein in crude fractions was determined spectrophotometrically from absorbances at 280 and 260 mμ (E. Layne, this series, Vol. 3 [73]).

change in conformation and charge of the protein molecule. Experimental evidence for such a conformational change with liver FDPase has been found in the enhanced affinity of the enzyme for AMP,[9] in the decreased fluoresence emission of the enzyme with the hydrophobic probe, 1-anilinonaphthalene-8-sulfonate,[10] and in the decreased absorbance of the enzyme at 295 mμ at alkaline pH,[11] all of which occur in the presence of substrate. It is of interest to note here that the substrate concentration during enzyme elution is more than sufficient to saturate the enzyme with FDP.[9]

Purifications of FDPases from a variety of different sources have been achieved in which a step employing substrate elution from substituted celluloses results in a large increase in specific activity.[12-14]

In summary, it may be noted that most of the published procedures for specific enzyme purification by substrate elution utilize anionic celluloses and eluants. However, there is no theoretical reason to limit this approach to any particular type of column material or eluant, since the principle of the technique appears to depend only on the production of a significant change in charge and/or conformation of the enzyme due to specific interaction with a substrate, activator, or inhibitor. The very successful use of neutral polysaccharides in enzyme purification[1] serves to illustrate this point.

Addendum. A recent report of the purification of yeast DPN+-specific isocitrate dehydrogenase is an example of specific elution from a cationic cellulose with an anionic eluant.[15] Homogeneous enzyme was isolated in high yield by a procedure including successive elutions from DEAE-cellulose and P-cellulose columns with citrate.

Also, a combination of specific adsorption on polymers of bovine gamma globulin containing covalently linked β-thiogalactosides followed by specific elution with either lactose or isopropylthiogalactoside was employed to obtain a high degree of purification of β-galactosidase from *Escherichia coli*.[16] In addition, the lactose repressor protein could be purified by specific adsorption followed by salt elution from such a polymer.

[9]M. G. Sarngadharan, A. Watanabe, and B. M. Pogell, *Biochemistry* 8, 1411 (1969).
[10]H. Aoe, M. G. Sarngadharan, and B. M. Pogell, *J. Biol. Chem.* 245, 6383 (1970).
[11]S. Pontremoli, E. Grazi, and A. Accorsi, *J. Biol. Chem.* 244, 6177 (1969).
[12]O. M. Rosen, S. M. Rosen, and B. L. Horecker, *Arch. Biochem. Biophys.* 112, 411 (1965).
[13]O. M. Rosen, *Arch. Biochem. Biophys.* 114, 31 (1966).
[14]J. Fernando, M. Enser, S. Pontremoli, and B. L. Horecker, *Arch. Biochem. Biophys.* 126, 599 (1968).
[15]G. D. Kuehn, L. D. Barnes, and D. E. Atkinson, *Fed. Proc.,* 29, 399 (1970) and unpublished observations.
[16]S. Tomino and K. Paigen, in "The Lactose Operon" (J. R. Beckwith and D. Zipser, eds.), p. 233. Cold Spring Harbor Lab., New York, 1970.

Section VII

Electrophoretic Procedures

Articles Related to Section VII

[33] Isoelectric Focusing of Proteins

By OLOF VESTERBERG

The method of isoelectric focusing has already found numerous applications in the study of proteins from various sources.[1-4] The method requires a stable pH gradient between the anode and the cathode in an electrolysis cell. This is obtained by the electrolysis of a water solution of a mixture of suitable low molecular weight ampholytes, called carrier ampholytes. When proteins are put into such a system each protein will migrate to, and focus at, its isoelectric point, p*I*. This makes the following possible:

1. *Separation of proteins* at a high degree of resolution for analytical as well as for preparative purposes. Proteins differing in isoelectric point with as little as 0.01 of a pH unit may be separated.[5,6] The resolving power is often of the same order or somewhat better than that obtainable by carefully conducted electrophoresis in polyacrylamide gel.[6]

2. *Characterization of each protein by its isoelectric point.* The p*I* values are simply obtained by measurement of pH at the maximum concentration of each protein after focusing and fractionation. These p*I* values can be determined with a high degree of reproducibility,[7] and are very useful in comparative studies of proteins. The minimum amount of protein necessary is limited only by the sensitivity of the method for the detection of the protein. The method is outstanding in this respect for p*I* determinations.

The technique requires a simple apparatus and is easy to handle. The principle allows great flexibility in performance. This is partly due to the focusing principle, which implies that every protein will migrate to and focus at its respective isoelectric point, independent of where it was put into the apparatus at the start. Therefore, the volume of the protein sample solution to be applied is not critical and even dilute protein solu-

[1]H. Haglund, *Sci. Tools* **14**, 17 (1968).
[2]O. Vesterberg, *Sv. Kem. Tidskr.* **80**, 213 (1968).
[3]O. Vesterberg, *in* "Methods in Microbiology" (J. R. Norris and D. W. Ribbons, eds.). Academic Press, New York, 1970.
[4]H. Haglund, *in* "Methods of Biochemical Analysis" (D. Glick, ed.), Vol. 19, p. 1, Wiley (Interscience), New York, 1971.
[5]O. Vesterberg, *Acta Chem. Scand.* **21**, 206 (1967).
[6]A. Carlström and O. Vesterberg, *Acta Chem. Scand.* **21**, 271 (1967).
[7]O. Vesterberg and H. Svensson, *Acta Chem. Scand.* **20**, 820 (1966).

tions of comparatively large volumes can be applied. This is a clear advantage over many conventional chromatographic and electrophoretic methods.

Isoelectric focusing like electrophoresis requires some means to counteract convective remixing of the separating components. At present three stabilizing techniques are used for this purpose: (1) density gradient, (2) polyacrylamide gel, (3) zone convection electrofocusing. The first technique is the most common, and an apparatus for 110 ml and another for 440 ml are commercially available.[8b] The second is of rapidly growing importance and is especially valuable for analytical separations and comparative work[8a] (see this volume [38]). The third technique is very simple, as it is in principle performed in a serpentine channel and does not require such additives as sucrose or gel.[9] Development of the last technique has been delayed by difficulties in fabricating a suitable apparatus with efficient cooling, but the problem has now been solved.[8b]

Basic Principle

Although isoelectric focusing of proteins has been described elsewhere,[1-4] a brief introduction is presented here. The net charge of a protein molecule in an acidic solution is positive because most amino groups carry a positive net charge and most carboxylic groups are protonated and electrically uncharged. If the pH is gradually increased, the number of carboxylic groups which carry a negative charge will increase, and the number of positively charged groups will decrease. At a certain pH value, the isoionic point, the net charge of the protein molecule, is zero. The isoionic point of a molecule is thus determined by the number and types of proteolytic groups and their dissociation constants. Whereas proteins show considerable variation in isoionic point, they are generally in the pH range of 3–11. It is also accepted that the positively charged anode in an electrolysis cell attracts negative ions and repels positive ions; the opposite is the case for the cathode. In conventional electrophoresis there is a constant pH between anode and cathode so that positively charged ions, e.g., proteins, migrate to the cathode while negatively charged ions migrate to the anode. In electrofocusing, a stable pH gradient is arranged; i.e., pH increases successively from anode to cathode. If a protein is put into this system at a pH lower than the isoionic point, the net charge of the molecule will be positive and it will migrate in the direction of the cathode. Due to

[8a]N. Catsimpoolas, *Separ. Sci.* 5, 523 (1970).
[8b]LKB-Produkter AB, S-161 25 Bromma, Sweden.
[9]E. Valmet, *Sci. Tools* 15, 8 (1968).

the presence of the pH gradient, the protein will migrate to an environment of successively higher pH values which, in turn, will influence the ionization and net charge of the molecule. The protein will eventually reach a pH where its net charge is zero and will stop migrating. This is the isoelectric point of the protein. The pH here is thus equal to the isoelectric point, which is very close to, or equal to, the isoionic point.[7] The consequence of this is that every protein will migrate to and focus at its respective isoelectric point in a stable pH gradient, irrespective of its origin in the apparatus at the time the current was applied. Therefore, the point of application and the volume of the protein solution are not critical. Once a final, stable focusing is reached, the resolution will be retained if the experiment is continued for a long time, since the focusing effect works against diffusion. This is contrary to conventional electrophoresis wherein diffusion is an obstacle.

From the facts mentioned it is understandable that the electrofocusing principle can be used as a powerful separation method. Furthermore, the isoelectric point can be determined simply by measuring the pH where the proteins are focused; pI values can be ascertained with a high degree of reproducibility,[7] which is often of the order of the reproducibility of ordinary laboratory pH meters. It should, however, be remembered that pI values are dependent on temperature in that they usually decrease with increasing temperature.[7] Therefore the pH must be measured at a constant temperature, usually the temperature of focusing. In a publication, the temperature should be stated.

When the pI of a protein determined by electrofocusing is compared with that ascertained by electrophoresis, often it is found that the former value is higher.[7,10] It is well known that the value of the isoelectric point of a protein determined by electrophoresis is influenced by the kind of buffer and the ionic strength used.[11] This phenomenon, in many cases, is caused by complex formation between the protein and buffer ions present at electrophoresis. Generally, the lower the ionic strength, the higher are the values. If the values are extrapolated to a very low ionic strength, the resulting pI has been found to approach or to be equal to that obtained by isoelectric focusing. This isoelectric point is often very close to the isoionic point.[2,7] This situation excludes complex formation.[7] A precise determination of the isoionic point by conventional methods is not very simple.

To sum up, it can be said that the isoionic point is valuable because it is a characteristic of the intrinsic acidity of the protein. The isoelectric point, pI, is the experimentally determinable value which is important for characterization and comparative purposes.

[10]H. Svensson, *Arch. Biochem. Biophys.*, Suppl. 1, 132 (1962).
[11]A. Tiselius and H. Svensson, *Trans. Faraday Soc.* **36**, 16 (1940).

The pH Gradient and the Carrier Ampholytes

The pH gradients are obtained by isoelectric focusing of special buffer substances which are ampholytes, and are called carrier ampholytes, here abbreviated as CA. As was clearly described by Svensson,[12,13] they must fulfill certain criteria. A certain buffering capacity of the ampholyte at pH values close to the isoelectric point is necessary because the CA must dictate the pH course. Furthermore, the CA must have a certain conductance at their pI, and this property is correlated with good buffering capacity. The CA should be of rather low molecular weight so as to facilitate their removal from proteins after isoelectric focusing by simple procedures such as molecular sieving or dialysis. To permit general use, numerous CA with pI values distributed over the pH interval between 3 and 11 should be available to cover the pH range in which most proteins are isoelectric. The exact number of CA necessary per unit of pH depends on several factors, including buffering capacity, electrophoretic mobility of the CA, and, above all, on the desired value of the pH gradient and the degree of protein resolution desired. Furthermore, the CA must have good solubility in water; hydrophilic character is also important so as to avoid adsorption to the hydrophobic sites of proteins. Light absorption of the CA above 260 nm should be low in order to permit detection of proteins after focusing by measurements at 280 nm. From the above, it follows that it is an advantage that the CA are isomers and homologs of aliphatic polyamino polycarboxylic acids. Such chemicals have been developed and synthesized by the author.[14] The general formula for the CA is,

$$R-N-(CH_2)_n-N-(CH_2)_n-COOH$$

where R can be

$$-(CH_2)_n-COOH, \text{ H, or } -(CH_2)_n-N-R$$
$$R$$

and $n < 5$. The synthesis has been developed further by LKB-Produkter, Stockholm, Sweden, and such CA are commercially available under the trade name Ampholine®. These chemicals fulfill the criteria of good carrier ampholytes as verified by numerous applications with a variety of proteins.[2-4]

The selection of a suitable pH range is important. The CA are sold as mixtures for several specific pH ranges, e.g., 3–10, 6–8. This means

[12]H. Svensson, *Acta Chem. Scand.* 15, 325 (1961).
[13]H. Svensson, *Acta Chem. Scand.* 16, 456 (1962).
[14]O. Vesterberg, *Acta Chem. Scand.* 23, 2653 (1969).

that most of the CA in a given sample have their p*I* in the correspond-ing pH range and are thus able to yield a pH gradient covering this range. The total pH range actually obtained will always extend some-what beyond the nominal pH range,[15] depending on the preparative procedure and also partly on the solutions used at the electrodes (cf. page 405). For protein separation and focusing, the pH range of the CA should be chosen so that the proteins under study have their p*I* values in this range. When the p*I* values are unknown, a wide pH range should be used. If only p*I* values obtained by electrophoresis are at hand, the dependence of such values on the ionic strength must be taken into account.

When the pH range is outside 6–8, it is advisable to add ampholytes of pH 6–8 in an amount equal to about 10% of the total in order to obtain a more even distribution of conductivity between the electrodes. Generally, an amount of carrier ampholytes is used which gives an aver-age concentration of 1% (weight/volume) in the density gradient columns and also in the zone convection electrofocusing apparatus. It is advantageous to use a higher concentration of CA for proteins with lower solubility[9]; the principle is comparable to the solubilizing effect of amino acids on proteins.[16] For electrofocusing in gel, an average concentration of 2% (w/v) is recommended since a higher concentra-tion counteracts electroendosmosis.[17]

Stabilization against Convection

As is the case for all electrophoretic methods, there is a need for stabilization of the separating protein zones against convective flow in the solution. For electrofocusing, three ways are presently used: density gradient, gel, and zone convection electrofocusing. Isoelectric focusing in gels is described in this volume [38]. Some general remarks on the other methods are made here; practical instruction is provided in the Manual which is an appendix to this article.

Isoelectric Focusing in Density Gradients

Experiments are conducted in vertical columns of which two types are presently available: one has a capacity of 110 ml, and the other of 440 ml (LKB-Produkter, Stockholm, Sweden) (Fig. 1). Density gradients suitable for electrofocusing can be made with many uncharged solutes which are dissolvable in water to a concentration that will increase the

---

[15]H. Davies, *in* "Protides of the Biological Fluids" (H. Peeters, ed.), Vol. 17, p. 389, Per-gamon, London, 1970.
[16]A. Grönwall, *C. R. Lab. Carlsberg* **24**, 185 (1942).
[17]Z. L. Awdeh, A. R. Williamson, and B. A. Askonas, *Nature (London)* **219**, 66 (1968).

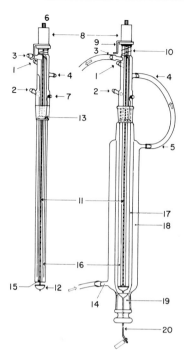

Fig. 1.   Electrofocusing column of 110-ml capacity. The outer cooling jacket (*18*) has an inlet at *14*, and an outlet at *5*. From the outer jacket the water flows through a tube into the central cooling jacket at *4*, and leaves the column at *3*. Two platinum electrodes are used. One electrode *13*, in contact with the plug *7*, is in the upper part of the column. The gas formed at this electrode escapes at *2*. The other electrode, in contact with the plug *6*, is wound on a rigid Teflon bar *11*. The gas formed at this electrode escapes at *1*. A piece of Teflon, *12*, is connected by the rigid Teflon bar, *11*, to the handle, *8*. The central tube is open at *15*, when the handle *8* is kept in the down position with the aid of the hook *9*. Before draining the column, the central tube is closed by lifting the plug *12*, which has a rubber ring gasket on the upper surface. It is kept closed with the aid of the spring *10*. Isoelectric focusing takes part in compartment *17*, which is filled through the tube *2*. At the bottom of the column there is a plug *19* with an attachment for a capillary tube to enable fractions to be sampled.

density sufficiently. The compounds must, of course, be harmless to proteins. A high purity and, especially, a low content of heavy metals are important. In most cases, sucrose has been used with 50% (w/v) as the densest solution. Sucrose is mild to proteins and can even have a protective effect.[2] Generally, approximately linear density gradients are prepared with the maximum solute concentration at the bottom of the column; there is a linear decrease of the concentration as a function of column height. When it is desired to increase the stabilization or the capacity of the gradient in order to carry a dense protein zone at a

certain level, nonlinear gradients may be used.[18] One should then remember that a certain strength of the density gradient must remain
also in other parts of the column to prevent convection disturbances
caused by the electrical heating effect. Glycerol has also been used and
is valuable for the pH range above pH 10; sucrose is partially dissociated at the higher pH and might interfere with p*I* determinations.[19]
Because the contribution to the density of an aqueous solution is less
for glycerol than for sucrose, glycerol is generally used with 55% w/v
as the highest concentration.[2] The same is true for ethylene glycol,
which also has been used. Mannitol and sorbitol give a higher density
contribution than glycerol and ethylene glycol, and can also be used.
Other useful agents include polyglycans, such as Dextran® and Ficoll®
available from Pharmacia, Uppsala, Sweden. Density gradients may be
prepared either manually, as described by Vesterberg and Svensson,[7] or
with the aid of a special gradient mixers[2,20] (Fig. 2). Another method of
preparation of a density gradient involves the use of a peristaltic pump
with three channels.[21]

Zone Convection Electrofocusing

The apparatus consists of two rectangular boxes, one of which serves
as a cover[9] (Fig. 3). The upper surface of the lower box is corrugated
with transverse ridges, with a height of about 10 mm, separated by depressions, and is facing the lid which has corresponding ridges. When
the halves are in position the ridges of the bottom part fit into the depressions of the lid so that a space of a few millimeters remains. From
one end to the other there will be a narrow wave-formed channel between the two parts which can be described as series of interconnected
broad U-tubes. The carrier ampholyte solution fills this space, and the
electrodes are situated at the ends. Both the lid and the bottom part are
cooled from the inside of each part by circulating liquid of constant
temperature ($+4°$). When the current is on, a density gradient is formed
in each depression by the solute. In the beginning this is due to thermal
diffusion (Ludvig-Soret effect[22]). When proteins focus, the density increases locally, making the proteins collect in the depression of the
bottom part. Proteins which have a tendency to precipitate at iso-

[18]H. Svensson, *in* "Laboratory Manual of Analytical Methods in Protein Chemistry Including Polypeptides" (P. Alexander and R. J. Block, eds.), pp. 195–244, Pergamon, London,
1960.
[19]T. Flatmark and O. Vesterberg, *Acta Chem. Scand.* **20**, 1497 (1966).
[20]H. Svensson and S. Pettersson, *Separ. Sci.* **3**, 209 (1968).
[21]S. R. Ayad, R. W. Bonsall, and S. Hunt, *Sci. Tools* **14**, 40 (1967).
[22]C. Ludwig, *Sitzber. Acad. Wiss. Wien* **20**, 539 (1856).

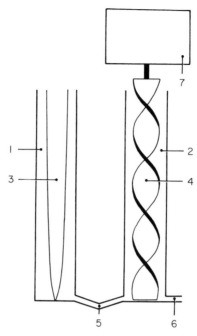

Fig. 2. Gradient mixer, *1* and *2*, compartments for the less dense and the dense solution, respectively; *3*, specially shaped plunger; *4*, spiral for stirring; *5*, communication channel; *6*, outlet; *7*, stirring motor.

electric focusing collect in the depressions, and do not interfere in the same way as they sometimes do by settling in density gradient columns. When electrofocusing is completed, the cover is raised, lifting up the ridges of the cover, which had been projecting into the depressions of the bottom part. The liquid level is thereby lowered, and the liquid collects in the depressions. Each one will contain a fraction separated by a ridge. Each fraction can now be collected without risk of contamination by neighboring fraction. The resolving power achieved by electrofocusing will be determined by the pH range and the number of depressions, provided sufficient voltage and cooling are applied. A good compromise is to have about 80 depressions. The zone convection principle offers the following advantages:

1. Stabilizing agents such as sucrose and gel are unnecessary; thus a source of possible difficulty is eliminated.

2. Filling of and collection from the apparatus are very simple.

3. The pH can be measured directly in the apparatus.

4. The principle offers excellent conditions for separation of proteins on a preparative scale.[23a] High protein concentrations will merely im-

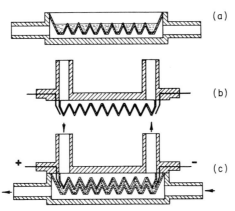

Fig. 3. Schematic drawing of a zone convection electrofocusing apparatus: (a) the bottom part filled as before (or after) the experiment; (b) the lid; (c) lid and bottom part assembled as during the experiment. By courtesy of Mr. E. Valmet.

prove the stabilizing conditions. Valmet reported loading 1 g or more of protein in an apparatus of 50-ml capacity.[23] The instrument is also valuable for fractionation of carrier ampholytes, e.g., for preparation of limited pH ranges. For example, Valmet[9] reported preparation of carrier ampholyte fractions covering only 0.2 pH units in one run from a 4% solution of ampholytes with the pH range 3–10.

5. The volume of the protein solution is not critical. It can be almost the same as the volume of the apparatus. The proteins are concentrated during separation.

6. Even rather crude protein solutions may be applied without prior purification. The salt concentration in the protein sample may be somewhat higher than in density gradient columns. Furthermore, precipitates will not disturb the separation. After fractionation, precipitates can be collected by centrifugation.

A drawback of the method is that it will need almost the same time for focusing as that needed for density gradient columns. Although the method is very attractive and offers many possibilities, especially for preparative purposes, it has not yet been used widely although it has been known for some years. The general use has been delayed solely by difficulties in constructing a version of the apparatus with efficient cooling. Valmet, who developed the principle, has arrived at a suitable apparatus after intensive and fruitful work together with the LKB company.

[23] E. Valmet, in "Protides of the Biological Fluids" (H. Peeters, ed.), Vol. 17, p. 401–408, Pergamon, London, 1970.
[23a] E. Valmet, Sci. Tools 16, 8 (1969).

The Resolving Power

It is often of general interest to calculate the resolving power of a separation method. In the case of the electrofocusing method this means the smallest pI difference, ΔpI, of proteins which still permits their separability. Using an equation presented by Svensson,[12] describing the concentration distribution of a focused ampholyte, it has been possible to develop an equation for the resolving power in terms of the pI difference, ΔpI, needed for separation of two proteins.[7]

$$\Delta \text{p}I = K\sqrt{\frac{D(\text{dpH}/\text{d}x)}{-E(\text{d}u/\text{dpH})}}$$

Here, K is a factor defined by the criterion for separation; D = diffusion coefficient, in $\text{cm}^2 \text{ sec}^{-1}$ of the protein; x = coordinate along the direction of the current; E = the field strength at the point of focusing, in volts/cm; u = electric mobility at pH close to the pI of the protein, in $\text{cm}^2 \text{ volt}^{-1} \text{ sec}^{-1}$. For different criteria of separation, Svensson[24] recommended a numerical value of $K=3$. The derivative dpH/dx is the value of the pH gradient, which is determinable by the experimenter in the selection of a certain carrier ampholyte pH range. Thus, a narrow pH range will yield a high resolving power. A pH range of about 1 pH unit allows the possibility of separating proteins with a ΔpI of about 0.02 pH unit; this has been confirmed in many applications.[5-7] Proteins with such a small pI difference are indeed very difficult to separate by other methods. A minimum field strength is necessary to get a good and rapid focusing and separation. Because a higher voltage also means a greater Joule heat effect, the maximum voltage is limited by the efficiency of the stabilization principle and cooling system. If heating is too great at a particular point in the density gradient this will result in thermal convections which may disturb the zones of focusing protein.

The resolution is also a function of the method used for analysis of the degree of focusing and separation. With the density gradient columns some resolution is lost while draining the column as described by Vesterberg and Svensson.[7] A special column of quartz glass has been constructed by Rilbe, permitting direct analysis of the degree of focusing and separation of protein zones in the column by densitometric or photograph recording using ultraviolet light[25] (see Fig. 4). This is especially valuable for analysis of the number and proportions of proteins in a mixture. However, when the separated components are to be iso-

[24]H. Svensson, *J. Chromatogr.* 25, 266 (1966).
[25]H. Rilbe, *in* "Protides of the Biological Fluids" (H. Peeters, ed.), Vol. 17, p. 369. Pergamon, London, 1970.

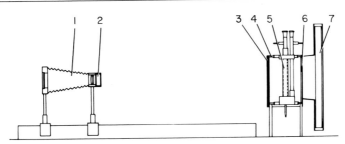

Fig. 4.   A special arrangement for photographic recording of protein zones. *1*, bellows; *2*, liquid filter; *3*, quartz-glass window; *4*, cell holder; *5*, quartz cell; *6*, Jena Filter-UG5; *7*, mercury lamp. By courtesy of Professor H. Rilbe.

lated or when the p*I* is to be determined, the contents of the column must be fractionated so that some of the resolution will be lost. As has been pointed out, the conditions for analysis and for fractionation of proteins focused in gels are favorable. The resolving power of a zone convection apparatus with 80 compartments is probably somewhat less than that obtainable with a density gradient or with a gel.

The Protein Solution

Is it necessary to use a prepurification step? It has been emphasized that the method of isoelectric focusing is a powerful one for separation. It possesses a high resolving power so that even crude mixtures of proteins can be separated. For analytical purposes, e.g., for determination of the relative proportions of proteins in a mixture, this is valuable. However, a given protein may comprise only a minor fraction of the total amount of protein in the sample. Therefore, for large-scale purification, it is often advantageous to use other methods either before or after isoelectric focusing. Some proteins, or impurities such as nucleic acids, will precipitate during isoelectric focusing. These compounds may often be removed beforehand by other methods. They may also be removed in a preliminary electrofocusing experiment; the zone convection electrofocusing principle seems to be valuable for this purpose.

Amount of Protein and Volume of Protein Solution

Capacity in the density gradient is limited merely by the amount of protein of interest in a zone and of components focusing close to it. When a crude mixture is at hand, the total charge of protein can be larger. The density gradient can carry a protein zone, provided its concentration does not exceed a certain value.[18] With the 110-ml column, about 5 mg of protein can be allowed in one zone for analytical purposes, and up to 25 mg for preparative purposes. When working with a wide pH range it is necessary to get narrow, well focused protein zones in order to obtain good separation. In a narrower pH range much broader zones will allow good resolving power. This is valuable because, by this means, very large protein concentrations which might otherwise result in precipitation of protein or local overloading in the density gradient columns can be avoided. The degree of focusing can be regulated by the voltage.

In both gel and zone convection electrofocusing, the maximum concentration of protein in a zone may be much greater than in density gradient systems. The limit is often set by the solubility of the protein. Gram quantities can be handled by zone convection electrofocusing.

Due to the focusing principle, the maximum volume of the protein solution to be applied may be almost as large as that of the apparatus in the case of the density gradient column and in the zone convection electrofocusing apparatus. With the gel technique in its present version, the volume is quite limited although this does not represent a problem since gels are used mostly for analytical purposes.

Generally, the point of application is not critical. Protein may be either distributed in the entire volume of the apparatus, or at a preselected position.

Preparation of the Protein Solution

During electrofocusing, salt ions migrate to the electrodes. This means that if the total amount of salt is very high at the start, the current density will be too high, even at a moderate voltage, and thereby may result in excessive electrical heating. The density gradient is especially sensitive to local temperature differences. Furthermore, the salt will produce acid at the anode and base at the cathode. The extreme pH values may occupy a significant part of the apparatus if the total amount of salt is high. Therefore, the total amount of salt in the protein solution should be limited. The salt concentration can be reduced by dialysis against a 1% water solution of glycine or of carrier ampholytes. Salt can also be removed by dialysis against a volatile buffer followed by lyophilization.

Preparation of Narrow pH Ranges

As has been pointed out earlier, narrow pH ranges of carrier ampholytes are useful in increasing the resolving power and in permitting application of a larger amount of protein. At present, carrier ampholyte mixtures with a minimum range of 0.5 pH unit are available commercially.[8b] The individual investigator can prepare without difficulty a desired narrow pH range. For this purpose carrier ampholytes of a suitable commercial pH range are focused in one experiment. After fractionation and pH measurement, those fractions with suitable pH values are selected. When such fractions are used as ampholytes in a new experiment, a narrow pH range is obtained. Such experiments are easy to perform in density gradient columns and with zone convection electrofocusing. Because the ampholytes of the second experiment comprise only a fraction of those used in the first run, a rather high ampholyte concentration, i.e., about 4–10%, should be used at the start in order to obtain sufficient material for the second experiment. Furthermore, when working at pH values outside the point of neutrality, it is recommended that ampholytes of pH 6–8 be added in an amount about 10% of the total in order to obtain a more even conductivity distribution between the electrodes.

Special Cases

Isoelectric Focusing of Less Soluble Proteins.

As is true of amino acids, carrier ampholytes have a significant solubilizing and protecting effect on many proteins.[2,16] However, during electrofocusing, especially with crude protein mixtures, precipitates are sometimes seen that may cause trouble in density gradient columns. It is not taken for granted that electrofocusing should be abandoned if precipitation occurs since, as noted below, there are many remedies. First, the precipitates may be caused by impurities. Second, if the protein of interest precipitates, the effect need not lead to denaturation since the carrier ampholytes have a protective effect, and the biological activity can very often be regained.[26] Precipitates, when they do occur, are generally easy to solubilize after fractionation by simply adjusting the pH.[26,27] One way to minimize precipitation is to decrease the amount of protein used.

Electrofocusing in gel and zone convection electrofocusing result in much less difficulty with precipitating proteins than in the density gradient systems. If the precipitates are caused by impurities, one could try to remove them before electrofocusing, or use the first focusing as a prepurification step. The zone convection technique is excellent for this purpose. If a density gradient column is used, it is sometimes good to select the polarity of the electrodes so that the precipitates collect in the lower part of the column, below the protein of interest. If the precipitrates do not come out nicely while draining the column, it is preferable to pump in a dense sucrose solution at the bottom and to extrude the contents of the column at the top. Care should then be taken to avoid air bubbles coming in before, or together with, the dense solution. Obviously, rising air bubbles in the column will disturb the zones of protein.

[26]W. M. Mitchell, *Biochim. Biophys. Acta* **178**, 194 (1969).
[27]O. Vesterberg, *in* "Protides of the Biological Fluids" (H. Peeters, ed.), Vol. 17, 383. Pergamon, London, 1970.

When precipitates occur, they usually appear slowly as the components reach their pI values. In some cases a sufficient separation may be obtained if the sample is placed into a pH gradient at a specific point, and the components are allowed to migrate in the direction of their pI's. The experiment is then interrupted either before or just at the time when precipitates appear.

Another approach is to use additives.[27] It is known that urea at a concentration of 2 and up to 4 M and formamide can have a solubilizing effect on proteins; urea has been used successfully.[28-30] In a density gradient column these agents can be added in equal amounts to both the dense and the less dense solution before preparation of the density gradient. It is also well known that nonionic detergents have a solubilizing effect on some proteins, e.g., membrane bound enzymes. For the latter purpose, they are used in concentrations ranging from 0.1 to 5%. For electrofocusing the compounds used should be nonionic and carry no net charge. Many such detergents are now available under various trade names, e.g., Tween 80®, Emasol®, Brij 39®, Triton X-100® (Sigma Chemical Co., St. Louis, Missouri).[27] Successful use of Triton X-100 has been reported.[31] Serum lipoproteins have been maintained in solution by using a water solution of 33% ethylene glycol.[32]

Susceptible Proteins and the Use of Thiol Compounds.

It has been observed that some sensitive proteins are more stable if focusing is carried out in the presence of a low concentration (about 10^{-4} M) of thiol compounds.[33] Promising results have been obtained with dithioerythritol.[34] The compound used should carry no net charge at the pH where proteins are focused. If the compound used has a negative net charge, the usual case for ordinary thiol compounds above pH 9.5, special remedies must be used. When working with alkaline proteins a small amount of mercaptan may be added at the cathode slowly and continuously in a small volume, or at intervals of some hours.[33] However, when using thiol compounds, especially at higher concentrations, cleavage of disulfide groups may occur. For studies of protein subunits, isoelectric focusing can be carried out in urea and mercaptoethanol.[35]

[28]I. Björk, *Acta Chem. Scand.* 22, 1355 (1968).
[29]J.-O. Jeppson, *Biochim. Biophys. Acta* 140, 468 (1967).
[30]J. G. Schoenmakers and H. Bloemendal, *Nature* (*London*) 220, 790 (1968).
[31]R. Möllby and T. Wadström, *in* "Protides of the Biological Fluids" (H. Peeters, ed.), Vol. 17, p. 465. Pergamon, London, 1970.
[32]G. Kostner, W. Albert, and A. Holasek, *Hoppe-Seyler's Z. Physiol. Chem.* 350, 1347 (1969).
[33]F. Wadström and K. Hisatsune, *Biochem. J.* 120, 725 (1970).
[34]W. W. Cleland, *Biochemistry* 3, 480 (1964).
[35]N. Catsimpoolas, *Fed. Eur. Biol. Soc. Lett.* 4, 259 (1969).

Appendix: Manual

Preparation of Solutions for Isoelectric Focusing in Density Gradient Columns and Performance of an Experiment

For equipment and chemicals see Tables I and II.

Data for isoelectric focusing in density gradient columns are summed up in Tables I, II, and III.

Table I

NECESSARY EQUIPMENT AND CHEMICALS FOR ISOELECTRIC FOCUSING IN DENSITY GRADIENT COLUMNS

Equipment and Chemicals	Comments
Column[a]	At present two types are available with a capacity of 110 and 440 ml, respectively. See Fig. 1
Ampholine[a]	Carrier ampholytes. These are available as 40% (w/v) water solutions of different pH ranges: 3–10, 3–6, 5–8, 7–10, and 3–5, 4–6, 5–7, 6–8, 7–9, 8–10. Recently also half pH ranges from pH 3.5 up to pH 10 are available on special order. For selection of pH range cf. page 405
Sucrose or glycerol	Analytical purity
Electrode solutions	Usually a dilute solution of NaOH is used at the cathode, and a dilute solution of H_2SO_4 or H_3PO_4 is used at the anode
Power supply and cables[a]	This instrument should have a variable voltage in the range 0–1000 V, giving a current of 0–20 mA, preferable with an ammeter readable at every 0.1 mA
Cooling device[a]	Although preliminary experiments may be made with running cold tap water in the cooling mantle of the column, there are many reasons for using a constant-temperature cooling bath. Because of the sensitivity of some proteins to higher temperatures, +4°C is recommended, The pI is dependent on temperature and therefore the same temperature should be used during focusing as at pH measurement. When a cooling bath is available, the water may be circulated through the cooling mantle of the column and then through a metal coil in a small bath. After draining of the column, the fractions are placed in this bath. pH measurement is carried out here, after 20 minutes for equilibration
pH meter	To determine pI of proteins, a pH meter is necessary. To permit a high accuracy, a pH meter with an accuracy of ± 0.01 pH units is recommended. The electrode should fit a test tube, so that not more than 1 ml of a sample solution is necessary.

[a]Available through LKB-Produkter AB, S-161 25 Bromma, Sweden.

TABLE II

AUXILIARY EQUIPMENT AND CHEMICALS FOR ISOELECTRIC FOCUSING IN DENSITY
GRADIENT COLUMNS

Equipment	Comments
Gradient mixer[a]	The density gradients can be prepared either manually or with the aid of an automatic gradient mixer. Two types, which give approximately linear density gradients, can be obtained, one for the 110-ml column and the other for the 440-ml column
Peristaltic pump[a]	Facilitates filling and emptying of the columns. The flow rate should be variable between 1 ml/min and 4 ml/min. A pump is suitable when a gradient mixer, and/or a photometer with a flowthrough measuring cell is used
Photometer with a flowthrough cell[a]	Although the proteins may be detected by measurements at 280 mμ on each fraction after the column is drained, it is more convenient to use a light absorption monitor with a flowthrough cell and a recorder (Uvicord, LKB)
Fraction collector[a]	Especially when using automatic emptying controlled by a peristaltic pump it is advantageous to use a fraction collector. The draining of the columns is usually done at a flow rate of 1 and up to 4 ml/minute. This means that the fraction collector should be adjustable to shift at 1 minute intervals.
Sephadex and columns	For separation of proteins from carrier ampholytes cf. page 410

[a]Available through LKB-Produkter AB, S-161 25 Bromma, Sweden.

TABLE III

DATA FOR ISOELECTRIC FOCUSING IN DENSITY GRADIENT COLUMNS

Datum	LKB 8101	LKB 8102
Volume	110 ml	440 ml
Maximum protein quantity	5–25 mg/zone	20–100 mg/zone
Maximum volume of protein solution	80 ml	320 ml
Maximum salt content in the protein solution	0.5 mmole	1.5 mmole
Ampholine quantity[a]	1–10 g	4–40 g
Filling rate: manually	3 ml/min	5 ml/min
with gradient mixer	5 ml/min	5 ml/min
Maximum power[b] at the start	3 W	6 W
Maximum voltage[b]	350–800 V	350-800 V
Electrofocusing time[b]	24–72 hours	24–72 hours
Emptying rate	2 ml/min	4 ml/min

[a]The lower figure is usually used. Higher concentrations can be used if precipitates of proteins occur (cf. page 393).
[b]cf. Appendix section: Voltage, Maximum Power, Focusing Time.

Selection of Carrier Ampholyte pH Range

The figures given below refer to the smaller column of 110-ml capacity, and the figures given in parentheses refer to the 440-ml column. When working outside the pH range 5–8, it is recommended that 0.25 ml (1.0 ml) of the Ampholine (40% w/v) pH 6–8 be added to prevent the development of a very low conductivity in this pH range.

Solutions to Prepare the Density Gradient

The following recipe can be followed when a gradient mixer is used. If the density gradient is to be prepared manually, all figures should be increased by a factor of 1.2, because of the inevitable loss in the burette and test tubes.

	110 ml column		440 ml column	
	Dense	Less dense solution	Dense	Less dense solution
Carrier ampholytes, 40% w/v	1.9 ml	0.6 ml	7.6 ml	2.4 ml
Distilled water and protein solution	37 ml	51.5 ml	148 ml	206 ml
Sucrose	26 g	—	104 g	—

After sucrose is dissolved, the solutions may be transferred directly to the respective vessels for the gradient mixer. Sucrose can be replaced by another nonionic solute. When glycerol is used to replace sucrose, the dense solution should contain 55% w/v of glycerol.

Choice of Electrode Polarity and Electrode Solutions

To compensate for differences in conductivity in different pH ranges, the central tube should be anodic (plus) for pH ranges below 6 and cathodic (minus) for pH ranges above 6. When protein precipitates occur in the higher half of the column and cause difficulty due to settling of precipitates, the assignment of electrodes may be reversed.

The following recipe shows the maximum quantities of acid and sodium hydroxide. The amount may often be decreased to one fifth.

Anode solutions		
	Anode in the central tube	Anode at the top of the column
Phosphoric or sulfuric acid	0.05 (0.2) ml of the concentrated acid, or 1 (4) ml of a 1 M solution	0.025 (0.1) ml of the concentrated acid, or 0.5 (2) ml of a 1 M solution
Distilled water	15 (56) ml	5 (25) ml
Sucrose	11 (44) g	—

	Cathode solutions	
	Cathode in the central tube	Cathode at the top of the column
Sodium hydroxide	0.1 (0.4) g, or 1 (4) ml of a 2 M solution	0.05 (0.2) g or 0.5 (2) ml of a 2 M solution
Distilled water	14 (56) ml	5 (25) ml
Sucrose	11 (44) g	—

Filling the Column

Position the column absolutely vertically. Connect the cooling water and close the clamp on the capillary tube at the bottom. With a pump, or with the aid of a funnel and a tube, the electrode solution for the central tube is run through the nipple of the central tube. The bottom valve of the central tube should be open to permit this solution to collect at the bottom of the column. Then the solution is pumped from the gradient mixer through the other nipple at the top of the column. It will layer over the electrode solution. The latter solution is then slowly pushed into the central tube as the level rises. However, some of the electrode solution remains at the bottom of the column to fill up the dead space beneath the central tube.

After the less dense and dense solutions have passed into the column, the top electrode solution is gently added to the column with the aid of a funnel and a tube. Check that this solution reaches above the top electrode. If necessary, add a few milliliters of distilled water.

Connect the power supply and use a suitable voltage (see discussion of voltage below). Remember that the voltage used can be lethal. Switch off the power supply before touching or fractionating the column.

Arrangement of the Density Gradient Manually

With this procedure the protein solution may replace the light solution in any desired fraction. The corresponding volumes from the burette are then discharged.

1. Label 24 (46 for the 440-ml column) test tubes of about 12 ml capacity, and place them in a rack.

2. Prepare the dense solution as described on page 405 and use a burette and drain (or pipette manually) according to Table IV into the labeled test tubes.

3. Prepare the less dense solution as described on page 405 and use a burette and fractionate (or pipette manually) according to Tables IVA and IVB into the labeled test tubes.

4. Mix the contents of each test tube carefully.

TABLE IVA

VOLUMETRIC AID FOR THE PREPARATION OF A CONSTANT DENSITY GRADIENT IN A 110 ML COLUMN

Fraction No.	Dense solution		Less dense solution	
	ml	Sum ml	ml	Sum ml
1	4.6	4.6	0.0	0.0
2	4.4	9.0	0.2	0.2
3	4.2	13.2	0.4	0.6
4	4.0	17.2	0.6	1.2
5	3.8	21.0	0.8	2.0
6	3.6	24.6	1.0	3.0
7	3.4	28.0	1.2	4.2
8	3.2	31.2	1.4	5.6
9	3.0	34.2	1.6	7.2
10	2.8	37.0	1.8	9.0
11	2.6	39.6	2.0	11.0
12	2.4	42.0	2.2	13.2
13	2.2	44.2	2.4	15.6
14	2.0	46.2	2.6	18.2
15	1.8	48.0	2.8	21.0
16	1.6	49.6	3.0	24.0
17	1.4	51.0	3.2	27.2
18	1.2	52.2	3.4	30.6
19	1.0	53.2	3.6	34.2
20	0.8	54.0	3.8	38.0
21	0.6	54.6	4.0	42.0
22	0.4	55.0	4.2	46.2
23	0.2	55.2	4.4	50.6
24	0.0	55.2	4.6	55.2

TABLE IVB

VOLUMETRIC AID FOR THE PREPARATION OF A CONSTANT DENSITY GRADIENT IN A 440-ML COLUMN[a]

Fraction No.	Dense solution		Less dense solution	
	ml	Sum ml	ml	Sum ml
1	9[b]	—	0	0
2	8.8	8.8	0.2	0.2
3	8.6	17.4	0.4	0.6
4	8.4	25.8	0.6	1.2
5	8.2	34.0	0.8	2.0
6	8.0	42.0	1.0	3.0
7	7.8	49.8	1.2	4.2
8	7.6	57.4	1.4	5.6

TABLE IV B (*continued*)

Volumetric Aid for the Preparation of a Constant Density Gradient in a 440-ml Column[a]

Fraction No.	Dense solution		Less dense solution	
	ml	Sum ml	ml	Sum ml
9	7.4	64.8	1.6	7.2
10	7.2	72.0	1.8	9.0
11	7.0	79.0	2.0	11.0
12	6.8	85.8	2.2	13.2
13	6.6	92.4	2.4	15.6
14	6.4	98.8[c]	2.6	18.2
15	6.2	6.2	2.8	21.0
16	6.0	12.2	3.0	24.0
17	5.8	18.0	3.2	27.2
18	5.6	23.6	3.4	30.6
19	5.4	29.0	3.6	34.2
20	5.2	34.2	3.8	38.0
21	5.0	39.2	4.0	42.0
22	4.8	44.0	4.2	46.2
23	4.6	48.6	4.4	50.6
24	4.4	53.0	4.6	55.2
25	4.2	57.2	4.8	60.0
26	4.0	61.2	5.0	65.0
27	3.8	65.0	5.2	70.2
28	3.6	68.6	5.4	75.6
29	3.4	72.0	5.6	81.2
30	3.2	75.2	5.8	87.0
31	3.0	78.2	6.0	93.0
32	2.8	81.0	6.2	99.2[c]
33	2.6	83.6	6.4	6.4
34	2.4	86.0	6.6	13.0
35	2.2	88.2	6.8	19.8
36	2.0	90.2	7.0	26.8
37	1.8	92.0	7.2	34.0
38	1.6	93.6	7.4	41.4
39	1.4	95.0	7.6	49.0
40	1.2	96.2	7.8	56.8
41	1.0	97.2	8.0	64.8
42	0.8	98.0	8.2	73.0
43	0.6	98.6	8.4	81.4
44	0.4	99.0	8.6	90.0
45	0.2	99.2	8.8	98.8
46	0	–	9[b]	–

[a]Fill tubes nos. 1 and 46 with the aid of a pipette. The dense solution is then filled into a burette. The successive burette readings on a 110-ml burette are given in the columns "Sum ml." The burette is to be refilled after every 100 ml.
[b]With a pipette.
[c]Refill the burette after these fractions.

These fractions can be transferred into the column through a nipple at the top of the column with the aid of a funnel attached to capillary tubing. The funnel should only be a few centimeters above the level of the nipple, and the flow rate should be adjusted to about 3 ml/min (5 ml/min). The fractions are added in numerical order.

Voltage, Maximum Power, and Focusing Time

The Joule heat is most pronounced in the region of the column with the lowest conductivity. If the local heat becomes too high it causes convection and disturbances in the density gradient. If this occurs, it is observable as refractive fringes when looking through a well illuminated column, and especially against a crosshatched background. During the first 12 hours of a run, the total power can be higher because it takes some time for the carrier ampholytes to be transported and become focused. This means that the conductivity is more evenly distributed at the start.

During focusing, conductance will decrease and vary in different parts of the column because of the following factors: (a) As the carrier ampholytes focus and reach their pI, their conductance is lowered. (b) The resulting conductance is determined by the conductance of the individual ampholytes and their relative amount. The conductance is not the same in all parts of the pH gradient. (c) The conductance will be lower at higher sucrose concentrations. In order to compensate for this, the part of the pH gradient having the best conductivity, i.e., either the most acidic or the most alkaline pH range, is arranged at the bottom of the column.

When the column has been filled with a gradient mixer, the voltage can usually be set at the final value at the beginning. When the column has been filled manually, the voltage should be set 150 V below the final value during the first few hours.

The following voltage and focusing times serve as a rough guide for 1% Ampholine solutions

pH range	Voltage	Time in days
3–10	400 V	2
3–5	500 V	1–2
4–6	600 V	1–2
5–7	700 V	1–2
6–8	700 V	1–2
7–9	600 V	2
8–10	500 V	2

The approach to final focusing is indicated by a constant current at a constant voltage. Always allow the experiment to continue for several hours beyond the point at which constant voltage is reached for final focusing. Ranges of pH narrower than 2 pH units will require longer periods than those noted above.

Collection of Fractions

Turn off the power supply and close the valve at the bottom of the central tube. Suck out the electrode solution in the central tube with the aid of a capillary tube and a syringe; this procedure reduces the risk of this solution mixing with the effluent during draining.

Open the pinch clamp on the capillary tube at the bottom, and drain either with the aid of a pump or by gravity. A flow rate of 1–2 ml/min (4–6 ml/min) and a fraction volume of 2–5 ml are adequate in most cases. If desired, a photometer with a flowthrough cell and a recorder may be used. A fraction collector is helpful. Usually biological activities can be determined on suitable dilutions in the presence of sucrose and carrier ampholytes. Fractions of interest may be submitted to a second focusing in a narrow pH range, often by utilizing the carrier ampholytes present in the sample.

Zone Convection Electrofocusing[35a]

1. Fill the anode compartment with an acid, e.g., 0.1 M phosphoric acid, and the cathode compartment with a base, e.g., 0.1 M sodium hydroxide.

2. Fill the other compartments with a water solution of the carrier ampholytes and protein up to a level slightly below the ridges. Alternatively the protein may be placed in one or a few preselected compartments.

3. Connect cooling and power supply, and use a suitable voltage for a sufficient time according to the recommendations of the manufacturer.

4. Switch off the power supply.

5. Raise the cover and collect the fractions. If necessary, separate protein precipitates by centrifugation of the fractions.

Separation of Proteins from Carrier Ampholytes

Although many experiments, such as determination of isoelectric points, homogeneity, and comparative studies, can be performed with-

[35a]Shortly an apparatus will be commercially available through LKB-Produkter AB, 161 25 Bromma, Sweden.

out removing the carrier ampholytes, the purification of proteins usually demands such a step.

Since proteins will have a molecular weight above 10,000, whereas the average molecular weight of the Ampholine® carrier ampholytes has been estimated to 800, the difference in size makes molecular sieving an attractive method for separation. Dialysis against a buffer must be regarded as a slow procedure, especially if it is desired to remove the last traces of carrier ampholytes. However, if the dialysis system is well stirred, and the buffer is changed frequently, it is possible to remove 99% of the carrier ampholytes by dialysis for about 15 hours. In a series of experiments, a column (2.5 × 20 cm) of Sephadex G-50, Fine (Pharmacia, Uppsala, Sweden) has been used in studies on the separation of mixtures of proteins and labeled carrier ampholytes.[36] Complete separation can be obtained with sample volumes of up to about 12 ml at a flow rate of the buffer of 1.2 ml/min. The proteins are then eluted in the void volume of the gel within less than 1 hour. The dilution factor is then only about 1.5, and, if desired, a buffer such as ammonium bicarbonate or ammonium acetate may be used which can later be removed by lyophilization. Fractions taken from density gradient columns should not contain more than 35% (w/v) of sucrose. If the sucrose concentration in a fraction is higher, the fraction could be diluted with a buffer or submitted to dialysis for 1 or 2 hours to reduce the viscosity.

Another method for separation of proteins from carrier ampholytes is based on precipitation of the proteins with ammonium sulfate and subsequent washing of the precipitate with an ammonium sulfate solution.[37]

When ammonium sulfate is dissolved in water, an acidic pH usually develops due to the evaporation of ammonia. The highest yield of a protein is often obtained at a pH close to its isoelectric point. One should then remember that the p*I* obtained by isoelectric focusing refers to an ionic strength close to zero and that the p*I* value decreases as ionic strength increases. The degree of saturation that produces a good yield of the protein in the precipitate is selected. After addition of ammonium sulfate, the solution is allowed to stand for some hours. The precipitate is then centrifuged and suspended in a fresh solution of ammonium sulfate of the same concentration, but four times the volume, as used in the first precipitation. The protein is then again centrifuged. By repeating the procedure three times it has been possible to remove more than 99.99% of the carrier ampholytes present at the beginning.[37] If excess

[36]O. Vesterberg, *Sci. Tools* **16**, 24 (1969).
[37]P. Nilsson, T. Wadström, and O. Vesterberg, *Biochim. Biophys. Acta* **221**, 146 (1970).

sucrose is present at the start of the first precipitation, the concentration must first be reduced; otherwise it might be difficult to collect the protein precipitate after centrifugation. The carrier ampholytes can also be separated from proteins by electrophoresis.[38-40] Using electrophoresis in gel it is easy to illustrate the separation of proteins from the Ampholine chemicals by using Amido Black for staining without prewashing of the gel. The Ampholine will then appear as blue spots, with a higher mobility than the proteins.[38-40] This simple procedure may be used to test the efficiency of any separation procedure.

Other methods for separation of proteins from carrier ampholytes include the use of ion exchangers[41] and partition chromatography by counter current. These two methods are especially valuable for the separation of peptides from carrier ampholytes. In partition chromatography in organic solvent-water systems, it could be expected that the carrier ampholytes would be more hydrophilic than most peptides due to the abundance of amino- and carboxylic groups in the carrier ampholyte molecules.

[38]C. Wrigley, *Biochem. Genetics* 4, 509 (1970).
[39]V. Macko and H. Stegemann, *Hoppe-Seyler's Z. Physiol. Chem.* 350, 917 (1969).
[40]G. Dale and A. L. Latner, *Clin. Chim. Acta* 24, 61 (1969).
[41]P. Wallén and B. Wiman, *Biochim. Biophys. Acta* (1971).

[34] Preparative Acrylamide Gel Electrophoresis: Continuous and Disc Techniques

By Louis Shuster

Zonal electrophoresis in polyacrylamide gels gives very good resolution of charged macromolecules by separating them according to both charge and size. The process has been called "electrophoretic molecular sieving."[1] Acrylamide gels are easier to work with than starch gels, and have the additional advantage that pore size can readily be varied. A general description of acrylamide gel electrophoresis may be found in several recent monographs[2,3] and the report of a symposium.[4]

[1]S. Hjertén, S. Jerstedt, and A. Tiselius, *Anal. Biochem.* 27, 108 (1969).
[2]I. Smith, "Chromatographic and Electrophoretic Techniques," Vol. II, "Zone Electrophoresis." Wiley, New York, 1968.
[3]G. Zweig and J. P. Whitaker, "Paper Chromatography and Electrophoresis," Vol. I, "Electrophoresis in Stabilizing Media." Academic Press, New York, 1967.
[4]*Ann. N. Y. Acad. Sci.* 121 (2), 305-650 (1964).

This chapter deals with methods for continuous and discontinuous preparative electrophoresis. The same apparatus can be used for both techniques, and the possible advantages and disadvantages of each are pointed out in appropriate sections.

The capacity of most preparative acrylamide gel systems is rather small—often 100 mg of protein or less, although one recently described column has a capacity of 1 g.[1] Preparative electrophoresis is therefore not generally used as an initial step in purification, but rather as one of the last steps.

Preliminary Experiments

The most important requirement for carrying out successful preparative gel electrophoresis is extensive experience with analytical gel electrophoresis. In the following description it is assumed that the reader already has such experience, and the emphasis is on differences between preparative and analytical electrophoresis of proteins.

In order to choose an appropriate gel porosity and buffer system, it will be necessary to have some idea of the size and charge of the protein that is being purified. The molecular weight of an enzyme in a crude extract can be estimated by gel filtration[5] or by ultracentrifugation in a sucrose gradient.[6] The charge can be deduced by testing the adsorption of the enzyme on different ion-exchanger celluloses.

The next step is to work with analytical gel electrophoresis until a clear pattern of stained bands is obtained. Analytical electrophoresis should be used to monitor all the preliminary purification steps. In many cases this monitoring may indicate which protein is being selectively concentrated during the purification. If it does not, some other way has to be found to establish which band in the analytical gel pattern represents the protein that is being purified. For some enzymes, especially dehydrogenases and hydrolases, modified histochemical techniques can be used to detect the enzyme within the gel.[2,3,7] These tests may also reveal the presence of isozymes.

If no suitable staining method is available, it will be necessary to attempt to elute the enzyme from the gel. Preliminary experiments should be run to show that the enzyme is stable in the buffers used and at the pH values encountered within the gel during electrophoresis. For discontinuous systems the running pH of acrylamide gels is about 0.5 pH unit above the nominal pH value for alkaline buffers, and about 0.5 pH unit below it for acidic buffers.[8] Different amounts of crude or

[5] P. Andrews, *Biochem. J.* 96, 595 (1965).
[6] R. G. Martin and B. N. Ames, *J. Biol. Chem.* 236, 1372 (1961).
[7] O. Gabriel, this volume [40].
[8] B. J. Davis, *Ann. N. Y. Acad. Sci.* 121, 404 (1964).

partially purified enzyme should be applied to several analytical gels. After electrophoresis for a short time (10 or 15 minutes) the gel is cut at the dye front, and the whole section that contains proteins is homogenized in a suitable buffer. This procedure will determine whether any enzyme activity can survive the conditions of electrophoresis, and also how much enzyme will have to be applied in order to yield enough activity for detection. It is advisable to assay the whole homogenate rather than an extract. Even if the enzyme is not eluted readily, it may still act on substrates that diffuse into and yield a product that diffuses out from, the gel particles.

If measurable activity is obtained in this manner, the next step is to run a regular analytical electrophoresis. The unstained gel is sliced into sections 1–2 mm thick and each section is broken up and assayed separately. For greater sensitivity the sections can be broken up in the incubation mixture. Sometimes freezing the gel beforehand helps in sectioning and extraction. Various slicing devices have been described, and several are available commercially.[9]

At this point the investigator should know the mobility of the enzyme relative to the dye front or another marker. Additional experiments can now be carried out in order to determine which gel pore size and buffer system give the best separation of the desired protein from impurities. Conditions should be chosen so that the mobility of the protein relative to that of the dye front is at least 0.2, and preferably greater than 0.3.

Scaling Up

Before proceeding to the more complicated commercial apparatus, it is very helpful to work initially with a simple column such as the one shown in Fig. 1. A useful column can be readily made from a 2 × 15 cm test tube with a flared top. The tube is placed in a test tube rack and a pipette or file is dropped into it several times in order to make a good-sized hole in the bottom. The bottom is then made even by filing it with wire gauze and ground smooth on emery cloth or sandpaper. The upper chamber consists of a quart-size polyethylene bottle with most of the bottom cut out. The column is kept in place with a rubber stopper. Electrodes can be made of a few turns of 30-gauge platinum wire wrapped around glass or plastic. Inexpensive electrodes can be obtained from Fisher Scientific Co. (No. 9-460). For casting the gel, the flared end of the column is covered with Saran wrap or dialysis tubing that is stretched taut with a rubber band. If dialysis tubing is used, it can remain on the column during electrophoresis.

[9]R. F. Goldberger, *Anal. Biochem.* 25, 46 (1968).

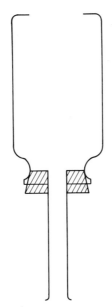

Fɪɢ. 1.   A simple column for scaling-up experiments.

For cooling, the beaker of lower buffer may be placed in a transparent ice bath—such as a plastic mouse cage—and the lower buffer can be mixed with a magnetic stirrer. Alternatively, electrophoresis may be carried out in a cold room.

Colored proteins should be used for these preliminary runs—e.g., hemoglobin, and serum albumin stained with bromophenol blue dye for anionic systems; cytochrome c and serum albumin stained with bromophenol blue dye for cationic systems (down to pH 4). Thus, for the widely used pH 8.3 Tris·glycine buffer system a suitable sample might contain 2 mg of bovine serum albumin, 2 mg of hemoglobin, and 0.1 ml of 0.05% bromophenol blue (made up in 0.005 N NaOH) in 1 ml of 10% sucrose or glycerol. A series of runs can be carried out, at the rate of 1 or 2 a day. The amount of protein applied should be increased in steps to the point of frank overloading. Gel height should be varied, from 1 or 2 cm to perhaps 8 cm. The investigator should systematically establish the effect of omitting spacer and sample gels, of overheating by raising the current or omitting cooling, of adding salt, and of distorting the gel surface. In some cases, he should electrophorese the colored proteins right off the bottom of the gel in order to watch their behavior during elution.

For detection of the oligomeric bands of serum albumin it will probably be necessary to fix and stain a small section of the gel, employing the usual techniques for analytical gels. Staining should also be used for complex mixtures, such as serum. Electrophoresis of serum in amounts ranging up to 0.2 or 0.3 ml should illustrate the problems that result from clogging of the gel surface by high molecular weight proteins in a complex mixture.

Finally, the actual sample that is being purified should be tested under conditions close to those that will be used in the preparative apparatus. In order to obtain some estimate of recovery, the enzyme can be eluted electrophoretically into a small dialysis bag filled with lower buffer and attached to the bottom of the column with a rubber band.

At this point, the investigator should no longer feel hesitant about assembling and using a rather complicated piece of apparatus. There are many laboratories where such apparatus is gathering dust on the shelf simply because enthusiastic use without proper preparation has led to initial failures and disappointment.

Preparation of Gels and Buffers

Purification of Monomers

The purity of acrylamide and bisacrylamide is more important for preparative than for analytical electrophoresis because of the problems that can result from inhomogeneities or from swelling of the gel. Some suppliers (Eastman, Canalco) now sell special grades of acrylamide for preparative electrophoresis. Standard grades (e.g., Eastman No. 5521) can be purified by one of the following recrystallization procedures.

It is important to remember that both monomers are *neurotoxic*, and contact with the dry powders or with solutions or inhalation of the dust should be avoided. These procedures should be carried out in a well-ventilated hood.

Recrystallization of Acrylamide.[10] Dissolve 70 g of acrylamide in 1 liter of reagent grade chloroform at 50°. Filter while hot without suction. Cool the filtrate to −20° and filter off the crystals on a Büchner funnel. Wash the crystals briefly in cold chloroform and dry in a vacuum desiccator. Store in a dark bottle.

Recrystallization of Bisacrylamide. Dissolve 12 g of bisacrylamide in 1 liter of reagent grade acetone at 40°–50° and filter while hot. Cool slowly to −20°. Filter or centrifuge to collect the crystals, wash with cold acetone, and dry in a vacuum desiccator. Store in a dark bottle.

Preparation of Gel Solutions

It is advisable to use freshly prepared monomer solutions for making preparative gels. For analytical gels and preliminary scaling up, the

[10]V. E. Loening, *Biochem. J.* 102, 251 (1967).

acrylamide-bisacrylamide solution can be stored at 4° in a brown bottle for at least 1 month. Solutions of ammonium persulfate are stable for 2 weeks at 4°.

The composition of the separating gel will vary, depending on the buffer system and the size of the macromolecules being separated. Some guidelines are given in Table I.

It is obvious that the short columns of dilute gel that are necessary in order to allow the elution of high molecular weight proteins within a reasonable time will require some mechanical support. This may be achieved by using nylon mesh[11] or a porous plastic disc or ring.[12]

The height of the separating gel column or slab is related to the mobility and to the method of elution. If elution is by electrophoresis from the end of the column, the gel height should be sufficient to give good *temporal* separation between protein peaks in the eluate—e.g., they should emerge from the column separated by an interval of at least 15 minutes. This may occur even though the protein bands are very close together in an analytical gel.

If the separated bands are to be cut out of the gel before elution, then a greater length of gel should be used so that *spatial* separation of at least 3 mm between bands can be achieved by increasing the length of time for which current is applied.

Gel Formulations

A few of the commonly used gel formulations are given here. In *each case* the stock solution is made up to a final volume of 100 ml with distilled water. For convenience, the three stock solutions are prepared so that the final gel solution can be made by mixing: 1 part monomer

TABLE I
SOME GUIDELINES FOR CHOOSING SEPARATING GELS

Molecular weight of protein to be separated	Acrylamide in separating gel (%)	Height of separating gel (cm)
10,000– 40,000	15–20	3–6
40,000–100,000	10–15	2–4
100,000–300,000	5–10	2–3
300,000–500,000	5	0.3–1
>500,000	2–5	0.3–1

[a]Adapted from the Canalco Prep-Disc Instruction Manual, Canal Industrial Corporation, Rockville, Maryland, 1968.

[11]Prep-Disc Instruction Manual, Canal Industrial Corporation, Rockville, Maryland, 1968.
[12]J. K. Smith and D. W. Moss, *Anal. Biochem.* 25, 500 (1968).

solution, 1 part buffer solution, 2 parts ammonium persulfate solution (Fisher No. A-682) (usually 0.14 g/100 ml).

For Separating Gels

Some manufacturers recommend the use of acrylamide and bis-acrylamide in the same proportions as for analytical gels. The Canalco procedure[11] calls for much less bisacrylamide in order to increase mobility and to decrease any tendency for the gel to pull away from the walls of the column. The concentration of acrylamide in the gel is given in the second column of Table II as a straight percentage value. The third column of Table II uses the notation of Hjertén,[13] where T = total monomer concentration (acrylamide + bisacrylamide) and C = the concentration of bisacrylamide expressed as a percentage of T. The relationship of electrophoretic mobility to T and C has been described by Morris[14] and by Hjertén et al.[1]

Buffers for Separating Gels

A. For pH 8.9 gels, to be used with the discontinuous buffer system of Davis[8]

HCl, 1.0 N	24.0 ml
Tris	18.15 g
TEMED[15]	0.23 ml

TABLE II
MONOMER SOLUTIONS FOR SEPARATING GELS

Acrylamide (g)	Bisacrylamide (g)	Final acrylamide Conc. (%)	T (%)	C (%)	Reference
40.0	0.12	10	10	0.3	a
30.0	0.8	7.5	7.7	2.6	b
20.0	0.68	5	5.2	3.3	c
19.0	1.0	4.75	5	5	d
22.8	1.2	5.7	6	5	e

[a] Prep-Disc Instruction Manual, Canal Industrial Corporation, Rockville, Maryland, 1968.
[b] Buchler Poly-Prep Instruction Manual, Buchler Instruments, Inc., Fort Lee, New Jersey, 1968.
[c] LKB 7900 Uniphor Instruction Manual, LKB-Produckter AB, Stockholm, Sweden, 1968.
[d] S. Raymond, in "Methods in Immunology and Immunochemistry" (C. A. Williams and M. W. Chase, eds.), Vol. 2, p. 47. Academic Press, New York, 1968.
[e] S. Hjertén, S. Jerstedt, and A. Tiselius, Anal. Biochem. 27, 108 (1969).

[13] S. Hjertén, Arch. Biochem. Biophys. 98, Suppl. 1, 147 (1962).
[14] C. J. O. R. Morris, Protides Biol. Fluids, Proc. Colloq. 14, 543 (1966).
[15] Abbreviation for tetramethylethylenediamine (MCB No. 8563).

This amount of TEMED will result in a gelling time of 1 to 2 hours at 0°. In order to decrease the gelling time to between 20 and 30 minutes at this temperature, the amount of TEMED is increased to 0.40 ml, and the concentration of ammonium persulfate stock solution is raised to 0.40 g/100 ml.

B. For pH 4.3 gels to be used with the discontinuous buffer system of Reisfield, Lewis, and Williams[16]

KOH, 1.0 N	24 ml
Glacial acetic acid	11.2 ml
TEMED	2.3 ml

Here, too, the concentration of the ammonium persulfate stock solution should be increased to 0.4 g/100 ml.

C. For pH 8.0 gels to be used with the continuous buffer system of Hjertén et al.[1]

Acetic acid, 1.0 N	10.0 ml
Tris	2.4 g
TEMED	0.4 ml

The higher ammonium sulfate concentration is recommended in order to decrease gelling time.

D. For a pH 8.4 gel to be used with the continuous buffer system of Raymond.[17]

Tris	0.43 g
Na₂EDTA	0.37 g
Boric acid	2.2 g
TEMED	0.4 ml

Buffers for Spacer Gels

A. For use with the discontinuous buffer system of Davis

1.	1.0 N HCl	24 ml	
	Tris	3.0 g	pH 6.8
	TEMED	0.23 ml	

or

2.	1.0 M H₃PO₄	12.8 ml	
	Tris	2.85 g	pH 7.2
	TEMED	0.1 ml	

For polymerization mix 1 part monomer solution with 1 part buffer and 2 parts of riboflavin solution, 1 mg/100 ml, and expose to fluorescent light.

[16]R. A. Reisfield, V. J. Lewis, and D. E. Williams, *Nature (London)* **195**, 281 (1963).
[17]S. Raymond, *in* "Methods in Immunology and Immunochemistry" (C. A. Williams and M. W. Chase, eds.), Vol. 2, p. 47. Academic Press, New York, 1968.

The catalyst solution for photopolymerization consists of riboflavin, 1 mg/100 ml, instead of ammonium persulfate. (Buchler[18] recommends adding 20% sucrose to the riboflavin solution to provide a stronger spacer.)

B. For use with the discontinuous buffer system at pH 4.3

1 N KOH	24.0 ml	
Glacial acetic acid	1.44 ml	pH 5.8
TEMED	0.23 ml	

Urea can be added to the various gel components to a final concentration of 6 M (36 g/100 ml). Because this amount of urea increases the volume of solutions considerably, it may be necessary to use a more concentrated solution of acid in making up the buffers. Urea also gives a tighter gel, so that the concentration of acrylamide in the separating gel can be decreased by 1 or 2%.

Electrode and Eluting Buffers

A discontinuous buffer system is most commonly used for analytical gels. That is, the composition, pH, and ionic strength of the buffer in the separating gel is different from that of the electrode buffers. The advantage of a discontinuous system is that it provides a very sharp starting zone, even with fairly large sample volumes. The theoretical basis for discontinuous buffers has been described by Ornstein.[19] However, very good resolution can also be obtained with continuous buffer systems, i.e., those in which the electrode and gel buffers are the same although these usually require careful application of a small volume of concentrated sample. A continuous system has the advantage of removing limitations on the length of the prerun (see below) or on the recirculation of buffer between the electrode chambers.

TABLE III
MONOMER SOLUTIONS FOR SPACER (LARGE PORE) GELS

Acrylamide (g)	Bisacrylamide (g)	Acrylamide (conc. %)	T (%)	C (%)	Reference
10.0	2.0	2.5	3.0	16.6	a
14.0	0.25	3.5	3.5	1.8	b

[a] Buchler Poly-Prep Instruction Manual, Buchler Instruments, Inc., Fort Lee, New Jersey, 1968.
[b] Prep Disc Instruction Manual, Canal Industrial Corporation, Rockville, Maryland, 1968.

[18] Buchler Poly-Prep Instruction Manual, Buchler Instruments, Inc., Fort Lee, New Jersey 1968.
[19] L. Ornstein, *Ann. N. Y. Acad. Sci.* 121, 321 (1964).

The electrode buffer is the same in both chambers for discontinuous analytical electrophoresis. In preparative discontinuous electrophoresis the composition of the buffer in the lower electrode chamber may resemble the separating gel rather than the upper buffer. Some examples are given below. All formulations are for a final volume of 1 liter. Each electrode chamber should contain at least 1 liter of buffer.

For discontinuous systems[11]

  1. Canalco System

Tris	3.0 g	
Glycine	14.4 g	0.025 M, pH 8.3

This is used for both the upper and lower electrode buffers. The elution buffer consists of a ⅛ dilution of the same Tris·HCl buffer used in the separating gel—i.e., 0.047 M final, pH 8.8. The reason for these differences will be explained in the description of the eluting chamber.

  2. Alkaline System[18]

Upper buffer	Tris	6.32 g	
	Glycine	3.94 g	0.05 M, pH 8.9
Lower and elution buffer	Tris	12.1 g	
	1.0 N HCl	50.0 ml	0.1 M, pH 8.1

  3. Acid System[16]

β-Alanine	31.2 g	
Glacial acetic acid	8.0 ml	pH 4.5

In order to preserve enzyme activity, it is possible to add sucrose or glycerol to the eluting buffer in concentrations of up to 20%, as well as 1 mM EDTA and mercaptoethanol or dithiothreitol (e.g., see Kohn and Jakoby[20]).

Power Supply and Electrodes

A well-regulated DC power supply that can provide up to 500 V at 0–100 mA is usually sufficient. A more sophisticated power supply has been described that features intermittent pulsing; pulsing at a rate of 75–300 pulses per second is claimed to reduce heating.[21]

The electrodes should be of platinum wire. These are usually supported on plastic for strength and placed symmetrically with respect to the gel surfaces.

Because of the high voltages used in electrophoresis, it is most desirable to have a safety interlock on the power supply so that the current

[20]L. D. Kohn and W. B. Jakoby, *J. Biol. Chem.* 243, 2486 (1968).
[21]Ortec Model 4200 Electrophoresis System User's Manual, Oak Ridge, Tennessee, 1969.

must be interrupted for access to the electrodes or buffer chambers. Where the electrophoresis apparatus is set up in a laboratory hood, a pressure-operated switch can be attached to the sash so that the power supply is on only when the hood window is down. It is also important to post a sign pointing out the danger whenever the power supply is in use. Safety recommendations for the use of high voltage power include proper grounding of the power supply.[22]

Cooling Requirements

Heat dissipation can be a problem in the scaling up of gel electrophoresis. The temperature rise will be greatest at the center of the gel, with the result that protein bands will be bowed downward. This may happen even though the dye front appears to be quite straight. For this reason, many preparative columns have a cold finger in the center as well as an outside jacket. With a short, wide column of gel, enough cooling can be achieved by recirculating the buffers.[23] The column may be cooled with ice water flowing at a rate of several liters a minute. For keeping the temperature of the gel close to $0°$ during electrophoresis, it is necessary to use a cooling solution at -2 to $-4°$.

The best indication that cooling is uniform and sufficient is the absence of distortion in protein bands. The simplest way to decrease any central distortion is to reduce the applied current.

Pumps

In addition to a pump for circulating cooling fluid it is very useful to have one or more well-controlled peristaltic pumps. These can be used for layering water over the gel during casting, for layering the sample evenly on the surface of the gel, for recirculating the buffer between electrode chambers in a continuous system, and for elution of bands as they emerge from the gel.

UV Monitor and Recorder

These may be used to measure the optical density at 280 mμ of protein fractions as they emerge from the eluting chamber on their way to the fraction collector.

Casting the Separating Gel

The column should be cleaned with dilute detergent, rinsed with distilled water, and air dried. The bottom of the gel column is closed off with a leakproof support that has a flush, flat surface, and the column

[22]E. W. Spencer, V. M. Ingram, and C. Levinthal, *Science* 152, 1722 (1966).
[23]A. D. Brownstone, *Anal. Biochem.* 27, 25 (1969).

is leveled. The stock solutions are cooled to the temperature at which the column will be run, and mixed in the proper proportions. The gel solution should be degassed under vacuum in a heavy-walled filter flask for a few minutes, preferably with magnetic stirring. This procedure will prevent the formation of bubbles in the gel.

If a porous plastic support is to be fixed to the bottom of the gel column, it should be saturated with the gel solution, preferably by vacuum infiltration, before the gel column is cast.

The amount of gel solution should be sufficient to give the required height of gel plus an additional 10 or 15% to allow for the volume of gel solution that does not polymerize at the top of the gel. The gel is poured down the side of the column to avoid bubble formation.

About 0.5 cm of water is carefully layered over the gel. The addition of a small amount of detergent to the water helps to achieve a smooth gel surface. The water is added very carefully through a small-bore polyethylene tube touching the wall of the chamber a few millimeters above the surface of the gel solution, by means of a syringe or a peristaltic pump. Distortion of the gel surface may result if the water is dropped into the gel solution or applied too rapidly. The gel is allowed to set at the temperature that will be used for electrophoresis. Setting may require 1 or 2 hours at 0° unless extra TEMED and ammonium persulfate are added to the standard formula for analytical gels. It has been recommended that the time for polymerization be kept to 20 or 30 minutes in order to produce a uniform gel.[1,18] Because the polymerization is exothermic, cooling solution should be circulated through the column during the casting and setting of the gel.

When polymerization of the separating gel is complete, a sharp refractile line will appear a few mm below the original level of the gel solution. The overlying liquid should be removed carefully with polyethylene tubing attached to a peristaltic pump or a hypodermic syringe. Care should be taken not to touch the gel surface. The surface of the gel should be washed twice with 10 ml of the spacer gel solution or with distilled water ia a spacer gel is not used.

Casting the Spacer Gel

It may be possible, especially with small volumes of fairly simple protein mixtures, to omit spacer gel and apply the sample directly to the surface of the separating gel. However, the use of spacer gel provides several advantages.

1. In some preparative columns (e.g., Canalco) the total gel length must be at least 4 cm in order to give the necessary mechanical strength. Yet in many cases it is undesirable to have a separating gel longer than

1 or 2 cm so that the desired fractions will elute within a reasonable time and without too much spreading by diffusion. The difference can be made up with a loose spacer gel.

2. With a high protein load or a mixture containing very large protein molecules there may be complexing and precipitation as the protein concentrates at the surface of the gel. The precipitate may clog the separating gel and cause the upper surface to shrink so that the edges pull away from the chamber and bands are distorted. The use of a spacer gel will prevent clogging of the surface of the separating gel. Distortion of the surface of the spacer gel is less likely to cause problems in separation.

3. The use of a spacer gel permits the application of fairly large sample volumes (up to 125 ml in the Buchler Polyprep Column).[18] The reason is that the spacer gel permits the slowest-moving protein bands to catch up with the fastest-moving bands before they enter the separating gel.

4. A spacer gel decreases interference by salt in the sample. Salt delays sharpening of bands and may decrease separation.

5. The use of a spacer gel eliminates the need for a prerun. The spacer gel is polymerized with riboflavin, and any residual ammonium persulfate in the separating gel is well out of the way before the protein bands arrive.

The volume of the spacer gel should equal the volume of the sample. The stock solutions are cooled and mixed, and the mixture is deaerated. The spacer gel solution is carefully applied over the separating gel and overlayed with about 0.5 cm of water. The gel solution is then illuminated with 2 vertical fluorescent lamps until polymerization is complete —about 30–45 minutes. The overlying liquid is then removed, and the gel surface is washed twice with about 10 ml of upper buffer. The washing must be done carefully because the spacer gel is rather fragile.

Preparation and Application of Sample

The sample should be concentrated to a volume of about 5 ml for continuous systems and about 10–20 ml for discontinuous systems. The capacity of the column will depend on the cross sectional area of the gel, and may range up to 5 mg of protein per square centimeter. The initial salt content should be less than 0.05 M, and the sample should be dialyzed against or diluted with the buffer used in the spacer gel. If spacer gel is omitted, the sample should be dialyzed against a ⅛ dilution of the separating gel buffer—e.g., 0.05 M Tris·HCl, pH 8.3, for the Tris·glycine system. A drop of 0.05% bromophenol blue dye is added to the sample in order to provide a dye front, and also to make it easier to watch the layering process. If there is any turbidity in the sample, it should be removed by high speed centrifugation or filtration through a Millipore filter. Otherwise there may be clogging of the gel surface. Sucrose or glycerol is added to a final concentration of 5 or 10% in order to make the sample dense enough to layer on the surface of the gel.

The sample may be concentrated by ultrafiltration or by dialysis against Ficoll, solid Carbowax, dry Sephadex or dry sucrose. However, it is inadvisable to have more than 20% sucrose in the sample because high concentrations of sucrose may cause shrinking of the gel surface and distorted bands.

It is well worth the effort to run a preliminary analytical electrophoresis on the protein sample in its final form, using the same gel system and a proportional load. The sample should enter the gel in 5–10 minutes, with zone sharpening from the front of the sample solution. Slow entry and zone sharpening from the top of the sample solution is a strong indication that there is too much salt. This problem seems to be more common when phosphate buffers are used.

The sample can also be incorporated into a loose gel that is polymerized on the spacer gel, provided that the conditions for photopolymerization do not inactivate the protein that is being purified. The stock solutions used for the Tris·glycine discontinuous system are:

Monomer:	Acrylamide	14.0 g
	Bisacrylamide	0.25 g
	Water to 100 ml	
Buffer:	NCl, 1.0 N	48.0 ml
	Tris	5.98 g
	TEMED	0.46 ml
	Water to 100 ml	
Catalyst:	Riboflavin, 4.0 mg/100 ml	

Four parts of sample is mixed with 1 part of buffer, 2 parts monomer, 1 part catalyst and a drop of 0.05% bromophenol blue dye. The gel is overlayered with water and polymerized by exposure to fluorescent light for 30–45 minutes.[11]

The advantage of a sample gel is that it prevents diffusion of the sample by convection at the start of electrophoresis and gives sharper starting zones. Most workers do not use a sample gel in preparative electrophoresis.

The upper electrode chamber is partially or completely filled with electrode buffer, and the sample is layered on the surface of the gel by means of fine (23 gauge) polyethylene tubing attached to a hypodermic syringe or a peristaltic pump.

All connections are checked, assembly of the apparatus is completed, and the power supply is turned on. A low current is used at the start of electrophoresis so that the sample will enter the top of the gel smoothly and with minimal heating or convection. This may take 1 hour or more. After the sample has entered the gel the current may be increased in order to give a reasonable rate of separation consistent with the need to prevent overheating. The rates obtained in analytical electrophoresis—i.e., separation within 5 cm of gel in 1 hour or less—cannot be achieved in preparative electrophoresis, and should not be attempted.

Preparative separations commonly require from 5 to 30 hours.

Collection of Separated Proteins

By Elution of Gel Sections

If it has been possible to extract the protein in good yield on a small scale, the same procedure may be used with large gel columns or slabs. In order to locate the protein band to be extracted, a small section or wedge (about 5 mm thick) is cut from the length of the gel and stained for 30–60 minutes. The usual stains are amido black (buffalo black NBR or naphthol blue black) or procion brilliant blue (Colab), 0.5% in 7.5% acetic acid. The remainder of the gel is covered in Saran wrap and frozen or stored at 4° until destaining has been completed. For destaining, the gel section is left overnight in a beaker containing about 100 ml of 7.5% acetic acid and a little Dowex 1-chloride resin. The resin adsorbs the dye as it leaches from the gel and speeds destaining.

A more rapid method for locating protein bands in acrylamide gels makes use of a fluorescent dye.[24] The gel is immersed for a few minutes in 0.1 M phosphate buffer, pH 6.8 containing 0.003% 1-analino-8-naphthalene sulfonate (magnesium salt, Eastman Organic Chemicals). This dye binds strongly to native albumin and to all denatured proteins to give a highly fluorescent complex. There is usually enough surface denaturation to visualize many protein bands. The sensitivity may be enhanced by exposing the gel to air for a few minutes or by immersing it for a few seconds in 2 N HCl. When exposed to longwave ultraviolet light, the bands fluoresce yellow or yellow-green. So little dye is taken up in this procedure that it is not likely to interfere in subsequent assays. If desired, the dyed portions may be shaved from the surface of the gel before the band is extracted.

The gel section is homogenized with about 10 volumes of buffer, either in a laboratory homogenizer or by repeated forcing through a hypodermic syringe. The homogenate is allowed to extract overnight

---

[24]B. K. Hartman and S. Udenfriend, *Anal. Biochem.* **30**, 391 (1969).

at 4°, with shaking if necessary. The gel particles are removed by centrifugation and washed once with a little buffer.

In many cases such extraction will remove only a small portion of the protein in the gel band. An alternate method of elution employs electrophoresis to move the protein out of the sectioned gel. A useful apparatus has been described by Sulitzeanu *et al.* for eluting the protein from a section of 2.8 cm diameter gel in about 4 hours.[25] The protein is eluted into about 4 ml of buffer contained in an elution chamber that is sealed off with a semipermeable membrane.

The column shown in Fig. 1 can also be used for electrophoretic elution. The gel section is homogenized in spacer gel buffer and centrifuged. The supernatant solution is decanted, and the broken up gel is mixed with an equal volume of spacer gel solution. The mixture is poured into the column and photopolymerized. A small dialysis bag filled with electrode buffer is attached to the lower end of the gel column with a rubber band and the electrophoresis is carried out for 2 or 3 hours with cooling. An elution chamber can be made by using a rubber stopper instead of Saran wrap to close off the column before the gel is poured. After polymerization the stopper is removed carefully to leave a space 3 or 4 mm deep. This space is filled with electrode buffer or Sephadex G-25 suspended in electrode buffer, and closed off with dialysis tubing to form an elution chamber. The use of a glass microscope slide to move the dialysis tubing into place will help to prevent air bubbles being trapped in the chamber.

Separated protein bands can also be eluted electrophoretically from a gel slab without sectioning. This is done by turning the slab 90° and placing it between 2 electrode grids that are contained in dialysis tubing.[26] The proteins are eluted by electrophoresis through the narrowest dimension of the gel slab. As they emerge from the gel the proteins encounter the dialysis membrane and sink to the bottom of the chamber, where they are collected in a series of narrow reservoirs filled with buffer. A 40% recovery of serum proteins has been reported for this method.[27]

By Elution during Electrophoretic Separation

In most preparative systems the electrophoresis is continued until protein zones migrate from the end of the gel. The emerging proteins are trapped within a small elution chamber that is separated from the lower electrode buffer by a semipermeable membrane.

[25]D. Sulitzeanu, M. Slavin, and E. Yecheskeli, *Anal. Biochem.* 21, 57 (1967).
[26]S. Raymond and E. M. Jordan, *Separ. Sci.* 1, 95 (1966).
[27]J. Broome and E. M. Jordan, *Separ. Sci.* 1, 319 (1966).

The use of semipermeable membranes presents certain problems that have been discussed by Hjertén et al.[1] Electrophoresis can change the pH and salt concentration at the surface of the membrane, and the emerging proteins may become denatured or adsorbed to the membrane. One possible solution to this problem is to design the apparatus so that the eluting buffer flows along the undersurface of the membrane, where ionic changes are opposite to those on the upper surface, before it enters the elution chamber.[28]

Various strategies may be employed for slowing the protein zones as they enter the eluting chamber, thereby decreasing the opportunity for convection and contact with the membrane. These include increasing the ionic strength and decreasing the pH (in an alkaline system) of the eluting buffer, adding sucrose to the eluting buffer, or packing the elution chamber with granular 4% agarose gel.[1] Increasing the flow rate of eluting buffer will also decrease adsorption to the membrane but results in very dilute solutions of protein.

The volume of the elution chamber is kept small in order to decrease dilution; the distance between the membrane and the end of the gel is often only 1–2 mm. Because the membrane is somewhat flexible, and the lower gel surface becomes convex during electrophoresis, there is a danger that the elution slit may become blocked. For this reason some columns have provisions for changing the position of the membrane.[11,18] In some cases the membrane is fixed on a rigid, porous support. A rigid membrane of semipermeable glass has also been used.[13] The bottom surface of the gel may be stabilized by using a porous polyethylene disc or nylon mesh for support.

A crosswise flow of buffer may be used to sweep proteins from the elution chamber. Because the temperature of the gel is higher than that of the buffer, there is a temperature gradient in the elution chamber that causes proteins to float just below the lower surface of the gel. A more efficient arrangement that is commonly employed is to have the buffer flow inward from several directions and then up a central channel that passes through the gel.

Discontinuous collection of eluted fractions can be achieved by using a solenoid valve to empty the elution chamber completely and refill it with fresh eluting buffer at programmed intervals. This procedure decreases dilution and the opportunity for contamination between zones. An automatic switch for turning off the current before the elution chamber is emptied may also be incorporated in the system. Two commercially available preparative columns employ discontinuous

[28]H. Hochstrasser, L. T. Skeggs, Jr., K. E. Lentz, and J. R. Kahn, Anal. Biochem. 6, 13 (1963).

collection.[23,29] The rate of collection can be programmed so that slow-moving bands may be collected in small, concentrated fractions just as easily as fast-moving bands, and with little or no mixing.

Analysis of Fractions

One consideration in the selection of gel and buffer systems for preparative electrophoresis is the kind of assays that will be carried out on the collected fractions. The most common tests are for protein content and biological activity, and for homogeneity—usually by analytical gel electrophoresis. Some elution buffers can interfere with these tests. Thus, Tris buffers in concentrations above 0.03 M can interfere with both the biuret reaction and the Folin-Lowry method. Often the optional density at 280 mμ will give sufficient indication of where proteins are eluted. Several ways of removing substances that interfere with the Lowry assay are described in this volume [35].

Because the eluting buffer is often similar in composition to the buffer in the separating gel or the upper electrode buffer, it may be possible to carry out analytical electrophoresis without any additional manipulations. For example, if the eluting buffer is Tris·glycine or Tris·HCl, the samples may be layered on separating gels containing Tris·HCl as in the usual Davis system. However, the amount of salt in the eluate may be sufficient to interfere and prior dialysis may be required. Electrophoresis of individual fractions will indicate the degree of separation that has been achieved in preparative electrophoresis. In order to establish that the purified fractions are homogeneous, it is usually necessary to combine active fractions, dialyze and concentrate, and assay for protein content. Samples ranging from 10 to about 400 μg of protein are then examined by analytical electrophoresis in order to reveal the presence of small amounts of impurities.

Disassembly

At the completion of preparative electrophoresis, the gel column is removed and its exact height is measured. A thin section is stained in order to determine which bands have remained in the column. Staining may also reveal inhomogeneity and surface clogging by precipitated or complexed proteins. These precipitates take the form of a strongly stained crust.

The whole gel column can be stained and destained electrophoretically while in place.[11] Both upper and lower buffer are replaced with 7% acetic acid. About 2 cm of 0.5% amido black or procion blue dye

[29]I. Schenkein, M. Levy, and P. Weis, *Anal. Biochem.* 25, 387 (1968).

in 15% acetic acid is layered on the gel, and electrophoresis is continued until the dye moves through the column in the direction of the cathode, leaving stained bands in a clear gel.

Where a discontinuous buffer is employed the gel cannot be re-used. In the case of a continuous buffer system the gel is not suitable for re-use because of residual slow-moving protein bands, surface clogging, and distortion of the gel.

Some Typical Separations

A few typical enzyme purifications by preparative gel electrophoresis are shown in Table IV. The buffer in all cases was the discontinuous Tris·glycine system of Davis.[8] Most of these experiments used commercial apparatus such as that listed in Table V.

TABLE IV

SOME TYPICAL ENZYME PURIFICATIONS BY PREPARATIVE ACRYLAMIDE GEL ELECTROPHORESIS

Starting material	Protein load Mg	Protein load Ml	Apparatus	Gel height Sep. gel (cm)	Gel height Spacer gel (cm)	Current and/or Voltage	Length of run (hours)	Yield (%)	Purification (fold)	Reference
Tartrate dehydrogenase from *Pseudomonas putida* (step 6)	18	4	Canalco Prep-Disc	3	1	15 mA	7	70	1.25	a
Oxaloglycolate reductive decarboxylase from *Pseudomonas putida* (step 5)	76	30	Buchler Poly-prep	6	—	50 mA	33	74	10.7	b
Thioredoxin reductase from *Escherichia coli*	22	50	Buchler Poly-prep	5	2.5	85 mA 200 V	8	90	1.3	c
Thymidylate kinase from: *E. coli*	85	5	Canalco Prep-Disc	3	1.5	15 mA	18	62	3.5	d
	15	5	Canalco Prep-Disc	3 (15%)	—	15 mA	18	55	7.5	d
Tryptophan oxygenase from *Pseudomonas acidovorans*	13.5	12.5	2.6 cm diameter column	12		300 V	18	32 (Extraction of gel section)	3.3	e

TABLE IV (*Continued*)

SOME TYPICAL ENZYME PURIFICATIONS BY PREPARATIVE ACRYLAMIDE GEL ELECTROPHORESIS

Starting material	Protein load Mg	Protein load Ml	Apparatus	Gel height Sep. gel (cm)	Gel height Spacer gel (cm)	Current and/or Voltage	Length of run (hours)	Yield (%)	Purification (fold)	Reference
Inorganic pyrophosphatase from *E. coli*	160	8	Buchler Poly-prep	6.9	2.5	40 mA	4	100	28	*f*
Amylase from *Bacillus macerans*	175	11.5	Buchler Poly-prep	3	1	50 mA	10	62	4.6	*g*
Asparaginase from *E. coli*	2	4	Canalco Prep-Disc	6	2	4 mA	10	90	2	*h*

[a] L. D. Kohn, P. M. Packman, R. N. Allen, and W. B. Jakoby, *J. Biol. Chem.* **243**, 2479 (1968).

[b] L. D. Kohn and W. B. Jakoby, *J. Biol. Chem.* **243**, 2486 (1968).

[c] L. Thelander, *J. Biol. Chem.* **242**, 852 (1967).

[d] D. J. Nelson and C. E. Carter, *J. Biol. Chem.* **244**, 5254 (1969).

[e] W. N. Poillon, H. Maeno, K. Koike, and P. Feigelson, *J. Biol. Chem.* **244**, 3447 (1969).

[f] H. Tono and A. Kornberg, *J. Biol. Chem.* **242**, 2375 (1967).

[g] J. A. DePinto and L. L. Campbell, *Biochemistry* **7**, 114 (1968).

[h] H. A. Whelan and J. C. Wriston, Jr., *Biochemistry* **8**, 2386 (1969).

TABLE V

FEATURES OF SOME COMMERCIALLY AVAILABLE APPARATUS FOR PREPARATIVE ACRYLAMIDE GEL ELECTROPHORESIS

Name	Manufacturer	Nominal capacity (mg protein)	Elution system	Special features	Reference
Prep-Disc	Canalco, Rockville, Md.	100	Continuous	Adjustable elution slit	a
Poly-Prep	Buchler, Fort Lee, N.J.	100	Continuous	Glass membrane; adjustable elution slit	b
Fractophorator	Buchler, Fort Lee, N.J.	25	Discontinuous	Sectioning of gel to give better lower surface for elution	c
Preparative acrylamide gel apparatus	Shandon, Sewickley, Pa.	60	Continuous	Porous plastic support attached to gel	d
Uniphor	LKB, Rockville, Md.	50	Continuous	Outlet tubing passes through dialysis membrane	e
Elution-convection apparatus	E C Apparatus, Philadelphia, Pa.	100	Electrophoretic convection	Electrophoresis through slab at right angles to direction of separation	f
Prep-P.A.G.E.	Quickfit, Fairfield, N.J.	100	Continuous	Gel supported by nylon cloth	g
Preparative disc apparatus	Mechanical Workshop, Hadassah Medical School, Jerusalem	100 mg per column	Electrophoresis of 3 mm-thick gel section	4 columns in 1 chamber	h

[a]Prep-Disc Instruction Manual, Canal Industrial Corporation, Rockville, Maryland, 1968.
[b]T. Jovin, A. Chrambach, and M. A. Naughton, Anal. Biochem. 9, 351 (1964).
[c]I. Schenkein, M. Levy, and P. Weis, Anal. Biochem. 25, 387 (1968).
[d]I. Smith, "Chromatographic and Electrophoretic Techniques" Vol. II, "Zone Electrophoresis," p. 475. Wiley, New York, 1968.
[e]B. Bergham, Sci. Tools 14, 34 (1967).
[f]S. Raymond and E. M. Jordan, Separ. Sci. 1, 95 (1966).
[g]Prep-P.A.G.E. User's Instruction Manual, Quickfit, Inc., Fairfield, New Jersey.
[h]D. Sulitzeanu, M. Slavin, and E. Yecheskeli, Anal. Biochem. 21, 57 (1967).

[35] Preparative Gel-Density Gradient Electrophoresis

by Louis Shuster

This technique combines the high resolution of acrylamide gel electrophoresis with the ease of recovery that characterizes electrophoresis in density gradients.[1] Separated proteins are recovered with very little dilution.

The apparatus is simple. The electrophoresis chamber[2] consists of a modified U-tube made from 25 mm diameter Pyrex tubing (Fig. 1). The tube is filled to the sidearms with a linear gradient of 60% to 20% sucrose in the same buffer that is used in the separating gel. The gradient is conveniently made with a 3-veined peristaltic pump.[3] The mixing chamber contains 20% sucrose, and the feeding tube from the mixing chamber reaches to the bottom of the U-tube in order to ensure the formation of the same gradient in both limbs of the U-tube. The buffered gradient (about 80 ml) is allowed to flow in until it enters the capillary sidearm (*A* in Fig. 1) connected to each limb of the chamber. The tubing attached to each side arm is closed off; the openings at the top of the right limb and at the right end are closed off with rubber stopper. About 15 ml of acrylamide gel (7.5%) in 3.75 *M* Tris· HCl, pH 8.8, is prepared according to Davis[4] (see this volume [34]). 5.5. ml of this gel is carefully layered on top of the sucrose gradient in the left arm of the chamber. About 1 ml of distilled water is layered on the gel. Polymerization should take 15–20 minutes at room temperature. Care should be taken to avoid entry of the gel into the capillary side arm. The remainder of the gel solution is kept on ice. If the gradient and the gel solutions are precooled, then the amount of TEMED and ammonium persulfate in the gel solution should be doubled (see this volume [34]). After the gel in the left limb has polymerized, the stoppers are removed and 5.5 ml of separating gel is layered on the sucrose in the right limb of the chamber. When the separating gel has hardened, overlying water is removed from both gels and their upper surfaces are washed with about 5 ml of spacer gel solution. Spacer gel (4 ml on each side) is applied, overlaid with water and polymerized with fluorescent light (see this volume [34]).

[1] L. Shuster and B. K. Schrier, *Anal. Biochem.* 19, 280 (1967).
[2] Manufactured by Metalloglass, Boston, Massachusetts.
[3] S. R. Ayad, R. W. Bonsall, and S. Hunt, *Anal. Biochem.* 22, 533 (1968).
[4] B. J. Davis, *Ann. N. Y. Acad. Sci.* 121, 404 (1964).

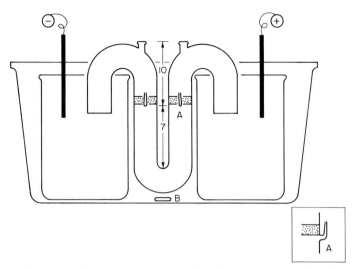

Fig. 1.   U-tube chamber for preparative acrylamide gel-sucrose gradient electrophoresis. The acrylamide gel is indicated by a stippled zone. The inset is a side view of sidearm *A*. The magnetic stirring bar is indicated by *B*. Distances are marked in centimeters.

The final gel column consists of 1.2 cm separating gel and about 1 cm of spacer gel. Gel height may be varied according to the guidelines described in this volume [34].

The chamber is suspended between two 1-liter Berzelius beakers filled with 0.025 M Tris·glycine buffer, pH 8.3. The beakers are placed in a clear plastic chamber, such as a mouse cage, which is filled with ice water. The top openings of the chamber are closed with serum stoppers. The space above the gel is completely filled on either side with the same Tris·glycine buffer as in the beakers. This is done by withdrawing trapped air through the serum stoppers by means of a hypodermic needle attached to a 50-ml syringe. The chamber is allowed to sit for 1 or 2 hours in the ice bath to allow for temperature equilibration. Cooling is facilitated by stirring the bath with a magnetic stirrer.

The sample of up to 50 mg of protein is prepared in the same buffer as that used for making the spacer gel, together with 10% sucrose or glycerol and 0.1 ml of 0.05% bromophenol blue in a total volume of 2–12 ml. The sample is carefully layered over the gel in the left arm of the chamber by means of a 3-inch hypodermic needle inserted through the serum stopper above the gel.

Electrophoresis is carried out at 10–15 mA (about 500 V) for 2–5 hours. This time is usually sufficient for the dye front to move a distance of 3–6 cm into the sucrose gradient. Overheating of the gradient will cause distortion of the dye front and, if allowed to continue, the formation of one or more discontinuities which permit convection. If the dye front becomes wavy or begins to form an upward plume at the center, the current should be decreased by one half.

At the end of the run, the power supply is turned off, and the electrodes are removed. The rubber tubing leading to both capillary sidearms is opened. A solution of 60% sucrose in water is slowly injected into the right capillary sidearm from a 20-ml hypodermic syringe or a peristaltic pump. The heavy sucrose solution flows down to the bottom of the U-tube and forces the gradient in the left side of the chamber out through the left sidearm to a fraction collector. It is important to have the bottom of the gel just above the sidearm in order to avoid a dead space that could permit mixing of the proteins in different fractions. Collection of fractions can also be intermittent. That is, electrophoresis can be interrupted and the fastest-moving proteins can be removed by injecting a few milliliters of gel buffer in 60% sucrose. Electrophoresis is then continued until later fractions emerge.

The usual size of each fraction is 0.5 to 1.0 ml. Because Tris interferes with the usual Lowry assay for protein, this assay cannot be applied directly. One solution to the problem is to precipitate the protein in a 0.1- or 0.2-ml aliquot of each fraction with 10% perchloric acid and 1% phosphotungstic acid, after which the tubes are held on ice for 1 hour. Precipitated protein is removed by centrifugation, washed once with 1 ml of 10% perchloric acid containing 1% phosphotungstic acid, and then dissolved in 1 ml of 0.1 N NaOH. The amount of protein is then determined by the phenol method of Lowry et al.[5]

Alternatively, the protein may be precipitated in a small glass tube at 0° for 30 minutes with 3.0 ml of 5% trichloroacetic acid. The protein suspension is filtered through a fine Millipore filter (No. EHWG-025). The tube and filter are washed twice with 3 ml of cold 5% trichloroacetic acid. The filter is transferred to a test tube containing 1.0 ml of 0.1 N NaOH for 2 minutes with occasional shaking in order to dissolve the protein from the filter. Aliquots of this solution are then used for the Lowry protein assay. Both methods give good recoveries down to at least 20 μg of protein.

Acrylamide gel-sucrose gradient electrophoresis has been used for several enzymes with recoveries of 60–90%.[1] Recovery of 70% of rat liver microsomal NADPH-cytochrome c reductase and an increase

[5] O. H. Lowry, N. J. Rosebrough, A. L. Farr, and R. J. Randall, J. Biol. Chem. 193, 265 (1951).

in specific activity of 3- to 6-fold has been reported by Ragnotti *et al.*[6] as a final step in the purification of the enzyme. The chamber used by these workers consists of a straight glass tube, 1 × 15 cm. The bottom is sealed with 2 ml of 7.5% acrylamide gel. The 60% to 20% sucrose gradient (5 ml) is overlaid with 2 ml of 7.5% gel and the tube is mounted vertically between two tanks of electrode buffer. Electrophoresis is carried out at a current of 5 mA in a cold room. Samples are collected by means of a hypodermic needle inserted between the upper gel and the wall of the tube. A chamber that could be used for separating both negatively and positively charged proteins at the same time on acrylamide gel-sucrose gradients has been described by Racusen and Foote.[7]

[6]G. Ragnotti, G. R. Lawford, and P. N. Campbell, *Biochem. J.* 112, 139 (1969).
[7]D. Racusen and M. Foote, *Anal. Biochem.* 25, 164 (1968).

Section VIII

Large-Scale Methods

[36] Large-Scale Growth of Bacteria*

By E. F. PHARES

Introduction

Scale-up of the purification of enzymes from microorganisms almost always requires a larger supply of the organism, which has the desired activities identical in character to the original, and in satisfactory concentrations. Often it is necessary for the biochemist and his colleagues to perform this task in the laboratory or related pilot plant without the assistance of personnel experienced in the fermentation art.

We have chosen to briefly discuss here some of the general methodology of fermentation, and to describe in particular some practical aspects of media and air sterilization. New or novel techniques in pH control and turbidity monitoring also are described. These descriptions should be used by the biochemist in conjunction with recent treatises and reviews on fermentation.[1-7]

Workers in this field must keep in mind the limitations of the physical controls of fermentation for enhancement of cell products. The control systems of the cell itself offer unlimited possibilities in this area, a subject alluded to elsewhere in this volume.[8]

*Research sponsored by U. S. Atomic Energy Commission under contract with Union Carbide Nuclear Corp.

[1]S. Aiba, A. E. Humphrey, and N. F. Millis, "Biochemical Engineering." Academic Press, New York, 1965.
[2]N. Blakebrough, "Biochemical and Biological Engineering Science," Vols. I and II. Academic Press, New York, 1967, 1969.
[3]I. Malek and Z. Fencl, eds., "Theoretical and Methodological Basis of Continuous Culture of Microorganisms." Academic Press, New York, 1966.
[4]J. R. Norris, D. W. Ribbons, eds., "Methods in Microbiology" Vols. I, II, and IIIA. Academic Press, New York, 1969-1970.
[5]G. L. Solomons, "Materials and Methods in Fermentation." Academic Press, New York, 1969.
[6]R. Steel and T. L. Miller, *Advan. Appl. Microbiol.* **12,** 153 (1970).
[7]F. C. Webb, "Biochemical Engineering." Van Nostrand, London, 1964.
[8]A. L. Demain, this Volume [12].

Equipment

General

The design and size of fermentors needed by a biochemical laboratory or pilot plant depends primarily on the nature of research being supported; available funds may be more of a factor in the choice of instrumentation and auxiliary equipment because these may represent several times the cost of the basic fermentor units.

Small units of 1-liter capacity or less, but with the ability to accommodate instrumentation, such as pH control and dissolved oxygen analysis systems, are highly desirable; these small units permit better development early in a project. Two such units are commercially available.[9] Working volume sizes of 3–10 liters are standard for development, usually in multiple units, and are also useful for small production batches. The 30–50-liter sizes serve a similar purpose but are usually used singly. They often are highly instrumented, because they have sufficient volume for many development experiments. The larger sizes, 100 to several hundred liters, normally are the "large-scale production" models for biochemists.

A variety of techniques have been used to furnish air and agitation for fermentation. In order of increasing complexity, the basic types of systems include the following: sparged tank, air lift, vortex, agitated and sparged (baffled), and draft tube fermentors. The vortex principle is used for moderate aeration (surface) with low foaming tendencies. The draft tube is similar but has a central tube to enhance the vortex; aeration is by surface exchange of gas or through a hollow shaft and impeller tip. The agitated and sparged tank is the basic design of the field, since it can best meet the oxygen transfer needs. Custom and commercial systems are available for easy interconversion among designs.[10] In fact general versatility is one of the most desireable features in equipment for the small pilot plant.

It has been suggested that the best way to select suitable equipment is to visit various laboratories and watch operations in progress.

Solomons has reviewed the characteristics of small[11] and large[12] pilot-scale fermentors. Solomons,[5] in an excellent book, and Steel and Miller[6] have further elaborated on equipment and methods.

---

[9]Biotech, Inc., Rockville, Maryland; Fermentation Design Inc., Allentown, Pennsylvania.
[10]Chemapec, Inc. Hoboken, New Jersey.
[11]G. L. Solomons, *Process Biochem.* 2 (3), 7 (1967).
[12]G. L. Solomons, *Process Biochem.* 3 (8), 17 (1968).

A 400-Liter System

The fermentor illustrated in Fig. 1 is an example of a versatile system which was designed to meet the needs of a small pilot plant in a research institution. The approaches applied in this unit are aimed at providing the same basic functions used in all sizes and designs of stirred and sparged fermentors. These basic functions are to provide: a sterile tank and medium; sufficient sterile air and agitation to meet the needs of the culture; and the means to carry out aseptically the necessary operations of inoculation, sampling, and control.

Ours is a 400-liter unit resembling larger production models.[13] The unit has a capacity of 350 liters when used with vigorous aeration. The

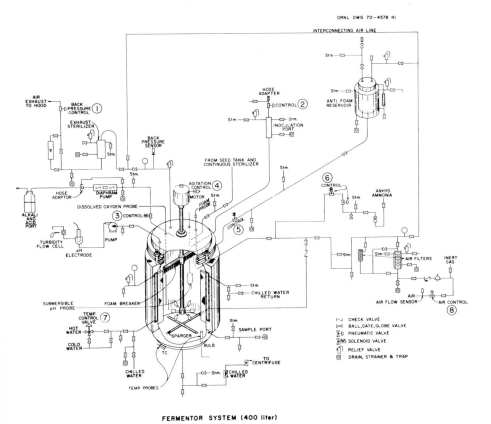

FERMENTOR SYSTEM (400 liter)

Fig. 1. Pilot-scale fermentor system (400 liters). For circled numbers, see text.

[13]Stainless and Steel Products Co., St. Paul, Minnesota.

details of tank and branch piping systems are mostly conventional. The open-turbine with variable pitch blades and a pipe sparger are considered less efficient than the vaned disk and single orifice design. Some other features and major auxiliary equipment are mentioned in later sections.

A fairly extensive control system, primarily based on conventional pneumatic equipment operates at points indicated by the circled numbers in the figure: (1) back pressure, Foxboro recorder-controller[14] to proportional valve; (2) inoculation port control, "Data trak" Programmer[15] to pneumatic diaphram valve. (3) acid and alkali port, L & N recorder-controller[16] to solenoid valve; (4) agitation, Foxboro recorder-controller to Reeves "Airtrol"[17]; (5) foam probe, to anti-foam solenoid valve; (6) anhydrous ammonia, L & N recorder controller to proportional valve and solenoid valve; (7) temperature controller, Foxboro recorder-controller to Foxboro mixing valve; (8) air-flow controller, Foxboro recorder-controller to proportional valve.

Other probes and monitors include those for thieve streams[18]: turbidity, visible wavelength (absorption) recorder (Isco)[19]; turbidity, fiber optics (nephelometer) recorder; pH electrode to L & N recorder-controller. The following tank probes are used: dissolved oxygen (Borkowski-Johnson)[20] to Foxboro recorder; pH probe, Foxboro glass electrode, Beckman "Lazaran" reference electrode to above L & N recorder-controller. Air flow, agitation, temperature, and pH may be controlled from line-following programmers. One single-function or two functions in a constant ratio are controlled by each programmer. As many as six inputs may be programmed by each instrument in an on-off fashion.

The control, recording, and programming systems were designed with three objectives: *control* of factors directly influencing the fermentation to give improved product yields; *documentation* of these and the noncontrolling functions for analysis and aid in reproducibility; *programming* of inoculation, aeration, agitation, pH, and temperature to allow less off-hours attention from a limited number of personnel.

Completely automatic sterilizing cycles now are available[9] which allow

---

[14]The Foxboro Co., Foxboro, Massachusetts.
[15]Data-trak, R I Controls, Division of Research, Inc., Minneapolis, Minnesota.
[16]Leeds and Northrup Co., Philadelphia, Pennsylvania.
[17]Moore Products Co., Spring House, Pennsylvania.
[18]A thieve stream represents an outlet system of extremely small capacity that allows sampling. The larcenous intent derives from the fact that the material is not returned to the main tank and is, in effect, stolen from it.
[19]Instrumentation Specialties Co., Lincoln, Nebraska.
[20]New Brunswick Scientific Co., New Brunswick, New Jersey.

untrained personnel to accomplish uniform operation. Experimental units with computer control interfaces have been produced commercially.[20] The success of these advanced systems remains to be evaluated.

Operation

In-Tank Sterilization Procedure and Related Techniques

Fermentor sterilization and media sterilization are most conveniently carried out simultaneously, unless medium is sterilized by a continuous process. Batch sterilization of the 400-liter fermentor diagrammed in Fig. 1 is described for a typical aerobic fermentation of 350 liters, final volume, using a minimal medium of inorganic salts and glucose. The presentation, in rather tedious detail, is aimed at illustrating the complexity of this type of system.

If the fermentation tank was not thoroughly cleaned after the last operation, a preliminary wash and short sterilizing cycle is required with dump valve or sample valve steam for 10 minutes at 120°. The clean tank is then filled with 315 liters of deionized water. Tap water may be used to advantage when compatibility with the media and organism has been demonstrated. Moreover, additional trace elements may not be necessary, although the mineral content of tap water can be variable. The use of hot water facilitates solution of salts and saves heating time, particularly in large tanks. With moderate agitator speed, about 125 rpm, the salts, trace elements, and antifoam agent are added. Occasionally a precipitate of $MgNH_3PO_4$ is formed when magnesium and phosphates are both present during sterilization; although no deleterious effects have been observed to result, magnesium salts are usually sterilized separately.

With the salts in the tank, the hand hole and exhaust valve are closed and steam at 40 psi is allowed into the tank jacket. The jacket drain is left open until the flow of condensate diminishes, then the drain is closed and the jacket steam-trap valve is opened. The inoculation port is opened, to vent tank air and to sterilize the valve and hose connection under the cap, which has a small bleed hole. The top air valve is opened and the tank air-supply valve is closed. Steam is applied to the dump valve (tank drain), sampling valve, seed-inlet valve, antifoam inlet, alkali-acid line, and anhydrous ammonia line; traps are opened and lines without traps are drained frequently or "cracked" and bled continuously. Steam is introduced to the air line above the air filter and is led in two directions: down through the filter to a trap, and through the interconnecting air line (I. C.) to the exhaust line trap. The air-line drain valve and the antifoam I. C. valve also are cracked. The anhydrous ammonia manifold is sterilized from the throttle valve, through the pneumatic control valve to the solenoid valve.

After about 30 minutes, the tank temperature has reached 120°. Steam pressure is dropped to between 22 and 24 psi. At this point, the I. C. valve, at the exhaust line and the antifoam tank, and the inoculation-port valve are closed. The tank exhaust is opened. Steam lines to seed-inlet, alkali, acid, antifoam, and ammonia ports are closed, and the corresponding tank valves are opened, allowing tank steam to pass. The thieve stream is bled slowly at the point of attachment of the pump tubing. The tank air valve is opened occasionally during this period. When the desired holding time is reached, 15 minutes in this case, jacket steam and trap valves are closed; jacket overflow and cooling water valves (tap or chilled) are opened. Air line steam is turned off, tank air valve is opened, and solenoids, bleed valves, and steam traps are closed. When ammonia manifold steam and control valves are turned off, ammonia is admitted and the drain is closed. When tank pressure has dropped to about 5 psi, air is admitted through the filter to maintain that pressure. Dogma requires complete drying of fiber glass filters before full operation is commenced, and presterilizing and predrying are recommended. However, on the theory that the first layers dry in a few minutes,[21] particularly in a heated filter, it is not a great risk to use low air as soon as required for maintaining positive tank pressure without flowthrough. The bulk of the drying air is not passed through the fermentor.

After a 5-minute resteaming of the hose adapter for the inoculation port, an aseptic connection is made with a previously sterilized, 40-liter carboy containing 15.4 kg of glucose monohydrate (Cerelose, Corn Products) in 32 liters. Before autoclaving, the carboy had been fitted with a connecting hose and dip tube and a vent with a glass-wool filter. The hose previously had been clamped off and the end wrapped with paper or foil. The connecting hose is fitted into a finger pump and, when the temperature has dropped to about 50°, the sugar solution is pumped into the tank. The back pressure should be dropped to about 2 psi during this operation. With smaller, heavy-walled vessels, solutions are conveniently blown into the fermentor with about 5 lbs air pressure applied through a small filter. This procedure may be used on large carboys if they have been inspected for possible damage, and if pressures are very carefully controlled. Wrapping the vessels would be prudent.

Temporary connections are conveniently accomplished with com-mercially available "Steri-Connectors"[8] or by puncture of resealing

[21]W. D. Maxon and E. L. Gaden, *Ind. Eng. Chem.* **48**, 2177 (1956).

rubber diaphrams[8-10] with large needles. Pumping 32 liters of glucose solution (Cerelose) will take about 30 minutes through a 0.5-inch i.d. hose with a finger pump having a heavy-duty speed changer. In the mean time, temperature control is set at 37°. The thieve stream is attached to the nutating tubing pump, and the medium is introduced to the pH electrode flow cell and to the turbidity monitor. Inoculation may be performed, when operating temperature is reached, by the same techniques used for adding the sugar. Small volumes are conveniently pumped in, after making certain that the port is sufficiently cool. Pumping the culture directly from a shake flask requires installing a dip tube, but this procedure probably is safer than pouring into a pressure flask for blowing in. After inoculation the control systems are set for normal operation.

Direct-steam injection has some advantages over jacket or other heat-exchange systems, especially in large installations. One must allow for considerable dilution of media, and, unless a secondary heat exchanger is used, objectionable materials in filtered steam must be accepted.

Sterilization of Media with Heat

Batch Sterilization

The logarithmic nature of the thermal death of vegetative cells and spores has long been known.[22] The relation was expressed by the familiar integrated form for a first-order reaction. Thus, for a pure culture of microorganisms, at a constant temperature,

$$\ln \frac{N_0}{N} = \int_0^t k\,dt = kt \tag{1}$$

where N_0 is the original number of organisms, N is the number surviving after heating during time t (min), and k is the specific reaction rate constant or thermal death rate constant. A plot of $\ln N_0/N$ against t gives k as the slope of a straight line; when k is known, survival values or sterilizing times may be calculated.

The rate constant is a function of temperature, and the Arrhenius equation has been used to describe this relation.[23]

$$k = Ae^{-E/RT}, \text{ and } \log k = 60 \log A + \frac{-E}{2.3\,RT} \text{ min}^{-1} \tag{2}$$

where E is the apparent energy of activation for spore destruction (calories/gram mole), R is the gas constant (calories/gram mole degree),

[22]O. Rahn, *Bacteriol. Rev.* 9, 1 (1945).
[23]M. J. Johnson, "Factors Involved in Rapid Sterilization of Fermentation Media," Univ. of Wisconsin Lecture Notes (1949).

T is absolute temperature, and A is a proportionality constant. E can be evaluated as the slope of a plot of log k against $1/T$ at two or more temperatures, and A may be determined experimentally at a known value of k.

Deindoerfer[24] determined values of E and A as 67,000 cal/mole and $1 \times 10^{36.3}$ sec^{-1}, respectively, for the destruction of spores of *Bacillus stearothermophilus*, strain 1518, which are typical heat-resistant spores used as test organisms in the food industry. He then was able to calculate kt for the rising and falling portions of a fermentor sterilization cycle by determining an average value of k by graphic integration. The necessary holding time at constant temperature for the required degree of spore reduction could then be calculated. This method required replotting the temperature-time curve as a velocity constant-time curve for the integration.

Deindoerfer and Humphrey[25] performed the rather complicated analytical integrations for various shaped sterilization curves. They used the del symbol, ∇, to denote spore reduction, ln N_0/N.

A method for rapid calculations of heat sterilization was reported by Richards.[26] Using Deindoerfer's calculations of E and A for *B. stearothermophilus*, he prepared a table of values for k (min^{-1}) from 100° to 125° and for the corresponding *cumulative* value of k from 100°; the latter then are $kt = $ ln $N_0/N = \nabla$ for a temperature curve with a slope of 1°/minute, resulting from a stepwise integration. Table I is an extension of Richards' table; tabulations have been changed from natural logarithms to $\log_{10}$ functions, which give spore reduction directly in $\log_{10}$ cycles. The Arrhenius equation (Eq. 2) for log k was used as follows:

$$\log k = 60 \times \log 36.2 + \frac{-67.7 \times 10^3}{2.302 \times 1.98 \ (273 + °C)} \ \text{min}^{-1}$$

$$= 37.9782 - \frac{14853.4}{273 + °C}$$

The table may be used for rapid analysis of heat-sterilization processes, even as they are in progress, and for the design of the processes when heating and cooling characteristics are known. The tabulated values are based on a linear temperature change of 1° per minute; a nonlinear curve is approximated with a linear estimate or is broken up into two or more linear portions. Spore destruction is negligible under 100°, and some workers disregard all temperatures below about 105°. Therefore,

[24]F. H. Deindoerfer, *Appl. Microbiol.* 5, 221 (1957).
[25]F. H. Deindoerfer and A. E. Humphrey, *Appl. Microbiol.* 7, 256 (1959).
[26]J. W. Richards, *Brit. Chem. Eng.* 10, 166 (1965).

TABLE I
TABULATION OF $k/2.3$ AND $\nabla/2.3$[a] FOR SPORE DESTRUCTION,[b] 100°–140°

| Temperature | | $k/2.3$ | $\nabla/2.3$ | Temperature | | $k/2.3$ | $\nabla/2.3$ | Temperature | | $k/2.3$ | $\nabla/2.3$ |
°F	°C			°F	°C			°F	°C		
212	100	0.006		238.8	115	0.216	1.03	265.8	130	5.75	29.6
213.8	101	0.008	0.01	240.6	116	0.271	1.29	267.6	131	7.09	36.7
215.6	102	0.010	0.02	242.4	117	0.339	1.63	269.4	132	8.74	45.5
217.4	103	0.013	0.04	244.2	118	0.425	2.06	271.2	133	10.8	56.2
219.2	104	0.016	0.05	246.0	119	0.531	2.59	273.0	134	13.2	69.4
221.0	105	0.021	0.07	247.8	120	0.663	3.25	274.8	135	16.3	85.7
222.8	106	0.027	0.10	249.6	121	0.827	4.08	276.7	136	20.4	106
224.6	107	0.034	0.14	251.4	122	1.03	5.10	278.4	137	24.5	131
226.4	108	0.043	0.18	253.2	123	1.28	6.39	280.2	138	30.6	161
228.2	109	0.054	0.23	255.0	124	1.59	7.98	282.0	139	36.7	198
230.0	110	0.068	0.30	256.8	125	1.98	9.96	283.8	140	44.9	243
231.8	111	0.086	0.39	258.6	126	2.45	12.4				
233.6	112	0.109	0.50	260.4	127	3.04	15.5				
235.2	113	0.137	0.63	262.2	128	3.76	19.2				
237.0	114	0.172	0.80	264.0	129	4.65	23.9				

[a]$\nabla/2.3 = \log 10\, N_0/N$, which is equal to the integrated $\log_{10}$ cycles of spore reduction for a time-temperature curve with a slope of 1°/minute.
[b]Modified and extended data of J. W. Richards, *Brit. Chem. Eng.* **10**, 166 (1965).

when approximating a nonlinear curve, large areas above the lower end of the curve are required to compensate for small segments of the higher end. Note that $K/2.3$ and $\nabla/2.3$ increase roughly 10-fold for a 10° rise in temperature, but that the rate of increase drops off with increased temperature.

Figure 2A is a time–temperature curve for batch sterilization of a fermentor containing 300 liters of salts medium with 1% tryptone. For integration of the heating and cooling portions, the linear approximation for each is selected, and the ratio of the span in minutes per degree is multipled by the appropriate $\nabla/2.3$ from the table. The holding time is multipled by $k/2.3$.

Heating: 110 to 119½° in 17 min; $\nabla/2.3 = \sim 2.9 \times 17/19½ = 2.5$
Holding: 119½°, 16 min; $\nabla/2.3 = 0.59 \times 16 = 9.5$
Cooling: 119½° to 100° in 7 min; $\nabla/2.3 = 2.9 \times 7/19½ = 0.8$
 Total $\nabla/2.3 = 2.5 + 9.5 + 0.8 = 12.8$ log cycles spore reduction.

Figure 2B is the curve for autoclave sterilization of a 40-liter glass carboy containing 15.4 kg of glucose in 32 liters. The heating portion is treated in two portions:

$$\text{Heating:} \qquad 0 \text{ to } 110° = 0.30 \times \frac{20}{10} = 0.6$$

$$110 \text{ to } 118° = (2.06 - 0.3) \times \frac{32}{8} = 7.04$$

$$\text{Cooling:} \quad 118 \text{ to } 100° = 2.06 \times \frac{25}{18} = 2.85$$

Total $\nabla/2.3 = 0.6 + 7.04 + 2.85 = 10.5$ log cycles reduction

An estimation of the necessary heating conditions for a particular medium is made from the product of the contamination level, volume, and a safety factor. The latter is usually 1×10^2 or 1×10^3 which means a probability of one contaminant in no more than 1 in 100 or 1000 tanks. Sterility is only probable, rather than absolute, because of the statistical nature of the thermal process. Table II lists analysis of a few representative media components used in our pilot plant. The 300 liters of medium used in the example contained 1% tryptone, i.e., 200 organisms per milliliter. With a safety factor of 1×10^3, the fermentor requires $300 \times 10^3 \times 200$ (organisms/ml) $\times 10^3 = 6 \times 10^{10} = 10.8$ log cycles spore reduction. In the example 12.8 cycles were obtained, and so holding time could safely have been 3 minutes shorter. In a similar manner, the autoclaved carboy of glucose (Fig. 2B), required a calculated 8.5 log cycles of spore reduction; 10.5 cycles were obtained. This is a 5-minute margin at a k value of 0.4 min^{-1}.

TABLE II

CONTAMINATION LEVELS OF SOME COMMON MEDIA COMPONENTS[a]

Media component	Organisms (plate counts)
M-9 Salts	< 1/ml
Cerelose (glucose) (4% w/v)	< 1/ml
Tryptone (1%)	
Brand A (lot 1)	~ 60/ml
(lot 2)	~ 200/ml
Brand B	< 1/ml
Brand C	< 1/ml
Peptone (1%)	
Brand A (lot 1)	< 1/ml
(lot 2)	~ 3/ml
Nutrient broth (1%)	
Brand A	~ 10/ml
Brand B	~ 5/ml
Yeast extract (1%)	
Brand A	< 1/ml
Typical commercial crude media	1×10^4 to 1×10^5/ml

[a]Random samples from stock supply.

An additional safety factor obtains in the assumption that all the organisms have the characteristics of the heat-resistant spores of *B. stearothermophilus*, strain 1518. Actually only about 10% of the contaminants in the tryptone survived at 100° for 10 minutes. Precise use of this approach would require that the velocity constant of the most resistant spores in each type of medium be determined and conditions adjusted accordingly, especially if there is any question of heat damage. Not many workers in small pilot plants can afford time for this kind of program, and that is why the heat-resistant organism is used as a guide.

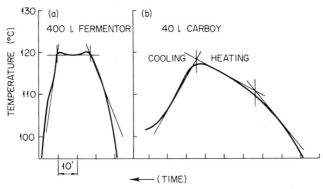

FIG. 2. Media sterilization time-temperature profile: (a) 400-liter fermentor; (b) 40-liter carboy.

One danger of reducing holding times to the limit of the formula for very clean media is the possibility of a dirty, or at least poorly designed, fermentor, which might easily contain many orders of magnitude more contaminants than does the medium. A short period, 5–10 minutes of direct steam sterilization immediately after clean-up from a previous run, will prevent vegetative cells from becoming spores. This procedure is aimed mainly at blind cavities which do not clean well; it does not replace the thorough cleaning of the fermentor. Particles also protect spores, and crude media which contain particles, e.g., commercial mashes, dirt, or even nonsoluble starch, require the introduction of an additional safety factor. Generation time and inoculum size are important considerations. In the case of an organism with a short generation time and with the use of a large inoculum, contamination is not a serious problem and smaller safety factors may be acceptable. In the alternative case, for a slow-growing organism and a small inoculum, careful sterilization is required.

Media, either sterile or nonsterile, should not be stored at favorable growth temperatures longer than necessary. This is particularly pertinent in continuous sterilization, where sufficient media may be mixed for several hours of operation of the sterilizer.

Autoclaving of media, small fermentors, glassware, etc., is an important part of the biochemical pilot plant, but many operations are carried out with little knowledge of the effect of various conditions and without any monitor of the efficiency of the process. Some aspects of the problem have been discussed in detail.[5,27] A variety of temperature-sensing aids for determination of efficiency of autoclaving have been used, especially in hospitals. Devices which may be read only after opening the autoclave include color tubes (chemical indicators), paints, pencils, compounds with appropriate melting points, tapes, maximum-reading thermometers, and spore strips. Immersible probes, for monitoring externally, are metal bulbs (usually liquid filled), metal resistance thermometers, thermistors, and thermocouples.

Although the described spore techniques may be the ultimate test of how an autoclave performs, to arrive at conclusive results requires many days. A new, self-contained, spore assay device is claimed to have shortened this time considerably.[28] The tapes,[28] while requiring steam exposure as well as sterilizing temperatures for color change, seem to have an indefinite end point of ~ 5 to 25 minutes at 120° although there is an intensification of color over this range. The color tube indication is generally based on hydrolysis reactions and has a temperature and time

[27] G. R. Wilkinson and L. G. Baker, *Progr. Ind. Microbiol.* 5, 233 (1964).
[28] Minnesota Mining and Mfg. Co., St. Paul, Minnesota.

relationship. A recent addition to the temperature indicator list is a small tab, for which accuracy of ± 1% is claimed.[29] For continuous, accurate analysis the probes with an external indicator or recorder appear to be the only satisfactory method available, especially in sterilization of liquids. Thermocouples are the choice of the probes because of their small size, fast response, and low cost. For temporary use, thermocouple leads No. 28 and smaller may be slipped past the autoclave door without permanent gasket damage. Sealing the bare wires in slits across the gasket is not aesthetic but may be done semipermanently. For a permanent installation, a ¼-inch Conax fitting may be used legally in the pressure wall, or the lead may be brought in through the steam-trap line. A pyrometer type instrument with a 150° scale, which can be read quite accurately, may be bought for less than 100 dollars; a recording potentiometer type will be more accurate and convenient and will cost several times as much. Resistance thermometers are very stable and accurate, but more bulky and considerably more expensive than thermocouples; hence they are used less frequently.

Factors affecting the observed rate of heating of a liquid in an autoclave include container material, weight, shape, size, liquid volume, type of liquid, position of thermocouple, and autoclave temperature. Table III illustrates most of these effects with tabulation of the time period required for spore reduction times of about 8 log cycles.

Heat transfer is very poor in polypropylene, and the use of this material as a container for liquids in the autoclave should be avoided. However, heat transfer is excellent in containers made of thin metals, which should be used more often for larger volumes. Glass conducts heat poorly enough that vessel weight should be considered in calculating autoclave time. Vessel size and liquid volume are larger factors than are vessel shape and the nature of the medium, although the latter have large effects when extreme. The commercial fermentor requires longer heating and cooling times for comparable liquid volumes because of the large mass of metal. Additional spore reduction is accomplished in a short period when the temperature of the medium approaches 120°. Thus higher autoclave temperatures shorten heating periods considerably, but they are more difficult to control for small volumes without either over or under sterilization.

With proper caution, Table III could be used as a guide to autoclaving times for a number of types of vessels. Table III is not recommended as a replacement for a thermocouple, but, since so few workers attempt any sort of monitoring, even a little help is better than none.

---

[29] William Wahl Corp., Los Angeles, California.

TABLE III

AUTOCLAVE STERILIZATION OF LIQUIDS[f]

Type[a]	Vessel Volume (liters)	Weight (kg)	Liquid[b] volume (liters)	Autoclave control T (°C)	Vessel heating[c] Profile (min) ∇/2.3 = 8			Approximate timer setting[d] and max T				Timing delay (min)	Exhaust delay[e] (min)
					25–100°	100° to max.	max. to 100°	∇/2.3 = 8 Min	T	∇/2.3 = 12 Min	T		
Carboy	40	11.9	32	121	61	54	18	110	116	123	117	3	2
Carboy	40	11.9	32 G	121	63	52	20	108	116	–	–	4	3
Carboy	40	11.9	32	128	50	40	19	77	119	84	120	9	4
Carboy	40	11.9	32 G	128	53	40	24	79	118	–	–	10	4
Fermentor	36		26	128	35	25	15	47	121	–	–	10	3
Kettle, S. S.	30	4.7	24	128	10	8	16	7	122-1/2	10	124-1/2	9	2
Kettle, S. S.	30	4.7		121	12	15	15	23	120	28	121	4	1
Carboy, poly-propylene	19	1.7	15	121	112	90	12	196	115	–	–	2	4
Carboy	20	4.6	16	121	31	33	15	59	118	68	118-1/2	3	2
Carboy	20	4.6	9	121	27	33	13	55	118	–	–	4	1
Carboy	20	4.6	16 G	121	34	40	14	68	117-1/2	–	–	4	2
Carboy	20	4.6	16	128	25	23	12	37	120-1/2	42	122-1/2	8	3
Carboy	20	4.6	9	128	23	22	13	36	121	–	–	9	2
Carboy	20	4.6	16 G	128	30	22	15	38	120	–	–	10	4
Carboy	20	6.5	16	121	35	44	13	74	117	85	118	3	2

Carboy	20	6.5	16	128	35	28	15	50	120	—	—	10	3
New Brunswick Fermentor	14	4.5	9	121	44	40	14	78	118	—	—	3	3
New Brunswick Fermentor	14	4.5	9	128	40	28	13	55	119-1/2	—	—	10	3
Carboy	12	4.35	9	128	30	26	11	43	120-1/2	—	—	10	3
Florence Fl.	12	2.35	9	128	18	18	11	24	121-1/2	28	122	10	2
Florence Fl., 5 flasks	12	2.35	9 (×5)	121	24	30	18	47	118	—	—	4	3
Florence Fl.	12	2.35	9	121	23	32	12	50	118-1/2	58	120	3	2
Erlenmeyer	6	1.25	5	121	12	21	12	29	119	36	120	3	1
Erlenmeyer	6	1.25	1	121	7	16	10	23	120-1/2	28	121	1	1
Erlenmeyer	4	1.07	3	121	9	25	12	31	119	—	—	2	1
Erlenmeyer	4	1.07	1	121	9	18	10	25	120-1/2	30	121	1	1
Erlenmeyer	2	0.5	1	121	8	16	10	22	120-1/2	25	121	1	1

[a] Glass unless otherwise indicated; S. S., stainless steel, 0.045-inch wall.

[b] Water or dilute aqueous media except those with 44% (w/v) glucose, marked G.

[c] Vessel temperature determined with chromel-alumel thermocouple positioned 1 cm from bottom of vessel and read with potentiometric recorder (Honeywell Type K, 0–150°).

[d] Timer setting = minutes: (25°–100°) + (100° to max) − (timing delay + exhaust delay). These delay times may vary and are a potential source of error when automatic operation is used.

[e] Time from start of autoclave exhaust stage to start of cooling of liquid (thermocouple).

[f] Data obtained with American Sterilizer autoclave Model 57 CR.

Continuous Sterilization

Continuous-heat sterilizers have some advantages over the batch-type primarily as a result of their unique characteristics of rapid heating, short holding, and rapid cooling. Medium is pumped, by positive displacement, through a heater, which may be a heat exchanger of the plate type or of the shell-and-tube type, or which may operate by direct steam injection. The holding section is a coil or section of pipe designed for minimal axial dispersion of medium, i.e., "plug flow." Cooling is by heat exchange, flashing, or quenching in a cold tank. The hot medium must be kept under sufficient pressure to prevent steam formation. Design and operation of these units have been described by several authors.[1,30-32]

The preservation of labile nutrients is the most significant feature of continuous sterilization. The destruction of growth factors, such as vitamins, and the degradation of sugars have lower activation energies (lower slope of the function log k vs. $1/T$) than has the killing of spores, and k for the small molecule does not increase as rapidly with increased temperatures. At ordinary sterilizing temperatures, rates of destruction for the two types are similar, but at higher temperatures, 140°–150°, and with holding times of a minute down to several seconds, spores are destroyed while appreciable amounts of nutrients are retained.[24] There are, of course, practical limits to the short heating, holding, and cooling steps.

The Maillard reaction,[32] the reaction of proteins, amino acids, or ammonium salts with sugars to form colored compounds (browning), occurs in most growth media in which sugars, especially glucose, have been sterilized by batch process in the presence of ammonia or amines. These Maillard compounds, as well as the large variety of reactants which can be produced by the heating of compounds in complex media, may be inhibitory to microorganisms. Nevertheless, most complete media can be sterilized continuously at 133° with a holding time of 1 minute, without these effects. In the situation where components cannot be mixed, they may be sterilized individually in concentrated solution in the continuous sterilizer and added to the main batch at a later time. Other advantages to the continuous process are the more optimal use of steam generating facilities and uniform scale-up conditions. The application to continuous culture is obvious, but the use of a holding tank for the sterile medium, rather than direct feed to the fermentor, is advisable.

[30]V. F. Pfeifer and C. Vojnovich, *Ind. Eng. Chem.* 44, 1940 (1952).
[31]F. H. Deindoerfer and A. E. Humphrey, *Appl. Microbiol.* 7, 264 (1959).
[32]N. Blakebrough, "Biochemical and Biological Engineering Science," Vol. II, p. 29. Academic Press, New York.

Calculations of necessary holding times in continuous sterilization may be carried out in the same manner as for batch treatment, except that heating and cooling contributions are usually neglected and only the k value for the spore at the holding temperature is required. The total volume to be sterilized for the particular run is used in determination of total spore numbers. Pumping rates, holding coil sizes, and temperature may be varied according to the characteristics of the individual sterilizer components. Particles in the medium are a more critical complication than in the batch process and should be removed by filtration. In the case of mash-type media, smaller k values should be used. A flow pattern in the holding coil which differs greatly from plug flow may lead to some over sterilization.[1] In straight, round pipe the mean velocity for laminar (viscous) flow is half the maximum, while for turbulent flow the mean is 0.82 of maximum. Thus a portion of the medium is oversterilized since the holding time normally should be calculated for maximum velocity.

Air Sterilization

General

Fermentor air, and most other gases, are commonly sterilized by filtration. Dry heat is used in smaller installations where absolute protection is required, such as for exhaust air containing pathogens and for protection from phage. Ultraviolet and ionizing radiation are not practical for large-scale work. Electrostatic precipitation and heat sterilization by adiabatic compression of air have been used but have not been developed as primary methods, although the principles involved probably contribute in some degree to most air-sterilization processes.

Solomons[5] and Elsworth[33] have recently described sterilization of gases by heat, and also by filtration. Two commercial incineration systems for small- to intermediate-size fermentors have been marketed.[20,34] The trend, however, is away from the use of heat as more efficient filters are developed, and some of the developments bear discussion.

Sterilization by Filtration

Sterilization of fermentation gases was first carried out by the familiar cotton wool filters, and a number of fibrous, granular, porous, and membrane materials have followed. The charcoal bed (activated carbon) has been the last survivor of the granular filters; considerable data show it to be less effective than even the most inefficient fibers,[1] and

[33]R. Elsworth, in "Methods in Microbiology" (J. R. Norris and D. W. Ribbons, eds.), Vol. I, p. 123. Academic Press, New York, 1969.
[34]The American Sterilizer Co., Erie, Pennsylvania.

its use certainly will not continue long. Porous materials including ceramics, fritted glass, and sintered metals have frequent application where the rather high pressure drops can be tolerated. Stainless steel is often used for steam and other corrosive services. The ceramics are used as standard equipment by at least one fermentor manufacturer for inlet and outlet filters. The proponents of this type of filter[10] point to the ruggedness and performance through long fermentation runs.

Fibrous Filters

Research on mechanisms of aerosol collection by fibrous materials, started in World War II, was extended in several laboratories.[35-38] This work resulted in semitheoretical relations to explain filter action, which Friedlander[39,40] and Aiba[1] further extended to similar general expressions; these are compatible with most of the available experimental data. They are based on the three primary mechanisms of particle collection by fibers which Friedlander has incorporated in the fiber efficiency component, N, of this basic equation:[41]

$$L = \frac{\pi}{4} \frac{d_F}{\alpha \eta} \ln \frac{N_1}{N_2} \qquad (3)$$

η = the effective individual fiber efficiency = diffusion + interception + impaction; L = bed depth (cm); $\bar{d}_F$ = fiber diameter (cm), mean or average; d_P = Particle diameter (cm); N_1 = Number of particles entering filter; N_2 = Number of particles leaving filter; α = Fiber volume fraction of filter.

Table IV lists calculated efficiencies (bed depths) for a range of particle sizes, fiber diameters, and air velocities, using the Friedlander equation. The striking effect of fiber diameter and the interplay of the mechanisms in the low velocity region are demonstrated. With fibers 1 μ in diameter, the diffusion mechanism predominates for particle diameters of 0.1 μ and below, except at higher velocities. Interception takes over for larger particles and is independent of velocity. The effect of impaction is noticeable above about 30 cm/second for 1-μ particles, and all parameters decrease rapidly with increase in fiber size.

Humphrey and Gaden[38] developed a high-efficiency filter design for spore removal based on high velocity operation in the impaction

[35] C. N. Davies, *Proc. Inst. Mech. Eng.* 1B, 185 (1952).
[36] C. Y. Chen, *Chem. Rev.* 55, 595 (1955).
[37] J. B. Wong and H. F. Johnstone, "Collection of Aerosols by Fiber Mats," *Ill., Univ. Eng. Exp. Sta. Tech. Rep.* 11, (1953).
[38] A. E. Humphrey and E. L. Gaden, Jr., *Ind. Eng. Chem.* 47, 924 (1955).
[39] S. K. Friedlander, *Ind. Eng. Chem.* 30, 1161 (1958).
[40] S. K. Friedlander, in "Biochemical and Biological Engineering Science," (N. Blakebrough, ed.) Vol. I, p. 49, Academic Press, New York, 1967.
[41] This equation is the same as that used by Wong and Johnstone.[37]

TABLE IV
FILTRATION PARAMETERS AND EFFICIENCIES FOR SMALL FIBERS[a]

Particle size (μ)	Fiber size (μ)	Velocity (cm/sec)	Parameters ($\times 10^2$)			Fiber efficiency, η	Bed for $10^{-4}\%$ penetration[b] (cm)
			Diffusion	Interception	Impaction		
0.05	1.0	2.5	24.2	0.2	—	0.24	0.1
0.08	1.0	10.0	7.1	0.6	—	0.077	0.36
0.1	1.0	2.5	11.6	0.7	—	0.123	0.2
0.1	1.0	10.0	4.6	0.7	—	0.053	0.5
0.1	1.0	20.0	2.8	0.7	—	0.035	0.8
0.1	1.0	200.0	0.4	0.7	—	0.011	3.0
0.2	1.0	10.0	2.1	2.8	—	0.049	0.54
0.2	1.0	20.0	1.3	2.8	—	0.041	0.66
1.0	1.0	10.0	0.5	70.0	—	0.70	0.04
1.0	1.0	40.0	—	70.0	9	0.79	<0.04
1.0	1.0	200.0	—	70.0	(50)	>>0.7	<<0.04
0.1	1.25	10.0	4.0	0.5	—	0.045	0.78
0.2	1.25	10.0	1.7	1.8	—	0.035	0.97
1.0	1.25	10.0	0.4	45.0	—	0.45	0.07
0.2	2.0	10.0	1.3	0.7	—	0.02	2.7
1.0	2.0	10.0	—	17.5	—	0.175	0.3
1.0	7.0	200.0	—	—	7.5	0.075	2.4

[a]Calculated from Eq. (3).
[b]Bed density = 0.1 g/cc; bed solids fraction = 0.04.

region, using 8-μ to 16-μ fibers. This type of filter has been widely used in large-scale installations for many years, but its design is inefficient for small particles.[38] None of the three collection mechanisms works well with small particles at high speed with large fibers. Of particular concern are some of the bacteriophage of the T-odd series which are stable in the dry state.

Nearly twenty years ago Blasewitz and Judson[42] developed high-efficiency, low-velocity aerosol filters, using standard AA glass fibers (Owens-Corning,[43] mean fiber diameter 1.3 μ), but their significance to fermentation was not generally recognized. Recently Elsworth,[33] working from data of Dorman,[44] has recommended thin beds of bonded AA fiber (Johns-Manville Ltd.[45] fiber diameter 0.6–6.0 μ), for use with small fermentors. At 0.08 g/cc density the penetration is < $10^{-5}\%$ for spores with 15 cm/sec velocity, and only a little greater for phage at 5 cm/second. For greater safety a 2-inch layer of fiber in a 6-inch length of glass pipe of the appropriate diameter is suggested; the bed is held

[42]A. G. Blasewitz and B. F. Judson, *Chem. Eng. Progr.* 51, 6-j (1955).
[43]Owens-Corning Fiberglas Corp., Toledo, Ohio.
[44]R. G. Dorman, *Chem. Ind.* p. 1946 (1967).
[45]Johns-Manville Co. Ltd, London.

in compression with stainless steel wire mesh pads and rubber stoppers.

Candle-type Fibrous Filters. The wrapped candle is a convenient form for pilot-scale filters. Relatively large areas can be achieved in small tubes, and edge leakage is no problem. We have used a candle of stainless steel screen, 3 cm × 20 cm, wrapped with 12 layers of glass fiber mat to a density of 0.17 g/cc. The glass, supplied by Johns-Manville Co.,[46] was AA grade bonded fiber (Microlite), 0.5 inch thick with a density of 0.6 lb/cu ft; fiber diameter was rated at 0.75–1.4 μ. The fiber was rolled wet, with firm pressure, and the ends were sealed with several turns of wire or string. The outside was wrapped with insulator's glass screen, and the filter was oven-dried for several hours. The dry filter, with a diameter of 5.0 cm and a glass bed depth of 0.8–0.9 cm, was installed in a vertically mounted milk-filter body with upward air flow. Air volume at 10 cm/second face velocity (inner) was 100 liters/minute, with a pressure drop 1.6 psi.

At densities over 0.15 g/cc, the soft, 1-μ fibers make a fairly stable bed. After in-line steaming, gentle pressure, 0.5–1 psi, for half an hour, followed by 1–2 hours at operating pressure, dries the filter if the supplied air is quite dry. As with high velocity filters, drying air probably should not go to the fermentor. A drain to atmosphere for the center of the candle aids in preventing condensate build-up while steaming and drying. A steam jacket or steam tracing is the quickest and best method of drying the filter.

The theoretical performance of the filter is demonstrated from a design example. For removal of all organisms, those with the highest penetration should be the criterion. Table IV shows highest penetration for 0.2-μ particles. Although there are not many organisms of this size, smaller ones may ride with dirt particles, which exist in huge concentrations and in all dimensions. Calculation of the required filter efficiency is made from contamination load and a safety factor. Air contamination varies from less than one to hundreds of organisms per cubic foot at typical air-compressor intake locations. One organism per liter is a convenient figure when real data are lacking. For a fermentation using 100 liters of air per minute for 7 days' operation,

$$\log 10 \, \frac{N_1}{N_2}$$

$$= \log \frac{1(\text{organism/liter}) \times 100(\text{liters/min}) \times 10^4(\text{min}) \times 10^3 \,(\text{safety factor})}{1.0}$$

$$= 9, \text{ and allowable penetration} = 1 \times 10^{-7}\%$$

[46]Johns-Manville Products Corp., New York, New York.

Bed depth for this particular filter material then may be calculated from Eq. (3) or estimated from Table IV. Fiber and particle diameters of 1.25 μ and 0.2 μ, respectively, are assumed. The bed is equivalent to 1.5 cm of 1.0 g/cc density or 0.4 volume fraction. At 10 cm/second face velocity, Table IV lists the bed depth for $10^{-4}\%$ penetration of a filter with these characteristics as $\sim$ 1.0 cm. This filter then would have penetration of $10^{-7}\%$, which was the design requirement.

The unit has operated successfully at 10 cm/second velocity for a 2-week period without bacterial or spore penetration, but no test for phage was attempted. It should be possible to mount such a filter into existing housings more than 2 inches in diameter. A 1-inch i.d. candle, 24 inches long will supply about 300 liters/minute at 10 cm/second. The two domestic glass-fiber manufacturers offer a number of fine fibers with low velocity filtration potential. Johns-Manville,[46] in addition to the AA bonded fiber used above offers loose fiber, mats, and webs of down to 0.7 μ and 0.4 μ diameters. Owens-Corning[43] offers: AA fiber, 0.75–1.25 μ, loose, bonded, or treated with uncured resin for customer molding; an AAA fiber of 0.625 μ; and a filtration medium, FM003, bonded, similar to the AA mats, with $\sim 1\mu$ fibers.

High-Efficiency Commercial Filters

There are now a number of different types of commercial filters on the market which are recommended for air sterilization; some of the properties of these, including that of the wrapped candle described above, are listed in Table V. The true membrane filters are "absolute" above a rated pore size, and they are obtainable in pore sizes to 0.01 μ. This is certainly smaller than any organism. Although flow rates at these limits are very poor, grades of 0.2 μ are quite satisfactory in the latter respect. Nevertheless, many workers do not trust a membrane because of the potential effects of an undetected pinhole. The Pall, Biotech, and Cox filters may be called modified membranes in that they are somewhat thicker and less vulnerable. The Cox product is very rugged and operates with some depth-filter action. The others are primarily depth filters, although the Whatman design permits liquid filtration at the 0.3 μ level with 98% removal. Ceramic air filters have long pores that usually are larger than the particles and thus may be to be considered of the depth type. Fiberglass filters rate well in flow rate versus pressure. They may be the most difficult to dry but also the most economical to operate. All fibrous depth filters are more or less sensitive to moisture, and air for them may need to be dried by heating to a few degrees above ambient.[38]

The choice of these depends on the particular application and

TABLE V
MEMBRANE AND SHALLOW-BED FILTERS[a]

Filter	Material and type	Penetration, various pore and particle sizes	Rate (ι/min/cm^2, $\Delta P = 5$ psig)	Remarks
Millipore GC[c]	Cellulose nitrate membrane	Absolute $>0.22\ \mu$	0.9	Ster. 120°,disks, tubes
Gelman AN 200[d]	Cellulose nitrate membrane	Absolute $>0.2\ \mu$	1.4	Ster. 120°
General Electric[e] Nucleopore	Polycarbonate membrane	Absolute for $0.5\ \mu$	5.6	Ster. 140°, disks, plates
Biotech[f]	Glass fiber ($\sim 1\ \mu$) paper disk	$2 \times 10^{-3}\%$ for $0.3\ \mu$ DOP, at 14.2 cm/sec		Max. 12 ι/min
Pall[g] Ultipor 0.12	Epoxy-bonded fiber mem-brane	Absolute $>0.35\ \mu$ in liquid	0.22	Pleated cartridge; from 1 ft^2 (disposable) up
Cox M-780[h] (AA-20)	Glass and asbestos, 780 μ thick	$<10^{-4}\%$, $0.3\ \mu$, *Pseudomonas*	0.25	"Tapered" pores; 47 mm dia. up
Whatman[i] Gamma 12-003	Bonded glass ~ 3 mm bed	$<10^{-4}\%$, NaCl 10 cm/sec	0.45–0.9	1 x 2¼ inch candle; also 2 x 10 inches and up
Domnick-Hunter[j] C-2ME	Bonded $0.5\ \mu$ glass fiber	Absolute, spores $<10^{-3}\%$, $0.4\ \mu$ DOP, methylene blue	1.1	3¼ x 1⅞ inches and up
Echo[k]	Polyvinyl alcohol 3–5 mm thick	$< 10^{-3}\%$, 80 cm/sec; $10^{-4}\%$, *Serratia* 5 cm/sec 5 mm thick[b]	1.6[b] (27 cm/sec.)	Ster. 120°, plates, all sizes
Selas,[l] Ser.C 01 Porosity	Ceramic, 3 μ pore 5mm thick	Absolute, liquid, $1.5\ \mu$	0.3	Cleaned by incineration
Wrapped Candle J-M, AA fiber	$1.25\ \mu$ (mean) diameter fiber (est.) 9 mm, 0.17 g/cc	$0.1\ \mu$ particle $10^{-10}\%$ $0.2\ \mu$ particle $10^{-7}\%$ $1.0\ \mu$ particle $10^{-100}\%$ 10 cm/sec	1.8 (30 cm/sec)	—

[a]Data from manufacturer's specifications; R. Steel and T. L. Miller, *Advan. Appl. Microbiol.* **12**, 153 (1970), or calculated from other available information.
[b]Calculated from data on PVA filters, 20-μ pore. *In* S. Aiba, A. E. Humphrey, and N. F. Millis, "Biochemical Engineering," Academic Press, New York, 1965.

perhaps on prejudice. Several years of general experience are needed to definitely point up the relative merits of the various designs. One fundamental question is whether a 0.2–0.5 μ pore will protect against contamination or whether a depth filter which works at all particle sizes, but on a probability basis, would be more suitable.

Aeration and Agitation (Mass Transfer)

Aeration

Mass transfer in fermentation is concerned with transferring gaseous oxygen from the bubbles of incoming air into dissolved oxygen in the medium, where it can be utilized by the cell. The nature and degree of the aeration and agitation, the physical properties of the medium, and cell demand determine the dissolved oxygen concentration. Maintaining a *proper* dissolved oxygen concentration is a major task of fermentation engineering.

The difficulty of keeping the dissolved oxygen concentration above the critical level, 0.1–0.2 ppm for *Escherichia coli,* arises from the low solubility of oxygen in aqueous solutions, 7 ppm at 37° in water and less in culture media. This represents only about a 1-minute supply for an *E. coli* culture with a doubling time of 1 hour at a cell density about 12 g/liter (wet).

Oxygen starvation leads to anaerobic metabolism, which is characterized by low cell yields and more acid products. Cultures which have suffered limiting oxygen after early log growth phase (a few g/liter) often never recover complete aerobic capacity. McDaniel, Bailey, and Zimmerli[47] demonstrated a 4-fold increase in yield of *E. coli* from a 16-fold increase in oxygen-absorption rate with less than doubling in glucose consumption. Certain processes suffer from too much oxygen for which the formation of vitamin B_{12} is a classic example.

A useful term, the volumetric oxygen transfer coefficient, $K_L a$, may be derived from the overall absorption rate process:

[47]L. E. McDaniel, E. G. Bailey, and A. Zimmerli, *Appl. Microbiol.* 13, 115 (1965).

---

[c]Millipore Filter Corp., Bedford, Massachusetts.
[d]Gelman Instrument Co., Ann Arbor, Michigan.
[e]General Electric Corp., Schenectady, New York.
[f]Biotech, Inc. (Sweden), 12221 Parkway Drive, Rockville, Maryland.
[g]Pall Trinity Micro Corp., Cortland, New York.
[h]Cox Instrument Div., Lynch Corp., 15300 Fullerton Ave., Detroit, Michigan.
[i]Reeve Angel, 9 Bridewell Place, Clifton, New Jersey. Large sizes, Sethco, Freeport, New York.
[j]Domnick-Hunter (Engineers) Ltd., Albert Road, Washington Co., Durham, England.
[k]Eikoh Kasei Co. Ltd., No. 3,2-chome, Kanda-Misakicho, Chiyodoka, Tokyo, Japan.
[l]Selas Flotronics, Spring House, Pennsylvania.

$$K_L a = \frac{\text{demand}}{\text{driving force}} = \frac{\text{millimoles } O_2 \text{ absorbed}}{(C^* - C_L)_{\text{mean}}} \ \text{min}^{-1} \qquad (4)$$

where K_L is the mass transfer coefficient for the liquid, a is the area of the bubble surface per liter, O_2 absorption is in liters per minute, and C^* and C_L are the oxygen concentrations per liter, of the liquid in equilibrium with the gas, and in the bulk liquid, respectively.

The coefficient is a fundamental term which may be used to compare aeration of an organism in any type vessel. Theoretically a $K_L a$ of 1 in a 10,000-gallon tank should represent exactly the same aeration as a $K_L a$ of 1 in a shake flask; thus it is a primary guide in fermentation scale-up.

The value of $K_L a$ can be estimated for a vessel by measuring the copper-catalyzed oxidation rate of sulfite to sulfate.[48] This method gives $K_L a C_G$, from which a value for $K_L a$ can be estimated.[49] However, it is usually reported as a "sulfite number," the maximum absorption or performance capability under those conditions. It is useful in comparing fermentor capabilities, but it is rarely the same as one determined from a biological experiment under the same conditions. Schultz and Gaden have pointed out some limitations of the reaction.[50]

A better determination of $K_L a$ is made during actual fermentation from an oxygen balance or directly, using a fast response oxygen electrode.[51] The determination and significance of these values have been reviewed.[49]

Having $K_L a$ values from fermentations and from sulfite numbers is useful for a general knowledge of equipment comparison. From a practical approach, however, the most useful information is the concentration of dissolved oxygen in the medium during a fermentation. By experience, and from a growth curve, most workers can perceive the change from logarithmic to linear growth when the critical oxygen level is reached. The dissolved oxygen electrode is a much better guide, and can furnish a continuous picture of the vital changes in cell physiology and the oxygen transfer equipment. Sterilizable galvanic-type electrodes developed by Borkowski and Johnson[52] may be fabricated in the laboratory, and several others are available commercially.[20]

[48]C. M. Cooper, G. A. Fernstrom, and S. A. Miller, *Ind. Eng. Chem.* **36,** 504 (1944).

[49]R. K. Finn, *in* "Biochemical and Biological Engineering Science" (N. Blakebrough, ed.), Vol. I, p. 69. Academic Press, New York, 1967.

[50]J. S. Schultz and E. L. Gaden, Jr., *Ind. Eng. Chem.* **48,** 2209 (1956).

[51]B. Bandyopadhyay, A. E. Humphrey, and H. Taguchi, *Biotechnol. Bioeng.* **9,** 533 (1967).

[52]J. D. Borkowski and M. J. Johnson, *Biotechnol. Bioeng.* **9,** 635 (1967).

Agitation

Agitation, the second parameter of mass transfer, is more of a design than an operational consideration. Tank, baffle, and impeller geometries are fairly standard for the stirred, sparged system. The design, including the power rating, should allow for transfer coefficients of up to 4 or 5 per minute. One practical point in design or operation is that tall tanks, or high liquid levels, are more efficient in air utilization but give lower overall transfer coefficients. Calderbank, (53) and Oldshue, (54) have discussed the agitation and scale-up aspects of mass transfer in detail.

Control of pH

The advantages of pH control in fermentation sometimes are not sufficiently considered by biochemists. Improved yields of products often are obtained by maintaining optimal conditions in all stages of growth, and by extending these stages far past the capability of buffers. However, many operations continue to be limited by buffer capacity simply because pH control is not introduced in the early stages of the project, and the necessary testing during scale-up is considered to be too costly in time. Thus it is important to study pH effects as early as possible. Several domestic and foreign manufacturers offer sterilizable electrodes with a variety of mountings for all sizes of equipment. One of the most successful products is the Ingold electrode[10]; this is a combination electrode with a pressurized electrolyte reservoir, which is guaranteed for over 200 sterilizing cycles.

Control from a Thieve Stream

The main source of trouble with electrodes involves contamination of the reference half-cell through the liquid junction. One remedy we have used to overcome this problem is an externally mounted flow cell, supplied by a miniature thieve stream; neither sterilization nor pressure is then required for electrodes since the stream is not returned to the tank. The same stream is monitored for turbidity, and the outflow is available for further analyses. It is an old idea with some advantages which justify its consideration in various applications. The sampling line for the 400-liter fermentor is a $\frac{1}{8}$ inch or $\frac{3}{16}$ inch o.d. by $\frac{1}{16}$ inch i.d. Teflon tube which penetrates the upper part of the tank through a stain-

---

[53]P. H. Calderbank, *in* "Biochemical and Biological Engineering Science" (N. Blakebrough, ed.), Vol. I, p. 102. Academic Press, New York, 1967.
[54]J. Y. Oldshue, *Biotechnol. Bioeng.* **8,** 3 (1966).

less steel tubing fitting and extends about 60 cm into the culture. It is conveniently fastened to a baffle. From the sampling tube the stream flows through a Teflon stopcock to a nutating (Sigmamotor AL-2)[55] tubing pump and to a miniature glass flow cell containing the pH electrodes. Latex tubing (Sigmamotor) ⅛ inch o.d. by ¹⁄₁₆ inch i.d. or a slightly heavier-walled silicone tubing (LKB No. 10251) is used. The pump is powered by an Electrocraft[56] motor and speed controller (E-500-4) with a 30:1 speed reducer. The pH cell effluent goes to a recording turbidity monitor (ISCO).[19] Electrodes have an indefinite life in this system. For two years we used a pair of miniature electrodes (Beckman)[57], which were changed, in good condition, for a miniature combination electrode unit.

The volume of the system up to the pH cell is less than 2 ml. With a flow rate of 2 ml/minute, the response time is about 1 minute. Two control systems have been used which allow a constancy of ± 0.1 pH unit. An on-off controller (L & N) operated a solenoid value for anhydrous ammonia or a pump and valve for alkali or acid. This control was through a timer, controlling 1–2 minutes out of a 5-minute cycle. The second instrument (L & N) with proportional band, rate time, and reset, operates a solenoid and a pneumatic metering valve for ammonia; or a valve and diaphragm pump for alkali or acid. Control is smoother but with about the same variation as with the on-off system.

Thirty-liter and 10-liter fermentors operate satisfactorily with this system at pumping rates of 0.5–1 ml/minute for the thieve stream. Since the runs seldom last more than 24 hours, not much medium is lost. On-off control is used, without a timer, to operate a tubing pump (Sigmamotor AL-4).[55] This system works, then, with adequate response and without excessive culture loss in the 10–400 liter fermentation volume range. As with most stream monitors, fouling of the flow cell can be a problem; some salts may precipitate, and old bacterial cultures tend to settle out, which can result in drifting of the water-control point. Mycelial cultures can be sampled if a small, coarse, ceramic filter is used over the inlet sample tube, although the response is thereby reduced further.

Nonpressurized, Submersible Electrodes

Two U. S. companies recently have announced submersible reference electrodes with no liquid reservoir and thus no pressure requirements. In fact, they can stand several atmospheres of pressure without inter-

[55]Sigmamotor Inc., Middleport, New York.
[56]Electro-Craft Corp., Division of Napco Industries, Hopkins, Minnesota.
[57]Beckman Instruments, Inc., Fullerton, California.

ference to operation. Beckman Instruments[57] uses a "polymer composite material" in conjunction with a silver-silver chloride half-cell; the result is called the "Lazaran" electrode. It is rated for extended operation at temperatures as high as 110°, and it has been extensively tested at much higher temperatures. Temperature damage results only during dry conditions, including dry steam. For periodic sterilization at 120° or above, submerged or in wet steam, the manufacturers expect a long lifetime from the electrode.

We have heated one of these electrodes in the autoclave for many hours at 130°, submerged in water, and at 120° in the fermentor, submerged in medium. In between times, over a period of about 6 months it has been operating at normal temperatures with excellent stability. The electrode is stored wet. For fermentor operation, the body of the electrode was machined to ½-inch diameter with a shoulder for mounting in a Leeds & Northrup immersion unit. A Foxboro[14] electrode was selected for measuring.

The other reference electrode, marketed by Universal Interloc,[58] contains "a year's supply of crystalline potassium chloride," which, with the silver-silver chloride half-cell, is replaceable. The submersible model is a combination electrode with a preamplifier; the latter no doubt contributes to its rather high cost and temperature limit of "120° or lower." It seems likely that a reference electrode similar to these will become standard in the near future for submerged applications in fermentation.

Foam Control

Uncontrolled foaming can ruin a highly aerated culture growing in enriched medium or one which is growing to high cell concentrations on minimal medium. Certain anaerobic organisms, e.g., *Clostridium pasteurianum*, actually produce so much gas, so vigorously, that foam is a problem. Most manufacturers of fermentors incorporate automatic antifoam addition systems which are controlled by a probe at the top of the tank. Mechanical foam breakers include paddles, centrifuges, nozzles, sonic devices, and combinations of these. Solomons,[5] and Bryant[59] have described these recently.

One of the most effective mechanical defoamers is a version of the rotating cone[10]; it is essentially an open, disk-type centrifuge in which foam is pushed between the rotating stacked cones, broken, and rejected, while gases pass out the top to the exhaust. In small fermentors,

[58]Universal Interloc, Inc., Santa Ana, California.
[59]J. Bryant, *in* "Methods in Microbiology" (J. R. Norris and D. W. Ribbons eds.), Vol. II, p. 187. Academic Press, New York, 1970.

the shaft-mounted unit requires high shaft speeds for effective operation.

Chemical antifoam agents may be dispensed automatically during the fermentation, may be added to the media before sterilization, or may be added from a preselected program. The proponents of the last of these systems note that it is easier to prevent foam than to destroy it. A variety of surface-active materials are sold as antifoam agents; of these, the biochemists have preferred the silicone emulsions because of their well established inertness in biological systems. Recently, polypropylene glycol (Dow, P-2000)[60] has gained favor because its action does not tend to decrease during the course of fermentation; nor, because it is a liquid, does it "clump" during sterilization. We have noticed toxicity to only one organism, *Photobacterium fischeri*. The polyglycol is a viscous oil, which does not emulsify well. Therefore, for dispensing purposes, in some laboratories it has been used as a 25% solution in ethanol.

It has been our experience that polypropylene glycol, 0.05 ml/liter to 0.1 ml/liter, in conjunction with a simple mechanical breaker, will control foam for an overnight bacterial fermentation in a salts and 4% glucose medium, in which cell yields of 40–60 g/liter, wet weight, are expected. For enriched media, at least 0.1 ml of antifoam is required per liter. The breaker is a two-bladed, ¼-inch mesh screen, supported in a three-sided tubing frame and fastened about 6 inches above the level of medium. The air exhaust has been moved to the center of the tank, just above the breaker, where foam is thrown away from it. The breaker also has another beneficial effect, that of rapid and uniform distribution of added antifoam. If foam overrides such an arrangement when an automatic addition system is in operation, the foam-probe makes contact every time a blade passes. Fatigue of the antifoam valve can be avoided with a delay relay which requires a signal of a few seconds' duration for the circuit to be energized. Foaming often may be reduced, without lowering of dissolved oxygen levels, by raising back-pressure and lowering the agitation rate or air flow. In the case of a shaft-mounted breaker, a high agitation rate usually is maintained.

Turbidity measurement by Split Fiber Optics

Turbidity monitors which use the absorption principle need special electronics to yield a linear readout, and this linearity is lost at cell concentrations of greater than about 1 g (wet weight) per liter. Conventional nephelometers provide direct linear electrical outputs pro-

[60]The Dow Chemical Co., Midland, Michigan.

portional to cell concentration. However, at high cell concentration, the cells begin to mask the light source from the detector, with resultant reversal to a downward slope.

We have been testing a prototype instrument[61] which uses a split fiber optics to furnish both light source and detector for a nephelometer (180°). These functions are intermixed in the $\frac{1}{8}$-inch diameter bundle, containing hundreds of fibers. When the optical system "looks" at a cell suspension, the light is reflected from the first cells it strikes back to the receptor, enabling the process to operate with light paths of only a few microns without being blinded. In fact, it is more efficient at high cell densities. With cell concentrations between zero and 60 g per liter, the slope increases about 10%. It is a nearly uniform rate of change and is easy to compensate.

The optics are mounted in a flow cell through which the effluent from the fermentor thieve stream is pumped after debubbling. A jet washes the face of the optics with a timed, short, water pulse. The signal is read on a print-type recorder.

Continuous Culture

The theory and advanced techniques of continuous culture have been widely described.[1,62] However, as Tempest has pointed out, one need not know all about the theory of continuous culture in order to use it, anymore than one need master the theory of internal combustion engines in order to operate an automobile. Tempest and his colleagues at Porton, who have pioneered both in theory and practice of this technique, have used it routinely for small-scale production.[63,64] An example is described here of continuous culture in the 400-liter fermentor for production of cells in the late logarithmic growth stage. Over 2000 liters of culture were produced in two working days with only two full-time operators, and a part-time assistant to freeze the harvested cells. A small continuous sterilizer was required in addition to batch-culture equipment. Several of the filters listed in Table V also can serve in this capacity, particularly where bacteriophage is not a problem.

[61]L. H. Thacker and W. F. Johnson, Instruments and Controls Division, Oak Ridge National Laboratory, Oak Ridge, Tennessee.
[62]I. Malek and Z. Fencl, "Theoretical and Methodical Basis of Continuous Culture of Microorganisms," Academic Press, New York (1966).
[63]R. Elsworth, G. A. Miller, A. R. Whitaker, D. Kitching, and P. D. Sayer, *J. Appl. Chem.* 17, 157 (1968).
[64]D. W. Tempest, *Biotechnol. Bioeng.* 7, 377 (1965).

Continuous Anaerobic Fermentation: Clostridium sticklandii as Example

Clostridium sticklandii is an obligate anaerobe from a group of organisms which carry out unique metabolic reactions.[65] The techniques described here apply to most anaerobic organisms. Less difficulty usually results from traces of oxygen on a large scale than is the case with small cultures. The opportunity of capping the fermentor with an anaerobic atmosphere after sterilization, and the ability of the larger inocula to create their own anaerobic conditions, facilitate operations. Often, the reducing agents which are ordinarily used on a small scale may be unnecessary. Some organisms benefit from ventilation with a slow sweep of inert gas while, with others, gas production is so vigorous that foam control is required and evolved hydrogen must be vented to an efficient hood.

Inoculum. Two 10-ml tubes of culture, grown overnight in an anaerobic jar, were used to inoculate 10 liters of medium in a 14-liter New Brunswick fermentor maintained under nitrogen. After 12 hours of growth at 30°, the turbidity read undiluted in a Klett-Summerson colorimeter at 540 nm, against water, was about 325 units.[66] Blank readings on the sulfide-containing medium are rather high and variable, so that water provides a more constant standard.

Medium Components. The following components were sterilized together in the fermentor, (concentrations are expressed in grams/liter): tryptone(N-Zcase, Sheffield Farms), 20; yeast extract (Difco), 10; sodium formate, 1.5; K_2HPO_4, 1.75; L-lysine·HCl, 1.0. Sterilized separately was $Na_2S·9H_2O$, 3 ml of a 10% solution being used for each liter of medium.

Continuous Sterilizer. A sterilizer, fabricated from stainless steel tubing, as described by Lempe,[67] was used. Heating and cooling sections were coils containing 50 feet of $\frac{5}{16}$ inch o.d. steel inside of $\frac{1}{2}$ inch o.d. tubing; the holding section was 24 ft of 1 inch o.d. tubing which provides a holding time of approximately 2 minutes with 1.5 liters/minute flow. A variable speed,[68] positive-displacement gear pump was used to feed the sterilizer, and two valves served to control pressure on the system at 40–50 psi.

Sterilization of Medium. Sterilization was carried out as described in a previous section for tank operation. All components, except sodium

[65]T. C. Stadtman, *Arch. Biochem. Biophys.* 125, 226 (1968).

[66]On the Klett scale, 500 units is an optical density of 1.00.

[67]J. Lempe, PP Div. Report No. 60-3442-2, 1959, U.S. Army Biological Laboratories, Fort Detrick, Frederick, Maryland.

[68]Gear-chem Model 70R5, ECO Engineering Co. Newark, New Jersey; U. S. Varidrive, 0.5 horsepower, U. S. Electric Motors, Milford, Connecticut.

sulfide, were made to a volume of 390 liters with deionized water, and sterilized for 30 minutes at 130°. These conditions are much more stringent than needed for spore destruction, but cells from the batch stage of the operation appear superior in enzyme activity to those from the continuous phase, in which the medium treatment was milder. The fermentor was kept under a positive pressure of 2–3 psi during cooling and throughout the run by introduction of nitrogen gas through the air supply inlet to the tank top. When the temperature approached the operating range, sodium sulfide solution was pumped in and the control was set at a temperature of 32°.

Fermentation. The fermentor was inoculated with the 10 liters of culture, described above at midnight and left overnight with slow agitation (45 rpm). At 9 AM, turbidity (Klett, 540 nm) was over 300 units, corresponding to a cell density of about 2.5 g/liter. The continuous sterilizer was steamed for about 30 minutes. Feeding of medium, which included the sodium sulfide, commenced at 9:45 AM at a rate of 1.5 liters/minute, with a holding temperature of 130° for 2 minutes. Harvesting, at the same rate, was alternated between two tubular bowl centrifuges.[69] A cooling coil was used between the fermentor and centrifuge to lower the harvest stream temperature to about 15°. Four 190-liter batches of medium were fed, and the tank contents were then cooled and harvested, except for 12 liters, which were removed aseptically as an inoculum for the next run. A total of 1160 liters yielded 3120 g of wet cells, from four centrifuge bowls, for a yield average of 2.7 g/liter.

The second day of operation produced 2800 g from 970 liters. All cells were frozen, as soon as they were taken from the centrifuge, by dropping small pellets of cells into liquid nitrogen.

Enzyme assays[70] of the seven individual centrifuge batches showed a progressive decrease in specific activity of cell suspensions for lysine fermentation and for glycine reductase during the first day of operation; activity was high again at the start of the second day of operation. This phenomenon demonstrates the transient nature of the cell physiology and the need for enzyme assays to detect changes in activities during fermentation in order to attempt to obtain optimum yields.

In this example of continuous culture, a medium component was not growth limiting, and cell density was determined by feed rate and temperature. For short runs, this is the most convenient, if not the most elegant, approach, and changes in physiology resulting from a limiting nutrient need not be investigated. Limiting components may result in either an increase or a decrease in the level of an enzyme.

[69]Sharples; Model 16, 15,000 rpm, 13,200 RCF; Model AS26, 15,000 rpm, 15,600 RCF. Pennsalt Chemicals Corp., Philadelphia, Pennsylvania.
[70]T. S. Stadtman, personal communication.

For larger-scale production, in the equipment described above but with a continuous, "desludging" centrifuge,[71] over 300 kg of *E. coli* was grown aerobically in 90 hours of continuous operation.

Semicontinuous Culture. Semicontinuous operation, which has no physiological resemblance to continuous culture, may be used to advantage where conditions for continuous culture are not applicable or convenient. The technique is very efficient when cells approaching the stationary growth stage are desired. For this procedure a holding tank for sterile medium, or a large-capacity sterilizer, and a holding tank for unharvested cells are required. The culture is grown, batchwise, to the required cell concentration and then is blown rapidly into the holding tank, which, preferably, has cooling facilities. A convenient amount of culture is retained for use as inoculum; this amount may vary from that remaining on the sides of the tank, up to as much as one-half of the tank contents. The fermentor is then refilled rapidly with fresh medium at the operating temperature.

An Example of an Aerobic System: Growth of *E. coli*

The regulation of glutamine synthetase synthesis in *E. coli* is a classic example of enzyme control.[72] One of the four major control mechanisms of the active enzyme level in *E. coli* W is the repression of synthesis by nitrogenous media, such as yeast extract and ammonia. Growth of cells on glutamate or in the presence of low levels of ammonia derepress synthesis and yield a 10- to 20-fold increase in the concentration of glutamine synthetase. This system serves as an example of an aerobic culture procedure.

Cells for purification of glutamine synthetase on a large scale were grown on a modification of a modified M-9 medium[73]:

Component	Grams/liter	
NH_4Cl (Allied Chemical, Tech.)		6.0
Na_2HPO_4 (Monsanto, Granular, Tech.)		6.0
KH_2PO_4 (Fisher, Purified)		3.0
Na_2SO_4 (Baker, Reagent)		1.1
$MgSO_4 \cdot 7H_2O$ (Mallinakrodt, AR)		0.2
Glycerol, 99% (GAF, USP)	50	ml/liter
Polypropylene glycol (Dow, P-2000)	0.05	ml/liter
Trace element mixture[74]	1.0	ml/liter

[71] Westphalia SAOOH-205, Centrico, Inc. Englewood, New Jersey.
[72] B. M. Shapiro and E. R. Stadtman, *Annu. Rev. Microbiol.* **24**, 501 (1970).
[73] B. Holmström and C. G. Hedén, *Biotechnol. Bioeng.* **6**, 419 (1964).
[74] According to Elsworth[63] and Tempest,[64] *E. coli* cannot assimilate EDTA complexes. However, the final concentration of EDTA in the medium used here is about 5% of that described by Tempest. We have observed no metal limitations at cell concentrations of over 50 g/liter.

Trace element mixture

Components	grams/liter
CaCl$_2$	0.5
FeCl$_2$·6H$_2$O	16.7
ZnSO$_4$·7H$_2$O	0.18
CuSO$_4$·5H$_2$O	0.16
CoCl$_2$·6H$_2$O	0.18
EDTA	20.1

The amount of ammonium chloride used limits growth to about 35 g/liter, wet weight. Sodium hydroxide was used to control pH; 22 liters of a 5N solution were sterilized for 90 minutes at a setting of 121° in the autoclave.

All medium components, except for magnesium sulfate, were sterilized together in a volume of 350 liters, as described in a previous section. The fermentor was held at 120° for 30 minutes. When the medium had cooled, magnesium sulfate was added and temperature control was set at 37°. The carboy containing the alkali was connected to the pump, and the pH controller was set for 7.0.

At 1 PM, the tank was inoculated with 1 liter of a shake flask culture which had been stored at 4° overnight. The following control settings were made: agitation, 120 rpm; airflow, 200 liters/minute; back pressure, 5 psi. Air and stirring were set to increase to 450 liters/minute and 260 rpm, respectively, after 7 hours. Heavy consumption of alkali ceased at 7:30 AM, and cell growth leveled off 45 minutes later. Aeration and agitation each were reduced by about half at this point. Data on derepression in this organism indicate nearly maximum synthetase activity at 5-6 hours after growth turnover, with the peak activity at 8-9 hours. Harvest was started in this run at 3 PM, about 6 hours after turnover; growth conditions were maintained during the 75-minute harvest with three centrifuges (two Sharples, Model AS-26, one Sharples, Model 16). The centrifuge cake, collected on Mylar sheets, was scraped off and frozen immediately at −20°.

With larger inocula this operation can be carried out on a 24-hour schedule. The growth level to which the culture is limited by ammonia does not appear critical. However, the carbon source must be in excess and pH should be maintained near neutrality. The dissolved oxygen concentration under the conditions described approaches the critical level at about 30 g/liter, and nearly linear growth is maintained for a short period. The sulfite oxidation number for this fermentor at the described level of aeration is about 3 mmoles O$_2$ per liter per minute. The increase in specific activity of glutamine synthetase under the above conditions is 9- to 10-fold, with a peak of about 0.8 μmoles γ-glutamyl hydroxamate formed per minute per milligram of protein.[75]

[75]E. R. Stadtman, personal communication.

The above examples are typical pilot plant procedures. In Table VI are listed other representative large-scale fermentations that have been carried out in our laboratory. The variety of organisms and range of yields reflect the versatility required in such equipment.

Improvised Methods

Usually large-scale equipment is not available for special, or even standard, procedures in most biochemical laboratories. Continuous culture is one method of scaling up in small equipment. We have also grown anaerobic organisms in 200-liter plastic drums which were sterilized with ethylene oxide and filled through the continuous sterilizer. Gentle agitation with magnetic stirring, and temperature control from heat lamps, or an internal coil or heater, are beneficial. Nonsterile growth is possible, particularly with fast-growing anaerobes, by the use of selective media and large inocula; a good example is the growth of *Clostridium kluyveri* in an oil drum.[76] *Galerina marginata* (Table VI) was cultured for the production of α-amanitin, in fifteen 20-liter carboys; a nonclogging sparger of nylon tubing with 0.01-inch diameter holes was used for runs lasting 3–5 weeks.

Photosynthetic organisms (Table VI) were grown in the 400-liter fermentor which was fitted with two external 1000-W halogen-cycle photographic lights[77]; the extra light port was fitted over the hand-hole opening of the fermentor, and water cells protected the glass ports from heat damage. Masking of light by foam and by spatter on the ports were the primary complications.

Many workers are skeptical of the long-range benefits of improvised or "homemade" equipment. However, we wish to mention a fermentor, of 28-liter capacity when used aerobically, which was fabricated from a 36-liter, bottom-tubulated, glass reactor vessel.[78] A commercial agitator shaft and bearing assembly was used; hollow baffles served for temperature control. The assembly was mounted in a platform from which the fermentor could be rolled directly into the autoclave. Full instrumentation and automatic controls were partially shared with the 400-liter unit. The size, ease of handling, and culture visibility have made this unit extremely versatile.

A unit similar to this, fitted with a magnetic agitator drive, has been

[76]E. R. Stadtman, Vol. I [84].
[77]Quartz-King 1000, Dual Light, Color Tran Industries, Burbank, California.
[78]Quick Fit VZ 30/12, QVF Limited, Staffordshire, England.

TABLE VI
Representative Large-Scale Fermentations

Organism	Volume (liters)	Yield (kg, wet weight)
Bacteria		
Aerobacter aerogenes	350	1.8
Azotobacter vinelandii	300	2.6
Bacillus subtilis 12 AT[+]	350	8.8
Bacillus thiaminolyticus	520[c]	6.0
"Cellulamonas"	350	4.0
Clostridium pasteurianum	400	5.0
Clostridium thermoaceticum	400	0.6
Escherichia coli B	1300[a]	26.6
Escherichia coli W	6600[a]	300
Escherichia coli CR 63	200	0.6 (cells)
+ T$_4$ Amber 44 N 82 phage		+ 1.5 x 10^{16} (phage particles)
Escherichia coli W 3110 (colicinogenic)	350	0.7
Escherichia coli K12, MO-7	350	17.0
Desulfovibrio desulfuricans	400	0.6
Desulfatomaculum orientis	400	0.3
Micrococcus lactilyticus	380	1.2
Streptococcus faecalis 10 Cl	1400[a]	6.7
Thiobacillus neopolitanus	400	0.4
Thiobacillus thioparus	380	0.6
Photobacterium fischeri	350	3.0
Photobacterium phosphorium	300	1.5
Fungi		
Galerina marginata	225[b]	0.7[d]
Neurospora crassa	350	22.0
Saccharomyces cerevisiae (mutant)	180	1.1
Saccharomyces cerevisiae (wild type)	350	5.6
Ustilago sphaerogena	300	10.0
Photosynthetic organisms		
Euglena B (heterotrophic, dark)	370	1.5
Euglena B (light)	370	0.3
Rhodospirillum rubrum	300	0.4
Rhodospirillum rubrum (autotrophic)	300	0.2

[a]Continuous culture.
[b]Fifteen 20-liter glass carboys, 15 liters each.
[c]Semicontinuous culture.
[d]Dry weight.

used for vaccine manufacture.[79] British laboratories use large-diameter glass pipe for smaller-size fermentors.[80]

[79]Personal communication, J. Cameron, Wellcome Research Laboratories, Beckenham, Kent, England.
[80]R. Elsworth, G. H. Capell, and R. C. Telling, *J. Appl. Bacteriol.* 21, 80 (1958).

[37] Scale-Up of Protein Isolation*,**

By STANLEY E. CHARM and CHARLES C. MATTEO

I. Introduction

Experience and information needed to solve the problems involved in the large-scale isolation of enzymes was unavailable several years ago. Success of scale-up operations required application of engineering principles, adaption of equipment from other areas of processing and development of new techniques.

The activities of the New England Enzyme Center upon which this chapter is largely based, include large-scale purification of macromolecules and subcellular structures from animal, plant, and microbiological sources, completion of several hundred fermentations, and preparation of bacterial viruses and virus-infected cells.

The operations presented are generally applicable to the processing of a wide range of biological materials. In many cases, specific examples and analytical solutions are given to the problems encountered in protein isolation.

II. The Pilot Plant Isolation Procedure

Thawing Frozen Material

With the exception of the continuous extraction of bacteria, the material to be extracted is usually accumulated in batches and frozen to prevent cellular deterioration. Thus, the first problem encountered is estimating the thawing time of frozen material. This is usually a trivial problem in laboratory-scale preparations, but large quantities are often

*The footnotes marked with an asterisk are references to a laboratory preparation. The unpublished pilot plant procedures developed at the New England Enzyme Center differ primarily in the quantities processed. However, important operation changes are usually made to accommodate the equipment available and to minimize overall processing time.
**The New England Enzyme Center is supported by PHS Grant No PO-7 FR 00346.

packed in drums or containers holding up to 100 lb of material. Thawing time increases with mass, and also varies considerably with geometry and thawing conditions employed. The time during which material remains thawed before it is used, is often critical. Material near the outside of a large container is in an unfrozen state longer than that in the center. The lower the thawing temperature, the greater the time difference. While a higher thawing temperature may diminish the time effect, it results in a steeper temperature gradient between the inner and outer regions. In general, since the amount of chemical reaction or change varies exponentially with temperature and linearly with time, the optimum thawing temperature that results in the least enzymatic change for the mass depends on geometry and thawing conditions.

The calculation of thawing time for finite objects is complex.[1] Values for several sizes and shapes of containers appear in Table I. It is seen that the thawing and freezing times depend on the heat transfer coefficient, a measure of the heat transfer conductance associated with the heating medium. For example, plunging one's hand into ice water at 0° certainly is more uncomfortable than placing one's hand in still air at 0°, even though both media are at the same temperature. This is due to the fact that the heat transfer coefficient for water is about ten times greater than that for still air. In Table II, heat transfer coefficients for air and water are noted for various conditions.

Preparation of Tissue

Removal from tissue of unwanted parts, e.g., connective tissue, blood vessels, and fat, is easily carried out in the laboratory by manual trimming. However, trimming is extremely time consuming when carried out by hand with large amounts employed in a pilot plant.

Failure to remove connective tissue (and membranes) can lead to the clogging of process lines and valves as material settles and is caught at these places. Erratic and variable flow results from blockage so that smooth operation is impossible. Lipid material presents a different problem. It may form complexes with proteins and can interfere with the sharpness of fractionation when precipitation techniques are used. An emulsion may form which interferes with settling and centrifugation of precipitates. Fat also may bind to filter materials and cause sharp decreases in flux rates. The same effect is exhibited in the flow through columns if material is retained on the surface.

While connective tissue is readily removed by centrifugation, it has been our experience that removal by screening is more desirable. This is primarily due to the large amount of solids in the crude extract and the difficulty with process flow control.

[1]Data calculated by computer and based on animal tissue containing 80% moisture.

TABLE I

Time[a] for Freezing and Thawing Infinite Cylinders and Slabs under Various Conditions[b]

Freezing (initial temp. = 25°)	Cylinder			Thawing (initial temp. = −20°C)	
	Freezing temp. (°C)			Thawing temp. (°C)	
	−17.8	−28.9	−40	4	25
Heat transfer coefficient cal/(cm² − °C − sec)	Time (hr)			Time (hr)	
Radius = Rm = 40 cm					
2.7×10^{-4} (still air)	440	230	100	460	135
1.1×10^{-3} (rapidly moving air)	210	110	82	300	100
2.7×10^{-3} (still water)	175	90	80	280	96
6.2×10^{-3} (rapidly moving water)	160	86	78	—	—
Rm = 20 cm					
2.7×10^{-4}	185	94	66	190	51
1.1×10^{-3}	74	43	30	96	30
2.7×10^{-3}	55	30	29	72	28.5
6.2×10^{-3}	41	25	28	—	—
Rm = 10 cm					
2.7×10^{-4}	82	46	33	73	21.5
1.1×10^{-3}	29	15	10.3	28	8.8
2.7×10^{-3}	16	9	6.4	23.5	6.8
6.2×10^{-3}	12.5	6.4	5.0	—	—
Rm = 5 cm					
2.7×10^{-4}	38	21	15.5	29	9.2
1.1×10^{-3}	11	6.4	4.0	9.5	3.3
2.7×10^{-3}	6	3.3	2.15	8.8	2.25
6.2×10^{-3}	42	2.15	1.5	—	—

Rm = 2 cm					
2.7×10^{-4}	14.5	7.1	5.4	11.5	3.4
1.1×10^{-3}	3.8	2.05	1.55	3.2	1.0
2.7×10^{-3}	1.8	1.0	0.7	1.65	5.8
6.2×10^{-3}	1.05	0.56	0.45	—	—
Slab ½ Thickness = Rm = 40 cm					
2.7×10^{-4}	670	330	270	800	255
1.1×10^{-3}	440	230	160	650	200
2.7×10^{-3}	400	215	150	610	195
6.2×10^{-3}	380	210	140	—	—
Rm = 20 cm					
2.7×10^{-4}	280	130	84	245	76
1.1×10^{-3}	125	64	45	180	58
2.7×10^{-3}	108	58	39	160	56
6.2×10^{-3}	95	48	37	—	—
Rm = 10 cm					
2.7×10^{-4}	94	52	38	89	32
1.1×10^{-3}	40	22	14.5	50	17.5
2.7×10^{-3}	29	16	11	41	14
6.2×10^{-3}	24	13.5	10	—	—
Rm = 5 cm					
2.7×10^{-4}	38	22	15.5	38	12
1.1×10^{-3}	14.5	8.2	5.5	16.5	5.3
2.7×10^{-3}	9.2	5.3	3.1	12.5	4.2
6.2×10^{-3}	7.7	3.9	2.6	—	—

TABLE I (continued)

TIME^a FOR FREEZING AND THAWING INFINITE CYLINDERS AND SLABS UNDER VARIOUS CONDITIONS^b

Heat transfer coefficient cal/(cm²−°C−sec) Freezing (initial temp. = 25°)	Cylinder Freezing temp. (°C)			Thawing (initial temp. = −20°C) Thawing temp. (°C)	
	−17.8	−28.9	−40	4	25
	Time (hr)			Time (hr)	
	Rm = 2 cm				
2.7×10^{-4}	14.0	8.4	5.7	13.0	4.0
1.1×10^{-3}	4.5	2.5	1.65	4.35	1.4
2.7×10^{-3}	2.4	1.30	0.88	2.6	.86
6.2×10^{-3}	1.5	0.84	0.60	–	–

^{a}Time for center to pass through change of state.

^{b}State change temperature, −5°C; latent heat, 70 cal/g.

²Thermal properties used in calculation

(Subscript 1 refers to frozen state, and subscript 2 to thawed state)

Freezing

Before change of state: $K_1 = 0.2 \times 10^{-2} \dfrac{\text{cal-cm}}{\text{sec-cm}^2 - °\text{C}}$ $C_1 = 0.9 \dfrac{\text{cal}}{\text{g-}°\text{C}}$ $\rho_1 = 0.9 \text{ g/cm}^3$

After change of state: $K_2 = 0.4 \times 10^{-2} \dfrac{\text{cal-cm}}{\text{sec-cm}^2 - °\text{C}}$ $C_2 = 0.4 \dfrac{\text{cal}}{\text{g-}°\text{C}}$ $\rho_2 = 1.0 \text{ g/cm}^3$

Thawing

Before change of state: $K_1 = 0.4 \times 10^{-2} \dfrac{\text{cal-cm}}{\text{sec-cm}^2 - °\text{C}}$ $C_1 = 0.4 \dfrac{\text{cal}}{\text{gm} - °\text{C}}$ $\rho_1 = 1.0 \text{ g/cm}^3$

After change of state: $K_2 = 0.4 \times 10^{-2} \dfrac{\text{cal-cm}}{\text{sec-cm}^2 - °\text{C}}$ $C_2 = 0.9 \dfrac{\text{cal}}{\text{gm} - °\text{C}}$ $\rho_2 = 0.9 \text{ g/cm}^3$

TABLE II
HEAT TRANSFER COEFFICIENTS FOR VARIOUS THAWING MEDIA

Medium	Heat transfer coefficient $(cal/cm^2 - °C\text{-sec})$
Still air 0°C	2.7×10^{-4}
Rapidly circulating air	1.1×10^{-3}
Still water	2.7×10^{-3}
Circulating water	6.2×10^{-3}

The choice of screen material is somewhat arbitrary and depends upon solid content and consistency of the suspension. For example, a suspension of ground spleen may be pumped through fiberglass window screening.[2] Both window screening and cheesecloth are convenient because they are disposable. Open mesh stainless steel has the advantage of being rigid but may be extremely difficult to clean. Continuous filtration on a rotary filter may be satisfactory in some cases, but is not useful if large amounts of lipid are present since this leads to clogging of the filter.

Fat can be often removed from a minced tissue by suspension in buffer. Particles of fat float to the top and are skimmed off. The problem is much more difficult once the tissue has been passed through a homogenizer, since colloidal suspensions or emulsions are formed. Manipulation of the purification steps may effectively eliminate this problem. If extraneous proteins are precipitated in the early stages of a preparation, filtration of the suspension usually removes the fat which becomes entrapped by the filter cake, leaving the filtrate fat free. Filter aid may be added to the suspension to maintain a reasonable filtration rate. This method is used in obtaining lactic dehydrogenase from dog fish muscle.

The desired protein may also be separated by precipitation. Sedimentation in a continuous centrifuge allows the fat to be collected with the supernatant liquid. If precipitation steps are not used, e.g., in the separation of microsomes from material such as liver, which involves a low speed and a high speed centrifugation, the DeLaval centrifuge may be used for liquid–liquid extraction. The lighter material is separated from the bulk of the supernatant fluid; microsomes are then pelleted by centrifugation in a continuous ultracentrifuge. Without treatment to eliminate fat, processing through the high speed centrifuge is hindered. Fat collects in globules which block both the process lines and the internal channels of the rotor.

[2]This material was used in the preparation of dipeptidyltransferase as cited in the Appendix.

A fourth way to remove fat is by solvent extraction. Preparation of an acetone powder, for example, removes lipids and dehydrates. Finely ground tissue is particularly important in the preparation of acetone powders to facilitate dispersion in the solvent and the subsequent extraction of water and fat. To obtain good cell breakage, a minimum amount of fluid is used during the grinding.

It also may be possible to effect a mechanical separation of fat and tissue based on particle flexibility and elasticity. A device employing the latter property is employed in the food industry for this purpose.[3] It consists of a perforated drum which rotates above a conveyor belt. The belt is held against the drum by tension. Those particles capable of being forced into the drum under the controlled tension are usually more flexible than those excluded. Separations are achieved at those temperatures where fat and tissues have different flexibilities.

Tissue Comminution and Cell Breakage

After thawing, cells and tissue are mechanically subdivided. The degree of disruption should be sufficient to release the desired enzyme, but not so severe as to release undesired cellular components. If material is too finely comminuted, mechanical difficulties may be encountered in removing solids.

Three types of comminuting equipment are employed at the Enzyme Center. A commercial meat grinder,[4] is used for mincing animal tissue. The Eppenbach Colloid Mill,[5] substitutes for the laboratory scale Waring Blendor. This device has a similar cutting action, and in addition the suspension is sheared by passage through adjustable openings after contacting the blades. The stream may be recycled through the hopper or the apparatus may be operated as a single-pass continuous system. Cooling fluid is circulated through a jacket to maintain low temperature. The Manton-Gaulin high pressure homogenizer is used for breaking microorganisms and operates by extrusion of a suspension through a small opening.[6]

Most animal tissue is first minced in a commercial meat grinder. This includes such varied tissue as muscle, heart, spleen, pancreas, and calf skin. The first stage of comminution may be used to precondition tissue for further homogenation. It is also a purification step, i.e., "soluble" enzymes are released, while other proteins remain entrapped in particulate organelles. If the mince is washed several times, soluble enzymes

[3]Bibun Co., Hiroshima, Japan.
[4]One horsepower "Butcherboy," Lasar Mfg. Co., Los Angeles, California.
[5]Gifford-Wood Co., Hudson, New York.
[6]Manton-Gaulin Mfg. Co., Inc., Everett, Massachusetts.

and much of the heme protein is removed. More severe comminution may release mitochondrial and microsomal material.[7]

The perfect comminutor imparts all its energy to dividing particles. From a thermodynamic view, the minimum energy required for comminuting would be the energy difference between the undivided and the divided states. In the ideal case, all the energy would be used to create new surface area. Thus, if the difference in surface area between state 1 and state 2 is A, the free energy difference is approximately $A\sigma$, where σ is the surface tension of the material. It can be shown that surface tension is related to tensile strength of the material. Consider a bar of material 1 cm² in cross section. If the bar can be pulled apart without deformation, the surface area is now 2 cm². If the force, F, pulling the material apart acted over the molecular distance D necessary for separation, then the work of separating the material is $F \times D$. The work of creating 2 cm² of new surface is 2σ, and

$$2\sigma = F \times D \qquad \text{or}$$

$$F = \frac{2\sigma}{D} \qquad (1)$$

The force, F, divided by the cross-sectional area over which it acted (in this case, 1 cm²) is the theoretical tensile strength of the material.

In practice, most of the comminution energy is consumed in deforming particles before they are reduced and in friction between particles after reduction. The latter energy ultimately appears as heat. Reduction of 1 g of tissue (1 cm³) to particles of 10^{-12} cm³ increases the total surface area to $10^{12} \times 6 \times 10^{-8}$ or 6×10^4 cm² (assuming cubic particles with sides of 10^{-4} cm). If the tensile strength of tissue is 10^5 dynes/cm², and the distance for separation of particles is of the order of 10^{-8} cm, the minimum theoretical work for this reduction is $(FD)/2 \times 6 \times 10^4$, or $(10^5 \times 10^{-8}/2 \times 6 \times 10^4 = 3 \times 10^1$ dyne-cm/g or 7×10^{-7} cal/g. Except for that energy required for movement, much of the energy added above this results in heating the material and must be removed to prevent a rise in temperature.

Breaking of Microorganisms. Intracellular substances may be extracted from microorganisms by a variety of methods. Large-scale techniques include mechanical methods such as extrusion of the cells under pressure or agitation with glass beads. In addition, the release of enzymes by osmotic shock, autolysis, or digestion of the cell by added enzymes such as lysozyme is often used. Sonication has not been successfully applied to processing large quantities of cells.

Extrusion under High Pressure. The Manton-Gaulin homogenizer has been found to be useful for disrupting large quantities of cells. Cells

---

[7] S. Cha and R. E. Parks, *J. Biol. Chem.* **239**, 1961 (1964).

are broken by extruding a suspension at pressures up to 8000 psi through a narrow channel (Fig. 1). The processing rate of 56 liters per hour (15 gal/hour) with the laboratory model allows up to 19 kg of cells in a typical suspension, i.e., 30% w/w, to be processed per hour. Larger versions are available with throughputs of 280 and 570 liters per hour. Gram-negative bacteria are easily ruptured by two passes through the device; about 80% breakage may be expected. The breakage of some cells such as micrococci, streptococci, yeast, and lactobacilli is more difficult. Breakage of high cell concentrations or the disruption of cells at the stationary phase of growth require modification of the procedure. This usually involves multiple passes or continuous recycling of the suspension. For example, *Lactobacillus casei* in a concentration of 10% (wet weight) required 5 passes for 90% breakage while *Escherichia coli* B at a concentration of 25% required 3 passes. Continuous recycling has been used for the breakage of *Aspergillis niger, Saccharomyces cerevisiae,* and *Pseudomonas aeruginosa,*[8] Baker's yeast may be brought to 90% breakage after two cycles by pretreatment with liquid nitrogen, i.e., freezing and thawing.

Cell breakage may be carried out using one of the three methods shown in Fig. 2. Equations describing the process assume a homogeneous sus-

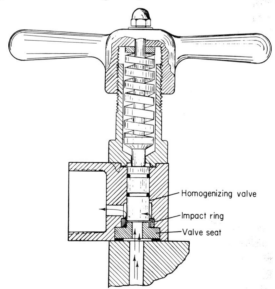

Homogenizing valve

Impact ring

Valve seat

Arrows show flow direction

FIG. 1.   Manton-Gaulin homogenizing valve used for cell disruption. (Courtesy Manton-Gaulin Co.)

[8]M. D. Lilly and P. Dunnill, *in* "Fermentation Advances" (D. Perlman, ed.), p. 225. Academic Press, New York, 1969.

pension in the reservoir, and assume that a constant fraction, D, of the whole cells entering the homogenizing valve survive unbroken.

Processing rates for each method as a function of K, (the fraction of cells surviving a single pass), are shown in Figs. 3 and 4 for 80% and for 90% cell breakage. Overall processing rates for all three methods decrease with increasing value of K. The single pass method has the highest throughput, but repeated passes are usually necessary for satisfactory breakage. Batch recycling approaches the output of the single pass method as the value of K increases. Efficiency of the other two methods in comparison to the single pass is illustrated in Fig. 5. For a given K value, the efficiency of the batch recycle is independent of the degree of breakage desired. The efficiency of the recycle and bleed technique decreases both as the values of K and the values for percentage of breakage increase. The extent to which some microorganisms are broken by passage through the Manton-Gaulin Mill is shown in Table III.

Time of exposure to high temperature ($10°$–$15°$ higher than entering temperature) is controlled by passing the effluent through a cooling coil or heat exchanger before recycling or further processing.

Two major precautions should be observed for proper use of the mill. The cell-buffer suspension, prepared by the use of a Waring Blendor or colloid mill, is passed through at least a single layer of cheesecloth and then through the Manton-Gaulin mill. This preliminary screening removes both large particles and foreign material which interfere with the pumping mechanism of the homogenizer. The second precaution is to avoid entrapped air in the mixture during suspension of the bacteria. Air accumulating in the homogenizer may result in an air lock which interferes with process flow.

The treatment of the extract which follows breakage often influences the breaking conditions. It is found, for example, that a single pass at the somewhat reduced pressure of 7000 psi yields optimal recovery of the enzyme DNA polymerase from *E. coli*. The enzyme is precipitated in the following step by streptomycin sulfate, presumably because of its association with nucleic acid. Since a second pass through the mill results in a greater shearing of the nucleic acid, the precipitation of the enzyme is considerably more difficult, A single pass yields only 60% cell breakage. Recovery of enzyme from the remaining intact cells may be effected by centrifugation and resuspension of the cells followed by a second pass through the Manton-Gaulin mill. Extracts from both passes are then combined for subsequent precipitation of the enzyme by streptomycin.

The presence of viscous nucleic acids is a nuisance. If, for example, the next step is centrifugation or filtration, the high viscosity hinders the removal of cell debris. When multiple passes through the homogenizer

(a) Single pass

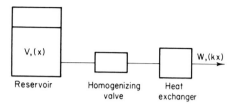

(b) Batch recycle

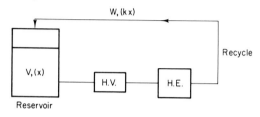

(c) Continuous recycle and bleed

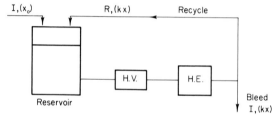

FIG. 2. Various arrangements of Manton-Gaulin homogenizer for cell disruption. The equations determining the whole cells remaining and the overall processing rate for each of the three methods are:

(a) single pass:

fraction unbroken, $\dfrac{X_f}{X_o} = K^n$

overall processing rate, $P = \dfrac{W}{n}$

(b) Batch recycle:

fraction unbroken, $\dfrac{X}{X_o} = e^{-\,W/v\,(1-K)t}$

overall processing rate $P = \dfrac{V}{t} = \dfrac{-W(1-K)}{\ln x/x_o}$

(c) Continuous recycle and bleed:

fraction unbroken, $\dfrac{X_f}{X_o} = \dfrac{1}{\dfrac{W}{I}(1/K - 1) + 1}$

overall processing rate $P = I = \dfrac{W(1-K)}{\left(\dfrac{X_o}{X_f} - 1\right)K}$

486

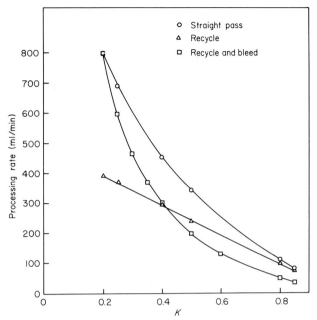

FIG. 3.   Processing rates in Manton-Gaulin for 80% cell breakage.

are not desirable, the problem of high viscosity may be eliminated by brief mechanical shearing in a Waring Blendor, colloid mill, or even by vigorous mechanical stirring of the suspension.

Cell Breakage with Glass Beads. Most other methods of mechanical disruption of cells use glass beads with a grinding mill.

In the Eppenbach mill, a suspension is subjected to strong shearing forces by circulation through a narrow gap between a rotor and stator. Glass beads (120 μ) are added to the suspension to a final concentration of 50–60 volume %. This equipment is normally used for batch volumes of approximately 10 liters although operation may be carried out on the continuous recycle and "bleed" method mentioned previously.

Temperature of the suspension rises 10° gradually in the 15 minutes required for disruption, even with cooling fluid in the jacket of the grinder. Higher bead concentrations intensify the heating problem

where I = bleed rate; K = fraction of cells surviving each pass; n = number of passes through machine; t = processing time; v = volume of suspension in reservoir; w = flow through machine; t = processing time; V = volume of suspension in reservoir; W = flow rate through homogenizing valve; X_0 = initial concentration of unbroken cells; X_f = concentration of unbroken cells in outlet stream. The equations can be used for other methods of cell breakage in which the homogenizing valve is replaced by another method of cell disruption.

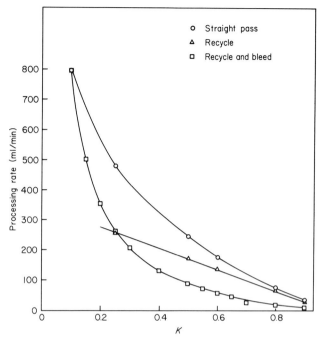

FIG. 4.   Processing rates in Manton-Gaulin for 90% cell breakage.

because the high viscosity, low thermal conductivity, and poor mixing result in less rapid heat transfer. With high bead concentrations, the suspension heats rapidly and has to be removed for cooling before breakage is complete. Other methods employing glass beads have been described. Among these is the use of a disintegrator with a rotary disc stirrer.[9] This was used as a model for constructing continuous production scale apparatus of 5-, 50-, and 200-liter capacity. The device described was capable of obtaining 100% breakage of *Saccharomyces cerevisiae* within 72 seconds at a concentration of 3.5 g (dry weight) /100 ml suspension. Increasing the concentration to 10.5 g/100 ml increased the breakage time to 96 seconds. A mill using larger glass beads (3 mm) has been used for breaking *Aspergillus niger*; the results have been compared to several other techniques.[10] Glass beads were used in a ratio of 6 kg to 3 liters of suspension. The continuous 5-liter mill required two passes at 10.8 liters/hour to achieve optimum breakage. Although Zetelaki[11] was unable to satisfactorily disrupt *Aspergillus*

[9] J. Rehacek, K. Beran, and V. Bicik, *Appl. Microbiol.* 17, 462 (1969).
[10] K. Zetelaki, *Process Biochem.* 12, 19 (1969).
[11] L. Edebo, *in* "Fermentation Advances" (D. Perlman, ed.), p. 249. Academic Press, New York, 1969.

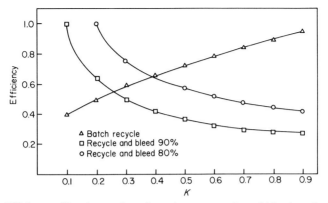

FIG. 5.   Efficiency of batch recycle and continuous recycle and bleed methods relative to single pass method. The relative efficiency is obtained by dividing the processing rate for each method by the processing rate for the single pass method.

niger in the Manton-Gaulin Mill, comparison with the experiments of Lilly indicates that the mycelium is disrupted if a batch recycling technique is used.[8]

Major disadvantages in using beads include their removal and the subsequent recovery of extract that adheres to them. Glass beads are separated from the disrupted cells by centrifugation, or by filtration in a Büchner funnel using Saran filter cloth. The beads must be resuspended in a wash buffer and filtered to remove adhered residue. Simple rinsing of the packed beads on the filter is much less efficient for recovering adhered material. In addition, the high viscosity of the suspension during processing leads to poor heat transfer and a rise in the temperature of the extract. Silicone antifoams, used during the growth of cells, have been found to interfere with breakage unless the cells have been washed.

TABLE III
TYPICAL K VALUES FOR BREAKAGE IN MANTON-GAULIN

	Concentration (g/l, wet weight)	Operating pressure	K^c	Percent breakage per pass
Aspergillus niger[a]	700	5000	0.76	24
Escherichia coli[b]	200	7000	0.5	50
Lactobacillus leichmanii	100	7000	—	—
Pseudomonas aeruginosa[a]	120	7000	0.35	65
Saccharomyces cerevisiae[a]	400	5000	—	21

[a]Calculated from data of Lilly and Dunnill[8].
[b]Stationary phase cells.
[c]K = fraction of entering whole cells broken by a pass through the homogenizing valve: % cell breakage for a pass through the valve = $(1-K)100$.

Mechanical Breakage with the X-Press. The X-press is used to disrupt cells by extrusion of the frozen material through a perforated disc. Rupture of cells is based on the changes in crystal structure and the shearing forces which occur upon flowing through the opening. A large unit capable of processing at least 100 g of frozen *Saccharomyces cerevisiae* per minute has been described.[11] With this system, cell breakage occurs at temperatures of about − 22°. The rapid change from liquid to phase II ice under pressure is thought to cause the breakage.

Breaking with Osmotic Shock. An osmotic shock procedure is useful for releasing certain hydrolytic enzymes and binding proteins from *E. coli* and other gram-negative bacteria such as *Salmonella typhimurium.*[12–14] Harvested cells are first washed with 30 mM Tris·HCl pH 7.0 and then suspended in 20% sucrose containing 30 mM Tris·HCl at pH 7.2 and 0.1 mM EDTA (1mM for stationary phase cells). After stirring for 10 minutes, the cells are removed by centrifugation in the DeLaval centrifuge. Cells are scraped from the bowl, cooled to 4°, and rapidly dispersed in cold water. Concentrated buffer may be added after 10 minutes if desired in order to stabilize enzymes or cells. The cells are then removed by centrifugation in a Sharples centrifuge. The supernatant liquid contains hydrolytic enzymes released by the sudden shift to low osmotic stength medium.[15] These enzymes are apparently located in the periplasmic region between the cell wall and the cytoplasmic membrane. Among the proteins released are the acid and alkaline phosphatases, asparaginase II,[16] ribonuclease I, and certain proteins involved in active transport. The technique is a gentle one, and cells remain viable after treatment. Only 4–7% of the total cell protein is released and the extraction technique yields a 14- to 20-fold purification over that obtained by disruption of the whole cell. The technique is also useful for removing such enzymes as phosphatase and asparaginase before the cells are extracted in a conventional manner. This greatly simplifies the subsequent purification of enzymes such as asparagine synthetase, the activity of which would be masked by the presence of the potent asparaginase.[15]

Enzymatic Breakage of Cells. Autolysis is commonly used for the extraction of cells, and the technique is easily adapted to large-scale operation. A suspension of cells is maintained at an elevated temperature (23°–37°) for several hours. After cooling, the cell extract is collected

[12]H. C. Neu and L. A. Heppel, *J. Biol. Chem.* 240, 3685 (1967).
[13]L. A. Heppel, *Science* 156, 1451 (1967).
[14]L. A. Heppel, *J. Gen. Physiol.* 54, 95 (1969).
[15]H. Cedar and J. H. Schwartz, *J. Biol. Chem.* 242, 3753 (1967).
[16]H. Cedar and J. H. Schwartz, *J. Biol. Chem.* 244, 4112 (1969).

by centrifugation. Among the enzymes extracted in this way are trans-aldolase from frozen *Candida utilis*,[17] UDP-Gal-4-epimerase from lyophilized *Candida pseudotropicalis*,[18] and tripeptide synthetase from air-dried bakers' yeast.[19]

Digestion of the cell wall may also be carried out by the addition of enzymes such as egg-white lysozyme. Through appropriate lysozyme concentration and control of digestion time, the degree of cell breakage may be controlled. Disadvantages of autolytic methods include the possible thermal inactivation of a desired enzyme or its destruction by cellular proteases.[20,21]

Criterion for Determining the Extent of Cell Breakage. A cell disruption process is designed to release intracellular material and the criterion for judging its completeness is the amount of product released. Microscopic or other observations may give data on the number of ruptured cells present, but not on the quantity of protein or enzyme released. However, the amount of enzyme activity may be correlated with secondary observations such as processing time, visual observation, or spin test data to give a convenient processing standard. Subsequent monitoring may be carried out by relying on these secondary criteria. They are especially helpful in continuous processing.

One of the simplest correlations is the use of graduated tubes in the DeLaval Gyrotest centrifuge.[22] A diluted sample of ruptured cell suspension is centrifuged for 3 minutes. Intact cells sediment first; above this is a layer of cell debris followed by the cytoplasmic components remaining in suspension (see Fig. 6). The relative quantity of each layer indicates the degree of breakage.

Extraction

Comminuted particles are frequently extracted batchwise in stirred jacketed vessels. Particle size determines not only the rate of extraction, but also the amount of extract which will be adsorbed on the particle surface. The larger the particle surface area the greater the amount of extract adsorbed to the particle and separated with the solids.

It is a well known general principle of extraction processes that, with a given quantity of extracting fluid, the use of many fractions of

[17]O. Tsolas and B. L. Horecker, *Arch. Biochem Biophys.* **136**, 287 (1970).
[18]R. A. Darrow and R. Rodstrom, *Biochemistry* **7**, 1645 (1968).
[19]J. E. Snoke, *J. Biol. Chem.* **213**, 813 (1955).
[20]I. T. Schulze and S. P. Colowick, J. Biochem. **244**, 2306 (1969).
[21]J. F. Clark and W. B. Jakoby, *J. Biol. Chem.* **245**, 6065 (1970).
[22]The DeLaval Separator Co., Poughkeepsie, New York.

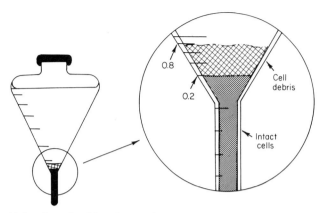

FIG. 6.   Estimation of cell breakage using graduated centrifuge tube.

the liquid leads to a more nearly complete removal of the desired solute than does the use of the entire liquid in a single extraction. This may be seen from the following formula,[23] which shows the relation between the volume of the material to be extracted, W, volume of extracting solvent used in each extraction, L, distribution ratio of desired solute between the solid and liquid phase, K, weight of solute initially associated with the solid phase, X_o, and the weight of substance remaining after the nth extraction, X_n.

$$X_n = X_o \left[\frac{KW}{KW+L} \right]^n \tag{2}$$

Consider the case where 10 k_g, or about 10 liters, of solids are extracted with 20 liters of buffer in 2 batches of 10 liters,
If $K = 0.1$

$$\frac{X_2}{X_o} = \left[\frac{(0.1)\,(10)}{(0.1)\,(10)+10} \right]^2 = \left[\frac{(1)}{11} \right]^2 = 0.0084 = \text{fraction of solute} \tag{3}$$
remaining unextracted after second extraction

If the extraction were carried out in one batch or $n = 1$ and $L = 20$,

$$\frac{X_1}{X_o} = \left[\frac{(0.1)\,(10)}{(0.1)\,(10)+20} \right] = \frac{1}{21} = 0.0476 = \text{fraction of solute} \tag{4}$$
remaining unextracted after single extraction

Further improvement in extraction efficiency may be obtained through arranging the staged extraction in countercurrent continuous flow, e.g., Fig. 7. In this method, the feed, i.e., the material to be ex-

[23]M. A. Joslyn, "Methods in Food Analysis," p. 50. Academic Press, New York, 1950.

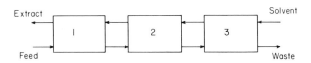

FIG. 7. Scheme of staged countercurrent extractor eliminate.

tracted, is not in contact with fresh solvent, but rather with solvent containing some solute. The fresh solvent is in contact with the partially extracted feed, and the concentration gradient favoring extraction is improved when it is needed most, i.e., when the solute concentration is lowest in the stream undergoing extraction. The larger the number of stages, the more thorough the extraction. The procedure for carrying out theoretical calculations for counter extraction processes is described in any elementary book in chemical engineering.[24]

The critical parameters for solid–liquid extraction are stage efficiency, i.e., how close solvent and solids come to equilibrium in a stage, number of stages, ratio of solvent to feed, and amount of solvent phase adsorbed to solids phase when separation of the phases occurs between stages and at the final stage. Stage efficiency is influenced by the character of agitation, fluid flow rate, and particle size. Separation of the extracted solids from the extract is brought about by centrifugation, filtration, or sedimentation, and these processes are also strongly influenced by particle size. Unfortunately, there are no quantitative data published on partition coefficients for proteins between solvent and solid phase.

Separation of Solids

Particles may be removed from suspension by centrifugation, settling, and filtration.

Centrifugation. Centrifugation is almost universally employed for solid–liquid separations in the laboratory and a wide array of centrifuges is available for batch use. Centrifugal forces up to 450,000 g may be developed although apparatus designed for the greater forces has a low capacity.

In the pilot plant, continuous centrifuges are necessary to handle the large volumes involved. Three types have been useful: disc-type centrifuge, hollow bowl type, e.g., Sharples, and high speed ultracentrifuges, e.g., Electronucleonics, R. K. All three permit continuous access to the liquid stream and at least intermittent access to solids.

The disc type centrifuge (Fig. 8) is readily available from several sources[22,25,26] and is suitable for clarifying suspensions with moderate

[24]G. Brown, "Unit Operations," p. 292. Wiley, New York, 1950.
[25]Westfalia Separator Div. Centrico Inc., Englewood, New Jersey.
[26]Bird Machine Co., South Walpole, Massachusetts.

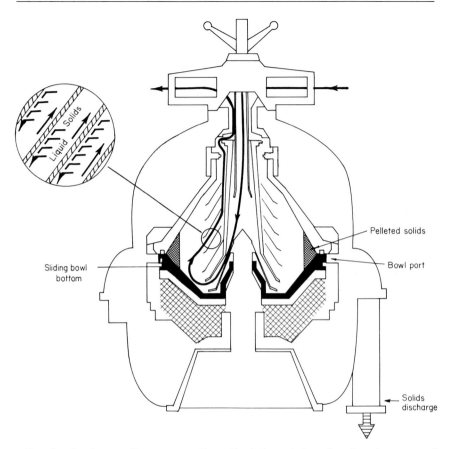

FIG. 8. Continuous disc-type centrifuge. Feed slurry is introduced at the center of the bowl and moves through the stack of discs. The stack is a nest of cones spaced a short distance apart. Solid particles in the feed, upon entering the narrow space between the discs, settle out radially under the influence of the centrifugal force. They impinge on the underside of the discs and slide outward to the bowl wall, where they compact before discharge. Fine particles can readily be captured at low g-force (approximately 8000) because of the short settling distance. Clarified effluent leaves at the inner edges of the discs and is discharged separately.

solids content (1–30%). The device has an infinite capacity for solids in that they may be discharged without stopping the centrifuge for cleaning. The DeLaval PX207 centrifuge used at the Enzyme Center has a solid capacity of 4.2 liters. When this amount has been collected, the contents of the bowl, including 4 liters of suspending fluid, are discharged. A certain amount of the extract is thus lost although the loss may be minimized by displacing the liquid in the bowl with buffer. Recent models of the PX type centrifuges permit discharge of only a portion of the material in the bowl. The resulting slurry of solids contains less suspending fluid. Much less extract is lost, and the need for displacement of the extract is eliminated. Rather than discharging the solids, it may be desirable to allow the bowl to come to rest, dismantle the centrifuge, and scrape the solids from the bowl. This results in a particularly dry product compared with the bowl discharge but adds to the total processing time. For this reason, it is usually employed when small batches of 50–100 liters are processed or, as in fermentation media, when the quantity of suspended solids is less than 2% of the total.

In addition to harvesting microbial cells, the DeLaval centrifuge is commonly used for removal of homogenized tissue from crude extracts and for separations of easily sedimented materials such as obtained by acetone or streptomycin precipitation. The centrifuge has also been used for liquid–liquid extraction.

It is possible to predict the efficiency of precipitate removal for a given flow rate in the continuous centrifuge by centrifuging 10 ml of the suspension in the *"gyrotester centrifuge."*

The DeLaval Company has correlated flow rate in the PX207 centrifuge with the time of centrifugation in the gyrotester. Conical centrifuge tubes are calibrated so that as little as 0.01% solids may be determined. The correlation between the gyrotester and continuous DeLaval centrifuge, unfortunately, had not been extended to the very low flow rates often used for enzyme preparation. Solids which settle in 3–5 minutes are processed in a Sharples Lab Super centrifuge while those not removed by the spin test are subjected to centrifugation in an ultracentrifuge.

Although the centrifuge is not refrigerated, low temperature operation is possible. The centrifuge may be precooled by passage of a refrigerated fluid through the machine prior to the product stream.[27] If the temperature of the supernatant liquid begins to rise, flow is momentarily interrupted, and the machine is cooled again. The increase in temperature upon passage through the machine is largely a function of the flow rate as illustrated in Fig. 9 for the DeLaval PX207. The increase in temperature is negligible at very high flow rates but many materials may be processed even at the lower rates. The effluent is passed through a heat exchanger for rapid cooling when operating at low flow rates.

Typical flow rates are the following: fermentation broth, 6 liters per minute; 30% acetone precipitate of protein (6% solids), 9 liters/minute; streptomycin precipitate, 12 liters/minute.

[27] H. E. Blair, S. E. Charm, D. Wallace, C. C. Matteo, and O. Tsolas, *Biotechnol. Bioeng.* 12, 321 (1970).

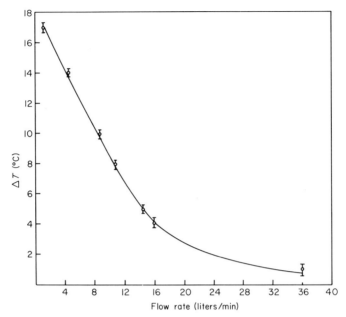

FIG. 9. Temperature rise as function of flow rate in a disc centrifuge.

Slightly higher *g* forces (15,000 *g*) are generated in tubular bowl centrifuges such as the Sharples A26.[28] The bowl capacity is about 4 liters, and the flow of liquid through the centrifuge is continuous. However, solid removal requires that the centrifuge be dismantled. In addition, the hollow bowl design has a much longer sedimentation path than a disc-type bowl, thus reducing the efficiency of the machine and requiring greater time for processing.

Among the centrifuges operating at higher *g* forces are the Sharples "Lab Super" centrifuge and zonal ultracentrifuges.[29,30]

Centrifugal forces of up to 60,000 *g* may be generated by the Sharples Lab Super centrifuges. The tubular bowl has a capacity of approximately 250 ml of solids and is easily assembled and operated. Because of their limited capacity, several machines are used simultaneously and are fed in parallel from a common, pressurized reservoir. Typical flow rates for each centrifuge range from 2 to 5 liters per hour. Although the centrifuge bowl itself is refrigerated, the effluent stream may heat at low flow rates.

These centrifuges are used for suspensions, such as ammonium sulfate precipitates, which require a higher *g* force for clarification than

[28]Sharples Co., New York, New York.
[29]Beckman Instruments, Inc., Palo Alto, California.
[30]Electro-Nucleonics, Inc., Fairfield, New Jersey.

obtainable in the DeLaval (about 8000 g). They are also useful for volumes too large for a batch centrifuge and too small for the DeLaval. They are not suitable for suspensions containing a high solids content because of the limited capacity of the bowl.

Continuous processing at forces approaching 100,000 g is possible with an ultracentrifuge. Although commonly used for gradient separations or isopycnic banding, they may be used to remove precipitates not easily separated by other centrifuges. The Beckman rotor, B VIII, has a 1-liter capacity (700 ml solids), but other centrifuges are available (Electronucleonics) with capacities of up to 6 liters. Typical flow rates for the Beckman are 2 liters per hour for bacteriophage, e.g., lambda phage, 1–2 liters per hour for ribosomes, and 10–20 liters per hour for ammonium sulfate precipitates.

One advantage of these centrifuges is that the system is sealed, thus eliminating foaming. Disadvantages include high cost for throughput, and corrosion sensitivity of rotors other than those made of titanium. They also must be dismantled to obtain the pelleted solids and have a low capacity for solids.

The simplest criterion for most centrifugation processes is the clarity of the effluent stream. Absolute clarity may not be necessary before the addition of such precipitating agents as streptomycin, protamine or ammonium sulfates. Solvent fractionation, pH precipitation or heat denaturation are other means for clarifying solutions, these result in the precipitation of small particles or cause formation of larger aggregates which are more easily removed.

It is possible that small particles will reappear upon resuspending the precipitate. If the desired protein is to be recovered from the precipitate, the problem obviously has not been overcome. This occurs in the preparation of glutamine synthetase from sheep brain acetone powder.[31]

It is possible to modify the original extraction procedure and eliminate the use of a colloid mill for the initial suspension in order to avoid fragmentation of the solid to small particles.

Pretreatment of the suspension often has a significant effect on the quality of the supernatant liquid. If the original extract were subjected to severe homogenization, extremely small fragments result. These are removed only with difficulty from the suspension. An autolysis carried out for a prolonged period has a similar effect. For example, the laboratory procedure for extraction of dipeptidyl transferase from beef spleen specified 22 hours of autolysis (see Appendix). The resulting suspension could not be clarified by centrifugation when the

[31]*V. Pamiljans, D. Krishnaswamy, and A. Meister, *Biochemistry* 1, 153 (1962).

operation was conducted on a large scale. In this case, it was found that a 3-hour autolysis released the same amount of enzyme, without the attendant fragmentation of the cell.

The hindered settling rate of spherical particles is given by the following equation.[32]

$$V = \frac{(\rho_P - \rho_L) \, g \, D^2}{18 \, \mu} \, F_s \qquad (5)$$

Where F_s is the correction factor for particle interactions in hindered settling and is given by:

$$F_s = \frac{X_L{}^2}{10^{1.82(1-X_L)}}$$

and D_p = diameter of particle; g = gravitational constant, 980.8 cm/(sec)^2 or 32.2 ft/(sec)^2; V = terminal settling velocity; X_L = volume fraction occupied by liquid; μ = viscosity of suspending fluid; ρ_L = density of liquid; ρ_P = density of particle.

The distance a particle moves in a gravitational field is $V\theta$, where θ is the time the field acts on the particle. In order for a centrifuge to remove a particle from suspension, it must impinge upon a surface. Those particles within a distance $V\theta$ of a surface will be removed in the centrifuge. The disc type centrifuge is thus more efficient than the hollow bowl because the additional surfaces between the center line and the bowl wall result in a decreased sedimentation path.

Because settling is dependent upon the density difference between the solid and the liquid, temperature affects the efficiency of separation. Most processes are carried out near 4°, the point of highest density for aqueous solutions. This results in the least density difference between solid and liquid. Lower temperatures also increase the viscosity of the suspending fluid thereby adding to a decreased sedimentation velocity.

The residence time, θ, for the continuous centrifuge is:

$$= \frac{\text{volume of centrifuge bowl}}{\text{flow rate through centrifuge}}$$

For a batch centrifuge, the residence time is simply the time the centrifuge is running.

For a continuous bowl type centrifuge (see Fig. 10), the allowable flow rate is given by:

$$Q = \frac{Y(D_p)^2 W^2 (\rho_p - \rho_L) \, (r_1{}^2 - r_0{}^2)}{36 \, (r_1 - r_0)^Z \mu} \, F_s \qquad (6)$$

[32]G. Brown, "Unit Operations," p. 78. Wiley, New York, 1950.

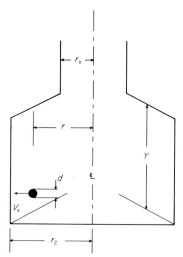

FIG. 10.   Diagram of continuous centrifuge.

where Y = length of bowl; W = angular volutional speed; r_1, r_0 = outside and inside radii of bowl, respectively.

To illustrate the calculation of flow rate in a continuous centrifuge, consider the following example:

EXAMPLE

Given: $Y = 10$ cm; $r_0 = 0.5$ cm; $r_1 = 3$ cm; $D_p = 0.005$ cm; $X_L = 0.85$; $\mu = 0.015$ poise; $\rho_p = 1.04$ g/cm^3; $\rho_L = 1.01$ g/cm^3. Speed $= 30,000$ rpm.
How fast may flow-through be?
Solution:

$$F_s = \frac{(0.85)^2}{10^{1.82(0.15)}} = 0.384$$

$$W = 2\pi \frac{30,000}{60} = 3142 \text{ radians/sec}$$

$$Q = \frac{(10)\,(5 \times 10^{-3})^2(3142)^2(1.04 - 1.01)(3^2 - (.5)^2)}{36(0.015)(3 - 0.5)^2}\,0.384$$

$$= 73 \text{ cm}^3/\text{sec}$$

Since the diameter and density of particles may vary in a suspension, there is a range of settling velocities. Frequently, several passes through a continuous centrifuge may be necessary before a complete separation of solids is achieved. It can be seen from Eq. 6 that, as the fraction of solids becomes less with each pass, the ease of separation improves because F_s becomes greater.

Settling may be thought of as sedimentation under a force of 1 g. Little equipment or labor is required, and it is easily adapted to large scale. A mixture is maintained for several hours in an unstirred vessel and allowed to separate into two layers. The clear layer may be removed by siphoning or pumping. Solids are then removed from the remaining volume by filtration or centrifugation. However, the volume to be processed may now only be as little as one-third of the original volume. For batch operation, it would be advantageous to plan a procedure that allows the preparation to settle overnight thereby permitting separation without additional labor.

The success of the method depends upon the specific properties of the suspension as well as on the temperature, ionic strength, and pH of the solvent insofar as they affect the aggregation of the particles. The main disadvantage of settling is that it is often too slow to be useful although the rate of settling may be increased by adding a flocculating agent. For example, acid-washed filter aid was used in the course of purifying dipeptidyltransferase from beef spleen (Appendix). Unfortunately, a systematic investigation of other flocculating agents for sedimentation has not been carried out. A difficulty, at least with solid additives, is the possible irreversible surface adsorption of the desired protein.

Filtration. Filtration is based upon the ability of a medium to permit the passage of the liquid phase through a porous structure and restrict passage of the solid phase. The important properties of the suspension to be filtered are those of viscosity of the liquid phase, particle size, and compressibility of the particles. Important properties of the filter medium include its porosity and compressibility. The driving force is the pressure difference across the filter medium.

In the laboratory, centrifugation is usually more convenient than filtration. However, as a process is scaled up, the relatively low capacities of higher speed centrifuges make filtration a more attractive method.

Continuous Rotary Filter. With a continuous rotary filter (Fig. 11), the filtering surface of a rotating drum is partially immersed in the suspension being filtered. A filter medium covers the drum and retains solids. Filtrate is drawn through the drum by an internal vacuum. As the partially submerged drum rotates, it is coated with the solids that are in turn removed at a point in the cycle by a scraper or "doctor" blade. Thus, the surface is renewed and the filter may be operated in a continuous manner. A variety of filter cloths and filter aids exist that must be matched to the characteristics of the suspension to be separated.

The continuous rotary filter may be used to separate the same type of solids as the DeLaval continuous centrifuge. The centrifuge is some-

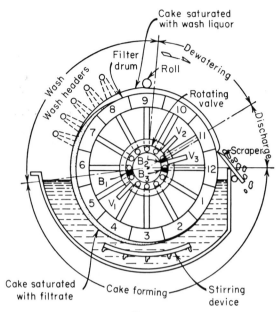

FIG. 11. Continuous rotary vacuum filter.

what more convenient to use because the filter must often be precoated. If the particle and suspending fluid densities are very close, centrifugation will be difficult; this condition has little effect on filtration. If solids settle rapidly, continuous centrifugation is difficult due to settled particles clogging lines and passage ways. This situation may arise in the acetone extraction of tissue. Settling is slow when substantial water is present, but in the presence of less water, settling of the tissue is rapid. Under the latter circumstances, continuous filtration is more convenient than centrifugation. When the volume fraction of solids is greater than 0.3, filtration is usually easier to apply.

Filter Press. The filter press is a batch type filter usually employed in pilot plant filtering and when the filter cake is valuable (Fig. 12). Although liquid flow through the filter press is continuous, solids accumulate, and the filter must be dismantled to remove them.

The conventional plate and frame filter press is not as convenient for solids separation as the continuous rotary filter because of the difficulty in removing solids from the former.

Modified Microporous Ultrafiltration. The difficulties in scaling up the capacities of high speed centrifuges are severe compared with those involved in scaling up the filtration of fine particles. Membrane filters with pore sizes ranging from 0.025 μ to 10 μ have been used for many

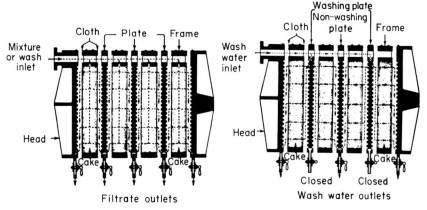

Fɪɢ. 12.   Filter press.

years for separation of fine particles in dilute solution, e.g., the series produced by Millipore Co.[33] and Gelman Co.[34] A diagram of microporous membranes is shown in Fig. 13. Fluid passes through an open pore membrane in much the same way it would pass through small pipes in laminar flow.

Particles that are difficult to centrifuge may lend themselves to filtration with open pore membranes. The problem in filtration occurs when particles rapidly clog the membrane although this difficulty may be overcome to some extent.

There are four general techniques available for renewing the membrane surface: (a) vibrating perforated plate above the membrane surface;[33,35] (b) high shear rate at the membrane surface, 10,000 sec^{-1};[36] (c) turbulent flow over the membrane surface and accompanying high shear rate (20,000 sec^{-1}); (d) continuous mechanical removal of solids, as in rotating drum filter.

Methods (a), (b), and (c) have hgh shear rates and shear stress associated with them. The vibrating perforated plate may have a shear rate in the order of 100,000 sec^{-1}, while the turbulent flow system noted in (c) has a shear rate of 20,000 sec^{-1}. Associated with high shear rates are high shear stresses. For example, for a Newtonian fluid with viscosity, μ, and undergoing a shear rate, γ, the shear stress, τ, is given by

$$\tau = \mu\gamma \qquad (7)$$

When μ is expressed in poises, and γ in sec^{-1}, τ is units of dynes/cm^2. The effectiveness of methods (a), (b), and (c) for clearing open pore

[33]Millipore Corporation, Bedford, Massachusetts.
[34]Gelman Instrument Co., Ann Arbor, Michigan.
[35]Chemapec, Inc., Hoboken, New Jersey.
[36]New England Enzyme Center, Boston, (experimental work).

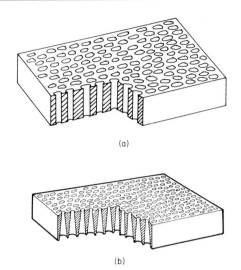

(a)

(b)

FIG. 13a.   Isotropic microporous membrane.

FIG. 13b.   Anisotropic microporous membrane.

membranes were tested using a suspension of casein precipitated with ammonium sulfate. The results, shown in Fig. 14, indicate that the vibrating perforated plate is most effective. The high shear-rate laminar-flow system is next, and the high shear-rate turbulent-flow system is less

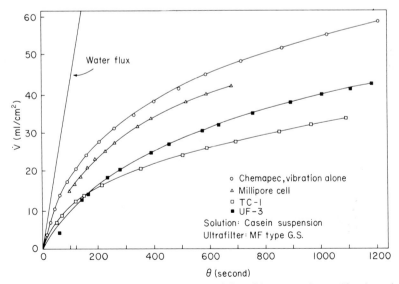

FIG. 14.   Ultrafiltration of casein micelles precipitated by ammonium sulfate in various ultrafiltration systems with a 0.22 μ membrane.

effective. In the latter two, the suspension is recycled continuously by the filter section and becomes increasingly more concentrated with each pass. The membrane pore size in each of these studies was 0.22 μ.

Increasing the pressure differential across the membrane up to about 70 psi improves the filtration rate. High pressures may actually decrease the rate by compressing the film of material that has accumulated on the membrane surface.

An effective low pressure filtration system for protein precipitates is obtained by using an open-pore membrane with a continuous rotary filter. With a filter surface of 3 square feet and a 0.22 μ membrane, it is possible to filter 13 liters per hour of casein precipitated by ammonium sulfate. Pressure across the membrane and rotating drum is about 12 psi. The backing for the membrane must be sufficiently rigid to maintain a smooth membrane surface when the drum is under vacuum; a porous polyethylene 1/8 inch thick is suitable as a backing.

Diffusion-Type Membranes. Proteins in solution may be removed by ultrafiltration with a diffusion type of membrane. A diffusion membrane has a filtering membrane surface supported by an amorphous open structure (Fig. 15). Filtration classification is initially due to the filtering membrane. However, as the solute is concentrated at the membrane surface, it forms an additional resistance to filtration. Of the various methods for clearing surfaces, the high shear rate with laminar flow is the most effective. An equilibrium exists between the concentrate resistance and the shear rate at the surface. Two enzyme systems (catalase and rennet) tested with these methods of ultrafiltration suffered various degrees of inactivation due to shear associated with cleaning the membranes. In Fig. 16a and Fig. 16b, the effect of shear rate on activity is shown as a function of time. It may be seen that at higher shear rates and with longer time, greater inactivation results. It is found that rennet

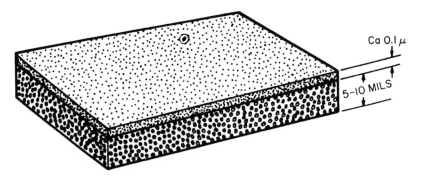

Ca 0.1 μ

5–10 MILS

Fig. 15. Anisotropic diffusive membrane.

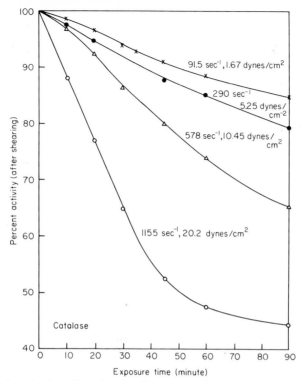

Fig. 16a. Inactivation of catalase by shearing at 4°.

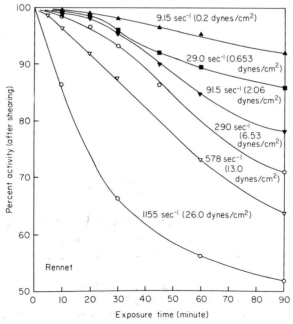

Fig. 16b. Inactivation of rennet by shearing at 4°.

is partially reactivated upon standing for 1–3 hours (Fig. 17). Catalase, on the other hand, is not reactivated after inactivation by shear. Shear tests on enzyme solutions were carried out in a Weissenberg Rheogoniometer[37] with a couette, or narrow-gap, concentric cylinder measuring system.[38]

Although ultrafiltration in a stirred cell has been applied successfully to the concentration of viruses as well as to enzymes,[39] inactivation of bacteriophage R17 during concentration in a thin channel system with high shear rate has been found. The results of a concentration experiment are tabulated below.

Sample	Titer/ml	Recovery %
Original	3.5×10^{12}	100
Concentrate, 2×	4.5×10^{12}	64
Concentrate, 3×	3×10^{12}	35

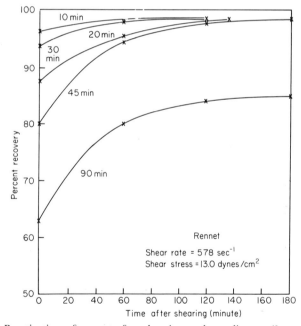

FIG. 17.   Reactivation of rennet after shearing and standing at 4°.

[37]Weissenberg Rheogoniometer, Farol Co., Bognor Regis, England.
[38]S. E. Charm and B. L. Wong, *Biotechnol. Bioeng.* **12**, 1113 (1970).
[39]D. I. C. Wang, T. Sonoyama, and R. I. Mateles, *Anal. Biochem.* **26**, 277 (1968).

It is assumed that inactivation is the result of shear. The technique would be useful if recirculation, which increases the shear effect, could be minimized.

It is a simple matter to construct an ultrafilter that employs dialysis tubing and is suitable for pilot plant scale concentration of enzymes, i.e., 50–100 liters per day. Dialysis tubing, 1 cm in diameter, is threaded through a smooth rayon cloth sleeve. The end of the dialysis tubing is fastened to the nozzle of a pressure can by a piece of Tygon tubing (Fig. 18). The dialysis tubing, backed by the cloth, forms a flexible rod when it has 30–40 psi within it. At this pressure, it is capable of being coiled without crimping and may be placed in the cylindrical container. The fluid to be filtered or concentrated is forced from the pressure can to the tubing and may be recycled through the filtering section by pump (see Fig. 19). In our laboratory, it was possible to coil 3500 cm² and to filter 5 liters/hour at 60 psi from a casein solution containing 1.6 mg/ml;

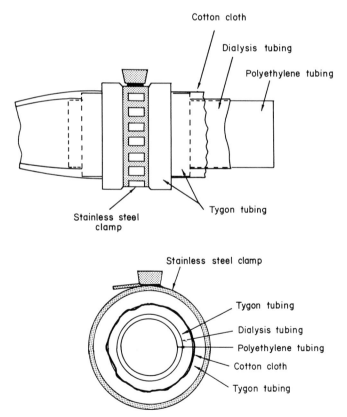

Fig. 18.   Arrangement of tubing for high pressure dialysis.

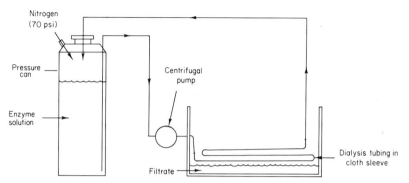

FIG. 19.   Apparatus for recycling solution during high pressure dialysis.

nearly 95% of the protein was retained. At completion of the process, dialysis tubing is removed from its cloth sleeve and the concentrated protein is removed by squeezing the tubing through a pair of rollers. By maintaining a constant volume in the system with appropriate fresh buffer during ultrafiltration, dialysis may be carried out rapidly.

Dialysis tubing withholds proteins with molecular weights of 20,000 and above, whereas the molecular weight retained by the various membranes is controlled more closely. Membranes are capable of fractionating proteins on the basis of molecular weights within groups of about 20,000.

Ultrafiltration of Proteins in Solution.   CONCENTRATION. Ultrafiltration of dissolved proteins is more difficult than ultrafiltration of protein precipitates. Ultrafiltration of proteins in solution is either for purposes of fractionation or concentration.

A catalase solution (molecular weight 225,000, 0.2 mg protein per milliliter) was ultrafiltered with a rotary filter using an Amicon membrane (PM-30).[40] The membrane usually retains proteins with molecular weights of 10,000–30,000. If catalase is allowed to filter dry on the membrane and is removed by the rubber doctor blade, about 50% of the activity is lost. However, by maintaining a concentrated protein solution on the membrane rather than the protein cake, losses in activity may be avoided. This is done by controlling the speed of the drum and the vacuum on the drying section of the filter. A suction system removes the protein concentrate from the drum. Catalase has been concentrated 15-fold in this manner without loss in activity.

FRACTIONATION. Fractionation with membranes of proteins in solutions is more difficult than concentration and the more concentrated the solution, the more difficult is fractionation. The major problem arises

[40]Amicon Corporation, Lexington, Massachusetts.

because proteins coat or polarize on the membrane and thereby restrict passage of other proteins. This occurs easily with membranes having pores of less than 0.2 μ in diameter. In such a situation only water passes through the membrane leaving behind a more concentrated protein solution without fractionation. By using a wiping cloth on the drum in the filter section of the cycle it is possible to alleviate the "polarization" problem (Fig. 20).

Concentrated protein solutions form protein micelles or aggregates thus making passage through membranes more difficult than with dilute protein solutions. It is possible by continuously diluting a concentrated protein solution, to wash the proteins through the membrane leaving behind the larger particles, e.g., ribosomes, viruses. The resulting dilute protein solution that passes the membrane may then be concentrated.

Plasma diluted 1:100 is filtered in the rotary filter with a 0.1 μ membrane. The filter is arranged so that the membrane is continuously wiped while filtering. The composition of the filtrate from diluted solutions that passed the membrane is 50% of the initial concentration. By maintaining a constant volume with saline, it is possible to wash through the plasma proteins (see Table IV).

In a similar test carried out with a recirculating filter (Amicon TCF) in which high shear at the surface was used to prevent the membrane from clogging, the initial concentration of protein solution passing through the membrane was 50% of the starting concentration as with the rotary filter. However, this eventually dropped to zero, i.e., only water passed through the membrane, (see Table IV). Similar results were obtained with solutions of casein at 5.7 mg/ml (Table V).

Reverse Osmosis. When ultrafiltration is applied to small molecules, e.g., molecular weight 500 or less, a substantial osmotic pressure acts against the filtering pressure. Under these circumstances, ultrafiltration is referred to as reverse osmosis.

Large-Scale Gradient Centrifugation vs. Ultrafiltration. Gradient centrifugation at 84,000 g is possible in a 6-liter continuous flow rotor (Electronucleonics Co.). This is the highest speed centrifuge available with this

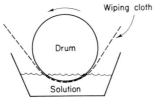

FIG. 20.   Wiping cloth on drum of rotary filter.

TABLE IV

COMPARISON OF PLASMA PROTEIN PASSAGE THROUGH 0.1 μ MEMBRANE ON ROTARY DRUM FILTER AND TCF-10 AMICON RECYCLE FILTER

Dilution	0	4	100	140	600	800	1000
Protein conc.[a]	65.61 mg/ml	16.62 mg/ml	0.723 mg/ml	0.477 mg/ml	0.108 mg/ml	0.082 mg/ml	0.068 mg/ml
Rotary drum filter with "wiping cloth"	10% steady state for 3 hours	10% steady	35% steady	35% steady	85% steady	100% steady	97% steady
TCF-10	10% → 5% after 3 hours	10% → 5% after 3 hours	14% → 9.3% after 3 hours	15% → 9.0% after 3 hours	—	50% → 0 after 3 hours	50% → 0 after 3 hours

[a] $\dfrac{\text{Protein conc. of effluent}}{\text{Protein conc. of feed solution}} \times 100\%$

TABLE V
COMPARISON OF CASEIN PASSAGE THROUGH 0.1 μ MEMBRANE IN
ROTARY DRUM FILTER AND TCF-10 AMICON RECYCLE FILTER

Protein conc.[a]	0.438 mg/ml	5.7 mg/ml
Rotary drum filter	40% steady	40% steady
TCF-10	40% steady	40% after 3 hours

[a] $\dfrac{\text{Protein conc. of effluent}}{\text{Protein conc. of solution}} \times 100\%$.

capacity. The advantage of this method of centrifugation is its ability to concentrate and to purify simultaneously. The procedure has been employed in plant-scale production of influenza vaccine. To carry out the same process with ultrafiltration would require an additonal purification step, either with fractionating membranes or by column chromatography.

Fractionation of Protein by Solubility

The selective precipitation of proteins by "salting out" is a widely used technique in the laboratory. Commonly used precipitating agents are ammonium sulfate, sodium sulfate, and streptomycin sulfate. Polymers such as polyethylene glycol[41] and organic solvents such as ethanol and acetone are also used as precipitants.[42]

The general equation describing the salting out process is[43]

$$\log s = B' - K_s' \frac{\Gamma}{2} \tag{8}$$

where s = solubility of protein in grams per liter of solution; B' = intercept constant; K_s' = intercept constant; $\dfrac{\Gamma}{2}$ = ionic strength (moles/liter).

The value of B' varies with pH, temperature, and the nature of the protein. It is independent of the salt used. The value of K_s' is independent of pH and temperature, and is a function of the protein and salt used.

A protein solution of concentration, s, will begin to precipitate at the ionic strength given by

$$\frac{\Gamma}{2} = \frac{B' - \log s}{K_s'} \tag{9}$$

There is no unique salt concentration for precipitating an enzyme. Rather, the limits of salt concentration needed to precipitate an enzyme

[41] This volume [23].
[42] This volume [22].
[43] A. A. Green and W. L. Hughes, this series, Vol. I, p. 72.

will vary with the concentration of enzyme.[44] Thus, enzymes may appear in different fractions varying not only with the stage of purification, but also with the activity of the starting material. A solubility curve of an enzyme is invaluable in the scaling-up process. Measurement of activity of a solution together with the plot would quickly give the optimum salt concentration required for recovery. Reproducible results in the precipitation of a protein require that the pH, temperature, and protein concentration be constant. In addition, because it may take several hours for a salted out solution to attain equilibrium, early separation of the precipitate may produce variable results. This point is particularly important in continuous systems, where the time alloted may be insufficient before centrifugation or filtration.

Ammonium sulfate is the most commonly used agent because of its high solubility, its low cost, and its protective effect on many enzymes. Organic solvents are capable of sharper fractionation, but denature most proteins unless all solutions are maintained at or below 0°. Salts must often be pulverized before addition in order to allow for rapid dissolving and for easier metering. Although liquids such as a saturated solution of ammonium sulfate are far easier to add, they necessarily dilute the suspension thereby changing the protein concentration. For high percent saturation this method also results in unwieldy volumes. Dialysis against a saturated solution has the marked advantages of simultaneous purification and concentration but has not had widespread application on a large scale.

Polyethylene glycol (PEG) has been used[41] in the fractionation of serum proteins,[45,46] in the precipitation of alfalfa mosaic virus,[47] and for the sedimentation of bacteriophage.[48] This technique is far more rapid than the use of an ultracentrifuge for the sedimentation of large volume of virus suspension.

Unlike inorganic salts, PEG can be separated from protein by adsorbing the protein to an ion exchange resin. The PEG is not adsorbed and can be washed through. Alternately, solvent precipitation of the protein may be employed. Since PEG is very soluble in such organic solvents as ethanol, it can readily be removed from the protein precipitate. A third method of removal is by ultrafiltration.

[44]M. Dixon and E. C. Webb, *Advan. Protein Chem.* 16, 197 (1962).
[45]A. Polson, G. M. Potgieter, J. F. Largier, G. E. F. Mears, and F. J. Joubert, *Biochim. Biophys Acta* 82, 463 (1964).
[46]P. W. Chun, M. Fried, and E. F. Ellis, *Anal. Biochem.* 19, 481 (1967).
[47]M. F. Clark, *J. Gen. Virol.* 3, 427 (1968).
[48]K. R. Yamamoto, B. M. Alberts, R. Benzinger, L. Hawthorne, and G. Treiber, *Virology* 40, 734 (1970).

Dilute protein solutions require higher salt concentration for precipitation of the protein. In addition to the larger volumes to be handled this factor can lead to increased difficulties in pilot plant operation. For example, centrifugation is less efficient because of the higher liquid density and a longer processing time is required.

Precipitation is an important tool for the concentration of proteins from solution, allowing further fractionation to be carried out with smaller volumes and leading to decreased processing time. Important to the large-scale operation are the physical characteristics of the final suspension which are often overlooked on the laboratory scale. The rate of addition of precipitating agent, for example, affects the final particle size. Rapid addition generally leads to smaller, more difficult to remove particles and may also denature protein or give inferior fractionation if localized high concentrations of precipitating agent exist. Removal of the heat of mixing is especially important when using organic solvents, and the type of agitation used has also had effects. A vibromixer, for example, which affords good mixing has lead to unexpected results in fractionation and even to differences in density of particles sufficient to cause them to float.

The high viscosity of many solutions such as a crude bacterial extract or of a solution of polythylene glycol, compounds the problems of adequate mixing and efficient heat removal. Recovery of precipitates by either centrifugation or filtration is hindered by high viscosity.

Scaling Up Differential Thermal Denaturation and Other Chemical Reactions

Occasionally, it is possible to purify a mixture of proteins by exposure to heat. If extraneous protein denatures more rapidly than the desired protein, it precipitates, leaving the desired protein in solution. The precipitated protein is then separated by filtration or centrifugation. It is possible from a consideration of the time-temperature relationships of this step on the laboratory scale to estimate what these parameters are on a large scale. Thermal denaturation of proteins appears to be a first-order reaction.[49] If the protein concentration of a given species remaining undenatured at time Θ is x, and if x_0 is the initial concentration, then the fraction in the native state at any time is

$$\ln x/x_0 = K\Theta \tag{10}$$

where K is the Arrhenius constant given by

$$K = A \, e^{-H/RT} \tag{11}$$

[49]M. Joly, "A Physico-Chemical Approach to the Denaturation of Proteins," p. 202. Academic Press, New York, 1965.

where A = constant; H = energy of activation for denaturation; R = gas constant; T = absolute temperature.

It follows that when this process is scaled up

$$\frac{x}{x_0} \text{ (lab scale)} = \frac{x}{x_0} \text{ (scale up)} \qquad (12)$$

The activation energy for denaturation for a particular enzyme may be determined from the rate of loss of activity at two different temperatures.

Thus:

$$\text{at } T_1: K_1 = \frac{\ln x/x_0}{\Theta}$$

$$\text{at } T_2: K_2 = \frac{\ln x/x_0}{\Theta}$$

From this, the value of H may be determined:

$$H = \frac{R \ln K_1/K_2}{\dfrac{1}{T_2} - \dfrac{1}{T_1}}$$

A mean value of H for several proteins is 82,000 cal/mol.[50]

In scaling up this step, the major differences occur for time-temperature relationships in heating and cooling the increased mass. For a stirred jacketed reactor, the time-temperature relationship is given by,[51]

$$\frac{UA\Theta}{MC} = \ln \frac{(T_m - T_1)}{(T_m - T)} \qquad (13)$$

where: U = overall heat transfer coefficient in vessel (see Table VI); A = heat transfer area; M = mass of fluid; C = specific heat; T_m = heating or cooling medium temperature; T = temperature of fluid at time Θ T_1 = initial temperature.

It is assumed in this case that the heating or cooling medium does not change temperature appreciably in the jacket of the stirred vessel. In the event that the heating or cooling medium does change temperature appreciably, an average between the inlet and outlet temperatures may be employed for the median temperature.

The overall heat transfer coefficient, U, is actually composed of several terms: the conductance (reciprocal of resistance) of the heating medium, vessel wall, and fluid in the vessel, all contribute to the total conductance or overall heat transfer coefficient as shown in Eq. (14):

[50]H. J. Morowitz, "Energy Flow in Biology," p. 114. Academic Press, New York, 1968.
[51]S. E. Charm, "Fundamentals of Food Engineering," AVI Publishing Co., Westport, Connecticut, 1970.

$$\frac{1}{U} = \frac{1}{h_i} + \frac{X_w}{K_w} + \frac{1}{h_o} \tag{14}$$

where h_i = heat transfer coefficient for the fluid inside the vessel; X_w = thickness of vessel wall; K_w = thermal conductivity of vessel wall; h_o = heat transfer coefficient for fluid in vessel jacket.

Equation (14) assumes a relatively thin-walled vessel, i.e., the thickness is small compared to the radius of the tank. Heat transfer coefficients vary with viscosity and circulation rate of the heating and heated fluids. Higher viscosity interferes with heat transfer; higher circulation rates enhance the transfer of heat from the fluid.

The overall heat transfer coefficient, U, may be obtained experimentally from a plot of $\ln (T_m - T_1)/(T_m - T)$ vs. Θ as shown by Eq. (13). The slope of the line is UA/MC from which U may be extracted. Overall heat transfer coefficients for water in various vessels at the New England Enzyme Center are noted in Table VI.

The use of Eq. (13) to scale up a chemical reaction such as thermal denaturation is illustrated in the following example.

EXAMPLE

Of a protein solution, initially at 5°, 1000 ml is placed in a stirred water bath at 75°. The solution is heated to 60°, held for 6 minutes at this

TABLE VI
OVERALL HEAT TRANSFER COEFFICIENTS FOR VARIOUS VESSELS
AT THE NEW ENGLAND ENZYME CENTER

Tank volume (liters)	Vessel material	Jacket medium	Jacket temperature (°C)	Agitator speed (rpm)	Overall heat transfer coefficient, Btu/(hr-ft²-°F)
100	Glass lined	Hot water	96		8.3
200	Stainless steel	Ethylene glycol	−17.5	150	12.9
200	Stainless steel	Ethylene glycol	−17.5	400	19.4
200	Stainless steel	Steam	115	133	103
200	Stainless steel	Steam	115	400	137
600	Stainless steel	Ethylene glycol	0	185	14.0
600	Stainless steel	Ethylene glycol	0	500	15.5
600	Stainless steel	Steam	100	200	60.0
600	Stainless steel	Steam	100	500	69.0
600, Fermentor	Stainless steel	Cold water	23.5	250	44
600, Fermentor	Stainless steel	Hot water	59	175	53
80, Fermentor	Stainless steel	Cold water	18.6	75	95
80, Fermentor	Stainless steel	Hot water	64	75	78.5

temperature and then cooled to 5° in a bath at 0°. The heating and cool-ing time–temperature relationships are shown in Fig. 21. It is now planned to carry this out with 90 liters in a jacketed glass-lined tank with hot water in the jacket at 65°. When the tank is cooled, water at 10° is passed through the jacket. It is essential that temperature of the product does not rise above 60°. An automatic control system limits the tank temperature so that it will not rise above 60°. How long should the prod-uct be held at 60° to be equivalent to the laboratory thermal denatura-tion process?

Solution:

1. Calculation of $(\ln \frac{x}{x_0})_{\text{lab}}$ from heating and cooling curve of beaker.

From Fig. 21, it is possible to obtain a plot of $e^{-82,000/2(273+T)}$ vs. Θ where $T = °C$ (see Fig. 22).

It is only after heating 0.6 hour that a significant effect on protein de-naturation occurs.

The fraction of protein unchanged in the laboratory process is pro-portional to the area under the curve in Fig. 22.

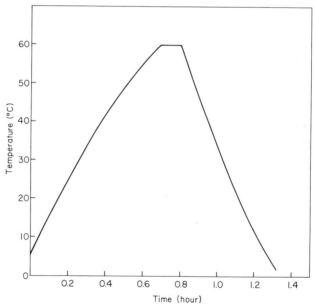

FIG. 21. Temperature vs. time for heating and cooling in a 1000-ml beaker.

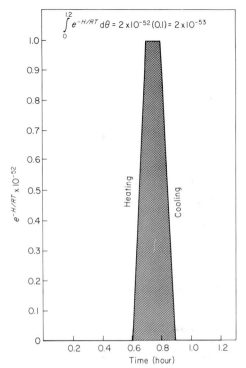

Fig. 22.   Integration to obtain fraction of undenatured protein by heating in a 1 liter beaker.

2. Calculation of time–temperature relationship for 90 liters in a glass-lined vessel[52]

$$\frac{UA\Theta}{MC} = 2.3 \log \frac{T_m - T_1}{T_m - T}$$

where $U = 45$ Btu/(hr) (sq. ft.) (°F); $M = $ lbs of fluid $= 198$; $C = 1$ Btu/(lb °F); $T_m = 65°$; $T_1 = 4°$; $A = 7.94$ sq. ft.; $(UA)/(MC) = 45\,(7.94)/(1.98)$ $(1)\,(2.3)$

The equation for a cooling curve is

$$0.72\Theta = \log \frac{10 - 60}{10 - T}$$

A plot of T vs. Θ is shown in Fig. 23.
From Fig. 23, a plot is made of $e^{-82,000/2(273+T)}$ vs. Θ as shown in Fig. 24.

[52] Although temperature is measured in °C and other parameters in English system units, both sides of Eq. (11) are dimensionless, making it possible to do this.

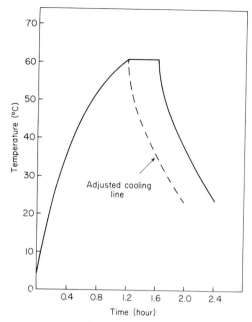

FIG. 23. Temperature vs. time for heating and cooling in 100-liter glass-lined tank.

The time at 60° is adjusted so that the area or number of units under this curve is the same as for the laboratory process.

In this case the effect of heating up and cooling down contribute the same number of units as the laboratory process, and no further time at 60° is required.

Scaling Up Column Chromatography

Solid–liquid chromatography is one of the most powerful tools available for the fractionation of proteins. The various types of packing materials and their application are described elsewhere in this volume,[53] but there are a number of points involved in scaling up the operation that deserve special mention.

Some changes are made in the standard preparation procedures for the packing material because of the longer time that may be required to prepare the resin. Since cellulose ion exchange resins are unstable in high concentrations of acid or base, the strength of hydrocholoric acids and sodium hydroxide solutions are kept below 0.1 N. Gels used for molecular–sieve chromatography are swollen at 60–80°, in order to shorten the time required for hydration to a few hours.

[53]See this volume, [27], [28], [30] and [31].

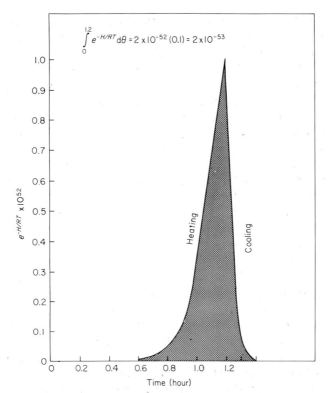

$$\int_0^{1.2} e^{-H/RT} d\theta = 2 \times 10^{-52} (0.1) = 2 \times 10^{-53}$$

FIG. 24.   Integration to obtain fraction of undenatured protein due to heating in 100-liter glass-lined tank.

Resins and gels of the smaller particle sizes are normally avoided in large-scale use because flow rates of the resultant packed columns are too slow. For the same reason, resins are extensively freed of "fines" before use. Column packing materials are stored at 4° as moist solids or in aqueous suspension adjusted to neutral pH. Bacterial growth is prevented by the addition of 0.03% toluene or 1% butanol to ion exchangers and of 0.02% sodium azide to suspensions of dextran gels.

When sample volumes of up to several hundred liters must be fractionated on a resin, the process of batch adsorption can save considerable time. The resin is suspended in the solution and removed after equilibration by filtration. Elution of the proteins may be carried out batchwise or, if finer resolution is desired, by packing a column with the resin and using a gradient elution technique. Batch adsorption should also be employed for solutions containing particulate matter. Adsorption onto a previously packed column is not recommended since the resin bed can act as a very effective filter, trapping the particles and thereby clogging the interstices of the column. The resultant decrease in flow rate can make processing of the preparation extremely time consuming, often leading to loss of enzyme activity.

Particular care must be taken to obtain uniform packing and to avoid the formation of channels in columns. The preferred method for packing is to add all resin as a slurry in one operation.[54,55] This method requires a column extension to accommodate the large suspension volume and can be cumbersome with large columns. As an alternative, a modification of the "aliquot method"[54] may be used. Buffer is added to one-third of the column height, and the slurry of packing material is added. When the resin has settled to a packed bed of approximately 2 inches, the outlet is opened. Additional suspension is then added as the liquid level in the column drops. The suspension must not be allowed to settle completely between each addition.

When the very porous dextran gels (e.g., Sephadex G-200) are used, the total hydrostatic head acting across the column must be limited to 10 cm of water. Sharply decreased flow rates result if this pressure is exceeded, and the column must be repacked for proper operation. The same result occurs if a bed height greater than 100 cm is used.

The problem is compounded with the use of dextran gel ion exchangers, e.g., DEAE-Sephadex A 50. The volume of packing is affected by ionic strength. Employing salt gradients of increasing concentration further compresses the bed. The compression coupled with the large elution volumes required for operation render these products extremely difficult to use in larger columns.[27]

Columns for large-scale chromatography can be fabricated relatively inexpensively from acrylic tubing. An adjustable seal is made by compressing a soft rubber gasket between two plastic disks (Fig. 25), thereby allowing a tight seal without the use of expensive precision-bore tubing. Although jacketed columns may be used to maintain low temperatures, it has been found more convenient to conduct the operation in a cold room, primarily because of the ease of maintaining the collected fractions cold.

Two fraction collectors suitable for collecting large fractions are the Durrum system,[56] and the Oak Ridge System.[57] The Durrum system is

[54]"Laboratory Methods of Column Packing," Whatman Data Sheet, Reeve-Angel Co., Clifton, New Jersey, 07014, 1968.

[55]"Sephadex-gel Filtration in Theory and Practice." Pharmacia Fine Chemicals Inc., Piscataway, New Jersey, 1967.

[56]Durrum Instrument, Palo Alto, California. This device is no longer produced by the company.

[57]C. W. Hancher, Biotechnol. Bioeng. 11, 1033 (1969).

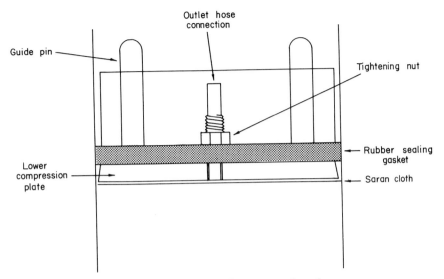

FIG. 25.   Schematic diagram of seal for chromatography column.

simpler in operation, cheaper and smaller. A synchronous motor rotates a distributor plate which is connected to the chromatography column by tubing. The plate rotates above a fixed plate containing adjacent nozzles connected to a fraction bottle by tubing. The time required for the distributor plate to traverse the diameter of a nozzle on the fixed plate determines the fraction size for a given flow rate through the column. A number of synchronous motors are available which are easily slipped in and out of the collector. Elution gradients of several hundred liters, collecting 3-liter fractions, have been carried out by this method.

The Oak Ridge system also works on a time basis but affords closer control between 1 and 150 minutes. When the fill-time interval for a fraction bottle is complete, a drive motor is started which causes a collector arm to rotate to the next position. The fluid continues to flow during the period of arm movement; as with the Durrum system, the edges of the collection nozzles touch one another so that no loss of sample occurs.

The system shown in Fig. 26 is used for large-scale gradients. Two buffer reservoirs are connected by a siphon. A pump connected to the mixing tank circulates buffer to a small elevated reservoir which feeds the column. The overflow from this container returns to the mixing tank.

Many complicated formulas have been devised for scaling up chromatography columns. However, for gel filtration a simple means of obtaining the same elution pattern, when increasing the scale of operations,

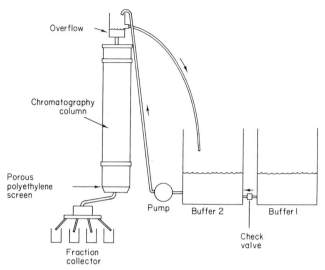

Fig. 26.   Arrangement for large-scale gradient elution.

was found to depend on maintaining dynamic similarity.[58] Dynamic similarity is obtained by arranging both large and small columns to be the same with respect to L/D and $DV\rho/\mu$, where L = length; V = flow rate ÷ cross-section area of column; D = column diameter; ρ = liquid density; μ = viscosity.

However, since ρ/μ is the same in both columns, it is actually DV that must be maintained constant. It is assumed that the packing density in the large and small column is the same. In addition, the ratio of the volume of protein solution to volume of packing is also maintained. Calculation of the dimensions and operating conditions for a large column from data for the small column is shown in the following example:

EXAMPLE

A Sephadex column, 1.5 cm in diameter and 40 cm in length, satisfactorily fractionates 1 ml of protein solution at a flow rate of 100 ml per hour. What are the column dimensions and flow rate necessary to fractionate 20 ml of solution?

Solution:
In small column

Volume of packing $= \dfrac{\pi(1.5)^2}{4}\, 40 = 70.5$ cm³

[58]S. E. Charm, C. C. Matteo, and R. Carlson, *Anal. Biochem.* 30, 1 (1969).

Cross section area $= 1.76$ cm^2

$$\frac{\text{Volume of protein solution}}{\text{Volume of packing}} = \frac{1}{70.5}$$

$$\frac{L}{D} = \frac{40}{1.5} = 26.6$$

$$DV = 1.5 \times \frac{100}{1.76} = 85$$

In large column

$$\frac{\text{Volume of protein solution}}{\text{Volume of packing}} = \frac{1}{70.5} = \frac{20}{\text{Volume of packing}}$$

Volume of packing $= 1400$ cm^3

$$\frac{L}{D} = 26.6 \qquad\qquad \text{or } L = 26.6D$$

Hence

$$\frac{\pi D^2}{4} \times 26.6D = 1400$$

$$20.9D^3 = 1400$$

$$D \cong 4 \text{ cm}, \ L = 106 \text{ cm}$$

The flow rate is then found:

$$DV = 85 = \frac{(D)(\text{flow rate})}{\frac{\pi}{4} D^2} = \frac{4 \times \text{flow rate}}{\frac{\pi(4)^2}{4}}$$

Flow rate $= 266$ cm^3 per hour.

Thus, even though the cross-sectional area is about 7 times greater, the flow rate is increased by a factor of only 2.7 to obtain the same elution pattern. Operating at a higher flow rate presents a risk of inferior separation.

Continuous Chromatography. The operation procedure for chromatography columns is essentially discontinuous. Sample application is followed sequentially by elution and regeneration of the bed. Some large-scale separations are carried out in commercial equipment using this cycling technique.

A truly continuous chromatography system has been described[59] by Fox *et al.* (Fig. 27). Sample solution is applied to a slowly rotating annular

[59] J. B. Fox, Jr., R. C. Calhoun, and W. J. Eglinton, *J. Chromatogr.* 43, 48 (1969).

FIG. 27. Continuous column chromatography process. From J. B. Fox, Jr., R. C. Calhoun, and W. J. Eglington, *J. Chromatogr.* 43, 48 (1969).

column. As the sample moves down the rotating bed, the components follow different helical paths. The various components leave the column at different points depending on their residence time in the column. Both sample application and the collection of fractions are performed continuously. The apparatus has been used for the separation of myoglobin and hemoglobin on Sephadex G-75 and the fractionation of skim milk on Sephadex G-25. Separation and yields are as good as those obtained with the usual column and the apparatus seems well suited for scaling up.[60]

[60]R. A. Nicholas and J. B. Fox, Jr., *J. Chromatogr.* 43, 61 (1969).

Concentration and Purification of Enzymes by Foaming

Foaming is a purification technique for the separation of soluble proteins based on differences in their effect on surface activity. If an inert gas is bubbled through a solution, the most surface-active components collect at the gas–liquid interface. The bubbles float to the top and are collected as a foam enriched in the solute. The method is based upon the fact that the amount of protein at an interface will change until the chemical potential at the surface, μ_s, is equal to the chemical potential in the bulk solution, μ_a. For surface-active materials, this results in a higher interfacial concentration. The analytical relationship between properties of the protein and surface concentration may be expressed through this principle.

At equilibrium

$$\mu_s = \mu_B \tag{15}$$

also

$$\mu_s = (\mu_o)_s + RT \ln C_s \tag{16}$$

$$\mu_B = (\mu_o)_B + RT \ln C_B \tag{17}$$

where C_s is the surface concentration, C_B is the bulk concentration of protein, $(\mu_o)_s$ and $(\mu_o)_B$ = standard state chemical potentials at surface and bulk, respectively, and

$$(\mu_o)_B - (\mu_o)_s = RT \ln \frac{C_s}{C_B} \tag{18}$$

and $(\mu_o)_B - (\mu_o)_s$ = heat of desorption of protein = λ.

Heats of desorption for protein have been found to be in the order of 5500 cal/mole.[61] The heat of desorption may be related to the surface tension through the Gibbs equation (Eq. 19), which is written to express the fact that, at equilibrium between a surface phase and a bulk phase, the free energies of the phases are equal.

$$\Gamma = \frac{-1}{RT} \frac{d\sigma}{d \ln c} \tag{19}$$

where Γ = surface excess; $(d\sigma)/(d \ln c)$ = change of surface tension with log concentration.

$$\frac{1}{tRT} \left(-\frac{d\sigma}{d(C_1)_B} \right) + 1 = e^{\lambda/RT} \tag{20}$$

where t = thickness of surface film

[61]M. Potash, Purification of proteins through foaming. M.S. thesis, Department of Chemical Engineering, Tufts University, Medford, Massachusetts, 1968.

Figure 28 is a typical surface tension–concentration diagram for proteins. Both at very low and relatively high concentrations, the value of the slope $(d\,\sigma)/(d\ln c)$ is zero. The surface film is enriched with a protein only for conditions where $(d\,\sigma)/(d\ln c)$ is not zero, and the components of a mixture can be separated only when the concentration of each is such as to yield differing values of $(d\,\sigma)/(d\ln c)$.

The separation of catalase and amylase is effected by controlling $(d\,\sigma)/(d\ln c)$ with ammonium sulfate.[62] In dilute concentrations, it is possible to effect a complete separation of these two enzymes when the procedure is carried out in a concentration range where $(d\,\sigma)/(d\ln c)$ for amylase is zero and $(d\,\sigma)/(d\ln c)$ for catalase is not zero.

The system for effecting the separation is shown in Fig. 29. The nitrogen bubbles continually form surface for concentrating protein from the bulk solution. The concentration gradient between the bulk phase and surface phase at zero time and at equilibrium is sketched in Fig. 30.

Foam fractionation may be used either for the removal of impurities from solution or for the collection of the desired protein in the foam. The addition of salts, alteration of pH, or the addition of surfactants is used to increase the selectivity of the process.[61] The method requires a low protein concentration for success and is therefore well suited for the dilute protein solutions existing in fermentation broths.[63]

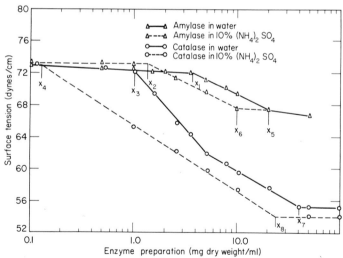

FIG. 28. Surface tension vs. log concentration for catalase and amylase solutions in water and 10% ammonium sulfate.

[62]S. E. Charm, J. Morningstar, C. C. Matteo, and B. Paltiel, *Anal. Biochem* **15**, 498 (1966).
[63]B. Holmstrom, *Biotechnol. Bioeng.* **10**, 551 (1968).

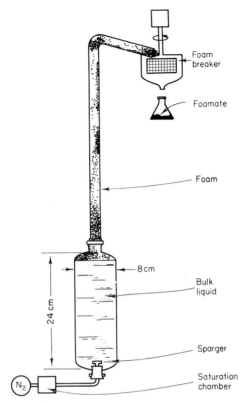

Fig. 29.   Foam separation apparatus.

Concentration of Enzymes through Crystallization of Water

Dilute protein solutions may be concentrated by freezing water and avoiding the occlusion of solute in the ice. When ice freezes sufficiently slowly, solutes diffuse from the ice–liquid interface and accumulate in the liquid phase. The continuous removal of pure solvent, as ice, results in concentration of solute. The simplest large-scale "freeze concentrator" is a jacketed tank. Refrigerant in the jacket freezes water on the tank wall. The agitator gently stirs the solution to aid the movement of solute away from the ice–liquid interface. The freezing rate is controlled by the temperature difference between the cooling medium and bulk solution. Generally, an initial difference of 3° is satisfactory. This is then increased as the thickness of ice on the walls of the vessel is increased. The method has the serious drawback that all solutes are concentrated, causing changes in ionic strength and pH.

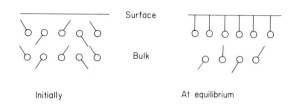

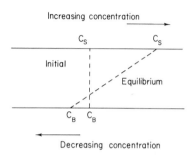

FIG. 30. Concentration of protein molecules in surface and bulk phases when surface is initially formed and at equilibrium.

Preservation of Enzymes

The major preservation methods for storing enzymes are freezing, drying, and refrigeration. The methods are aimed at inhibiting microbial and chemical activity.

Freezing. Water availability is substantially reduced in freezing. This in itself may have the effect of lowering microbial activity. In addition the rate of chemical reactions are slowed. However, at temperatures as low as −40° it is possible to have as much as 5% of water unfrozen. This is the water that is in close association with protein and thereby behaves differently from "free" water in this respect. In a study of frozen fish, it was found that only at −60°C is all the water frozen.[64,65]

Although many enzymes are stabilized in freezing, a number are not. Microbial cells may be particularly sensitive to freezing. Polyalcohols, e.g., propylene glycol or glycerol (15%), have been useful in reducing inactivation due to freezing. The rate of freezing, particularly in the case of cells, may also influence viability. During slow freezing, osmotic pressure gradients may be produced across cell walls by

[64]Sussman, M. V., "A study of the behavior of water in frozen fish tissue using nuclear magnetic resonance and X-ray powder diffraction techniques." Report to U.S. Bureau of Bureau of Commercial Fisheries, Technological Laboratories, Gloucester, Massachusetts, 1965.

[65]S. E. Charm and P. Moody, *ASHRAE J.*, April, 1969.

the crystallization of water with subsequent concentration of salts. In addition, large crystals may form, possibly causing mechanical rupture of the cell. In rapid freezing, the crystals are small and salts are immobilized before they can form the concentration gradients responsible for osmotic pressure changes.

The reasons for inactivation of certain enzymes by freezing are not clear.

Storage at Temperatures below °C without a Change of State. It is possible to store protein solutions without a change of state at temperatures down to −22°C by preventing the expansion of the solution which must accompany the change to ice I. This is easily carried out with a steel chamber filled with liquid and capped so as to exclude all air. The pressure–temperature relationship for this system is shown in Fig. 31. Several enzymes which are inactivated by freezing have been stored in this way without loss of activity. The enzymes tested were

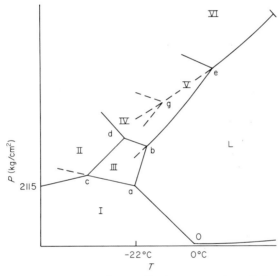

FIG. 31. Pressure–temperature diagram for ice-water system. Below −22°C it is not possible to have a liquid phase. Phase equilibrium pressure–temperature relationships:

Point	T° (C)	P (kg/cm²)	Phases
a	−22.0	2115	Ice I, ice III, L
b	−17.0	3530	Ice III, ice V, L
c	−34.7	2170	Ice I, ice II, ice III
d	−24.3	3530	Ice II, ice III, ice V
e	0.16	6380	Ice V, ice VI, L

catalase and carboxypeptidase A. Stability could also be achieved by mixing the enzymes with 15% glycerol at this low temperature.

Dehydration. Preservation by dehydration depends on the fact that chemical and microbial activity are very low at low moisture content or water activity levels. Microbial activity is generally inhibited at moisture concentrations of less than 10% while most chemical activity is inhibited at a level below 3%.

The time of drying must be sufficiently rapid so that reactions do not occur to any great extent while moisture is being reduced to the safe level. The extraction of water from tissue with a solvent is one method that is frequently employed for lowering moisture. This may be done with cold acetone, thus restricting to a minimum chemical reactions while moisture is being removed. After water is removed, fat may be extracted with warmer acetone. Acetone powders turn brown if moisture is not sufficiently removed due to the reaction of reducing sugar with amino acids.

Another method for removing moisture at low temperature is by freeze drying, and this represents a common method of enzyme preservation.[66] The material to be dried is first frozen and then subjected to high vacuum (less than 5 mm Hg pressure). Ice sublimes as heat is withdrawn from the surroundings. It has been noted that not all the water freezes, i.e., nonfrozen water exists at low temperature due to its close association with protein. This water evaporates at higher temperatures after the removal of ice, but it is essential that this moisture also be removed if the enzyme is to be stabilized. Residual moisture is essentially removed by vacuum drying.

Since, in freeze drying, the material to be dried is first frozen, all the hazards associated with freezing mentioned previously will be present. Therefore, the rate of freezing may influence the final freeze dried product as much as the dehydration step.

Vacuum drying permits a relatively low temperature removal of water at temperatures down to the freezing point. At any given pressure in the vacuum chamber, moisture having a vapor pressure equal to it will evaporate. The temperature at which water evaporates depends on its degree of association with the solid material.

A convenient, nondestructive method for measuring the moisture content of dried protein material is to determine the dielectric constant of the powder. This may be carried out even after packaging the dried product. The principle of the method is that the dielectric constant

---

[66]See this volume [5].

of water is quite high compared to most other materials. Commercial devices for this purpose are available.[67]

Inactivation of Enzymes Exposed to Shearing

Enzymes exposed to shearing for a sufficient length of time undergo inactivation.[38] When the shear rate (sec^{-1}) times the exposure time (sec) is greater than 10^4, inactivation becomes detectable. At 10^7, about 50% inactivation has been observed in solutions of catalase, rennet, and carboxypeptidase.[68] For a solution flowing in a circular tube, the mass average shear rate × time is $16/3$ (L/D) where $L =$ length and $D =$ diameter of the tube. For a rectangular thin channel with a narrow height and a relatively wide width it is 6 L/D, where D is the height of the channel.

In the ultrafiltration of enzyme solutions, recirculation at a high shear rate past the membrane surface is a common practice for maintaining a high flux through the membrane. If sufficient recirculation occurs, the shearing in this ultrafiltration system results in enzyme inactivation. Circulation through a Teflon tube by a finger pump for 24 hours also causes inactivation of catalase and rennet solutions.

A vibrating plate above the membrane surface has been employed in ultrafiltration systems and has been found to result in inactivation because of shearing.[69] In the processing of enzyme solutions, it is important to consider the shear rate × exposure time associated with such processes as agitation, pumping, and flow.

Monitoring of Unit Operations

It is essential that the various unit operations in an isolation process be monitored. The physical parameters most easily measured are time, temperature, turbidity, volume of solids, solid and liquid density, particle or aggregate size, electrical conductivity, refractive index, viscosity, pH, and absorbance. When disrupting cells, it is possible to establish the effectiveness of the operation by sedimenting a sample by centrifugation as previously explained, see Fig. 6. The percent solids settled as a function of time is a convenient method of "fingerprinting" any suspension. It is a function of suspension viscosity, density, particle size, and concentration. Absorbance or turbidity measurements, although convenient, are not as reliable for measuring solids as is the centrifugal sedimentation test.

---

[67]Forte Engineering Co., Norwood, Massachusetts.

[68]Note that (shear rate) × (time) is dimensionless.

[69]S. E. Charm and C. J. Lai, *Biotechnol. and Bioengin.* (in press) 1971.

The particle size distribution in a suspension may be determined with a Coulter counter or by means of sedimentation methods. The latter techniques require a rather prolonged period for a complete determination. It is not useful for particles lighter than their suspending fluid or for these sedimenting very slowly.

Use of the Coulter counter is more satisfactory for the determination of particle size. Supernatant fluid which has been filtered through a $1\ \mu$ or smaller pore membrane is used for diluting the sample. The suspension must be at constant temperature during the procedure in order to prevent changes in particle size either by further precipitation or by dissolving the precipitate.

The concentration of ammonium sulfate or salt in a suspension may be determined by conductivity, by solution density as determined with a hydrometer, by use of an "ion-specific" electrode or by refractometry. In dilute salt solutions, other solutes interfere with these measurements. Electrical conductivity, specific gravity, and refractive index as a function of ammonium sulfate concentration are shown in Figs. 32a–c.

The density of solids may be quickly established by employing a series of copper sulfate solutions differing in concentration of the salt.[70] The density of the solution in which a particle just remains suspended indicates the density of the particle.

Some of the above parameters may be monitored continuously, as necessary, in a fully continuous process.

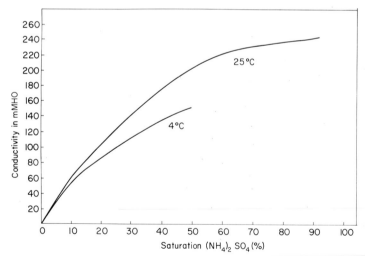

FIG. 32a. Conductivity vs. percent saturation $(NH_4)_2SO_4$ at 4° and 25°.

[70] R. A. Phillips, D. D. Van Slyke, P. B. Hamilton, V. P. Dole, and K. Emerson, Jr., *J. Biol. Chem.* **183**, 305 (1950).

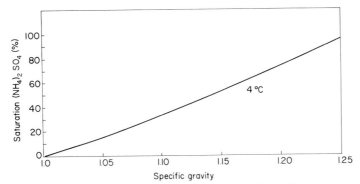

FIG. 32b.   Specific gravity vs. percent saturation $(NH_4)_2SO_4$ at 4°.

Laboratory vs. Scaled-Up Procedure

Unit operations for the isolation of protein are basically the same in the laboratory and in scaled-up operations. However, several problems peculiar to the large-scale preparation should be considered. These include foaming, control of temperature, high shear stresses, and extended processing times.

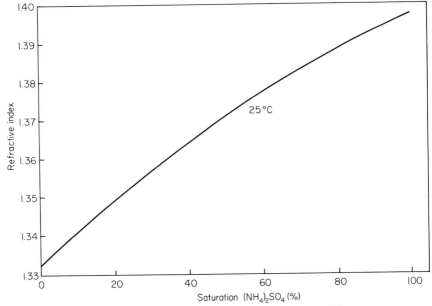

FIG. 32c.   Refractive index vs. percent saturation $(NH_4)_2SO_4$ at 25°.

Aeration of solutions during transfer, mixing, or centrifugation may generate considerable foam. Although in most cases, little enzyme activity is lost as a result of foaming,[62] foam does interfere with unit operations. For example, pH adjustment, and addition of solid and liquid reagents are hampered by the presence of excess foam. Similarly, accurate volume determinations must be calculated from the density and weight of a solution instead of reading directly from a graduated vessel. The addition of an antifoam, e.g., a silicone, usually minimizes the problem of foam, although some enzymes are suspected of being sensitive to silicone.

Low temperatures are maintained in large-scale equipment by circulation of coolant in jacketed tanks and by use of refrigerated centrifuges. Apparatus without provision for refrigeration, such as the DeLaval centrifuge or the Manton-Gaulin homogenizer, is precooled before use. If a cold feed is introduced, the large mass of the equipment usually prevents rapid temperature changes. If processing requires a long time, it may be necessary to interrupt operation periodically and recool the equipment.

The most frequent problem in scaling up is the control of process time. Nearly all isolation processors at present are batch or semicontinuous. Certain steps require a longer period to complete on a large scale than in the laboratory, and enzyme instability, which may not be apparent in the laboratory, becomes evident with the increased time required.

There have been over 300 different preparations scaled up at the New England Enzyme Center. These have all been carried out in a batch or semicontinuous manner. The laboratory and pilot plant yields of enzyme from several scaled-up preparations are shown in Table VII. It can be seen that the overall processing time for the large preparations and the recovery of enzyme are comparable to the respective laboratory procedure. Thus, it may be inferred that time limitations do not present an impasse to scale up. Once a particular step is recognized as time-sensitive, the preparation is either divided into several batches, or a continuous process is employed at that point. The ultimate minimization of process time may be obtained with the development and utilization of techniques for a completely continuous process.

A comparison between laboratory isolation and a scaled-up method of preparation is shown in Table VIII for transaldolase.[27] The major difference between the two occurs at the acetone fractionation steps. At this point, enzyme stability is critically dependent on time and temperature. A three-step procedure is employed in the laboratory involving fractions of 10–25%, 25–37%, and 37–53%. Trial experiments showed the enzyme to be unstable after the first precipitation. After 1 hour at 5°, activity is lost completely. Considerable time is saved in the pilot plant and loss of activity is minimized by taking only two fractions (0–33%, 33–50%). The scale-up fractionation process is limited by the capacity of the DeLaval PX207 since the total solids resulting from a larger volume at the first acetone precipitation would exceed the bowl capacity. A single pass is not possible, and the delay in processing would result in a concomitant loss in activity. The DeLaval used in the pilot plant preparation is maintained at low temperature by periodically cycling a 50% mixture of acetone and water at −10° through it. In this way, it is possible to maintain temperatures between −3° and −7° during the acetone fractionations.

Except for volumes, laboratory and large-scale procedures are the same in most of the remaining steps. The overall recoveries of enzyme from both the pilot plant and laboratory procedure are also comparable.

It is frequently possible to purify several enzymes in the same preparation. For example, in another preparation carried out at the New England Enzyme Center, DNA polymerase, polynucleotide ligase and a phospholipase are all obtained from a single run with 50 kg of *E. coli* B. This requires the services of four technicians for 5 days.

Continuous vs. Batch Enzyme Isolation Procedures

With techniques now available, it is possible to envision a general continuous system for enzyme isolation (Fig. 33a). The system should permit continuous access to solid as well as liquid streams. Such a process eliminates most high speed centrifuges since they must be dismantled to make solids accessible. However, ultrafiltration devices, e.g., a modified continuous rotary filter, lend themselves to continuous ultrafiltration. From the scheme noted in Fig. 33a, it may be seen that continuous processing also may be carried out up to column chromatography.[71]

Dunnill[72] reported an improvement in yield for a continuous process over a batch process in the extraction of prolyl-*t*-RNA synthetase from mung bean. The flow diagram for this process is shown in Fig. 33b.

It is much simpler to initiate a batch process than a continuous one. However, once in operation, the continuous process operates automatically, whereas the batch process requires constant attention, es-

---

[71]Although continuous separations based on molecular size are possible, gradient elution of proteins from an ion-exchange resin has not been carried out.

[72]P. M. Dunnill, P. Dunnill, A. Boddy, M. Houldsworth, and M. D. Lilly, *Biotechnol. Bioeng.* **9**, 343 (1967).

TABLE VII

YIELDS OF ENZYME FROM SEVERAL SCALED-UP PREPARATIONS

Enzyme	Net starting material, lab procedure	Yield from lab process	Time (days)	Total starting material, scale-up procedure	Yield from scaled-up procedure	Time (days)
Alkaline phosphatase	Human placenta, 370 g	75,000 units	8	30.8 kg	3.2×10^6 units	8
Arylpyruvate ketoenol tautomerase	Lamb kidney, 1 kg	400 units	6	38 kg	4×10^4 units	4
β-D-Galactosidase	Escherichia coli, 100 g	5 g	3–4	2.1 kg	100 g	4
β-Hydroxy-decanoate thioester dehydrase	Escherichia coli, 0.5 kg	0.03 g 3×10^4	10–14	11 kg	5.7 g	6
D-Amino acid oxidase	Hog kidney, 3.25 kg	units	5–10	100 kg	2.7×10^6 units	4
Dipeptidyl-transferase	Beef spleen, 5.5 kg	22,700 units	2	180 kg	335,300 units	3

Enzyme	Source	Starting amount			Final amount		
Diphosphopyridine nucleotidase	Bacillus subtilis, 1 kg	300 units	5	8.2 kg	2500 units		3
Diphosphopyridine nucleotidase	Horse brain, acetone powder, 0.24 kg	3300 units 1×10^5	5–10	2.4 kg	3×10^5 units 5.5×10^6		5
DNA polymerase	Escherichia coli B, 450 g	units 5000	3	29.5 kg	units 6×10^5		3
Glutamine synthetase	Sheep brain acetone powder, 300 g	units	1	6 kg	units		3
Lactic acid dehydrogenase	Dog fish muscle, 9 kg	2 g 5×10^4	3	105 kg	21 g 5×10^5		6
Thiophorase	Pig heart, 3.4 kg	units	5	40 kg	units		3
Thyroxine-binding inter-α-globulin	Cohn fraction IV-4, 250 g	250 mg	3	1.5 kg	1.5 g		3
Tripeptide synthetase	Dried yeast, 3.8 liter	40 mg	7	80 liters	500 mg		7
UDP-4-galactose epimerase	Freeze-dried Candida tropicalis, 0.5 kg	3.42×10^6 units	10	3 kg	3.01×10^8 units		5

TABLE VIII

PILOT PLANT VS. LABORATORY PROCEDURE FOR TRANSALDOLASE FROM *Candida utilis*

AUTOLYSIS
Thaw 23 kg of yeast and hold at 25° for 24 hours (LAB: thaw 900 g, rest the same)

EXTRACTION
Add 4.5 liters of 0.5 M NaHCO$_3$ and 42 liters water. Stir for 15 min; 4°C, 30 min (LAB: Buffer—add 60 ml and 533 ml water; rest the same)

CENTRIFUGE
DeLaval (continuous flow) 8000 g, 4°C, 2 hours (LAB: 4500 g, 30 min)

Supernatant   PPT discarded

Assay at this point is 100% activity and purification 1; all other assays relative to this (LAB: same).

ACIDIFICATION
Add 5 N acetic acid to pH 4.8, 4°C, 30 min (LAB: same)

ACETONE FRACTIONATION (0–33%)
Add 26 liters of −70°C acetone, 1° to 7°C 10 min (LAB: 0–25% 300 ml acetone, then 25–37% acetone fraction)

(NH$_4$)$_2$SO$_4$ FRACTIONATION (0–50%)
Adjust vol. to 300 units of transaldolase per ml, add (NH$_4$)$_2$SO$_4$ g/liter 4°C, 30 min (LAB: volume 167 ml; rest the same

CENTRIFUGE
8000 g, 4°C, 30 min (LAB: same)

Supernatant   PPT discarded
(3.6 liters)

(NH$_4$)$_2$SO$_4$ FRACTIONATION (50–75%)
CENTRIFUGE
8000 g 4°C 30 min (LAB: same)

PPT   Supernatant discarded

Dissolve in minimum volume of TEA-EDTA buffer (LAB: Repeat 2 times)

(720 ml)

(LAB: 81 ml)

Recovery 49% purification 12 × (LAB: recovery 57% purification 13 ×).

Add to a solution containing NAPO$_3$ KCl and H$_2$O. Final concentrations:

CRYSTALLIZATION OF TYPE III
Add 50% and 100% (NH$_4$)$_2$SO$_4$ solutions to 50% saturation and 1000 units/ml of transaldolase. Let stand 2 days, 4°C, 48 hours (LAB: same)

CENTRIFUGE
8000 g 4°C, 15 min (LAB: same)

PPT   Supernatant

1st CRYSTAL WASH
Suspend in 10 ml of cold distilled water and let stand for 10 min; 0°C, 15 min. (LAB: 1 ml water; rest the same)

CENTRIFUGE
5000 g, 0°C, 10 min

Supernatant   PPT

2nd CRYSTAL WASH
Suspend in 4.5 ml of distilled water and let stand for 10 min; 0°C, 15 min (LAB: suspend in 0.5 ml)

CENTRIFUGE
5000 g 0°C 10 min (LAB: same)

Supernatant   PPT

3rd CRYSTAL WASH

CENTRIFUGE
DeLaval 8000 g 1° to 7°C 10 min (LAB: −10°C, rest the same)

Supernatant   PPT discarded

ACETONE FRACTIONATION (33–50%)
Add 25 liters −70°C acetone. Let stand 10 min; −3°C to −7°C 10 min (LAB: 37–53%; add 500 ml acetone −10° 15 min)

CENTRIFUGE
DeLaval 8000 g, −7° to −3°C 15 min (LAB: 4500 g)

PPT   Supernatant discarded

Suspend in 2 liters of distilled water and adjust to pH 7.0 with 1 N NaOH 4°C, 30 min (LAB: suspend in 80 ml, 15 min)

CENTRIFUGE
8000 g 4°C, 30 min (LAB: same)

PPT   Supernatant

Suspend in 1 liter of distilled H$_2$O.

0.5 M NaPO$_3$, pH 6.5, 0.1 M KCl and 175 units transaldolase per ml, 4°C, 30 min. (LAB: same)

ACID (NH$_4$)$_2$SO$_4$ FRACTIONATION (0–60%)
Add (NH$_4$)$_2$SO$_4$ g/liter, 4°C, 30 min (LAB: same)

CENTRIFUGE
8000 g 4°C, 30 min

Supernatant   PPT discarded

ACID (NH$_4$)$_2$SO$_4$ FRACTIONATION (60–75%)   Adjust pH to 5.0 with 2 N acetic acid and add (NH$_4$)$_2$SO$_4$ g/liter, 4°C, 30 min (LAB: same)

CENTRIFUGE
8000 g 4°C, 30 min (LAB: same)

Supernatant discarded   PPT

Dissolve in 400 ml of distilled H$_2$O, 4°C, 10 min (LAB: same)

(510 ml)

Recovery 40% purification 21 × (LAB: 54 ml, 47% recovery, purification 21 ×)

Suspend in 4.5 ml of distilled water and let stand for 10 min; 0°C, 15 min (LAB: same)

Supernatant   Precipitate
                     Suspend in 250 ml of 60% (NH$_4$)$_2$SO$_4$ solution and store in cold 4°C, 10 min

Freeze (contains mostly types I and II)

Recovery 19% purification 17 ×   Recovery 11%, purification 172 ×

(1100 ml)   (250 ml)

(LAB: Supernatant discarded, ppt in 30 ml; recovery 13%, purification 191 ×)

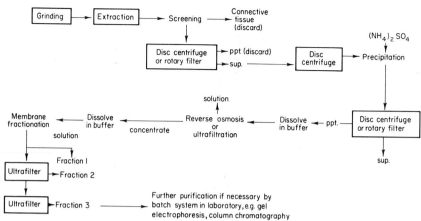

FIG. 33a.   General flow diagram for continuous process.

pecially during time-sensitive stages. If a large number of these are present, a large-scale batch process may be impractical, and the effort to develop a continuous process becomes desirable. Although most scaled-up enzyme preparations are batch processes, many steps, including those of cell breakage and centrifugation, are semicontinuous or continuous. Large volumes of material can be handled only when means for continuous centrifuging are available.

It is relatively simple to carry out an isolated process in a continuous manner. However, the development of a truly continuous process requires the coordination of every step so that a steady state exists, i.e., there is no depletion or accumulation at any point. The development and maintenance of the steady state is the major difficulty of the continuous process.

Major differences exist with respect to equipment requirements in batch and continuous processes. The same piece of equipment, e.g., a centrifuge, may be used at several stages of a batch process whereas separate pieces of apparatus are required for continuous processing. The continuous process also requires automatic controls.

Production of Microorganisms as a Source of Enzymes

Microorganisms are often used as the source material for enzyme purification. The intracellular concentration of enzymes may be controlled, and in many cases greatly increased, by preliminary laboratory work to isolate mutants and to determine optimum growth conditions. Growth of the organism is then carried out in the pilot plant. Aspects

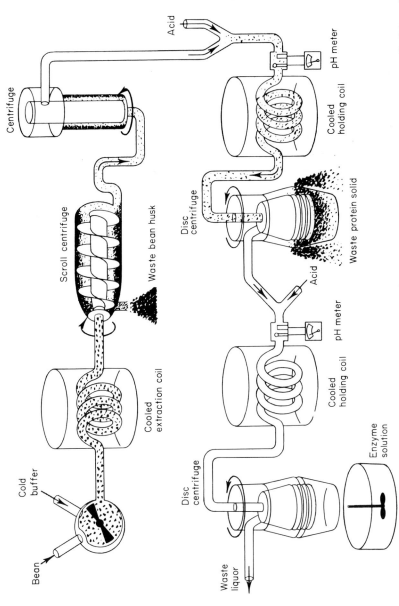

FIG. 33b.   Continuous isolation of enzyme prolyl-*t*RNA synthetase from mung bean. From M. D. Lilly and P. Dunnill, *Science J.* **5**, 59 (1969).

of the large-scale propagation of microorganisms are considered below, as well as elsewhere in this volume.[73]

Sterilization

Sterilization time is a function of the volume of the medium. To obtain the same level or probability of bacterial survival, 500 liters of medium will have to be held at an elevated temperature for a longer time than will 1 liter. Because of this, the fermentor medium may differ in composition from the smaller volume in a flask and may adversely affect growth if a limiting nutrient has been destroyed by more severe heating or if toxic thermal degradation products are produced. Therefore, in testing a medium for scale up, the small-scale system must be "oversterilized" to duplicate the time–temperature effects in the larger system.

A method similar to that used in calculating equivalent "thermal denaturation" processes may be applied to determining the effects of sterilization processes. The time-temperature relationships for heating various amounts of medium in an autoclave are shown in Figs. 34a and 34b; that for 550 liters of medium in a fermentor is shown in

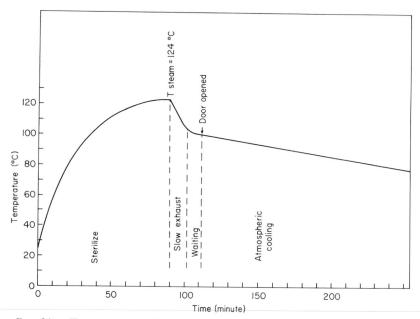

FIG. 34a.   Temperature vs. time for sterilization in an autoclave.

[73]See this volume [9].

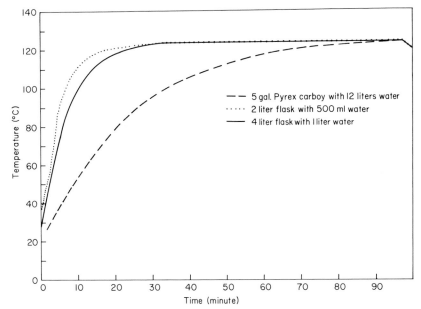

FIG. 34b. Time–temperature relationship for 10 liters in 20-liter carboy for sterilization and cooling.

Fig. 35. It can be seen that 12 liters of solution, when autoclaved in a 5-gallon glass carboy, requires 80 minutes to just reach 120°. Since a normal laboratory technique may allow only 45–60 minutes in an autoclave, it is probable that the sterilization temperature may never be attained in many cases. In contrast, the total time for heating, holding at temperature, and cooling for a 550-liter fermentation may be as long as 3 hours.

Components of the growth medium which are sensitive to heat, either by themselves or by their interaction with other components at elevated temperatures, are sterilized separately in an autoclave. Included in this group are glucose, amino acids, vitamins, and some trace elements. Membrane filtration may be employed for sterilization of heat-sensitive materials. The last method eliminates heating but introduces a higher risk of contamination because of the greater amount of handling required; it is not effective in eliminating viruses and phage.

The problems associated with the longer time required for sterilization in large-scale systems may be eliminated by the use of high temperature-short time sterilization. The thermal death time for a microorganism may be 1 minute at 240°F and 0.1 second at 290°F, sterilization may be accomplished in fractions of seconds rather than minutes at lower temperatures.

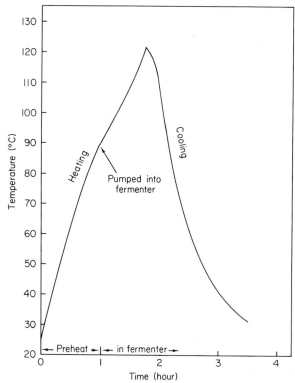

FIG. 35. Temperature vs. time for sterilization of 500 liters. Solution is preheated in a "make up" tank before introduction into the fermentor.

There are a number of continuous, high temperature–short time sterilizing systems. One employs a plate heat exchanger to heat the medium by high pressure steam. The heated fluid is then cooled in another section of the heat exchanger using the entering feed medium as the cooling agent. Other means of cooling may be used, e.g., propylene glycol or direct expansion of refrigerant into the plate heat exchanger.

Another such system employs the direct injection of steam into the medium and subsequent "flashing off" under vacuum of excess condensed water. The plate heat exchanger, although not as effective a heating system as the steam injection method, appears to be easier to control.

Inoculation

Medium components sterilized separately are transferred to the fermentor through a sterile port. They are followed by a seed culture.

The size of this inoculum usually ranges from 2% to 10% of the total volume of medium. There are instances, however, where the size of inoculum for a 500-liter culture may be as small as 200 ml or as large as 80 liters. A small inoculum may be used for an overnight culture to enable harvesting while the cells are still in log phase. Knowledge of the generation time permits projection to the harvest time. However, it should be noted that the growth rate may differ greatly from a shake flask to an aerated fermentor. This method of projecting the time required will not give accurate predictions if there is an indeterminate lag phase after inoculation of the large fermentor.

An inoculum that is in the logarithmic phase of growth, normally does not produce a lag or delay period in the fermentor. While the use of a culture in stationary phase may result in a short delay of an hour or so, there is a high probability that the delay will be considerably longer. In addition to occupying the fermentor for a needlessly long period, other complications may arise. For example, *Bacillus stearothermophilus* may be grown at 57° with vigorous aeration. A 500-liter fermentation is completed in only 4 hours if 10 liters of a rapidly growing inoculum are used. Using an inoculum in the stationary phase extends the time to 16 hours. Growth with aeration for long periods at high temperature results in significant evaporation of water from the medium (40% of the original volume over a 16-hour period). To maintain aeration at elevated temperatures for extended periods, it would be necessary to saturate the gas with water vapor at the growth temperature. Otherwise, the fermentation medium will be concentrated because of moisture removed by the dry air. Any filters used on the air lines would have to be maintained at a higher temperature to prevent condensation from reducing the efficiency of the filter.

Monitoring Growth

Progress of the fermentation is followed by measurement of the increase in cell mass. A 10-ml sample is centrifuged in the DeLaval Gyrotest centrifuge, and the percentage of solids is noted (Fig. 6). Although absorbance or Klett readings may be used, the percentage of solids method provides greater accuracy at a high cell density. When the medium itself has a high optical absorption, the centrifugation procedure may be the only accurate way to quickly measure growth.

Change in pH of the medium may be used as an indication of the growth of the culture. Observation of the rate of consumption of base (or acid) required to maintain the pH is used for fermentations of *Clostridium pasteurianum* and *Pseudomonas saccharophila*. The end of log phase is indicated by a marked decrease in the need for adjusting solution. Recording pH changes in the medium is also used to determine the state of a culture. The derepression of *Streptococcus faecalis* (ATCC 8043) on a glucose-arginine medium depends first upon the utilization of glucose. This is accompanied by a drop in pH. When the sugar has been depleted, utilization of arginine as a carbon source results in a rise in pH.

Other monitoring methods must be sought when the fermentation extends into stationary phase. Many compounds are produced in appreciable amounts only after cell division has stopped. Their presence may be detected by determination of metabolic products in the medium or in the cells themselves. Thus, the presence of intracellular enzymes involved in the biosynthesis of actinomycins from *Streptomyces antibioticus* (ATCC 14888) is indicated by the appearance of actinomycin in the medium. A sample of the culture is centrifuged and the supernatant liquid is extracted with an equal volume of ethyl acetate. A yellow color indicates the presence of actinomycins, and the amount is quantified by a simple colorimetric determination. Alternately, a small sample of the cells may be extracted to determine the amount of the desired product in the cells; this does not always require breakage of the cell. For example, growth of *Micrococcus denitrificans* is continued well into stationary phase for optimum production of cytochrome *c*. The culture is monitored by centrifuging and washing the cells in dilute buffer. The absorbance of the wash buffer before and after the addition of sodium dithionite provides a measure of the cytochrome *c* released.

The purity of the culture is routinely checked by staining of the culture and the inoculum. The gram stain is a rapid and sensitive technique for use in the course of a fermentation. Gross contamination is readily detected although the method is not effective in revealing low levels of contamination. Repetition of the test over a period of time is more significant if a contaminating strain is likely to overgrow the parent culture. Plating a culture and noting the characteristics of the colonies is desirable, but the time required for development of colonies makes it unsuitable for monitoring.

It is also useful to monitor the concentration of dissolved oxygen during the course of the fermentation. Variation in oxygen concentration is responsible for many of the discrepancies noted between flask cultures and fermentations. In addition to its effect on the overall growth rate, changes in dissolved oxygen affect the biosynthesis of many enzymes.[74] Certain enzymes, e.g. glycerol dehydrogenase, are actively destroyed by cells engaged in aerobic metabolism.[75]

[74] J. W. T. Wimpenny, *Process Biochem.* 6, 19 (1969).
[75] E. C. Lin, A. P. Levin, and B. Magasanik, *J. Biol. Chem.* 235, 1824 (1960).

Harvesting

Bacteria are usually collected by centrifugation at the end of a fermentation run. The DeLaval PX207 can process 500 liters in less than 2 hours. The spent medium remaining in the bowl with the pelleted cells is displaced with a suitable buffer, and the contents of the bowl are then discharged. Additional wash buffer is added, and the suspension is recentrifuged. After the required washes are completed, the centrifuge is stopped and disassembled. Most cells are collected as a paste from the walls of the bowl, but the remaining fluid in the bowl often consists of a suspension of cells which must be recentrifuged. This is usually done with a large-batch rotor or in a Sharples centrifuge.

Filtration may be used as a method of collection. *Neurospora crassa* has been filtered on cheesecloth, and *Streptomyces antibioticus* ATCC 14888 on a Büchner funnel. In the latter case, the mycelium accumulates at the bottom of the fermentation vessel and the fermentor has to be disassembled to obtain it for filtration.

Derepression

Application of the phenomena of derepression for large-scale procedures is primarily concerned with methods of decreasing the effector concentration in the medium. For example, synthesis of ribonucleotide reductase by *Lactobacillus lichmanii* (ATCC 7830) is stimulated when a culture growing in the presence of the repressor deoxyadenosine (AdR),[76] is diluted 5-fold with medium free of AdR. A culture of 80 liters was grown and transferred to 400 liters of medium in a second fermentor. This technique requires two growth vessels and is limited by the volume of the larger vessel.

If a bacterial mutant which has a nonfunctioning enzyme in the pathway is selected, the intracellular formation of end product is prevented. This aids the derepression, but may impose a requirement for an exogenous supply of the compound. Such is the case for the derepression of an *E. coli* mutant used in the preparation of aspartate transcarbamylase[77]; the strain requires uracil for growth. The concentration of uracil in the medium is chosen to allow rapid growth until there are approximately 3×10^8 bacteria per milliliter. At this point, depletion of uracil results in derepression of the operon for pyrimide synthesis. However, the growth rate is vastly decreased because of a lack of uracil, and the generation time increases to 8 hours.

The concentrations of effector used for a large-scale growth should be chosen to enable the culture to reach a high cell density before the

[76]*M. Goulian and W. S. Beck, *J. Biol. Chem.* 241, 4233 (1966).
[77]*J. C. Gerhart, and H. Holoubek, *J. Biol. Chem.* 242, 2886 (1967).

onset of derepression. If the cell population is low at that time, a decreased growth rate may result in the uneconomical prolonging of the fermentation in order to obtain a reasonable yield. The effector must not be supplied in such an excess that another nutrient becomes growth limiting before the onset of derepression. In such a case, the activity of the desired enzyme is greatly decreased or even nonexistent. This may occur in a large-scale fermentation because of the thermal degradation of vitamins and other trace nutrients during sterilization.

Laboratory techniques involving the centrifugation of a culture and resuspension in fresh medium are not well suited to large-scale growth. Because of the large volumes, cell metabolism is arrested for too long a period before resuspension. There is also a high probability of contamination during manipulation.

Processing of Bacteriophage

The lytic cycle is introduced by the addition of virulent phage to the culture or by the induction of lysogenic bacteria. The techniques for infection with a virulent phage are similar to those used on a laboratory scale.

Two techniques are suitable for induction in large volumes. The first is irradiation with ultraviolet light. A rapidly growing culture is subjected to an empirically determined radiation dosage from an immersed lamp that is too low to kill a normal cell. This dosage depends upon the medium and culture conditions, i.e., temperature and cell density. Any change in volume, degree of mixing, or vessel geometry requires further experimentation to determine the correct dosage. As the volume is increased, multiple radiation sources become necessary because of the limited penetration of radiation through the suspension, and equipment cost therefore rises rapidly. Alternatively, a flow-through radiation device may be used. Culture broth is pumped from the growth vessel through an annular irradiation chamber surrounding the lamp. Radiation dosage is controlled by varying the intensity of the lamp, the flow rate, or the thickness of the annulus. Multiple chambers may be operated in parallel if the maximum flow rate for a single unit is too low to process the desired volume. The apparatus has the advantage of supplying a uniform, reproducible dosage to the cell suspension. Disadvantages include the necessity for two culture vessels and heating of the culture broth as the result of low flow rates. It is well suited for the continuous system described below.

The second, and perhaps the most useful, induction method for large-scale use is "heat shifting." Lysogens of phage mutants producing a temperature-sensitive repressor can be induced simply by raising the temperature of the medium. For example, bacteria lysogenic for lambda phage with a point mutation in the lambda Cl region produce a thermolabile repressor.[78] The bacteria are grown at a temperature between 30° and 34°. At a cell density of 10^9 cells/ml, the temperature is raised to 42°, inactivating the repressor and causing induction of the prophage. After 15 minutes, the temperature is lowered to 37°. Incubation is continued with the phage in the vegetative state. Free phage are normally released into the medium if incubation is continued sufficiently long. The exact time will vary with the specific virus–host system and with such culture conditions as temperature, medium composition, and aeration rate. Bursting of the cell and release of phage may occur at any time between 20 minutes and several hours after infection or induction.

If phage-induced substances are to be isolated, the bacteria are not allowed to lyse. Lysis is usually prevented by rapid cooling of the culture, thereby arresting cell metabolism. This may be accomplished by adding the cell suspension to vessels containing ice, by pumping the medium through a heat exchanger, or by subjecting the medium to a high vacuum and a lower temperature by removal of latent heat. Although the last two methods retain sterility, the first is the easiest with large volumes. Crushed ice provides instantaneous cooling, limited only by the rate at which medium can be pumped from the fermentor. Disadvantages arise from the dilution of the medium, e.g., decreasing ionic strength and increasing volume.

Use of a heat exchanger requires either a large supply of coolant or a high-capacity compressor. The rate of cooling will be limited by the compressor capacity or surface area of the exchanger. Cooling by vacuum is limited by the rate at which the vacuum system removes the water vapor. A steam jet attached to the fermentor is the most efficient means of accomplishing this. This system has the possible disadvantages of degassing the culture medium and causing excessive foaming.

The total heat that must be removed is calculated from the equation:

$$MC_p(T_2 - T_1) = \text{cooling load}$$

where M = mass of culture; C_p = specific heat of culture; T_1 = initial temperature; T_2 = final temperature.

[78]F. Jacob, and J. Monod, J. Mol. Biol. 3, 318 (1961).

EXAMPLE

Medium, 100 liters, is to be cooled from 37° to 2°C. Determine the cooling load required. Determine the amount of ice that must be added to accomplish this.

Solution:

Assume that the medium has the same properties as water; $C_p = 1.0$ cal/g °C; $\rho = 1$ g/cm^3; total mass $= 100 \times 10^3 = 10^5$ g; $T_2 - T_1 = -35°$; $MC_p(T_2 - T_1) = -(10^5)(1)(35) = -35 \times 10^5$ cal, therefore, 35×10^5 cal must be removed.

The heat of fusion of ice is 80 cal/g, therefore

$$\frac{35 \times 10^5}{8 \times 10^4} = 43.75 \text{ kg} = 96.2 \text{ lb of crushed ice are required per 100 liters}$$

This is conveniently rounded off to the value of 1 lb of ice per liter of culture.

An estimate also may be made of the amount of water that must be evaporated. The heat of vaporization of water is approximately 540 cal/g. Thus, for every liter of water removed as vapor, 5.4×10^5 calories are also removed.

$$\frac{35 \quad \times 10^5}{5.4 \times 10^5} = 6.5 \text{ liters/100 liters}$$

The requirement for rapid cooling may be eliminated by the use of mutant phage that do not cause cell lysis. In this case, incubation can often be extended, resulting in higher yields of the desired product. For example, yields of a phage-directed exonuclease, present in quantities 30 times that of the wild-type level, are produced by λ T_{11}, a defective lysogenic phage[79]; infective phage are not produced by the cell.

Lysis inhibition, by addition of high concentrations of magnesium ions, has also led to increased phage yields in E. coli cultures.[80] A concentration of 0.2 M MgSO$_4$ can prevent lysis of E. coli infected with ΦX174, M-12, and QB. In the case of ΦX174 infected E. coli C, cells which normally lyse 35 minutes after infection remain intact after 3–4 hours. Phage yields are 5- to 10-fold higher than controls lacking magnesium. Similar effects occur with M-12 and QB infected cells. Normally infected cells start to lyse at 50–70 minutes. The addition of magnesium prevents lysis and results in a 2- to 3-fold higher phage yield.

The intact cells containing whole phage or phage-induced substances,

[79]C. M. Radding, J. Mol. Biol. 18, 235 (1966).
[80]H. H. Gschwender and P. N. Hofschneider, Biochim. Biophys. Acta 190, 454 (1969).

are recovered by low speed centrifugation. The cells are resuspended in a volume from one-tenth to one-hundredth of the original culture volume. Rupture of the cells may be accomplished in several ways.

1. Freezing of the intact cells followed by suspension in a blendor. This technique is not used to recover infective phage because they are destroyed by freezing.

2. Lysis effected by the addition of chloroform (25 ml/liter), lysozyme (0.15 g/liter), and EDTA (ml of 0.5 M EDTA per liter).

3. Resuspending the cells in EDTA. Cells protected from lysis by the addition of magnesium are ruptured when added to 50 mM EDTA and subjected to a blendor.

4. Rupturing the cells by the mechanical methods described previously. Phage-infected cells are easier to break than uninfected cells and these more severe methods are not usually necessary.

A suspension of ruptured cells is extremely viscous, and subsequent processing is impaired unless nucleic acids are degraded. This may be achieved mechanically by the use of a blendor. However, severe shearing inactivates phage with resultant loss of infectivity. Nucleic acids may be digested by incubation with added deoxyribonuclease. Ribonuclease may also be effective, but is usually omitted when isolating RNA phages.

Cell debris is removed by centrifugation at 10,000–50,000 g, and the pellet may be washed and recentrifuged for complete recovery of the supernatant liquid. Phage are isolated by the methods described below.

The major problem is that of collection of virus from a large volume of lysate. There are three ways by which this has been accomplished on a large scale: high speed centrifugation, ultrafiltration, and precipitation of the phage followed by centrifugation at moderate g forces. Each of these methods has presented some difficulties.

Centrifugation of the crude lysate at 90,000 g is an effective means of removing phage, but a typical flow rate for removal of 95% of lambda phage through an ultracentrifuge is 2 liters per hour. Although the rate may be increased somewhat for larger phage, such as the T-even series, it is, nevertheless, low, and only small volumes can be processed. Even with the larger K series rotors (Electronucleonics), processing volume is not much greater than 100 liters per day.

Ultrafiltration may be useful, but inactivation has been observed with lambda and R17. In addition, the remaining concentrated suspension must be further purified, usually by centrifugation to remove proteins, ribosomes, and other material retained by the membrane. Precipitation with ammonium sulfate has been used for separating RNA phage; salt is added to the crude lysate (320 g/liter), and the suspension is centrifuged first in the DeLaval and then in the Sharples between 3 and 10 hours after addition of the salt. This procedure requires as long as 10 hours for 500 liters, considerably faster than the other methods but with loss of phage activity when ammonium sulfate is present in the precipitate. The precipitated phage must therefore be processed immediately.

Polyethylene glycol (PEG) may be used to concentrate a wide variety of phage from lysates.[48] Phage may be collected by brief low speed centrifugation or by allowing the suspension to settle overnight if sodium chloride is added to 0.5 M and the PEG concentration is adjusted to between 2 and 10%. Nearly quantitative recovery of infectivity is obtained over a range of phage titers from 10^5 to 10^8 plaque-forming units/ml. The mechanism of separation appears to be by phase partition rather than by precipitation. Evidence for this comes from the observation that the fraction of infectivity removed from solution by a fixed concentration of PEG is not a function of phage concentration over a 10^8-fold range.

Difficulties with this method arise from the simultaneous sedimentation of ribosomes and nucleic acids after the addition of PEG, and the sedimentation of a large fraction of DNA by PEG concentrations greater than 5%. Minimal medium is generally used to prepare the lysates. Naturally occurring materials, e.g., yeast extract, form a precipitate with PEG under the conditions used to sediment phage and are an additional source of contamination.

Continuous Production of Phage-Infected Cells

Large quantities of infected cells for use as a source of phage-induced substances can be produced in a continuous system similar to that outlined in Fig. 36. As an illustration of the high productivity, *Escherichia coli*, infected with T_2 phage, has been produced at the rate of 100 g per hour using a growth vessel containing 10 liters of medium and an infectection vessel of 4 liters.[81] Fresh medium is supplied at the rate of 10.4 liters per hour as a proportioning pump meters a fixed cell phage ratio into the second vessel. The infected cells have a mean residence time of 22 minutes before cooling and harvesting.

The system can also be used for the induction of a lysogen, either by ultraviolet radiation within the flow chamber described above, or by temperature induction of the culture. In the latter case, the culture may be heated to 42° and incubated in a third vessel for 15 minutes. The temperature in the "infection vessel" is then lowered to 37°.

[81] E. C. Short, Jr., and J. F. Koerner, *J. Biol. Chem.* 244, 1487 (1969).

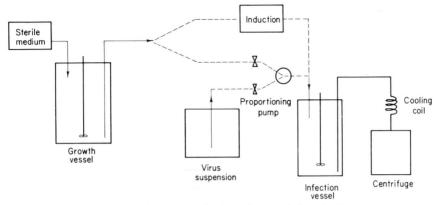

Fig. 36.   Continuous system for production of phage-infected cells.

Appendix: Examples of Scaled-Up Preparations

The scaled-up procedure for the isolation of dipeptidyltransferase (cathepsin C) from beef spleen is given.[82] This is followed by a discussion of some of the factors considered in the development of the pilot plant process. Yields obtained from the laboratory and pilot plant procedures are compared in Table IX.

Procedure

1. Partially frozen beef spleens were passed through a meat grinder twice, initially using a plate with 0.5-inch openings and then a plate with 0.25-inch openings. The total weight of the ground spleens was 180 kg.

TABLE IX
ISOLATION OF DIPEPTIDYLTRANSFERASE: LABORATORY AND PILOT PLANT

	Laboratory		Pilot plant	
	Volume (liters)	Activity (units)	Volume (liters)	Activity (units)
Acid extract	12.18	3.4×10^4	372	11×10^5
0.40% AS supernatant	—	—	380	8.6×10^5
40–70% AS fractions	0.194	2.27×10^4	5.16	6.85×10^5
Overall recovery	67%		61.5%	

[82]Reference for laboratory procedure: R. M. Metrione, A. G. Neves, and J. S. Fruton, *Biochemistry* **5**, 1597 (1966).

2. Two volumes of distilled water containing 0.6 g EDTA per liter were added to the spleens, and the mixture was brought to room temperature.

3. The suspension was adjusted to pH 3.5 by the slow addition of 5.5 liters of 6 N sulfuric acid. After 1 hour of stirring, two additional liters of 6 N sulfuric acid were added to readjust the suspension to pH 2.5.

4. The temperature was brought to 37° and the mixture was stirred for 3 hours. Connective tissue was removed during this period by pumping the suspension through fiber glass window screening.

5. The extract was centrifuged in the DeLavel PX207. The supernatant fluid was displaced before discharging the bowl. Five passes lowered the content of solids from 12% to 1%.

6. The supernatant fluid (372 liters) was cooled to 4° and ammonium sulfate was added to 40% saturation. Acid-washed filter acid[83] (Hi-Flo Super-Cel, Johns-Manville), 1.7 kg, was added and the mixture was left to settle overnight in the cold room.

7. The clear upper layer was decanted (330 liters). The remaining mixture was filtered on Büchner funnels using Schleicher & Schuell No. 576 paper.

8. The filtrate and the decanted supernatant from step 7 were brought to 70% saturation with ammonium sulfate and left for 1 hour in the cold.

9. The 70% precipitate was removed by centrifugation in the De-Laval. The precipitate was scraped from the bowl, suspended in a minimum volume of water, and dialyzed for 24 hours against three changes of 0.9% NaCl.

10. The supernatant from step 9 was saved and allowed to settle overnight. Precipitate which settled out was collected by centrifugation in the Sharples and dialyzed as above.

11. Insoluble material, present after dialysis, was removed by centrifugation and the supernatant liquid from step 9 (3350 ml) and step 10 (810 ml) were combined.

Analysis of Large Scale Preparation

1a. Manual trimming was carried out in the laboratory to remove the outer membrane from fresh spleen. The organs were then cut into strips and frozen. These steps were eliminated in the pilot plant because dissection was too time-consuming. The spleens were only partially

---

[83]Acid-washed filter aid: Dry material was weighed and suspended in HCl. The suspension was filtered, and the solids were washed with distilled water until the suspension was above pH 5.5.

thawed before grinding because the second pass through the grinder was much more difficult with completely thawed material.

3a. Sulfuric acid was added through a piece of tubing fed by an aspirator bottle. Over 0.5 hour was required to adjust the pH; electrodes were immersed in the tank and had to be cleaned intermittently of accumulated tissue during pH adjustment.

4a. The laboratory procedure called for 22 hours at 38°. Severe degradation of particulate matter occurred in the pilot plant under these conditions. The resulting suspension could not be easily centrifuged because of the small particle size. Monitoring of the autolysis showed that maximum enzyme activity was released after 3 hours.

The suspension was pumped through a layer of fiber glass window screening stretched over a second tank. This removed the membranes that were not trimmed in step 1. It was more economical to discard the screening material after use than to clean it. A basket centrifuge could also have been used at the step to remove the coarser material.

5a. This was the point at which prolonged autolysis led to difficulty in centrifugation. Even with the shorter digestion time, the solid material was not easily removed. Five passes were used for two reasons: The efficiency of centrifugation increased as the volume fraction of solids decreased. A faster flow rate resulted in a smaller temperature rise in the product stream. When foaming was a problem, the DeLaval was operated under 25–30 psi of back pressure, thereby minimizing the entrapment of air in the exit stream.

Displacement of the bowl before discharging solids allowed recovery of most of the extract. Nevertheless, some of the extract was lost because of mixing or because it had a higher density than the displacing solution. Current models of the DeLaval PX207 allow partial discharge of the bowl contents and would allow almost complete recovery of extract.

6a. Settling at this step allowed an easy reduction of volume with little extra labor. The precipitate did not settle out overnight in the absence of filter aid.

8a. To save time, the decanted layer was brought to 70% saturation while the remainder was still filtering. The wait before separation of the precipitate was only 1 hour rather than overnight as in the laboratory preparation.

9a. Filtration, as used in the laboratory was too slow to be used at the step. A dry precipitate was recovered from the DeLaval by stopping the bowl and scraping out the pelleted solids. Less volume was thereby carried over to the dialysis.

10a. The layer containing the precipitate was drawn off and centrifuged at a higher g force. The presence of a precipitate possibly resulted from centrifugation before precipitation had reached equilibrium. Solid material still settled out at this step after overnight equilibration before step 9. This result could be due to the g force of the DeLaval which was relatively low for sedimentation of small particles.

Lactic Dehydrogenase from Dogfish Muscle[84]

1. Starting material (105 kg) was divided into three parts, each of which was prepared separately. Frozen dogfish muscle (35 kg) was ground in a commercial meat grinder and extracted in 95 liters of cold 1 mM EDTA containing 1 mM mercaptoethanol. The suspension, 111 liters, was stirred at 4° for 1 hour and centrifuged in a DeLaval PX207 centrifuge.

2. Solid ammonium sulfate was added with stirring to 45% of saturation. The suspension was left for 2 hours and filtered overnight through Canton flannel bags in a cold room. This step removed the precipitate and remaining fat. It was necessary to add a small quantity of filter aid to each bag so as to obtain a clear filtrate of 112 liters.

3. Solid ammonium sulfate was added to the filtrate to 70% of saturation and the suspension was filtered overnight through fluted filter paper on ten large plastic funnels set up on stands around two low plastic vessels.

4. The pink precipitate was scraped from the filter paper and dissolved in cold water.

5. Steps 1 through 4 were repeated using another equal amount of frozen muscle, and the two portions were combined. The total volume was 12 liters.

6. Half the suspension was dialyzed at 4° overnight against 5 mM Tris·HCl at pH 7.6.

7. The precipitate was centrifuged, and the supernatant liquid was charged onto a 20-liter column of DEAE-cellulose which had been equilibrated with 50 mM Tris·HCl at pH 7.6. Because of the large amount of protein present, two DEAE-cellulose columns were used. Fractions were eluted with the same buffer.

8. Fractions containing enzyme activity were pooled and solid ammonium sulfate added to 70% of saturation. The mixture was stored at 4° for several hours, after which the precipitate was centrifuged in the zonal centrifuge and dissolved in a small amount of cold, distilled water.

[84]Adapted from F. Stolzenbach, this series, Vol. 9, p. 278.

Section IX

Criteria for Homogeneity

[38] Gel Electrofocusing

by C. W. WRIGLEY

Gel electrofocusing provides rapid analytical fractionation of amphoteric compounds, such as proteins, on the basis of their isoelectric points. The high resolving power of density gradient electrofocusing (this volume [33]) is combined with the convenience and flexibility of gel electrophoresis (this volume [39]). The technique was originally described in a number of articles that were published independently and almost simultaneously in 1968.[1-9]

Gel electrofocusing is an extension of the technique of electrofocusing in a density gradient[10] and is based on the same principle of fractionation. Replacement of the density gradient by a gel as separation support makes the procedure more suitable for analytical fractionation of many small samples. A stable pH gradient is produced by applying a voltage to a mixture of carrier ampholytes in a column or slab of gel. The carrier ampholytes (polyaminopolycarboxylic acids having a broad spectrum of isoelectric points) become arranged in the gel in order of increasing isoelectric point from anode to cathode. They are confined to the gel by using dilute solutions of a strong acid and a strong base in anode and cathode compartments, respectively. Molecules of a protein migrate in the pH gradient until they become concentrated (focused) in the pH region, where the protein is isoelectric. The zone remains sharpened in this position as long as the pH gradient is stable and the voltage is maintained.

Samples

Proteins are well suited to fractionation by electrofocusing because of their generally low diffusion coefficients and steep pH-mobility curves near the isoelectric point. Smaller amphoteric molecules can

[1] Z. L. Awdeh, A. R. Williamson, and B. A. Askonas, *Nature (London)* **219**, 66 (1968).
[2] N. Catsimpoolas, *Anal. Biochem.* **26**, 480 (1968).
[3] N. Catsimpoolas, *Sci. Tools* **16**, 1 (1969).
[4] G. Dale and A. L. Latner, *Lancet* No. 7547, 847 (1968).
[5] J. S. Fawcett, *Fed. Eur. Biochem. Soc. Lett.* **1**, 81 (1968).
[6] D. H. Leaback and A. C. Rutter, *Biochem. Biophys. Res. Commun.* **32**, 447 (1968).
[7] R. F. Riley and M. K. Coleman, *J. Lab. Clin. Med.* **72**, 714 (1968).
[8] C. W. Wrigley, *J. Chromatogr.* **36**, 362 (1968).
[9] C. W. Wrigley, *Sci. Tools* **15**, 17 (1968).
[10] O. Vesterberg and H. Svensson, *Acta Chem. Scand.* **20**, 820 (1966).

also be electrofocused though results might not be as good as those obtainable with proteins.

Since proteins can be focused to their isoelectric pH values from any part of the pH gradient, it is not necessary to apply the sample as a sharp zone. Incorporation of the sample throughout the gel permits analysis of very dilute protein solutions (about 0.005%); however, possible modification of protein during polymerization is a disadvantage of this method. Alternatively, concentrated sample solution can be layered on top of the gel in the manner usually adopted for gel electrophoresis. A high concentration of salt in the sample should be avoided since electrofocusing is delayed during the electrophoretic removal of salt from the gel. Disaggregating substances such as urea may be added to the sample and the gel to maintain the solubility of otherwise insoluble samples, but possible variations in the effects of such compounds with pH should be borne in mind.

Nature of the Supporting Medium

In electrofocusing, the gel serves merely as an anticonvectant, not as a molecular sieve as it does in gel electrophoresis. For this reason, it is important that gel pore size should not be so small as to restrict unduly the focusing of proteins. The 7.5% polyacrylamide gel recommended in the above procedure provides good mechanical strength and allows adequate movement of globular proteins with a molecular weight of up to 100,000. For larger proteins, a lower concentration of acrylamide should be used, with the addition of agarose (0.5%), if necessary, to enhance gel strength.[11]

The rate of drift of the pH gradient due to endosmosis is low with polyacrylamide gel (1–2 cm in 16 hours[6]). This is therefore the medium nearly always adopted for gel electrofocusing. Agarose or starch gels are of limited value for gel electrofocusing as the pH gradient drifts considerably during prolonged running, especially at alkaline pH values. Electrofocusing on paper media[12] offers some advantages, but endosmosis is a difficulty in this case also.

Choice of pH Range

When dealing with proteins of unknown isoelectric point, it is usual to analyze first of all in the pH range 3 to 10. On the basis of this experiment, a narrower pH range can be chosen for further analysis if desired. The narrowest ranges of commercially available ampholytes span two pH units. For fractionation of proteins with very similar

[11]A. C. Peacock and C. W. Dingman, *Biochemistry* 7, 668 (1968).

isoelectric points, carrier ampholytes for ranges covering only 0.5 pH unit can be prepared by using the density gradient electrofocusing column.[12]

Materials

Apparatus.

Gel electrofocusing can be performed in conventional apparatus for gel electrophoresis using cylindrical[2-5, 7-9] or slab[1,6] gels. The procedure below is based on the use of 65 × 5 mm columns in disc electrophoresis apparatus.[13] This apparatus is available commercially or can be made up in the laboratory. However closer comparison of samples is obtained by analyzing them side by side on a rectangular gel. Comparison of samples on cylindrical gels can be improved by incorporating a dividing wall across the upper 10 mm of the tube so that two samples can be loaded side by side on top of the gel and compared by simultaneous electrofocusing in the respective halves of the same gel column. This technique is useful for comparing a purified preparation with the original crude extract.

Whereas the definition of zones separated by gel electrophoresis becomes reduced by diffusional spreading as fractionation proceeds, zones resolved by electrofocusing remain sharpened during prolonged fractionation. For this reason, the length of the gel used for electrofocusing is not as restricted as in the case of gel electrophoresis. Gels double the length of those above, or longer, can be used successfully for better zone separation.

Reagents

1. Carrier ampholyte solution (40% w/w), obtainable in 25-ml bottles in the pH ranges 3–10, 3–6, 5–8, 7–10, 3–5, 4–6, 5–7, 6–8, 7–9, and 8–10 from LKB-Produkter AB, S-161 25 Bromma 1, Sweden or from LKB Instruments, Inc., 12221 Parklawn Drive, Rockville, Maryland. Dispense the viscous solution into water from a pipette calibrated "to contain", and rinse the pipette with the diluted solution.
2. Acrylamide solution, containing 30 g acrylamide and 1 g N,N'-methylene bisacrylamide dissolved in water to 100 ml
3. Potassium persulfate, fresh 1% aqueous solution
3a. Riboflavin, 0.015% aqueous solution (store in a dark bottle)

[12]C. W. Wrigley, *in* "New Techniques in Amino Acid, Peptide and Protein Analysis" (A. Niederwieser and G. Pataki, eds.). Ann Arbor Science Publ. Inc., Ann Arbor, Michigan, in press.
[13]B. J. Davis, *Ann. N. Y. Acad. Sci.* 121, 404 (1964).

4. Anode solution, 0.2% sulfuric (or phosphoric) acid
5. Cathode solution, 0.4% ethanolamine (or ethylenediamine)

Procedure

The quantities given are for eight tubes (5 × 65 mm) in conventional disc electrophoresis apparatus.

Step 1. Gel Preparation. For chemical polymerization, mix reagents in the following proportions. The gel sets about 30 minutes after addition of the persulfate solution (at 20–25°): distilled water, 8.0 ml: carrier ampholyte solution, 0.3 ml; acrylamide solution, 3.0 ml; potassium persulfate, 0.7 ml. (Protein sample, if required[14].)

As an alternative to chemical polymerization, gels may be photopolymerized by replacing persulfate (reagent 3) in the above mixture with riboflavin (reagent 3a). In this case polymerization takes place about 20 minutes after exposure close to a fluorescent light, but generally it is not initiated by the normal intensity of laboratory lighting.

Remove dissolved air from the gel mixture by evacuation before filling running tubes to within 5–10 mm of the top. Overlayer carefully with water from a fine-tipped pipette and allow to polymerize.

Step 2. Fractionation. The polarity of the electrodes is optional and is often determined by such considerations as the pH stability of the sample. Pour the appropriate electrode solutions into the electrode vessels after mounting the gel tubes in the apparatus.[15] Cooling may be necessary during electrofocusing of labile enzymes. Raise the voltage gradually, maintaining a maximum current of 2 mA per tube, up to a maximum of about 400 V. Electrical resistance rises during formation of the pH gradient and becomes approximately constant when the gradient is established. This takes about 15 minutes with salt-free samples. Focusing of proteins can take from 30 to over 120 minutes, depending on the method of loading, gel pore size, the pH range, and the characteristics of the proteins. The time for electrofocusing of a particular sample is best determined experimentally by removing tubes at various times. It is important to ensure that electrofocusing is continued to completion. There is generally no harm in electrofocusing for an hour or so longer than the minimum time.

[14] If the sample is to be distributed throughout the gel, it should be included at this stage in the gel mixture. If necessary, reduce the volume of water to make allowance for the sample.

[15] Instead of setting the sample in the gel, the sample solution (in 10% sucrose) can be layered on top of the gel and under a protecting layer of 1% carrier ampholyte solution (in 5% sucrose). The sample can be applied after assembling tubes in the disc electrophoresis apparatus, either before switching on the current or after running for about 30 minutes to establish the pH gradient.

Step 3. Determination of the pH Gradient. After electrolysis, the course of the pH gradient in the set of gels can be determined. Cut a gel, identical to those to be stained, into 10–20 equal sections and measure the pH of a 1 ml of water extract of each piece.

Step 4. Detection of Proteins. Gels cannot be stained directly with the usual protein stains, since the stains are strongly bound to carrier ampholytes. Proteins can be detected either by precipitation or after the removal of carrier ampholytes.

a. PROTEIN FIXATION. Zones of most proteins appear quickly as white precipitation bands when electrofocused gels, removed from the running tubes, are immersed in 5% trichloroacetic acid. Precipitated zones should be photographed against a black background with side lighting.

b. REMOVAL OF CARRIER AMPHOLYTES. If staining is required, gently agitate gels in 5% trichloroacetic acid (10 ml per gel) and change the trichloroacetic acid at least 5 times at 1 to 2 hourly intervals. Removal of ampholytes can also be obtained by electrophoresis of the gels in 2.5% (10 mA per tube for 20 hours), using the procedures developed for electrophoretic destaining following disc electrophoresis.[13] Higher currents are possible with transverse electrophoresis.[16]

c. PROTEIN STAINING. After removal of carrier ampholytes, stain gels by immersing for 1 hour in amido black (1% in 7% acetic acid). Destain by electrophoresis or by washing in 7% acetic acid.

Other Methods of Detection. Virtually all the procedures devised for detecting specific compounds and enzymes after gel electrophoresis, can be applied to gel electrofocusing provided possible interference of carrier ampholytes is avoided.

Staining and fixing can be avoided if gels are run in quartz tubes and then scanned directly by ultraviolet light (280 nm) in a densitometer.[17] At this wavelength, the absorbance of carrier ampholytes is low, but variable along the gel. A blank scan should therefore be performed on a gel without sample.

Enzyme staining procedures (this volume [40]), have been used successfully with gel electrofocusing,[4,6] but there is the possibility of interference from carrier ampholytes because of their tendency to complex metal ions and because of variation in pH along the gel.

Immunological procedures have been described for detecting components after electrofocusing[3,7] (usually in agarose gel). Carrier ampholytes do not interfere.

[16]C. Schwabe, *Anal. Biochem.* 17, 201 (1966).

[17]A. R. Dravid, H. Freden, and S. Larsson, *J. Chromatogr.* 41, 53 (1969).

Validity of Fractionation

In contrast to electrophoresis, electrofocusing is likely to give an estimate of isoelectric point that is approximately equivalent to the isoionic point, since the electrofocusing environment is salt free and the ionic strength of the carrier ampholytes is low. So far there is no evidence to indicate that carrier ampholytes complex with proteins.

Heterogeneity due to characteristics other than isoelectric point can be observed with gel electrofocusing if care is not taken to ensure that zones focus to equilibrium. Gel electrofocusing of serum albumin and its oligomers[9] is an example of this pitfall. Artifactual heterogeneity can also arise from the use of persulfate for gel polymerization as observed, for example, in the gel electrofocusing of hemoglobin.[12] Preliminary electrolysis is not necessarily a satisfactory way of removing excess persulfate. In preference, photopolymerization is recommended whenever possible.

Combined Gel Electrofocusing and Electrophoresis

Both gel electrophoresis and gel electrofocusing are valuable tests of protein homogeneity. Neither replaces the other, since the criteria of fractionation are different in each case. The two procedures can be combined effectively for two-dimensional analysis.[9,12,18] Generally a narrower gel (3 mm) is used for electrofocusing in the first dimension. This gel is inserted in the sample position of a rectangular slab of acrylamide, starch, or agarose gel for electrophoresis in the second dimension. For many proteins it is possible to choose a pH for electrophoresis such that carrier ampholytes are separated from proteins during electrophoresis, thus making direct protein staining possible without prior removal of ampholytes.

Even if it is possible to demonstrate homogeneity for a preparation by as critical a procedure as combined electrophoresis and electrofocusing, it must be realized that the words "purity" and "homogeneity" are concepts that have "no meaning except with reference to the methods and assumptions used in studying the substances being discussed."[19] Major qualitative and minor quantitative differences in the composition of uncharged amino acids could still remain undetected.

[18]G. Dale and A. L. Latner, *Clin. Chim. Acta.* 24, 61 (1969).
[19]N. W. Pirie, *Biol. Rev. Biol. Proc. Cambridge Phil. Soc.* 15, 377 (1940).

[39] Analytical Disc Gel Electrophoresis[1]

By Othmar Gabriel

Disc electrophoresis (discontinuous electrophoresis) has attracted the attention of many investigators because of its well-defined, sharp separation boundaries, which can be achieved with only microgram quantities of protein. The theoretical basis for this method, which combines the principles of electrophoresis and gel filtration, is the Kohlrausch regulation principle. During electrophoresis a specific combination of ions forms a moving front of the "leading" ion followed by the "trailing" ion, thereby bracketing the sample protein between them. This permits concentration of the protein sample into an extremely sharp layer at the origin and yields the sharp resolution of individual boundaries. A further advantage of the method is offered by the use of polyacrylamide gel as anticonvection medium; the qualitative and quantitative selection of the gel components permits choice of a wide range of properties for the suporting medium. Thus, it is possible to anticipate the specific conditions for the best possible separation by proper selection of the gel components. More detailed information about the principles of the method have been described by Ornstein.[1a]

It is the purpose of this presentation to provide analytical systems specifically applicable to the analysis of enzymes. Purification of enzymes by conventional methods can and should be followed at each of the later stages of the purification procedure by analysis using disc gel electrophoresis. Similarly, conditions for preparative scale electrophoresis,[2] a method recently used successfully as a purification step for enzymes, can be defined only by preliminary experiments with the analytical system. Disc electrophoresis is also a powerful tool in establishing homogeneity of a protein although the results, as with all other methods used to establish purity of proteins, must be interpreted with caution. The present article is designed to familiarize the investigator with the practical aspects of the method as carried out with the simplest possible means and with specific emphasis on problems related to the separation of enzymes. The material is presented in two major complementary sections, the first of which deals with the electrophoretic procedure and

[1]The experimental results from the author's laboratory and preparation of this chapter were supported by a Grant of the Public Health Service AI-07241 from the National Institutes of Health.
[1a]L. Ornstein, *Ann. N. Y. Acad. Sci.* 121, 321 (1964).
[2]L. Shuster, this volume [34].

the second[3] with methods of staining required to localize enzymes in gels.

For readers wishing more information on all aspects of disc gel electrophoresis, several comprehensive treatments of the subject have appeared recently.[4-6] The use of disc gel electrophoresis for quantitative evaluation, the use of modified gels containing urea or detergents, or the use of methods for assessment of molecular weights of proteins are not considered in this presentation.

Electrophoretic Procedure

General Description

According to the original procedure of Davis,[7] disc gel electrophoresis is carried out in small glass tubes, 5 mm (i.d.) × 65 mm, in which three different gel preparations are layered contiguously.

The separating gel is prepared at the proper concentration of acrylamide and of the proper percentage of cross-linkage to provide a three-dimensional network optimal for effective "molecular sieving" of the sample components.

The low cross-linked stacking gel, differing in pH and composition from the separating gel, effectively concentrates or "stacks" the protein sample into a sharp starting zone.

The sample gel is frequently identical to the stacking gel and, as the name indicates, contains the protein sample. However, the use of a sample gel is not recommended for enzymes, as will be explained below.

The glass columns with the acrylamide gel in place are inserted into the electrophoretic chamber. The sample is applied last and the electrophoretic run is carried out. Immediately thereafter, procedures[3] permitting location of the separated components are applied.

Specific Requirements for Enzyme Protein Separations

The unique problems related to the separation of enzyme proteins by polyacrylamide disc electrophoresis result from the need to demonstrate a functional enzyme after electrophoresis. Therefore, conditions contributing to inactivation of the enzyme must be avoided.

Any electrophoretic procedure will evolve heat which is proportional to the product of the current and the potential gradient. In order to

[3] O. Gabriel, this volume [40].

[4] G. Zweig and J. R. Whitaker, "Paper Chromatography and Electrophoresis." Vol. I, p. 151. Academic Press, New York, 1967.

[5] H. Rainer Maurer, "Disk-Electrophorese." de Gruyter, Berlin, 1968.

[6] G. Reich, G. Hebestreit, and J. Winkler, Z. Chem. 6, 401 (1966).

[7] B. J. Davis, Ann. N. Y. Acad. Sci. 121, 404 (1964).

avoid denaturation of the enzyme, heat generated by the current must be dissipated by convection. For practical purposes, this can be accomplished by conducting the electrophoretic run in a cold room or a refrigerator at temperatures of 0–4°.[8] It should be mentioned here that the geometry of the glass columns, i.e., the surface to volume relationship, should not exceed the dimensions usually used (0.5 mm i.d.) if excessive heating inside the tube is to be avoided. A problem intimately associated with the heat generated during electrophoresis is the ionic strength of the buffer system. High ionic strength requires the use of a low voltage gradient (V/cm) which, in turn, results in an impractically long time for separation and consequent poor resolution due to diffusion. This consideration also imposes limitations upon the addition of coenzymes or substrates with ionic character; such compounds may be necessary for stabilization of the enzyme during electrophoresis. As a first approximation, the concentration of ionic low molecular components should not exceed 10^{-2} M.

One problem frequently encountered in the maintenance of a functional enzyme is the prevention of oxidation of protein sulfhydryl groups. Consideration has to be given to removal or counteraction of harmful components which remain after polymerization of the separating gel; ammonium persulfate is a particular nuisance.[9–12] Preelectrophoresis of the separating gel has been used successfully. Thus, a buffer system, identical in ionic strength with the one used for polymerization, is employed in both buffer compartments to maintain an identical ionic environment. After completion of preelectrophoresis of the separating gel, the stacking gel is polymerized in place. For the separation of enzymes, a sample gel is not recommended because of possible inactivation of the enzyme during the polymerization process. Instead, the sample is mixed with sucrose or glycerol (5–25%) and layered on top of the stacking gel. It is noteworthy that the best electrophoretic separations are achieved at pH values at which most proteins are either negatively charged, about pH 9, or positively charged, about pH 4. Surprisingly, most enzymes tested under these extreme conditions retain at least some of their catalytic activity. Rather poor electrophoretic separations are achieved at neutrality. From these observations it would appear that good separations are accomplished primarily because of differences in

---

[8] An effective way of facilitating convection is to fill the lower buffer compartment sufficiently high to immerse the individual gel columns in the buffer solution.

[9] W. M. Mitchell, *Biochim. Biophys. Acta* 147, 171 (1967).

[10] J. M. Brewer, *Science* 156, 256 (1967).

[11] H. Fantes and I. G. S. Furminger, *Nature (London)* 215, 750 (1967).

[12] A. Bennick, *Anal. Biochem.* 26, 453 (1968).

size and shape of the molecules rather than to differences in net charge.

Usually, tracking dyes are mixed into the sample or are added to the upper buffer compartment. Bromophenol blue (anion) may be used for systems at a basic pH, whereas methylene green (cation) is effective for systems at an acidic pH. Electrophoresis is terminated as the disc of the tracking dye is seen to approach the lower end of the separating gel. It is suggested that the power supply be disconnected completely before disassembling the electrode compartment. Individual gels are then removed from the apparatus and immediately subjected to the staining process.

Necessary Apparatus and Equipment

The apparatus for disc electrophoresis was described earlier in this series[13] and can be obtained commercially.[14,15] It usually consists of plastic containers which serve as electrode compartments and accommodates several glass tubes (0.5 cm, i.d.; 6–7 cm long) which contain the gel; several samples are subjected to electrophoresis simultaneously. A power supply which can deliver 50 mA at 500 V is satisfactory. It is desirable to have an instrument capable of delivering constant current.

A variety of additional equipment for disc electrophoresis is commercially available but not essential. There is, for example, equipment for destaining, cutting, and scanning of gels.

Preparation of Glass Columns. A glass tube of 5 mm i.d. is cut into smaller pieces (about 65 mm) of equal length. The sharp edges are removed mechanically with a fine Carborundum cloth. Fire polishing is not recommended for this purpose since the inside diameter must be constant for the entire length of the tube. The glass columns are cleaned with hot nitric acid and rinsed extensively with distilled water. After drying in a dust-free atmosphere, the tubes are immersed in a 0.5% solution of Photo Flow 600 (Eastman Kodak Company) and dried again. These columns are ready for accepting the polyacrylamide gel.

Gel Polymerization

The coated gel columns are closed at one end with a rubber stopper[16] and are placed on the laboratory bench in a vertical position. All the solutions necessary for polymerization are warmed to room temperature

[13]A. M. Altschul and W. J. Evans, *in* "Methods in Enzymology" (C. H. W. Hirs, ed.), Vol. XI, p. 179. Academic Press, New York, 1967.
[14]Canal Industrial Corporation, Rockville, Maryland.
[15]Buchler Instruments Inc., Fort Lee, New Jersey.
[16]B-D Vacutainer stoppers, Becton, Dickinson and Co., Columbus, Nebraska and Rutherford, New Jersey.

and are evacuated with an aspirator or vacuum pump; it is important to reduce the solutions to a low partial pressure of air since oxygen acts as chain terminator in the polymerization process. Reproducible results in the polymerization process can be obtained only when identical conditions for the polymerization process are maintained.

As indicated, the original method described by Davis[7] used three individual gel layers: sample gel, stacking gel, and separating gel, each polymerized on top of the other. In order to avoid exposure of the enzyme during polymerization of all three gels, the opposite sequence has been followed so that the separating gel is prepared first. The polymerization of the individual gel layers requires great care and manual dexterity. The objective is to produce a completely flat interphase between the layers at both ends of the polymerized gels.

Separating Gel. In order to achieve a perfectly even surface at the lower end of the gel, 100 μl of a 40% sucrose solution are placed into the bottom of the glass tube which is in vertical position inserted in a rubber stopper. On top of this sucrose layer is placed the solution with the formulation for the separating gel. It is useful to mark the glass columns on the outside or to deliver identical volumes (1.2 ml solution, about 30 mm length) so as to have reproducible and identical lengths of separating gel. Immediately after this step, a 3–4 mm layer of water is placed on top of the gel solution. Mixing of water with the gel solution is avoided by using a water-filled hypodermic syringe (No. 23 needle) and delivering the water by touching the wall of the glass with the needle about 1 or 2 mm above the surface of the gel solution. Small droplets will form on tubes which are not cleaned or coated properly; these droplets drip into the gel solution and result in mixing. The technique of layering water on top of the gel solution, when properly carried out, results in a flat and even surface after polymerization and thereby avoids distortion of boundaries during electrophoresis.

If the reagents for the separating gel are freshly prepared, polymerization will not exceed 30 minutes and the progress of polymerization can be observed directly. After overlayering of the gel solution with water, the interphase is visible but will disappear; it again becomes visible after completion of polymerization. When polymerization is complete, each tube is turned 180° and drained of excess water, which is carefully removed with absorbent tissue.

Preelectrophoresis of Separating Gel. Residual peroxide, remaining in the gel after polymerization, has been shown to inactivate enzymes and lead to formation of other artifacts. If it is not necessary to carry out preelectrophoresis, one may proceed directly to the next step with stacking gel. Electrophoretic removal has been recommended and has produced good results when carried out immediately after polymerization. For preelectrophoresis, the rubber stopper closing one end of the tube is removed. Care must be taken to admit air when removing the stopper in order to avoid dislodging the gel from the surface of the glass. For this purpose one can deform the rubber stopper by squeezing it so that air can enter. Alternatively, a hypodermic needle may be used to permit entry of air. The lower end of gel is rinsed well to remove the sucrose solution and the glass columns are inserted into the electrophoresis apparatus. The buffer solution used for the upper and lower buffer compartments has a composition identical to that used in the polymerization reaction. Care is taken that air bubbles are not trapped in the glass columns. This preliminary electrophoresis is carried out for 2 hours at about 3 mA per tube. Thereafter, the tubes may be stored for several days at 4° in a beaker containing a small amount of the same buffer and covered with parafilm.

Stacking Gel. The column with the separating gel in place is dried with tissue to remove excess liquid from the area into which the stacking gel will be polymerized. The upper part of the column is rinsed twice with about 100 μl of a solution containing the components of the stacking gel. The column is again placed into a vertical position. (If the same type of rubber stopper is used as described above, it is advantageous to use a stopper with an opening in the center to prevent dislocation of the separating gel.) The appropriate amount of stacking gel solution, 0.2 ml (about 5 mm), is placed on top of the separating gel. Immediately thereafter, a layer of 3–4 mm of water is carefully placed on top of the polymerizing solution observing the precautions mentioned above. Without moving the tubes, a fluorescent light is placed 1–2 inches above the upper ends of the columns: photopolymerization occurs within about half an hour. The gel becomes slightly turbid, indicating the progress of photopolymerization. After polymerization, excess water is poured off and the upper part of the column is rinsed carefully with the buffer which will be used in the upper electrode compartment. The columns thus prepared are best used between one-half to one hour after completion of the process.

Application of the Sample. In the original procedure of Davis,[7] the sample was mixed with a solution identical in composition to the stacking gel. However, the polymerization process carried out in the presence of enzyme may denature the protein. Agents which may be desirable in the sample solution, e.g., mercaptans required to maintain the enzyme activity, will increase the time necessary to achieve polymerization. Because of the difficulty in retaining enzyme activity, it has become standard practice to apply the enzyme sample in glycerol (10%) or sucrose directly, without gelling the sample.[17] Sample application is best carried out after the individual gel columns are positioned in the electrophoresis apparatus. Gas bubbles trapped in the column or in the lower buffer compartment must be removed; a U-shaped Pasteur pipette is useful for this purpose. After sample application, buffer is carefully layered on top of each column filling it completely with buffer solution. Finally, the upper buffer compartment is filled with the same buffer. To 500 ml of buffer solution, about 2 ml of 0.001% bromophenol blue (anion) or methyl green (cation) are added.

The amount of sample applied to each column will depend on the number of proteins present. A practical range for samples with only a few components is 10–20 μg, whereas up to 200 μg of a crude sample may be applied. The sample should be low in inorganic salts and preferably dialyzed against the same buffer used in the upper compartment.

Electrophoresis is carried out in a refrigerator or in a cold room at 0–4°. Current at about 1–2 mA per tube or less is practical. It is useful to maintain constant current and to observe the change of voltage during the course of electrophoresis. A drastic change in the potential gradient is an indication of overheating. It should be noted that current and voltage depend directly on the ionic strength of the buffer employed. In some cases only 1 mA current will be permissible to avoid heat denaturation. The duration of the electrophoresis should not exceed 2 hours and the procedure is considered as completed when the disc of the tracking dye is seen to reach the lower end of the separating gel. Major protein bands also can be seen directly due to differences in the refractive index of the system at the point at which protein is concentrated.

After the run, the power supply is disconnected and the column samples are removed from the apparatus. Rimming of the gel with a syringe filled with water and a hypodermic needle (about 25 gauge) will permit extrusion of the gel from the column.[7] Even the somewhat hard gels, those containing 15% acrylamide gel or more, can be removed by this procedure. The harder gels can be removed more easily by using 10% glycerol solution in water instead of water itself. Great care has to be taken when handling the gels to get them out of the column without breaking. Several staining procedures used for the subsequent location of enzymatic activity may lead to spurious staining when gels are handled directly with fingers. In this instance, gloves or forceps should be used. It is also suggested that the gels be rinsed with ice cold buffer solution after extrusion from the glass tube in order to remove reagents used during electrophoresis. In some staining procedures the position of the

[17]S. Hjertén, S. Jerstedt, and A. Tiselius, *Anal. Biochem.* 11, 211, 219 (1965).

tracking dye which forms a convenient reference boundary, disappears. It was found helpful to insert into the gel, at the site of the tracking dye, a copper wire about 5 mm in length, so as to maintain this reference point for comparison with other gels.

Gel Systems

Component Chemicals

Acrylamide (CH_2=$CH \cdot CONH_2$) is polymerized with the bifunctional molecule N,N'-methylene bisacrylamide (CH_2=$CH \cdot CONH)_2 \cdot CH_2$ to result in a clear and transparent gel. The percentage of acrylamide and the degree of cross-linkage can be controlled by changing amounts and proportions of the two monomers used. The pore size of the gel best suited to fit the molecular dimensions of the components to be separated will be selected to achieve maximal separation.

The polymerization process is a free radical addition process which has to be initiated by appropriate catalysts. The process of polymerization should be carried out in the absence of oxygen, which acts as chain terminator. Ammonium persulfate is used as a catalyst for the preparation of the separating gel; photopolymerization with riboflavin as the catalyst is usually employed to prepare the stacking gel. In both systems, $N,N,N'N'$-tetramethylethylenediamine can be used to accelerate polymerization.

The purity of the individual monomers is of great importance for the reproducibility of the polymerization process. Commercial sources provide acrylamide and N,N'-methylene bisacrylamide in sufficient purity to permit its use without further action.[14] Some of the reagents can be obtained premixed and ready for polymerization (Canalco). Recently, highly purified components became available which permit the formation of gels transparent in the ultraviolet range between 260 and 280 nm[18] and allow direct scanning for proteins and for nucleic acids. The components must be stored in dark bottles under refrigeration. Once the monomers are dissolved, the shelf life under refrigeration is about 3 months. It is recommended that the pH of the solution be checked periodically since aging will result in an increase in pH. The requirement of additional time for polymerization also serves as an indication that monomers are deteriorating. It should be mentioned that acrylamide is a toxic substance and that contact with solutions of the compound can result in skin irritations.

---

[18]Bio-Rad Laboratories, Richmond, California.

Criteria for Proper Choice of Gel Systems

The various systems available today for disc gel electrophoresis are so numerous that a few general remarks are in order. Several properties of the enzyme to be separated should be known in order to arrive at a reasonable selection: (a) the molecular weight, (b) the isoelectric point of the protein, (c) the stability of the enzyme at pH 4–9.

In most instances an approximate knowledge of the above three parameters will permit the selection of a gel system.

It is recommended that the versatility of the method be exploited by including experiments under several different conditions. Usually, the composition of the stacking gel is not varied since it serves primarily as an anticonvection medium. As a first approximation, the volume of the sample should be in the proper proportion to the volume of the stacking gel and should not exceed that volume. The properties of the enzyme in terms of its net charge at the pH of the separating gel will help in deciding whether a system be chosen to operate in an alkaline or acidic milieu. The "standard" gels which have been used most successfully are the ones operating at a pH of about 9 (System I of the table). Sample proteins which have a negative net charge at this pH will migrate toward the anode. Usually, the sample columns, arranged in a vertical position, are attached to the power supply with the cathode connected to the upper buffer compartment and the anode to the lower buffer compartment. In this "anionic" electrophoretic system, bromophenol blue solution is added into the cathodic electrode compartment as a tracking dye. Conversely, for basic proteins, it is necessary to work at a pH of about 4 to obtain positively charged proteins that will migrate to the cathode (System III of the table). In the latter system, the cathode is connected to the lower buffer compartment and methyl green may be mixed directly with the sample as the tracking dye.

These rather extreme pH values are very effective in terms of separation but may be detrimental to enzymatic activity. A compromise solution is a gel system operating at neutrality or at a slightly alkaline pH (System II of the table).

On the basis of these simple guidelines, a selection for the pH of the gel system will be possible. For testing the homogeneity of an enzyme, the study of its mobility at different pH values is indicated.

One of the more effective ways of improving resolution into individual components is variation of the composition of the separating gel. Migration of proteins due to their net charge in the electrical potential applied along the gel will depend primarily on their retention by molecular sieving. Thus, proteins of high molecular weight will be separated best by using gels of larger pore size, whereas smaller proteins will separate better with gels of smaller pore size. A gel containing 7–7.5% acrylamide has the most desirable mechanical properties and is commonly used for proteins ranging in molecular weights from 10^4 to $10.^6$ For smaller proteins, below $10,^4$ gels containing as much as 15–30% acrylamide may be tried. Conversely, for species with very high molecular weights, gels with large pores, prepared with about 4% acrylamide, are used. The most suitable conditions for the separation are best defined by performing the procedure at different concentrations of acrylamide. The mechanical properties of the gel are of importance in their handling: gels of large pore size are very brittle whereas the gels of low pore size may be difficult to remove from the glass column. It is recommended that the amount of bisacrylamide be decreased as the concentration of acrylamide is increased in order to improve the mechanical properties of the gels. This step will also limit excessive heat formation during the polymerization process for the low pore-size gels.

A summary of the above principles is presented in the table. The numbers in the table refer to the formulations given below. Successful use of disc gel electrophoresis requires taking advantage of the flexibility and versatility of the method in that empirical variations of pH and pore size are required before a study can be considered to be complete.

Formulation of Gel Systems for Disc Electrophoresis

Only gel systems which have been used successfully for separation of enzymes are included. The formulations presented here have been reported previously by a number of investigators.[7,19–23]

Stock solutions may be prepared in advance but must be refrigerated and protected from light. The solutions are brought to room temperature, mixed in the proportions indicated and evacuated with an aspirator or a vacuum pump to reduce the oxygen content. The actual working solutions are prepared immediately prior to polymerization of gel, and only in amounts necessary for the day.

In the following listing, the formulations are presented for separating gels, stacking gels, and buffer solutions. The amounts indicated are for stock solutions of 100 ml unless indicated otherwise. The following abbreviations are used: TEMED for N,N,N',N'-tetramethylethylenediamine; Bis for N,N'-methylenebisacrylamide; and Tris for tris(hydroxymethyl)aminomethane.

[19] B. J. Davis: Preprint: "Disc Electrophoresis." Distillation Products, Division of Eastman Kodak Co., Rochester, New York, 1962.
[20] R. A. Reisfeld, V. J. Lewis, and D. E. Williams, *Nature (London)* 195, 281 (1962).
[21] D. E. Williams and R. A. Reisfeld, *Ann. N. Y. Acad. Sci.* 121, 373 (1964).
[22] Information supplied by Canal Industrial Corp., Rockville, Maryland.
[23] D. E. Williams and R. A. Reisfeld, *Ann. N. Y. Acad. Sci.* 121, 373 (1964).

SELECTION OF GEL SYSTEMS FOR DISC ELECTROPHORESIS
(Numbers refer to formulation of gel systems page 576)

| Range molecular weights approx | Percent acrylamide | Anionic Enzyme Sample (upper buffer compartment cathode; lower buffer compartment anode; tracking dye: methylene blue) | | | | | | Cationic Enzyme Sample (upper buffer compartment anode; lower buffer compartment cathode; tracking dye: methyl green) | | |
| | | System I | | | System II | | | System III | | |
		Separating gel, run at pH 9.5[a]	Stacking gel, run at pH 8.3[a]	Buffer system pH 8.3	Separating gel, run at pH 8.0	Stacking gel, run at pH 7.0	Buffer system pH 7.0	Separating gel, run at pH 3.8	Stacking gel, run at pH 5.0	Buffer system pH 4.5
$> 10^6$	3.75	4	2	3	—	—	—	—	—	—
	5.00	5	2	3	—	—	—	—	—	—
	7.0	1	2	3	—	—	—	—	—	—
10^4–10^6	7.5	6	2	3	11	12	13	14	15	16
	10	7	2	3	—	—	—	—	—	—
	15	8	2	3	—	—	—	17	15	16
$< 10^4$	22.5	9	2	3	—	—	—	—	—	—
	30	10	2	3	—	—	—	—	—	—

[a] pH values are given for 25°. Because of the temperature coefficient of Tris buffer, the pH of these gels at 0° is 10.2 and 9.6 respectively.

Separating Gel Formulations

(Numbers refer to the table)

(1) 7% Acrylamide, 0.18% Bis, pH 8.9
    Stock solutions: (a) 1 N HCl, 48 ml
                  Tris, 36.3 g
                  TEMED, 0.23 ml; resulting pH 8.8 to 9.0
            (b) Acrylamide, 28 g
                  Bis, 0.735 g
            (c) Ammonium persulfate, 0.14 g
  Working solution: 1 part (a), 1 part (b), 2 parts (c)
(4) 3.75% Acrylamide, 0.3% Bis, pH 8.9
    Stock solutions: (a) As in (1)
            (b) Acrylamide, 15 g
            (c) Riboflavin, 4.0 mg
  Working solution: 1   part (a), 2 parts (b), 1 part (c), 4 parts H_2O
Use photopolymerization.
(5) 5% Acrylamide, 0.25% Bis, pH 8.9
    Stock solutions: (a) As in (1)
            (b) Acrylamide, 20 g
                  Bis, 1.0 g
            (c) As in (4)
  Working solution: 1 part (a), 2 parts (b), 1 part (c), 4 parts H_2O
Use photopolymerization.
(6) 7.5% Acrylamide, 0.18% Bis, pH 8.9
    Stock solutions: (a) As in (1)
            (b) Acrylamide, 30 g
            (c) Bis, 0.735 g
  Working solution: 1 part (a), 1 part (b), 2 parts (c)
(7) 10% Acrylamide, 0.1% Bis, pH 8.9
    Stock solutions: (a) As in (1)
            (b) Acrylamide, 60 g
                  Bis, 0.6 g
            (c) As in (1)
  Working solution: 1 part (a), 1.34 parts (b), 4 parts (c), 1.66 parts H_2O
(8) 15% Acrylamide, 0.1% Bis, pH 8.9
    Stock solutions: (a) As in (1)
            (b) As in (7)
            (c) As in (1)
  Working solution: 1 part (a), 2 parts (b), 4 parts (c), 1 part H_2O
(9) 22.5% Acrylamide, 0.15% Bis, pH 8.9
    Stock solutions: (a) As in (1)
            (b) As in (7)
            (c) As in (1)
  Working solution: 1 part (a), 3 parts (b), 4 parts (c)
(10) 30% Acrylamide, 0.2% Bis, pH 8.9
    Stock solutions: (a) As in (1)
            (b) As in (7)
            (c) Ammonium persulfate, 0.18 g
  Working solution: 1 part (a), 4 parts (b), 3 parts (c)

(11) 7.5% Acrylamide, 0.2% Bis, pH 7.5
    Stock solutions: (a) 1 N HCl, 48 ml
                    Tris, 6.85 g
                    TEMED, 0.46 ml, pH 7.5
               (b) Acrylamide, 30 g
                    Bis, 0.8 g
               (c) As in (1)
    Working solution: 1 part (a), 2 parts (b), 4 parts (c), 1 part H_2O
(14) 7.5 Acrylamide, 0.2% Bis, pH 4.3
    Stock solutions: (a) 1 N KOH, 48 ml
                    Acetic acid, 17.2 ml
                    TEMED, 4.0 ml; resulting pH 4.3
               (b) As in (11)
               (c) Ammonium persulfate, 0.28 g
    Working solution: 1 part (a), 2 parts (b), 4 parts (c), 1 part H_2O
(17) 15% Acrylamide, 0.1% Bis, pH 4.3
    Stock solutions: (a) As in (14)
               (b) Acrylamide, 60 g
                    Bis, 0.4 g
               (c) As in (14)
    Working solution: 1 part (a), 2 parts (b), 4 parts (c), 1 part H_2O

Stacking Gel Formulations

(Numbers refer to the table)

Polymerization by exposure to fluorescent light
  (2) Stock solutions: (a) Tris, 5.98 g
                      TEMED, 0.46 ml
                      1 N HCl (about 48 ml) adjust to pH 6.7
                 (b) Acrylamide, 10 g
                      Bis, 2.5 g
                 (c) Riboflavin, 4.0 mg
                 (d) Sucrose, 40 g
    Working solution: 1 part (a), 1 part (b), 1 part H_2O, 1 part (c), 4 parts (d)
  (12) Stock solutions: (a) 1 M H_3PO_4, 39.0 ml
                      Tris, 4.95 g
                      TEMED, 0.46 ml; resulting pH 5.5
                 (b) As in (2)
                 (c) As in (2)
                 (d) As in (2)
    Working solution: 1 part (a), 2 parts (b), 1 part (c), 4 parts (d)
  (15) Stock solutions: (a) 1 N KOH, 48.0 ml
                      Acetic acid, 2.87 ml
                      TEMED, 0.46 ml; resulting pH 6.7
                 (b) As in (2)
                 (c) As in (2)
    Working solution: 1 part (a), 2 parts (b), 1 part (c), 4 parts H_2O

Buffer Formulations

(Numbers refer to the table)

The amounts of material indicated are for the preparation of 1 liter of buffer solution.

(3) Tris, 3.0 g
Glycine, 14.4 g, pH 8.3

(13) Diethylbarbituric acid, 5.52 g
Tris, 1.0 g, pH 7.0

(16) β-Alanine, 3.12 g
Acetic acid, 0.8 ml, pH 4.5

[40] Locating Enzymes on Gels[1]

By OTHMAR GABRIEL

The most important task remaining after successful separation of enzymes is the location of the resolved components. Ideally, one protein-staining band will match a specific locus exhibiting enzymatic activity. The precautions mentioned in this volume [39] will facilitate obtaining a functional enzyme protein. In most instances, a compromise is necessary between selection of appropriate conditions for localization of enzymatic activity and localization of protein. Very similar problems have been encountered in the localization of enzymes in tissues, and a vast body of valuable information can be obtained from the histochemical litera-ture.[1a–3] Consequently, histochemical methods may serve effectively as a guide and be adapted to the staining of enzymatic activity in acrylamide gels. As mentioned, the stability of the enzyme must be maintained dur-ing electrophoresis; equally important, the activity of the enzyme should be located promptly, prior to diffusion of the separated components. It is helpful to obtain the maximum amount of information possible con-cerning the properties of the enzyme to be separated, e.g., optimal con-ditions for enzymatic activity and necessary cofactors.

[1] The experimental results from the author's laboratory and preparation of this chapter were supported by a Grant of the Public Health Service AI-07241 from the National Institutes of Health.
[1a] T. Barka and P. J. Anderson, "Histochemistry," Harper & Row, New York, 1963.
[2] M. S. Burstone, "Enzymatic Histochemistry and Its Application in the Study of Neo-plasms." Academic Press, New York, 1962.
[3] A. G. E. Pearse, "Histochemistry." Little, Brown, Boston, Massachusetts, 1968.

Localization of Enzymes

Following successful electrophoretic separation, the process of diffusion continues after the current is turned off. Therefore, the enzyme assay should be carried out as quickly thereafter as possible. In cases where the enzyme can be frozen without loss of activity, the gel may be stored in a frozen state.

Several techniques may be used to locate enzymes.

1. Separate staining for protein and for enzymatic activity. Since most sensitive methods for the detection of protein require fixation in acid, a procedure which results in complete denaturation, a duplicate sample must be stained for enzymatic activity. It should be emphasized that polyacrylamide gels change their dimensions as a function of their ionic environment with consequent contraction or expansion of the gel. This should be kept in mind when comparing samples treated with different reagents.

2. Where applicable, utilization of a system allowing binding of a fluorophore to protein permits the localization of protein and assay of enzyme in the same gel. This method is a compromise in terms of sensitivity of both protein and enzyme detection.

3. Staining for enzymatic activity *in situ* in the acrylamide gel followed by staining for protein in the same gel.

4. Cutting of the extruded gel into segments, elution of the enzyme from the gel, and assay of the eluted material. The segments with positive enzymatic activity are compared with duplicate samples stained for protein. Recovery of enzyme varies widely as a function of the molecular size of the protein and the degree of cross-linkage of the polyacrylamide gel used for the separation. Crushing the gel in an appropriate buffer helps in releasing enzyme activity. If it is feasible to freeze the enzyme without loss of activity, freezing and thawing of the gel may facilitate release of enzyme.

In all staining methods, the penetration of buffers, substrates, and reagents will be influenced by the extent of cross-linkage and can become limiting factors both for the progress of the enzymatic reaction and for the staining process. Since all reagents must migrate from the outside of the gel toward the center, the appearance of "rings" rather than "discs" of stained material is generally observed. Thus, the staining process is frequently completed on the outside of the gel but remains incomplete near the center. Consequently, optical evaluation of the band density, determined by photometric scanning of the staining pattern, may lead to appreciable errors in quantitation. A more complex situation arises upon optical evaluation of the color intensity produced by an enzymatic reaction. The many factors determining enzymatic activity, e.g., pH, ionic environment, concentrations of cofactors, cannot be sufficiently controlled to permit meaningful quantitation of enzymatic activity in polyacrylamide gels. It also should be emphasized that staining for enzymatic activity must be carried out in the presence as well as in the absence of substrate. Obviously, only bands obtained in the presence of substrate reflect enzymatic activity. However, this caution is necessary since many stains will present as a spurious positive reaction at the junction of separating and stacking gel.

Discontinuity of pH or the concentration of components in the gel can also lead to staining artifacts. For example, a pH gradient or the presence of adventitious additives can markedly influence the enzymatic reaction.

Several major groups of methods, suitable for the localization of enzymes after polyacrylamide gel electrophoresis, can be distinguished. Most, if not all, of these represent adaptations of techniques originally described by histochemists: (a) simultaneous capture reactions, (b) postincubation coupling reactions, (c) autochromic reactions, (d) "indicator" gel methods. Some of the major features of these four groups of methods will be discussed since they provide an understanding of the principles involved and should enable the investigator to develop his own modification for a specific enzyme.

(a) *Simultaneous Capture Reactions.* The method is used whenever the enzyme is stable under the conditions necessary for the staining process. The principle is based on the release of a primary reaction product from the substrate which, in turn, couples with a reagent present in the incubation mixture to form an insoluble colored product. The method has been used extensively for the histochemical demonstration of hydrolytic enzymes resulting in the release of a coupling agent capable of reacting with a diazonium salt present in the incubation mixture. It is pertinent to mention the complexities involved here in order to understand the problems related to the staining of enzymes in general. The accuracy of enzyme localization depends on the kinetics of the enzymatic reaction releasing the primary reaction product. Once the reaction has occurred, diffusion of the low molecular weight compound will take place. The extent of diffusion will be a function of the kinetics of the capture reaction. Since all the events occur in the three-dimensional network of the polyacrylamide gel, and not in a homogeneous solution, these events will be further complicated by the diffusion rates of the individual components into the gel. It is desirable to recall these complexities when proceeding to more difficult systems (b, c, d).

(b) *Postincubation Coupling.* Ideally, the requirement here is for the primary reaction product to remain without diffusion, in the area where it is released by the enzyme. Unfortunately, this requirement rarely is met, and diffusion broadens the stained band. The component released as the result of the enzyme-catalyzed reaction will diffuse throughout the entire incubation period. For an assessment of the proper incubation time prior to coupling, valuable information can be obtained by substituting an autochromic substrate (c).

(c) *Autochromic Methods.* As the name implies, these methods are based on the utilization of substrates that permit direct visualization of the reaction. Included are methods that utilize changes in optical properties of the substrate or the product(s), such as fluorescence and ultraviolet light absorbance. The great advantage of these methods is the direct observation of the progress of the enzymatic reaction *in situ* as well as an assessment of the diffusion occurring during the incubation period. For the purpose of localization of the enzyme it is sufficient to observe the initial color and to note the appropriate location by inserting a marker.[4] The same gel subsequently may be stained to locate proteins. Thus, bands displaying enzymatic activity can be matched with the corresponding enzyme-protein in the same gel.

(d) *Sandwich-Type Incubation with "Indicator" Gel.* For many important classes of enzymes, e.g., transferases or catabolic enzymes with high molecular weight substrates, the first three methods are not applicable. The staining process for these enzymes necessitates the presence of high molecular weight substrates, acceptors, or auxiliary enzymes in the separating gel. Invariably, such components will have great influence on the mobility of the components to be separated. Nevertheless, several papers have been published suggesting the copolymerization of high molecular weight compounds directly into the separating gel.[5,6]

The "indicator" gel method avoids interference by the high molecular substances required for the staining process. The indicator gel is first prepared to contain all the components necessary for the staining process. For this purpose, high molecular weight substrates, acceptors, or auxiliary enzymes are included in the polymerization mixture usually employed for preparation of separating gels. The procedure as well as the dimensions for this preparation are identical to that for the separating gel. Upon completion of polymerization, the gel is cut longitudinally into three slices, e.g., with a Canalco gel slicer, and maintained under refrigeration in a moist chamber until used. Independently, the enzyme sample is subjected to electrophoresis in the usual way without any modi-

[4] It was found convenient to insert a 5 mm-long copper wire into the gel at the appropriate site.
[5] M. Lieflander and H. Stegemann, *Hoppe-Seglers Z. Physiol. Chem.* 349, 157 (1968).
[6] H. Stegemann, *Z. Anal. Chem.* 243, 573 (1968).

fications in the composition of the separating gel. Upon termination of electrophoresis, the gel is cut into three longitudinal slices. One of these is placed along an "indicator" gel so that the two gel surfaces are contiguous. This "sandwich" is then incubated in a moist chamber. Depending on the type of enzyme, the indicator gel will be subjected to one of the staining methods described above (a–c) to permit localization of enzyme.

Numerous modifications of the "indicator" gel method are possible. Most important of these is the use of agar, starch gel, or paper rather than polyacrylamide in preparing the "indicator" gel.

Precautions must be taken to avoid inactivation or denaturation of high molecular weight material by the polymerization procedure. It is therefore suggested that a drop of the enzyme sample be placed on a freshly prepared "indicator" gel to serve as a positive control.

Staining: Specific Methods

A selection of methods for the detection of enzymatic reactions is presented in two sections. Section A will be concerned with the methods used for the recognition of the enzyme as a protein. Section B deals with procedures designed to detect enzymatic activity. These methods have been found to work satisfactorily and are applicable not only to the specific enzymes indicated but to groups of enzymes catalyzing analogous reactions. The description of specific methods is followed by tables which are offered to provide guidance for the detection of individual enzymes; the enzymes presented in Tables I, II, and III are listed in alphabetical order.

A. Methods for Protein Staining

The most sensitive methods for the staining of proteins entail the fixation of the protein and result in irreversible denaturation. For the necessary correlation between protein stain and enzymatic activity, two individual gels or gel segments must be subjected separately to the staining processes for protein and enzymatic activity, respectively. The change in dimensions of the gel due to shrinking or swelling make it desirable to carry out both staining procedures on the same gel, which would therefore have to be sliced in half parallel to its long axis. However, several staining processes for enzymatic activity permit subsequent counterstaining for proteins. Alternatively, a compromise can be achieved by using fluorescent dyes for protein detection. Despite the decreased sensitivity of the latter, the method nevertheless provides sufficient residual enzymatic activity to permit detection on the same gel.

(1) *Amido Black 10B*[7] *(Amido Schwartz)*

Use of this stain is reliable but time consuming. Rapid destaining by electrolytic procedures is not advisable since loss of minor components has been reported.

Method. One gram of the Amido black dye is dissolved in 100 ml of 7.5% acetic acid. Insoluble material is decanted, and the supernatant liquid is used for the staining process. Place each gel into a test tube and immerse completely in the dye solution for 1–5 hours. For destaining, the solution is drained and replaced with 7.5% acetic acid. Slight agitation and repeated changes of the acetic acid will remove excess stain within 24–48 hours. The procedure is completed when the background of the gel is transparent. Store in 7.5% acetic acid.

(2) *Coomassie Blue*[8]

At present, this is the method of choice, although exploration of other methods is suggested.

Method. After completion of electrophoresis, the individual sample gels are placed into test tubes and immersed in 12.5% trichloroacetic acid. After 30 minutes, the gels are treated with a freshly prepared 1:20 dilution in 12.5% trichloroacetic acid of a 1% solution of Coomassie Blue (Coomassie Brilliant Blue R250: Colab Laboratories, Inc., Chicago Heights, Illinois). The gels must remain in the staining solution for 30 minutes to 1 hour. Destaining is achieved by transfer to 12.5% trichloroacetic acid. The procedure is much quicker than that with Amido black. For maximal sensitivity, the same gel may be subjected to a second round of staining. Trichloroacetic acid, 12.5–20%, is used for fixation and 0.05–0.1% solution of the dye is used in the same trichloroacetic acid concentration. Fixation and staining periods are at least doubled, although exact conditions for specific proteins must be found empirically.

(3) *Fluorescence Labeling.*[9]

The method is based on the binding of a fluorophore to the protein, thereby resulting in fluorescence. Enhancement of fluorescence is achieved by denaturation of the protein. For most purposes, ammonium persulfate remaining after polymerization in the gel, or surface denaturation caused by removal of the gel by extrusion from the glass

[7]E. H. Woeller, *Anal. Biochem.* 2, 508 (1961).
[8]A. Chrambach, R. A. Reisfeld, M. Wyckoff, and J. Zacchari, *Anal. Biochem.* 20, 150 (1967).
[9]B. K. Hartman and S. Udenfriend, *Anal. Biochem.* 30, 391 (1969).

column, will cause sufficient denaturation to allow binding of fluorophore to protein. If there is insufficient surface denaturation, enhancement of the fluorescence can be achieved by further deliberate denaturation, e.g., exposure of the gel to air, exposure to fumes of concentrated HCl or immersion in 3 N HCl for a few seconds to a few minutes. After each denaturation step, the gel is again placed into the buffered staining solution. As indicated above, the method is a compromise in terms of both sensitivity for the detection of protein and the location of enzymatic activity.

Anilinonaphthalene sulfonate (ANS) may be used as fluorophore. The magnesium salt of 1-anilino-8-naphthalene sulfonate[10] is recrystallized twice from hot water and the crystals are filtered and dried. An aqueous stock solution of 1 mg ANS per milliliter is prepared and remains stable in a refrigerator for several months. Immediately prior to use, a dilution of the dye to 0.003% with 0.1 N sodium phosphate buffer, pH 6.8, is prepared. The freshly extruded gels are placed into a petri dish and immersed in this solution. Fluorescence is observed by using a portable, longwave ultraviolet lamp. The amount of protein necessary for detection by this method is of the order of 100 μg but can be decreased to about 20 μg by denaturation.

B. Methods for Localization of Enzymatic Activity

Several methods used for the localization of enzymes on acrylamide gels are presented here. It is the objective of this section to provide examples for each of the main groups of enzymes and to place emphasis upon the demonstration of the different techniques used for enzyme detection. The letters (a), (b), (c), and (d) refer to the type reaction described above in the section "Localization of Enzymes," whereas the numbers are used to designate the specific technique described subsequently in this article. The following examples are chosen:

(1) Dehydrogenases
    (1.1) Lactate dehydrogenase. Simultaneous capture technique (a)
    (1.2) TDPG-oxidoreductase. Postincubation capture technique (b)

(2) Transferases
    (2.1) Phosphoglucomutase. Indicator gel method (d)
    (2.2) TDPG-pyrophosphorylase. Indicator gel method (d)
    (2.3) Sucrose phosphorylase. Postincubation capture technique (b)

(3) Hydrolases
    (3.1) (3.2) Carboxylic acid esterase. Postincubation capture technique (b)

[10]Eastman Organic Chemicals, Eastman Kodak Company, Rochester, New York 14605.

(3.3) Alkaline phosphatase. Simultaneous capture method (a)

(3.4) Alkaline phosphatase. Postincubation capture method (b)

(3.5) Acid phosphatase. Simultaneous capture method (a)

(3.6) Glycosidases. Autochromic method (c)

(3.7) Lipase. Simultaneous capture method (a)

(3.8) Leucine aminopeptidase. Postincubation capture technique (b)

(4) Lyases

(4.1) Aldolase. Simultaneous capture technique (a)

(4.2) Carbonic anhydrase. Autochromic method (c)

(4.3) Tryptophan synthetase. Simultaneous capture technique (a)

(5) Isomerases

(5.1) Maleate isomerase. Postincubation capture technique (b)

(5.2) Phosphoglucose isomerase. Simultaneous capture technique (a)

(5.3) Uridine diphosphogalactose-4-epimerase. Simultaneous capture technique (a)

(1) *Dehydrogenases*

A variety of methods is available among this group of enzymes, most of which use tetrazolium salts. The tetrazolium salt serves as the terminal electron acceptor leading to the reduced, colored formazan derivative. These methods have been widely applied to starch gel electrophoresis[11] and have been modified for acrylamide gels. Many of the enzymes studied with this technique are DPN and TPN requiring but the usefulness of the method has been extended beyond this requirement.

The use of tetrazolium salts as staining agents requires special care to prevent nonspecific staining. These salts are light sensitive and the incubation should be carried out in the absence of light. The staining process is pH and temperature dependent so that the incubation mixture must be prepared just prior to use. If preservation of the stained gel is anticipated, the residual reagent should be removed by extensive washing to avoid increased background coloration resulting from light-catalyzed color formation.

(1.1) *Lactate Dehydrogenase. Simultaneous Capture Technique* (a). The technique is illustrated with lactic acid dehydrogenase[12] but is applicable with minor modifications to many pyridine nucleotide-requiring enzymes.

STAINING PROCEDURE. After electrophoresis, the extruded gels are transferred to small test tubes. In order to assure the appropriate pH along the entire length, gels are immersed in ice cold 0.5 M Tris·chloride buffer, pH 7.5, for 15 minutes prior to staining.

[11]I. H. Fine and L. A. Costello, this series Vol. 6, p. 958 (1963).
[12]A. Dietz and T. Lubzant, *Anal. Biochem.* 20, 246 (1967).

STAINING SOLUTION.[13] The staining solution should be prepared immediately prior to use in quantities sufficient for the experiment.

Sodium lactate,[1] M	1.0 ml
DPN, 10 mg/ml	1.0 ml
NaCl, 0.1 M	1.0 ml
MgCl$_2$, 5 mM	1.0 ml
Tris·chloride buffer, 0.5 M, pH 7.4	2.5 ml
Nitroblue tetrazolium (NBT), 1 mg/ml	2.5 ml
Phenazine methosulfate (PMS), 1 mg/ml	0.25 ml

Total immersion of the gel in the staining solution is followed by incubation at room temperature. Periodically, the progress of staining is checked, although the gels are protected from light. After completion of the process, the solution is discarded and the gels are washed sequentially in water and 7.5% acetic acid.

COMMENTS. Demonstration of substrate-dependent stain is required since formation of reduced pyridine nucleotide in the absence of exogenous substrate ("nothing dehydrogenase") has been reported by numerous investigators.[14]

(1.2) *TDPG-Oxidoreductase.*[15] *Postincubation capture technique* (b). A second method, applicable to oxidoreductases as well as to other groups of enzymes such as transferases and hydrolases, uses the postincubation capture technique.[16] The method is based on the enzymatic conversion of a nonreducing substrate to products with reducing properties.

STAINING PROCEDURE. The reactivity of TTC (2,3,5-triphenyl tetrazolium chloride) depends on the redox potential of the reducing agent, the pH of the solution, the temperature at which the reaction takes place, and the time allowed for the reaction to proceed. Judicious selection of the reaction conditions provides a qualitative differentiation between different groups of reducing sugars. For example, keto sugars, such as fructose or a nucleotide-linked 4-keto intermediate, will react at room temperature, whereas aldohexoses require heating to 100°.

Procedure a: Localization of enzymes producing keto sugars

1. The extruded gel is incubated under the appropriate assay conditions. *Example*: TDPG-oxidoreductase catalyzes the conversion of the

[13]The same procedure may be modified for use with enzymes as indicated in Table I. Substrate, other components such as sulfhydryl group containing reagents, metal ions, DPN or TPN, and the pH of buffer have to be changed depending on the requirements of individual enzymes.
[14]S. H. Hori and T. Kawamura, *Acta Histochem. Cytophem.* 1, 95 (1968).
[15]This enzyme is classified as a lyase according to the rules of nomenclature presented by the International Union of Biochemistry.
[16]O. Gabriel and S. F. Wang, *Anal. Biochem.* 27, 545 (1969).

nonreducing sugar nucleotide to the reducing intermediate TDP-4-keto-6-deoxyglucose. The latter compound reacts with TTC. When enzyme substrate is the limiting factor, incubation is carried out in narrow test tubes wherein 0.5 ml liquid is sufficient to immerse the gel completely. The gel is kept for 20 minutes at 37° in 0.5 ml of a solution containing 3 μmoles of TDPG and 30 μmoles of Tris·chloride at pH 8.0.

2. Upon completion of the incubation period, the gel is rinsed with water and incubated at room temperature in a freshly prepared solution of 0.1% TTC in 1 N NaOH. The staining process is completed in 5–10 minutes.

3. The gel is washed in 7.5% acetic acid to remove excess reagent.

4. The same gel may be stained subsequently for protein using Coomassie Blue.

Procedure b: Staining aldose reaction products

1. As in procedure a according to the requirements of the enzyme; for example, see 3.6.

2. The gel is rinsed with water, then transferred to a test tube containing 0.1 M iodoacetamide for 5 minutes at room temperature. Upon removal, the gel is rinsed with distilled water.

3. The gel is immersed in a freshly prepared solution consisting of 0.1% TTC in 0.5 N NaOH and heated in a boiling water bath for 60–90 seconds; the tube should be gently agitated during this period. The exact heating time must be determined empirically, but heating should be terminated as soon as a pink background appears.

4. Immediately after heating, the gel is rinsed with water, washed, and stored in 7.5% acetic acid.

(2) *Transferases*

The transfer of a group from a donor to an acceptor molecule is characteristic for all transferases. Since this category represents enzymes transferring a wide variety of groups other than hydrogen and electrons, several different techniques for localization of these enzymes must be provided. One principle, which has been applied successfully, involves coupling with other enzymes or enzyme systems so that a sequence of enzymatic reactions leads to action by a dehydrogenase which, in turn, can be localized using the methods already described. However, this approach introduces the use of a high molecular weight "auxiliary" enzyme which can only enter the three-dimensional network of the gel to a limited extent.

A method which avoids some of these difficulties but requires greater manual dexterity is the indicator gel method (d) described below.

(2.1) *Phosphoglucomutase. Indicator Gel Method* (d). Phosphoglucomutase is subjected to electrophoresis on a 7.5% acrylamide separating gel at pH 8.9.[17] Upon completion of the procedure, the gel is extruded, rinsed in cold buffer, and sliced into three longitudinal segments. "Indicator" gel is prepared independently but has identical dimensions to the separating gel; polymerization is carried out in a glass column identical to that used in the electrophoretic procedure. Stock solutions for the indicator gel (7.5% acrylamide) have the following composition:

Stock solutions: (a) 48 ml 1 N HCl ⎫
                      36.3 g Tris ⎬ up to 100 ml with H_2O, pH 8.9
                      0.23 g TEMED ⎭
                 (b) 0.1 M Mercaptoethanol
                 (c) 30 g acrylamide ⎫ 100 ml
                      0.8 g Bis ⎬
Working solution: 1 part (a), 1 part (b), 2 parts (c), 4 parts (d)

The last addition to this working solution is glucose-6-phosphate dehydrogenase. The polymerization process is carried out and the gels are sliced longitudinally into three segments (Canalco gel slicer) which are placed into a solution (total volume, 2 ml) containing the following components: 100 μl 1 M Tris·chloride at pH 8.0; 200 μl 0.1 M $MgCl_2$; 200 μl 0.01 M EDTA; 100 μl glucose-1-phosphate (10 μmoles/ml); 2 μl glucose-1,6-diphosphate (10 μmoles/ml); 25 μl TPN (100 mg/ml); 10 μl PMS (20 μmoles/ml); 20 μl NBT (10 μmoles/ml).

This incubation should be performed at 4° for about 1 hour prior to use. The indicator gel is removed from the above solution, placed in close contact along its entire length with the separating gel obtained after electrophoresis, and incubated in a small moist chamber at room temperature. A petri dish with filter paper saturated with water serves well. Incubation is carried out in subdued light. In the areas corresponding to the position of enzymatic activity, the "indicator" gel will show color formation and is washed with water after which it may be stored in 7.5% acetic acid.

(2.2) *TDPG-Pyrophosphorylase. Indicator Gel Method* (d). A crude preparation of TDPG-pyrophosphorylase isolated from *Pseudomonas aeruginosa* is used. Disc electrophoresis was carried out in a 7.5% separating gel.[17] Upon completion of electrophoresis, the gel was rinsed with cold buffer, sliced, and used immediately for the assay.

INDICATOR GEL PREPARATION. An identical procedure is followed as described for phosphoglucomutase (2.1), except that both glucose-6-phosphate dehydrogenase and phosphoglucomutase are added to the

---

[17]O. Gabriel and I. Weemaes, unpublished data.

working solution. After polymerization, the indicator gel is sliced and placed into a solution with the following composition for 1 hour at 4°:

Total volume: 0.8 ml 50 M of Tris·chloride, pH 8.0, containing 10 mM MgCl$_2$ and 1 mM EDTA, 25 μl TPN (10 mg/ml), 1 μl glucose-1, 6-diphosphate (10 μmoles/ml), 25 μl TDPG (10 μmoles/ml), 50 μl inorganic pyrophosphate (10 μmoles/ml), 10 μl PMS (20 μmoles/ml), 50 μl NBT (10 μmoles/ml).

The indicator gel is removed from this solution and placed upon the freshly sliced separating gel from the electrophoretic run, in the same manner as described in (2.1). Incubation is at room temperature.

COMMENT. The incorporation of enzymes into the matrix of the indicator acrylamide gel and maintenance of such enzymatic activity was accomplished in the examples chosen by the addition of 12.5 mM mercaptoethanol as a protective agent. All the other low molecular weight components necessary for the enzymatic reaction and for the staining procedure are initially omitted from the working solution in order to avoid either inactivation of enzyme or inhibition of the polymerization reaction. After polymerization is completed, the low molecular weight components enter the gel by diffusion at 4°. The indicator gel is then ready for staining and is placed upon the sliced separating gel obtained after electrophoresis. Incubation is carried out at room temperature. In order to test the function of the indicator gel, a drop of enzyme may be placed directly onto a small segment of gel; a positive stain should develop.

In the examples of TDPG-pyrophosphorylase (2.2) two widely separated bands of activity were obtained, only one of which, that with the greater mobility, was dependent upon substrate (TDPG).

The above two examples were chosen because they are also demonstrable using the simultaneous capture technique. For the latter purpose, all the components necessary for the staining procedure, including the auxiliary enzymes, are included in the amounts indicated in a total volume of 0.8 ml of incubation mixture. The gel is totally immersed in the solution and incubated at room temperature.

(2.3) *Sucrose Phosphorylase.*[16] *Postincubation Capture Reaction* (b). The enzyme catalyzes the transfer of a glucose residue from sucrose to inorganic phosphate. Incubation is carried out for 20 minutes at 30° in 3 ml of 33 mM potassium phosphate, pH 6.9, containing 0.2 M sucrose. The rest of the procedure is described in (1.2).

(3) Hydrolases

A large number of methods are available for the study of hydrolases, thereby representing a reflection both of the early development of methods in histochemistry and of the relative stability of this class of enzymes. Most of the methods are postincubation capture reactions (b), and a variety of esterases of different specificity can be located. For example, an enzyme-catalyzed reaction, acting on a substrate such as α-naphthylacetate, releases a compound, α-naphthol, which, upon addition of a diazonium salt (fast blue RR) will lead to a colored insoluble product indicating the location of the enzyme. The following general classes of hydrolases will be presented with detailed methods: esterases, phosphatases, glycosidases, lipases, and peptidases. In addition, examples for the illustration of the principles for autochromic techniques (c) are provided.

(3.1, 3.2) *Carboxylic acid esterase activity*[18–20]

(3.1) *Simultaneous Capture Reaction* (a). The extruded gels are washed twice in 0.1 M Tris·chloride, pH 7.0, for 15 minutes. The diazonium salt reagent is prepared just prior to use with all solutions kept 0° and mixing carried out in an ice bath. Two drops of p-chloroaniline (36 mg/ml of 1 N HCl) are mixed with 2 drops of sodium nitrite (26 mg/ml) in 1 ml of ice water and shaken. This solution is added to 25 ml of 0.1 M Tris·chloride, pH 7.0, and 0.2 ml of α-naphthyl acetate (1% solution in acetone). The gels are incubated in this solution until stained bands appear.

(3.2) *Simultaneous Capture Reaction* (a). A choice of three different substrates, the acetyl, propionyl, or butyryl esters of α-naphthol, is provided in the procedure. Substrate components are as follows: 0.2 M Tris-chloride, pH 7.4, 2 ml; H_2O, 47 ml; α-naphthyl acetate, -propionate or -butyrate (1% in acetone), 1 ml; Fast Blue RR (diazonium salt), 25 mg.

(3.3)–(3.5) *Phosphatases*

(3.3) *Alkaline phosphatase.*[21] *Simultaneous Capture Method* (a). The solution for incubation is composed of 33 mM Tris·chloride, pH 9.5, 25 mM sodium α-naphthyl phosphate and 1 mg/ml of fast red TR (diazonium salt of 4-chloro-o-toluidine). After staining is completed, excess reagent is removed by washing with water.

(3.4) *Alkaline Phosphatase.*[21] *Postincubation Capture Method* (b). Incubation is carried out at 25° for 15 minutes in a solution containing 5 mM

[18]R. L. Hunter and E. A. Maynard, *J. Histochem. Cytochem.* 10, 677 (1962).
[19]C. L. Markert and R. L. Hunter, *J. Histochem. Cytochem.* 7, 42 (1959).
[20]B. J. Davis, Enzyme Analysis 3b, Canalco, Canal Industrial Corporation, Rockville, Maryland, 1963.
[21]J. M. Allen and G. J. Hyncik, *J. Histochem. Cytochem.* 11, 169 (1963).

sodium β-glycerophosphate, 15 mM CaCl$_2$, 33 mM Tris·chloride at pH 9.5. The gel is rinsed with water, placed into 3 mM lead nitrate in 80 mM Tris·maleate at pH 7.0 for 30 minutes. It is washed for 1 hour in water with frequent changes of water and finally placed into a 5% ammonium sulfide solution for 2 minutes. Thereafter, it is again washed with water.

(3.5) *Acid Phosphatases.*[1a,22,23] *Simultaneous Capture Method* (a). Simultaneous incubation coupling method (a) may be used with sodium α-naphthyl phosphate as substrate and rosaniline diazonium salt as the coupling agent. It should be emphasized that at least two incubations at 4° to 5° for 15 minutes each in 0.1 M acetate buffer, pH 5.0, are necessary prior to incubation with substrate in order to adjust the pH of the gel to an acidic milieu. If this step is not carried out, alkaline phosphatase will react. The following solutions need to be prepared: Solution 1: Equal volumes of 4% rosaniline·HCl in 2 N HCl and 4% NaNO$_2$; they are combined just prior to use. Solution 2: 20 mg of sodium α-naphthyl phosphate in 13 ml H$_2$O and 5 ml Veronal-acetate buffer (9.7 g sodium acetate·3H$_2$O and 14.7 g sodium barbital in 500 ml water). For use, 1.6 ml of solution 1 is mixed with 18 ml of solution 2. Adjust to pH 5.0 with 1 N NaOH and filter. Incubation is carried out for 12–18 hours at 4°–5°.

(3.6) *Glycosidases. Autochromic Method.* The localization of glycosidases was chosen to demonstrate the principle of autochromic methods (c). Synthetic chromogenic glycosides, e.g., *o*- or *p*-nitrophenylglycosides, have been widely used for purposes of assay, and these compounds are commercially available.[24] For example, the following procedures were employed for the demonstration of α-mannosidase and β-D-glucosaminidase, respectively[16]:

After electrophoresis, the gels are rinsed with cold water and incubated for 30 minutes at room temperature in 0.2 M potassium acetate at pH 4.9.

DEMONSTRATION OF α-MANNOSIDASE. The gel is incubated in 3 ml of a solution containing 5 mM *p*-nitrophenyl α-D-mannopyranoside[24] in 25 mM potassium citrate at pH 4.5 for 30 minutes at room temperature.

DEMONSTRATION OF β-D-GLUCOSAMINIDASE. The gel is incubated in 3 ml of a solution containing 5 mM *o*-nitrophenyl-*N*-acetyl-β-D-glucosaminide[24] in 25 mM potassium citrate, pH 5.5, for 30 minutes at room temperature. The enzyme-catalyzed cleavage results in the liberation of *o*- and *p*-nitrophenol, respectively. These may be observed

[22] T. Barka, *J. Histochem. Cytochem.* 9, 542 (1961).
[23] P. J. Anderson, S. K. Sing, and N. Christoff, *Proc. 4th Int. Congr. Neuropathol.* 1, 75 (1962).
[24] Pierce Chemical Company, Rockford, Illinois.

directly by their yellow color at an alkaline pH. Thus, the influence of diffusion on substrates during the incubation period can be assessed immediately. As a consequence, it is possible to mark the location of these enzymes by the insertion of a small piece of copper wire.[4] Thereafter, the gel is washed free of substrate, fixed in 12.5% trichloroacetic acid, and stained with Coomassie Blue.[8] The location of enzymatic activity and protein on the same gel obviates the need for the matching of separate samples. Alternatively, activity may be located by means of the reducing properties of the enzymatically released sugar moiety using the tetrazolium chloride method described above (1.2).

(3.7) *Lipase.*[25] *Simultaneous Capture Method* (a). Five milliliters of 0.4 M Tris·chloride, pH 7.4, 1.0 ml of 2.5% sodium taurocholate solution, and 3 ml of H_2O are thoroughly mixed with 0.1 ml of a 2% solution of α-naphthyl nonanoate[26] (-pelargonate) in dimethylacetamide and 10 mg of Fast Blue BB salt. The turbid solution is filtered, and the filtrate is used for incubation at 37° for 40–60 minutes. The gels are washed with water.

(3.8) *Leucine Aminopeptidase.*[27] A substrate stock solution is prepared by dissolving 24 mg of either L-leucyl-4 methoxy-β-naphthylamide (1)[26] or L-leucyl-4 methoxy-β-naphthylamide (11)[26] in a few drops of methanol. The solution is diluted with water to a final volume of 3 ml. The staining solution is prepared as follows: substrate (stock solution), 3.0 ml; sodium acetate (0.1 M, pH 6.5) 30.0 ml; sodium chloride (0.14 M), 24.0 ml; potassium cyanide (0.02 M), 3.0 ml; Fast Blue B salt, 30 mg.

Gels are incubated at 37° for 15 minutes to 2 hours, depending upon their activity. They are rinsed with 0.14 M sodium chloride and immersed for 2 minutes in 0.1 M $CuSO_4$. They are rinsed again in sodium chloride. Substrate I results in formation of a purplish-blue stain whereas Substrate II yields a red stain.

(4) Lyases

In this group of enzymes only a few procedures are available; some were originally described for electrophoresis using starch gel as the anticonvection medium.

(4.1) *Aldolase.*[28] A simultaneous capture technique was employed by coupling aldolase with glyceraldehyde-3-phosphate dehydrogenase. This permits use of the methods described earlier (1.1) based on

[25]M. Abe, S. P. Kramer, and A. M. Seligman, *J. Histochem. Cytochem.* 12, 364 (1964).
[26]Cyclo Chemical Corporation, Los Angeles, California.
[27]M. M. Nachlas, B. Morris, D. Rosenblatt, and A. M. Seligman, *J. Biophys. Biochem. Cytol.* 7, 261 (1960).
[28]D. C. Nicholas and H. S. Bachelard, *Biochem. J.* 112, 587 (1969).

formation of reduced pyridine nucleotide. The incubation is carried out at 37° in the following solution: 3 mM fructose 1,6-diphosphate, pH 8.25, 40 mM Na$_2$HAsO$_4$; 40 mM Na$_2$P$_2$O$_7$; 0.2 mM DPN; glyceraldehyde-3-phosphate dehydrogenase (0.2 units/ml); PMS, 0.01 mg/ml; NBT, 0.2 mg/ml in 60 mM glycine-NaOH at pH 8.25.

(4.2) *Carbonic Anhydrase.* Carbonic anhydrase was visualized by taking advantage of its esterase activity with β-naphthyl acetate as a substrate.[29]

Alternatively, an autochromic method[30] (c) may be applied, which utilizes the pH change occurring at the site of enzymatic activity:

The gel is immersed in 0.1 M Veronal acetate at pH 8.1 to which a few drops of 0.2% bromothymol blue in 50% ethanol are added. The gel is transferred to cold, CO$_2$-saturated water; a yellow band appears at the location of the enzyme.

(4.3) *Tryptophan Synthetase.*[31] *Simultaneous Capture Reaction* (a). This enzyme catalyzes the release of glyceraldehyde 3-phosphate from its substrate, indole glycerophosphate. Coupling with glyceraldehyde-3-phosphate dehydrogenase results in formation of DPNH and permits detection of the enzyme by the nitroblue tetrazolium method (1.1).

(5) *Isomerases*

(5.1) *Maleate Isomerase.*[32] *Postincubation Capture Reaction* (b). The enzyme is coupled with fumarase and malate dehydrogenase, leading to formation of DPNH. The postincubation method employed nitroblue tetrazolium as the staining agent. Incubation of the gels is carried out at room temperature for 20 minutes in the following assay mixture: 50 mM Tris·chloride at pH 8.4; 15 mM DPN; 5 mM thioglycerol; 20 mM potassium maleate; 150 units of fumarase per milliliter; 100 units of malate dehydrogenase per milliliter. After incubation, gels were rinsed with water and maintained in the dark, immersed in the staining solution which contained, per milliliter, either 40 μg of PMS and 500 μg of NBT or 50 μg of PMS and 2 μg of indophenol red.

(5.2) *Phosphoglucose Isomerase.*[33] The simultaneous capture reaction with nitroblue tetrazolium (1.1) is used by coupling the isomerase with glucose-6-phosphate dehydrogenase. Fructose 6-phosphate is used as the substrate. The method was originally suggested for starch gel electrophoresis.[34]

[29] J. E. A. McIntosh, *Biochem. J.* 114, 463 (1969).
[30] L. J. Edwards and R. L. Patton, *Stain Technol.* 4, 333 (1966).
[31] I. P. Crawford, J. Ito, and M. Hatanaka, *Ann. N.Y. Acad. Sci.* 151, 171 (1968).
[32] W. Scher and W. B. Jakoby, *J. Biol. Chem.* 244, 1878 (1969).
[33] L. I. Fitch, C. W. Parr, and S. G. Welch, *Biochem. J.* 110, 56P (1968).
[34] R. K. Scopes, *Biochem. J.* 107, 139 (1968).

TABLE I

DEHYDROGENASES

Enzyme	Substrate	Principle of method	Mode of detection and reference to method[a]	References
Alcohol dehydrogenase	Ethanol, 2-butanol	Formation of DPNH, PMS, NBT	(a) (1.1)	b, c
Amine dehydrogenase	Methylamine	PMS, 2, 6-dichlorophenolindophenol	(d)	d
Amino acid oxidase	L-Leucine	PMS, TTC, or iodonitrotetrazolium salt	(a) (1.1)	e, f
	D-Phenylalanine	PMS, TTC	(a) (1.1)	e
Aldehyde dehydrogenase	Glycolaldehyde	Formation of DPNH, PMS, NBT, 2-mercaptoethanol	(a) (1.1)	g
Catalase	H_2O_2	Incubation with KI releases I_2 which stains with starch	(b) (d)	h, i, j
Folate reductase	Dihydrofolate	Formation of tetrahydrofolate thiazolyl blue or 3(4, 5-dimethylthiazolyl-1, 2)2, 5-diphenyltetrazolium bromide	(a) (c)	k, l
Fucose dehydrogenase	L-Fucose	Formation of DPNH, PMS, NBT	(a) (1.1)	m
Glucose oxidase	D-Glucose	PMS, tetranitroblue tetrazolium	(a)	m
Glucose-6-phosphate dehydrogenase	Glucose 6-phosphate	Formation of TPNH, PMS, NBT	(a) (1.1)	o, p, q, r
Glutamate dehydrogenase	L-Glutamic acid	Formation of DPNH or TPNH, PMS, NBT	(a) (1.1)	r, s, t
Glyceraldehyde phosphate dehydrogenase	D-Glyceraldehyde phosphate	Formation of DPNH, PMS, NBT	(a) (1.1)	u
Glycerophosphate dehydrogenase	L-α-Glycerophosphate	Formation of DPNH, PMS, NBT	(a) (1.1)	u
Hexose-6-phosphate dehydrogenase	D-Galactose-6-phosphate	Formation of DPNH, PMS, NBT	(a) (1.1)	p
Histidinol dehydrogenase	Histidinol	Formation of DPNH, PMS, NBT	(a) (1.1)	v
Homoserine dehydrogenase	L-Homoserine	Formation of TPNH, PMS, NBT	(a) (1.1)	w
Hydroxysteroid dehydrogenase	α- and β-Hydroxysteroids	Formation of DPNH, PMS, NBT	(q) (1.1)	x

Isocitrate dehydrogenase	D-L-Isocitric acid	Formation of DPNH or NADPH, PMS, NBT	(a) (1.1)	r, y
α-Ketoglutaric semialdehyde dehydrogenase	α-Ketoglutaric semialdehyde	Formation of TPNH, PMS, NBT	(a) (1.1)	z
α-Ketoisocaproate dehydrogenase	αKetoisocaproic acid	Formation of DPNH, PMS, NBT	(a) (1.1)	aa
α-Keto-β-methyl valeriate dehydrogenase	α-Keto-β-methyl valeric acid	Formation of DPNH, PMS, NBT	(a) (1.1)	aa
Lactate dehydrogenase	Pyruvic acid	Utilization of DPNH, PMS, NBT	(b)	bb
	D, L-Lactic acid	Formation of DPNH, PMS, NBT	(a) (1.1)	cc, dd, ee
Lipoamide dehydrogenase	DPNH, lipoamide	Utilization of DPNH-2,6-dichloroindophenol	(b)	ff
Lipoyl dehydrogenase	DPNH, lipoamide	NBT, DPNH	(a)	gg, hh
Malate dehydrogenase	L-Malic acid	Formation of DPNH or TPNH, PMS, NBT	(a) (1.1)	r, y, cc, ii, jj
	β-Isopropyl malate	Formation of DPNH, PMS, NBT	—	kk
Melilotate hydroxylase	Melilotic acid	FAD, NBT	—	ll
Myeloperoxidase	H_2O_2	Benzidine hydrochloride	(a)	i, mm, nn, oo
Nitrate reductase	Nitrate (TPNH)	Formation of nitrite, N-(1-naphthyl) ethylenediamine hydrochloride, sulfanilamide, azo dye formation	(b)	pp
Oxidase, DPNH-, and TPNH-	DPNH, TPNH	Utilization of DPNH, NBT	(a)	r
Oxidase (ceruloplasmin)	o-Dianisidine, p-phenylenediamine	Oxidation to colored product	(a)	b, qq
6-Phosphogluconate dehydrogenase	6-Phosphogluconic acid	Formation of TPNH, PMS, NBT	(a) (1.1)	rr
Tartrate dehydrogenase	Mesotartrate	Formation of DPNH, PMS, NBT	(b)	ss
TDPG-Oxidoreductase	TDPG	Formation of reducing substance, TTC	(b) (1.2)	tt
Tyrosinase	p-Cresol, catechol	Incubation of substrates in presence of L-proline	(a)	uu, vv
UDPG-Dehydrogenase	UDPG	NADH formation, PMS, NBT	—	rr

footnotes to table 1

a Mode of detection: (a) simultaneous capture reaction, (b) postincubation capture reaction, (c) autochromic method, (d) "indicator" gel method.

The numbers refer to the methods described in Section B (Methods for Localization of

Enzymatic Activity). Modification of the reference method according to the specific requirements of the enzyme in terms of substrate and optimal conditions for enzymatic activity, will permit staining of the enzyme.

[b]E. H. Grell, K. B. Jacobson, and J. B. Murphy, *Ann. N. Y. Acad. Sci.* **151**, 441 (1968).

[c]W. Sofer and H. Ursprung, *J. Biol. Chem.* **243**, 3110 (1968).

[d]R. R. Eady and P. J. Large, *Biochem. J.* **106**, 245 (1968).

[e]M. B. Hayes and D. Wellner, *J. Biol. Chem.* **244**, 6636 (1969).

[f]B. Curti, Y. Massey, and M. Zmudka, *J. Biol. Chem.* **243**, 2306 (1968).

[g]R. G. von Tigerstrom and W. E. Razzell, *J. Biol. Chem.* **243**, 2691 (1968).

[h]This method requires incorporation of small amount of starch into separating gel, and was described originally for starch gel electrophoresis. J. G. Scandalios, *Ann. N. Y. Acad. Sci.* **151**, 274 (1968).

[i]Canalco. Information Bulletin Enzyme Analysis **8** (1963).

[j]Suggested visualization of enzyme with indicator gel (d) containing starch.

[k]P. F. Nixon and R. L. Blakley, *J. Biol. Chem.* **243**, 4722 (1968).

[l]A. M. Albrecht, F. K. Pearce, W. J. Suling, and D. J. Hutchison, *Biochemistry* **8**, 960 (1969).

[m]H. Schachter, J. Sarney, E. J. McGuire, and S. Roseman, *J. Biol. Chem.* **244**, 4785 (1969).

[n]J. Jos, J. Frezal, J. Rey, and M. Lamy, *Nature (London)* **213**, 516 (1967).

[o]S. H. Hori and S. I. Matsui, *J. Histochem. Cytochem.* **16**, 62 (1968).

[p]C. H. R. Shaw and A. L. Koen, *Ann. N. Y. Acad. Sci.* **151**, 149 (1968).

[q]W. E. Criss and K. W. McKerns, *Biochemistry* **7**, 125 (1968).

[r]S. L. Allen, *Ann. N. Y. Acad. Sci.* **151**, 190 (1968).

[s]L. Corman, L. M. Prescott, and N. O. Kaplan, *J. Biol. Chem.* **242**, 1383 (1967).

[t]H. B. Lé John, I. Suzuki, and J. A. Wright, *J. Biol. Chem.* **243**, 118 (1968).

[u]Suggested procedure: a similar procedure was used on starch gel electrophoresis. In this instance, fructose-1,6-diphosphate was used as substrate coupled with aldolase and triose-phosphate isomerase. R. K. Scopes, *Biochem. J.* **107**, 139 (1968).

[v]E. H. Creaser, D. J. Bennett, and R. B. Drysdale, *Biochem. J.* **103**, 36 (1967).

[w]J. W. Ogilvie, J. H. Sightler, and R. B. Clark, *Biochemistry* **8**, 3557 (1969).

[r] C. R. Roe and N. O. Kaplan, *Biochemistry* **8**, 5093 (1969).
[u] N. S. Henderson. *Ann. N. Y. Acad. Sci.* **151**, 429 (1968).
[z] E. Adams and G. Rosso, *J. Biol. Chem.* **242**, 1802 (1967).
[aa] J. A. Bowden and J. L. Connelly, *J. Biol. Chem* **243**, 3526 (1968).
[bb] E. M. Tarmy and N. O. Kaplan, *J. Biol. Chem.* **243**, 2579 (1968).
[cc] P. S. Chen, *J. Exp. Zool.* **168**, 337 (1968).
[dd] L. Schatz and H. L. Segal, *J. Biol. Chem.* **244**, 4393 (1969).
[ee] R. Cammack, *Biochem. J.* **115**, 55 (1969).
[ff] S. Ide, T. Hayakawa, K. Okabe, and M. Koike, *J. Biol. Chem.* **242**, 5460 (19¶¶).
[gg] S. A. Millard, A. Kubose, and E. M. Gal, *J. Biol. Chem.* **244**, 2511.
[hh] M. L. Cohn, L. Wang, W. Scouten, and I. R. McManus, *Biochim. Biophys. Acta* **159**, 182 (1968).
[ii] C. A. Villez, *Ann. N. Y. Acad. Sci.* **151**, 222 (1968).
[jj] W. H. Murphey, C. Barnaby, F. J. Lin, and N. O. Kaplan, *J. Biol. Chem.* **242**, 1548 (1967).
[kk] S. J. Parsons and R. O. Burns, *J. Biol. Chem.* **244**, 996 (1969).
[ll] C. C. Levy, *J. Biol. Chem.* **242**, 747 (1967).
[mm] N. Felberg and J. Schultz, *Anal. Biochem.* **23**, 241 (1968).
[nn] S. R. Himmelhoch, W. H. Evans, M. G. Mage, and E. A. Peterson, *Biochemistry* **8**, 914 (1969).
[oo] A. Novacky and R. E. Hampton, *Phytochemistry* **7**, 1143 (1968).
[pp] J. Ingle, *Biochem. J.* **108**, 715 (1968).
[qq] M. D. Poulik, *Ann. N. Y. Acad. Sci.* **151**, 476 (1968).
[rr] Suggested procedure.
[ss] L. D. Kohn, P. M. Packman, R. H. Allen, and W. B. Jakoby, *J. Biol. Chem.* **243**, 2479 (1968).
[tt] O. Gabriel and S. F. Wang, *Anal. Biochem.* **27**, 545 (1969).
[uu] R. L. Jolley, Jr., R. M. Nelson, and D. A. Robb, *J. Biol. Chem.* **244**, 3251 (1969).
[vv] R. L. Jolley, Jr., and H. S. Mason, *J. Biol. Chem.* **240**, 1489 (1965).

TABLE II
TRANSFERASES

Enzyme	Substrates	Coupling enzyme(s)	Staining procedure	Mode of detection and reference to method[a]	References
		Principle of methods			
ADPG-glycogen transglycosylase	ADPG, maltoheptaose	—	Iodine stain for glycogen	(b) (d)	b
Aspartate transaminase	2-Oxoglutarate		Reaction of diazomium salt with oxaloacetate	(d)	c,d
Cellobiose phosphorylase	D-Xylose, glucose 1-phosphate	—	Release of inorganic phosphate, visualization as the Ca salt	(b)	e
Creatine kinase	Creatine phosphate, ADP	Hexokinase and glucose-6-phosphate dehydrogenase	Formation of glucose 6-phosphate, TPNH, PMS, NBT	(a)	f
DNA-polymerase	Poly A (A-T) radioactive deoxyribonucleotide	—	Scan for radioactivity incorporated into polymer	(d)	g
Galactokinase	Galactose, ATP	Pyruvate kinase, lactic acid dehydrogenase	Disappearance of DPNH due to ATP generating system. Visualization in ultraviolet light	(c)	h
Galactose-1-phosphate uridyl transferase	Galactose-1-phosphate UDPG	Phosphoglucomutase, glucose-6-phosphate dehydrogenase	Indicator gel method using starch gel. TPNH formation observed in the ultraviolet light	(d)	i
Glycerol kinase	Glycerol, ATP	L-α-Glycerophosphate dehydrogenase	Formation of DPNH, PMS NBT	(a)	j
Glycogen phosphorylase	Glucose-1-phosphate maltoheptaose	—	Iodine stain for polymer or liberation of inorganic phosphate	(a) (b) (d)	b,k,l,m,n
Hexokinase	Glucose, ATP	Glucose-6-phosphate dehydrogenase	Formation of TPNH, PMS, NBT	(a)	n,o

Phosphofructokinase	Fructose-6-phosphate ATP	Aldolase, triosephosphate isomerase, glyceraldehyde-3-phosphate dehydrogenase	Formation of TPNH, PMS, NBT	(a)	p
Phosphoglucomutase	Glucose-1-phosphate	Glucose-6-phosphate dehydrogenase	Formation of TPNH, PMS, NBT	(a) (d) (2.1)	l,q,r
Polynucleotide phosphorylase	ApA, nucleoside diphosphate	—	Formation of TPNH, PMS, NBT	(b)	s,t,u
			Stain for polymer with acridine orange or methylene blue		v
Purine nucleoside pyrophosphorylase Sucrose Phosphorylase	Sucrose, inorganic phosphate	—	Formation of fructose, reducing properies, TTC stain	(b)(1.2)	w
Tyrosine aminotransferase	L-Tyrosine α-ketoglutarate	—	PMS, p-iodonitrotetrazolium violet	(a)	h
UDPG-glycogen transglucosylase	UDPG, glycogen maltoheptaose	—	Iodine stain for polymer	(d)	b

[a] See footnote a of Table I.

[b] J. F. Fredrick, Ann. N. Y. Acad. Sci. 151, 413 (1968).

[c] C. M. Michuda and M. Martinez-Carrion, Biochemistry 8, 1095 (1969).

[d] F. Fonnum, Biochem. J. 106, 401 (1968).

[e] J. K. Alexander, J. Biol. Chem. 243, 2899 (1968).

[f] R. H. Yue, H. K. Jacobs, K. Okabe, H. J. Keutel, and S. A. Kuby. Biochemistry 7, 4291 (1968).

[g] T. M. Jovin, P. T. Englund, and L. L. Bertsch, J. Biol. Chem. 244, 2996 (1969).

[h] D. G. Walker and H. H. Khan, Biochem. J. 108, 169 (1968).

[i] W. G. Ng, W. R. Bergren, M. Fields, and G. N. Donnell, Biochem. Biophys. Res. Commun. 37, 354 (1969).

[j] S. Hayashi and E. C. C. Lin, J. Biol. Chem. 242, 1030 (1967).

[k] S. A. Assaf and D. J. Graves, J. Biol. Chem. 244, 5544 (1969).

[l] H. Harris, D. A. Hopkinson, and J. Luffman, Ann. N. Y. Acad. Sci. 151, 232 (1968).

[m] C. H. Davis, L. H. Schliselfeld, D. P. Wolf, C. A. Leavitt, and E. G. Krebs, J. Biol. Chem. 242, 4824 (1967) (J. L. Hedrick, A. J. Smith, and G. E. Bruening, Biochemistry 8, 4012 (1969).

[n] R. T. Schimke and L. Grossbard, Ann. N. Y. Acad. Sci. 151, 332 (1968).

[o] H. M. Katzen, D. D. Soderman, and V. J. Cirillo, Ann. N. Y. Acad. Sci. 151, 351 (1968).

[p] D. J. H. Brock, Biochem. J. 113, 235 (1969).

[q] D. M. Dawson and A. Mitchell, Biochemistry 8, 609 (1969).

[r] O. Gabriel and I. Weemaes, in preparation.

[s] C. B. Klee, J. Biol. Chem. 242, 3579 (1967).

[t] P. S. Fitt, A. E. Fitt, and H. Wille, Biochem. J. 110, 475 (1968).

[u] C. B. Klee, J. Biol. Chem. 244, 2558 (1969).

[v] H. L. Engelbrecht and H. L. Sadoff, J. Biol. Chem. 244, 6228 (1969).

[w] O. Gabriel and S. F. Wang, Anal. Biochem. 27, 545 (1969).

[x] F. A. Valeriote, F. Auricchio, G. M. Tomkins, and D. Riley, J. Biol. Chem. 244, 3618 (1969).

TABLE III
Hydrolases

Enzyme	Substrate	Principle of detection	Mode of detection and reference to method[a]	References
L-Alanyl-β-naphthyl amidase	L-Alanyl-β-naphthyl amide	Post incubation coupling with Red B	(b)	b
Amylase	Soluble starch	Iodine stain for nondegraded starch	(b) (d)	c–f
Aminopepsidases } Arylamidases }	Leucyl- or arginyl-β-naphthylamines	Coupling of naphthylamine with fast Garnet GBC	(b)	g,h
Cholinesterase	Acetylthiocholine	Formation of copper thiocholine	(b)	i,j
Diphosphopyridine nucleosidase	DPN	Coupling with lactic dehydrogenase PMS, NBT	(b)	k
Esterases	α- or β-naphthyl acetate, α- or β-naphthyl butyrate	Coupling with diazo blue	(a) (3.1) (3.2)	l–n
Fucosidase	Glycoprotein	Reaction of released sugar with TTC	(b) (1.2), (d) (2.1)	o–q
Galactosidase	Lactose	Glucose oxidase, PMS, NBT	(d)	r
	6-bromo-2-naphthyl-β-D-glucosaminide	Coupling with Diazo Blue B	(b)	s
Glucosidaminidase	4-Methylumbelliferone glycoside	Fluorescence in the ultraviolet	(c)	t, u
	o-Nitrophenyl-N-acetyl-β-D-glucosaminide	Release of o-nitrophenol	(c)	
Glucosidase	4-Methylumbelliferone glycoside	Fluorescence in the ultraviolet	(c)	t, u
	Naphthol glucuronide	Coupling with diazopararosaniline	(a)	v
Glycosidase	Glycoside	Release of reducing sugar, reaction with TTC	(b) (1.2)	u,w
	o-, p-Nitrophenylglycoside	Release of o-, p-nitrophenol	(c) (3.6)	w

Guanase	Guanine	Coupling with xanthineoxidase, NBT	(a)	x
Histaminase	Putrescine	Pyrroline formation	(a)	y
Invertase	Sucrose	Reaction of fructose with TTC	(b) (1.2)	w
Lactamase	Cephaloridine	Reaction with iodine-starch	(d)	z
Lipase	α-Naphtholnonanoate	Coupling with Fast Blue BB salt	(a) (3.7)	aa
Pepsinogen	Hemoglobin	Amido Black 10B stain for protein	(d)	bb
Nucleosidetriphos-phatase	ATP, ITP, GTP	Release of inorganic phosphate, reaction with $Pb(NO_3)_2$ and $(NH_4)_2S$	(b) (3.4)	cc, dd
Nucleotidase	5'-Nucleotides	Release of inorganic phosphate, $Pb(NO_3)_2$, $(NH_4)_2S$	(b) (3.4)	ee,ff,gg
Phosphatase (acid)	α-Naphthyl phosphate	Coupling with diazopararosaniline	(a) (3.5)	gg,hh,ii,jj,kk
Phosphatase (alkaline)	α-Naphthyl phosphate	Coupling with Fast Red TR	(a) (3.3)	kk,ll,mm
	β-Glycerophosphate	Release of inorganic phosphate $Pb(NO_3)_2$, $(NH_4)_2S$	(b) (3.4)	nn
	p-Nitrophenyl phosphate	Release of p-nitrophenol	(c)	oo
Phosphodiesterase	TMP-5'-d-naphthyl ester	Coupling with diazonium salt	(b)	pp
	Bis-p-nitrophenyl-phosphate	Release of p-nitrophenol	(c)	pp,qq
Pyrophosphatase (inorganic)	Sodium pyrophosphate	Release of inorganic phosphate	(b) (3.4)	ll,mm,rr
Ribonuclease	Low molecular weight RNA	Stain for undegraded RNA	(b) (d)	ss,tt
Threonine deaminase	L-Threonine	Reaction of 2-oxobutyrate with 2,4-dinitro-phenylhydrazine	(b)	uu
Urease	Urea	pH change. NBT conversion to formazan	(a)	vv

[a] see footnote a of Table I.

[b] C. S. Beck, C. W. Hasinoff, and M. E. Smith, *J. Neurochem.* 15, 1297 (1968).

[c] Z. Ogita, *Ann. N. Y. Acad. Sci.* 151, 243 (1968).

[d] T. Friedman and C. J. Epstein, *J. Biol. Chem.* 242, 5131 (1967).

[e] P. J. Keller and B. J. Allan, *J. Biol. Chem.* 242, 281 (1967).

[f] H. Heller and R. G. Kulka, *Biochim. Biophys. Acta* **165**, 393 (1968).

[g] N. Marks, R. K. Datta, and A. Lajtha, *J. Biol. Chem.* **243**, 2882 (1968).

[h] L. A. Idahl and I. B. Taljedal, *Biochem. J.* **106**, 161 (1968).

[i] G. A. Davis and B. W. Agranoff, *Nature (London)* **220**, 277 (1968).

[j] P. K. Das and J. Liddell, *Biochem. J.* **116**, 875 (1970).

[k] N. I. Swislocki and N. O. Kaplan, *J. Biol. Chem.* **242**, 1083 (1967).

[l] R. L. Hunter, D. C. Bennet, and A. H. Dodge, *Ann. N. Y. Acad. Sci.* **151**, 594 (1968).

[m] W. J. Collins and A. J. Forgash, *J. Insect Physiol.* **14**, 1515 (1968).

[n] R. L. Hunter and E. A. Maynard, *J. Histochem. Cytochem.* **10**, 677 (1962).

[o] C. L. Markert and R. L. Hunter, *J. Histochem. Cytochem.* **7**, 42 (1959).

[p] B. J. Davis, Enzyme Analysis 3b, Canalco (1963).

[q] D. Aminoff and K. Furukawa, *J. Biol. Chem.* **245**, 1659 (1970). The authors incorporated the substrate into the gel prior to polymerization. Alternatively, the indicator gel method (d) can be used.

[r] D. H. Alpers, *J. Biol. Chem.* **244**, 1238 (1969).

[s] D. H. Alpers, E. Steers, Jr., S. Shifrin, and G. Tomkins, *Ann. N. Y. Acad. Sci.* **151**, 545 (1968).

[t] D. Robinson, R. G. Price, and N. Dance, *Biochem. J.* **102**, 525 (1961).

[u] R. G. Price and N. Dance, *Biochem. J.* **105**, 877 (1967).

[v] R. Ganschow and K. Paigen, *Genetics* **59**, 335 (1968).

[w] O. Gabriel and S. F. Wang, *Anal. Biochem.* **27**, 545 (1969). (N. P. Neumann and J. O. Lampen, *Biochemistry* **6**, 468 (1967).

[x] R. Currie, F. Bergel, and R. C. Bray, *Biochem. J.* **104**, 634 (1967).

[y] J. K. Smith, *Biochem. J.* **103**, 110 (1967).

[z] This method was originally described for starch gel electrophoresis: T. D. Hennessey and M. H. Richmond, *Biochem. J.* **109**, 469 (1968). The indicator gel method (d) containing soluble starch can be used.

[aa] M. M. Nachlas, B. Morris, D. Rosenblatt, and A. M. Seligman, *J. Biophys. Biochem. Cytol.* 7, 261 (1960).

[bb] D. Lee and A. P. Ryle, *Biochem. J.* 104, 735 (1967).

[cc] M. J. Selwin, *Biochem. J.* 105, 279 (1967).

[dd] A. Abrams and C. Baron, *Biochemistry* 6, 225 (1967).

[ee] H. F. Dvorak and L. A. Heppel, *J. Biol. Chem.* 243, 2647 (1968).

[ff] C. W. T. Pilcher and T. G. Scott, *Biochem. J.* 104, 41C (1967).

[gg] C. Arsenis and O. Touster, *J. Biol. Chem.* 243, 5702 (1968).

[hh] T. J. Barka, *J. Histochem. Cytochem.* 9, 542 (1961).

[ii] P. J. Anderson, S. K. Song, and N. Christoff, *Proc. Int. Congr. Neuropathol. 4th*, Vol 1, p. 75.

[jj] M. Igarashi and V. P. Hollander, *J. Biol. Chem.* 243, 6084 (1968).

[kk] P. T. Iype and C. H. Heidelberger, *Arch. Biochem. Biophys.* 128, 434 (1968).

[ll] D. W. Moss, R. H. Eaton, J. K. Smith, and L. G. Whitby, *Biochem. J.* 102, 53 (1967).

[mm] R. D. Cox, P. Gilbert, and M. J. Griffin, *Biochem. J.* 105, 155 (1967).

[nn] J. M. Allen and G. J. Hyncik, *J. Histochem. Cytochem.* 11, 169 (1963).

[oo] H. H. Sussman, P. A. Small, Jr., and E. Cotlove, *J. Biol. Chem.* 243, 160 (1968).

[pp] B. Lerch, *Experientia* 24, 889 (1968).

[qq] C. L. Harvey, K. C. Olson, and R. Wright, *Biochemistry* 9, 921 (1970).

[rr] H. Tond and A. Kornberg, *J. Biol. Chem.* 242, 2375 (1967).

[ss] G. Wolf, *Experientia* 24, 890 (1968).

[tt] S. Biswas and V. P. Hollander, *J. Biol. Chem.* 244, 4185 (1969).

[uu] According to the International Union of Biochemistry, this enzyme is classified as a deaminating lyase: G. W. Hatfield and H. E. Umbarger, *J. Biol. Chem.* 245, 1736 (1970).

[vv] W. N. Fishbein, *Proc. Int. Symp. Chromatogr. Electrophoresis, 5th, Ann Arbor, Michigan*, p. 238 (1969).

(5.3) *Uridinediphosphogalactose-4-epimerase.*[35] *Simultaneous Capture Technique* (a). UDP-Galactose is used as the substrate and coupled with UDPG-dehydrogenase, resulting in the formation of DPNH. The nitro-blue tetrazolium reaction (1.1) is applied for localization of the enzyme. A control without substrate is important here since bands appear in the absence of the substrate.

Documentation and Storage

Many of the stains both for protein and for enzymatic activity are sufficiently fixed onto the gel so that they may be stored for long periods of time when the gels are held at 4° in 7.5% acetic acid and protected from light. A permanent record, of course, is provided best by a photograph with high contrast film; Polaroid films are convenient for this purpose.

[35]R. A. Darrow and R. Rodstrom, *Biochemistry* 7, 1645 (1968).

Author Index

Numbers in parentheses are reference numbers and indicate that an author's work is referred to although his name is not cited in the text.

A

Abe, M., 592
Abrams, A., 206(*qq*), 207, 208, 214(5), 601(*dd*), 603
Accorsi, A., 385
Ackers, G. K., 290, 295
Adams, E., 595(z), 597
Adelberg, E. A., 57, 92
Adler, H. J., 15, 16(8), 17(8)
Adolph, K., 267
Agranoff, B. W., 600(*i*), 602
Aiba, S., 441, 456(1), 457(1), 458, 462, 469(1)
Aisen, P., 255
Åkeson, A., 261, 264 (19)
Alberghina, F. A. M., 81, 85(12)
Albert, W., 402
Alberts, B. M., 239, 512, 552(48)
Albertsson, P.-Å., 239, 240 (2, 3), 242(4)
Albrecht, A. M., 594(*l*), 596
Alexander, J. K., 598(*e*), 599
Alexander, R. F., 157
Algranati, I. D., 32
Allan, B. J., 600(*e*), 601
Allen, J. M., 138, 590, 601(*nn*), 603
Allen, R. H., 250, 595(*ss*), 597
Allen, R. N., 431(*a*), 432
Allen, S. L., 594(*r*), 595(*r*), 596
Allman, D. W., 229
Alpers, D. H., 600(*r, s*), 602
Alpert, N. L., 13
Altgelt, K. H., 295
Altschul, A. M., 568
Amarose, A. P., 165
Ambe, K. S., 205(*q*), 206
Ames, B. N., 89, 178, 413
Aminoff, D., 600(*q*), 602

Anacker, W. F., 327
Andersen, R. D., 262(*j*), 263
Anderson, N. G., 168, 175, 180(11), 192, 195
Anderson, P. J., 578, 591(1a), 601(*ii*), 603
Andersson, B., 376
Andrews, E. C., 228, 229(*p*), 230
Andrews, P., 290, 291, 308(9), 413
Anfinsen, C. B., 345, 347(1, 5, 6), 353, 355(1), 356, 357(1), 358, 360(25, 27), 375, 378(5), 379(1)
Aoe, H., 385
Araki, C., 294
Arima, K., 211
Arndt, W. F., 104
Aronson, A. I., 93
Arsenis, C., 347, 601(*gg*), 603
Asai, J., 229
Ashworth, J. N., 233
Askonas, B. A., 393, 559, 561(1)
Assaf, S. A., 598(*k*), 599
Astrin, K. H., 165
Atkinson, D. E., 385
Auerbach, J., 11
Auricchio, F., 599
Avey, H. P., 262(*i*), 263
Awdeh, Z. L., 393, 559, 561(1)
Ayad, S. R., 395, 434
Axén, R., 348

B

Babelay, E. F., 168
Babson, A. L., 8
Bachelard, H. S., 592
Bachman, B. J., 81
Bacon, J. S. D., 122
Bailey, E. G., 463

Subject Index

A

ACA-Automatic Clinical Analyzer, 13
Acetabularia, Golgi apparatus isolation
  from, 147
Acetic acid, buffer acid properties, 3
Acetohydroxy acid synthetase, production
  in *S. typhimurium,* 89
Acetone
  as protein precipitant, 511
  use in membrane protein extraction, 214
Acetonitrile, as reagent for protein crystal-
  lization, 261
N-Acetylamide, as inducer for amidase, 87
N-Acetyl-β-D-glucosaminidase, analysis of
  in multiple systems, 18–22
N-Acetylglucosaminyl transferase, reaction
  in Golgi apparatus, 142
Acid phosphatase
  analysis of
    continuous-flow method, 11–12
    in multiple systems, 18, 19, 22
  large-scale isolation of, 490
  staining on gels, 585, 591, 601
Acid, monoprotic, degree of dissociation
  related to pH, 4
Acrylamide, recrystallization of, 416
Acrylamide gel, for preparative gel-density
  gradient electrophoresis, 434–437
Acrylamide gel electrophoresis (preparative), 412–433
  applications of, 430–433
  buffers for, 418–421
  commercial equipment for, 433
  cooling requirements for, 422
  disassembly operation, 429–430
  fraction analysis, 429
  gel formulations and preparation,
    416–418
    casting of, 422–424
    guidelines for choosing, 417
  power supply and electrodes, 421–422
  preliminary experiments, 413–414

protein collection, 426–429
  by elution during electrophoresis,
    427–429
  by elution of gel sections, 426–427
  pumps for, 422
  sample preparation and application,
    424–426
  scaling up in, 414–416
  uv monitor and recorder for, 422
Adenine
  in derepression of thiamine pathway
    enzymes, 89
  use in enzyme induction, 92–93
ADPG-glycogen transglycosylase, staining
  on gels, 598
Adrenal glucose-6-phosphate dehydrogenase, purification of using nonionic
  polymers, 240, 246–248
Adsorbents, for ion-exchange chromatography, 274–282
Aeration, in large-scale growth of bacteria,
  463–465
Aerobacter aerogenes
  enzyme induction in, 93
  large-scale growth of, 475
Aerobic bacteria
  enrichment culture techniques for,
    61–64
    camphor-decomposing bacteria, 62
    filamentous blue-green algae, 63–64
    iron-oxidizing bacteria, 62–63
    Pseudomonas spp., 61–62
Aeromonas liquefaciens, enzyme induction in,
  94
Affinity chromatography, 345–378
  adsorption and elution of proteins,
    350–351
  applications of, 345–346, 355–362
  principles of, 345–351
  procedure for, 352–355
  protein and peptide coupling to agarose
    in, 373–378
  solid matrix support for, 346–347

in membrane protein extraction, 215, 226

Biocytin, use in affinity chromatography, 357–358, 360

Bio-gels, *see also* Polyacrylamide gels
in gel filtration, 294
physical characteristics of, 316–318

Biotech filter, 461, 462

Bisacrylamide, recrystallization of, 416

1,4-Bis-2(5-phenyloxazolyl)benzene, reagent for xanthine oxidase, 151

Bladder epithelial cells, plasma membrane isolation from, 123

Blood
fractionation of, 155
dextran-mixing method, 155–156
hypaque-dextran two-phase method, 156–159
leukocyte isolation from, 155–157
sample collection by filter paper technique, 153–154

Blue-green algae, filamentous type, enrichment culture techniques for, 63–64

Bromoacetyl-agarose, preparation and use in affinity chromatography, 366

Bromothymol blue, in enzyme stain, 593

Brushite, conversion to hydroxyapatite, 327

Bubble trap, for zonal centrifugation, 200

Buffer(s)
for anaerobic column chromatography, 322–325
for Golgi apparatus isolation, 131
preparation of, 3–5
for preparative acrylamide gel electrophoresis, 418–421

Buffer acids, for vitamin B_6 assay, 3

n-Butyl alcohol
in purification of lipoproteins, 225–227
use in membrane protein extraction, 214

B-XIV zonal rotor, 169, 175
diagrams of operation, 184–185
properties of, 179
schematic diagram of, 190

C

Cab-O-Sil, use in affinity chromatography, 349

Cacodylic acid, buffer acid properties, 3

Calam and Hockenhull's medium, composition of, 73, 77

Calcium acetate, reagent for plasma membrane preparation, 124–125

Calcium ion, protein complexes with effect on fractionation, 238

Calcium phosphate gel, deposited on cellulose, preparation, 339–342

Camphor-decomposing bacteria, enrichment culture technique for, 62

Canavanine, use in enzyme induction in microorganisms, 91

Candida pseudotropicalis, enzyme isolation from, large-scale, 491

Candida tropicalis, enzyme preparation from, 537

Candida utilis, enzyme isolation from, large-scale, 491, 538–539

Candle-type fibrous filters, in large-scale growth of bacteria, 460–461, 462

Capillaries, microdiffusion cells from, 255–258

Carbonic anhydrase, staining on gels, 585, 593

Carboxylic acid esterase, staining on gels, 584, 590

Carboxymethyl cellulose (CMC), enzyme elution from, 379–385

Carboxypeptidase
shearing effects on, 531
use in affinity chromatography, 349

Carboxypeptidase A, preservation of, 530

Casein, ultrafiltration of, 503
comparison to rotary filtration, 511

Catalase
inactivation-reactivation studies on, 505, 506, 531
induction in *Rhodopseudomonas spheroides*, 90
isolation
by foaming, 526
large-scale, 508
preservation of, 530
staining on gels, 594

Cathepsin C, *see* Dipeptidyltransferase

CA-type membranes, 42

Cauliflower, Golgi apparatus isolation from, 147

Cell breakage, for large-scale growth of bacteria, 484

G

Q

Quinolinate phosphoribosyltransferase, crystallization of, 250

R

Radish, Golgi apparatus isolation from, 147

Raulin-Thom medium with glucose, composition of, 73, 77

Razor blade chopping device, 146–147

Red blood cells, membrane studies on, 115

Rennet, in activation studies on, 505, 531

Reograd rotor, for zonal centrifugation, 202

Repressors, in enzyme production, 87–89

Restricted-diffusion chromatography, *see* Gel filtration

Reverse osmosis, in large-scale protein isolation, 509

Rhamnose, in yeast spheroplast preparation, 122

Rhodopseudomonas spheroides, catalase induction in, 90

Rhodospirillum rubrum, large-scale growth of, 475

Ribonuclease
  analysis of in multiple systems, 19
  crystallization of, 261
  purification
    by affinity chromatography, 361–362, 378
    by elution with substrate, 379
  staining on gels, 601
  use in bacterial membrane preparation, 106–107

Ribonuclease A, dialysis of, 24

Ribonuclease I, large-scale isolation of, 490

Ribonuclease-S-peptide, affinity chromatography of, 355, 378

Ribosomal ribonuclease, in bacterial membranes, 113

Ribosomes, preparation from *N. crassa*, 81, 84, 85

RNA, in bacterial membranes, 114

tRNA
  crystallization of, 263, 265
    micro technique, 266–267

preparation from *N. crassa*, 84
purification on modified cellulose, 347

Robot Chemist, 11

Rock and roll dialyzer, 28–29

Rocking dialyzer, 28–29

Rosaniline·HCl, in enzyme stain, 591

Rotary filter, in large-scale protein isolation, 500, 501

Rotating dialyzer, 29

Rubidium bromide, as gradient for zonal centrifugation, 174

Rubidium chloride, as gradient for zonal centrifugation, 174

Rubidium formate, as gradient for zonal centrifugation, 174

S

Saboraud-dextrose agar, composition of, 73

Saboraud-maltose agar, composition of, 73

Saccharomyces carlsbergensis, spheroplast preparation from, 122

Saccharomyces cerevisiae
  large-scale growth of, 475, 484, 488, 489
  spheroplast preparation from, 122

Salmonella typhimurium
  acetohydroxy·acid synthetase production in, 89
  histidine mutants of, 92
  large-scale growth of, 490
  membrane preparation from, 105, 110, 115
  sulfate reduction enzymes from, 89

Salting-out of proteins, on large-scale, 511–513

Salts, use in membrane protein extraction, 211–212

Schlieren effect, of enzymes, 251

Sephadex adsorbents
  in affinity chromatography, 347
  in gel filtration, 291–294
  in ion-exchange chromatography, 274–285
  in large-scale protein isolation, 520–524
  physical characteristics of, 315
  use in purification of enzymes by elution with substrate, 383–384

Sepharose, *see also* Agarose beads, Agarose gels